P9-ARG-839

# MARINE ECOLOGY

## A Comprehensive, Integrated Treatise on Life in Oceans and Coastal Waters

*Volume I* ENVIRONMENTAL FACTORS

*Volume II* PHYSIOLOGICAL MECHANISMS

*Volume III* CULTIVATION

*Volume IV* DYNAMICS

*Volume V* OCEAN MANAGEMENT

# MARINE ECOLOGY

## A Comprehensive, Integrated Treatise on Life in Oceans and Coastal Waters

Editor
OTTO KINNE
*Biologische Anstalt Helgoland*
*Hamburg, Federal Republic of Germany*

VOLUME V
Ocean Management
Part 2: Ecosystems and Organic Resources

*A Wiley–Interscience Publication*

1983
JOHN WILEY & SONS
Chichester · New York · Brisbane · Toronto · Singapore

***Library of Congress Cataloging in Publication Data:***

(Revised for vol. 5, pts. 2, 3 & 4)
Kinne, Otto.
Marine ecology.
Vol. 5 published: Chichester, New York, Wiley.
Includes bibliographies.
CONTENTS: v. 1. Environmental factors. 3 v.—v. 2. Physiological mechanisms. 2. v.—[etc.]—v. 5. Ocean management. 4. v.
1. Marine ecology—Collected works. I. Title.
QH541.5.S3K5 574.5′2636 79-121779

ISBN 0 471 90159 8

***British Library Cataloguing in Publication Data:***

Marine ecology.—(A Wiley-Interscience publication)
Vol. V: Ocean management
Part 2.
1. Marine ecology
I. Title II. Kinne, Otto
574.5′2636 QH541.5.S3

ISBN 0 471 90159 8

Typeset by Preface Ltd, Salisbury, Wiltshire
Printed by The Pitman Press, Bath, Avon.

# FOREWORD

## to

# VOLUME V: OCEAN MANAGEMENT

'Ocean Management', the last volume of *Marine Ecology*, describes and evaluates all the essential information available on structures and functions of interorganismic coexistence; on organic resources of the seas; on pollution of marine habitats and on the protection of life in oceans and coastal waters. The volume consists of four parts:

Part 1

**Zonations and Organismic Assemblages**

Chapter 1: Introduction to Part 1—Zonations and Organismic Assemblages
Chapter 2: Zonations
Chapter 3: General Features of Organismic Assemblages in the Pelagial and Benthal
Chapter 4: Structure and Dynamics of Assemblages in the Pelagial
Chapter 5: Structure and Dynamics of Assemblages in the Benthal
Chapter 6: Major Pelagic Assemblages
Chapter 7: Specific Pelagic Assemblages
Chapter 8: Major Benthic Assemblages
Chapter 9: Specific Benthic Assemblages

Part 2

**Ecosystems and Organic Resources**

Chapter 1: Introduction to Part 2—Ecosystems and Organic Resources
Chapter 2: Open-Ocean Ecosystems
Chapter 3: Coastal Ecosystems
3.1: Brackish Waters, Estuaries and Lagoons
3.2: Flow Patterns of Energy and Matter

Part 3

**Pollution and Protection of the Seas: Radioactive Materials, Heavy Metals, and Oil**



Part 4

**Pollution and Protection of the Seas: Pesticides, Domestic Wastes, and Thermal Deformations**



The culmination of *Marine Ecology*, Volume V has its roots in, and draws much of its basic substance from, the preceding volumes of the treatise:

Volume I ('Environmental Factors') is concerned with the most important environmental factors operating in oceans and coastal waters and their effects on microorganisms, plants and animals.

Volume II ('Physiological Mechanisms') reviews the information available on the mechanisms involved in the synthesis and transversion of organic material; in thermoregulation, ion- and osmoregulation; in evolution and population genetics; and in organismic orientation in space and time.

Volume III ('Cultivation') comprehensively assesses the art of maintaining, rearing, breeding and experimenting with marine organisms under environmental and nutritive conditions which are, to a considerable degree, controlled.

Volume IV ('Dynamics') summarizes and critically evaluates the knowledge available on the production, transformation and decomposition of organic matter in the marine environment, as well as on food webs and population dynamics.

Of necessity somewhat heterogeneous in concept and coverage, 'Ocean Management' introduces the readers to fields of applied marine ecology, i.e. to man's use of oceans and coastal waters for his own ends. In order to provide a solid basis for a sound assessment of the sea's man-supporting qualities, Parts 1 and 2 deal comprehensively with the basic multi-specific units encountered and with the resources which they constitute. After summarizing our current knowledge on the large variety of organismic groupings in the form of zonations and assemblages, and after considering the structures and functions of major marine ecosystems, the significance of these units and their components as resources utilizable for food or for raw materials are reviewed in depth. Parts 3 and 4, finally, evaluate man's potentially destructive impact. They focus on the different facets of pollution and critically assess measures—currently applied and considered practicable in the future—for protecting life in oceans and coastal waters from detrimental human influences.

Selected as early as 1965, the title 'Ocean Management' still seems too ambitious, if not somewhat misleading—even though, as anticipated, a host of new data and many new and important insights into the machinery of ecological systems have been brought to light in the meantime. Environmental management requires the concerted, judicious, responsible application of science and technology for the protection and control of those properties of ecosystems, species, resources, or areas that are regarded as absolute requirements for the continued support of civilized human societies*. Our knowledge has remained insufficient for an objective, exact definition of such properties. Hence, maintenance of a high degree of organismic and environmental diversity, and maximum possible conservation of natural conditions is deemed essential to avoid or to reduce irreversible long-term damage.

The means and aims of ocean management receive detailed attention in Part 2. Our capacity for true management is still restricted to narrowly-defined areas and to specific organisms, i.e. to a few heavily exploited or otherwise especially endangered species. Marine ecosystem management remains problematic; it is in need of much more basic ecological knowledge than is at present available.

The recent trend of using English as the international scientific language—in itself of great significance for international communication and cooperation—has often been disadvantageous to scientists with insufficient command of that language. It has frequently diminished the representativeness and distorted the emphasis of the total information actually available. While organizing and editing *Marine Ecology*, I have therefore attempted to include scientists from countries which made important contributions to the field of marine ecology, but whose scientists largely use non-English languages. In this way I wanted to underline the international status and significance of marine ecological research and the need to draw from different sources in order to provide the best possible representation of the state-of-the-art. Of course, I had to pay for this: an enormous amount of time and effort had to be invested in translation and manuscript improvement.

While I am writing this last foreword of the treatise, *Marine Ecology* is nearing completion. It is my sincere wish to thank again all those who have supported me during the many years of planning, carrying out and finalizing this *magnum opus*. From 1965 to 1981, the work on *Marine Ecology* has taken up most of my evenings, weekends and holidays. Who can blame me for feeling relieved?

As was the case with previous volumes of this treatise, I have received much help and advice while working on Volume V. With profound gratitude I acknowledge the close and fruitful cooperation with all contributors; the support, patience and confidence of the publishers; and, last but not least, the technical assistance of Monica Blake, Alice Langley, Julia Maxim, Seetha Murthy, Sherry Stansbury and Helga Witt.

O.K.

*KINNE, O. 1980: *14th European Marine Biology Symposium 'Protection of Life in the Sea'*: Summary of symposium papers and conclusions. *Helgoländer Meeresunters.*, **33**, 732–761.

# CONTENTS
# OF
# VOLUME V, PART 2

# CONTRIBUTORS
# TO
# VOLUME V, PART 2


FIELD, J. G., *University of Cape Town, Zoology Department, Rondebosch 7700, Republic of South Africa.*

GULLAND, J. A., *Marine Resources Service, Fishery Resources and Environmental Division, Food and Agriculture Organization of the United Nations, Via delle Terme di Caracalla, 00100 Rome, Italy.*

HEDGPETH, J. W., *5660, Montecito Avenue, Santa Rosa, California 95404, USA.*

KINNE, O., *Biologische Anstalt Helgoland (Zentrale), Notkestraße 31, 2000 Hamburg 52, Federal Republic of Germany.*

MICHANEK, G., *Göteborgs Universitet, Avd för Marin Botanik, Carl Skottsbergs Gata 22, S-413 19 Göteborg, Sweden.*

VINOGRADOV, M. E., *Academy of Sciences of the USSR, P. P. Shirshov Institute of Oceanology, 23, Krasikova, 117218 Moscow, USSR.*

# OCEAN MANAGEMENT

Marine Ecology Vol. V, Part 2
Edited by Otto Kinne


# 1. INTRODUCTION TO PART 2—ECOSYSTEMS AND ORGANIC RESOURCES

O. KINNE

## (1) General Aspects

While Part 1 of Volume V focused on the description, analysis, and evaluation of zonations and organismic assemblages, i.e. of groups of species living together in a defined space, Part 2 concentrates on interacting and interdependent biotic and abiotic components forced into a dynamic, functionally integrated supra-specific unit—the ecosystem—by co-ordinated patterns of energy flow, material cycling, and homeostatic control mechanisms. After reviewing ecosystems of open sea areas, coasts, and estuaries, Part 2 considers the resources which they constitute, i.e. their significance for supporting human life.

The term 'ecosystem' was introduced, interpreted, and defined by the English botanist A. G. TANSLEY (1935); he realized (p. 300) that ecosytems

> 'show organisation, which is the inevitable result of the interactions and consequent mutual adjustment of their components'.

In Volume IV of this treatise, 'ecosystem' has been defined as

> 'an integrated spatial entity of interacting and interdependent biotic and abiotic components which are linked by energy flow and material cycling, by exchange between biotic and abiotic and among biotic constituents, as well as by homeostatic control mechanisms' (KINNE, 1978, p.1).

Terms closely related to the concept of ecosystems are 'association' (first introduced in botany, and redefined by MARGALEF, 1977,1978), 'biocoenosis' (MOEBIUS, 1877; see also this volume, pp. 647, 740), 'community' (PETERSEN, 1911, 1914), and 'organismic assemblage' (Volume V: PERES, 1982). For details and definitions concerning these terms consult Volume IV, p. 1 and Volume V, pp. 4, 48, 49, 740–742.

Our treatment of the world-wide resources formed by marine organisms is subdivided into chapters devoted to benthic plants, planktonic plants, and fishery products. While the reviewers completed their assessments it turned out that the information presently available on, and the extent to which use is being made of, planktonic plants is insufficient for a comprehensive, critical review. The present status is, therefore, briefly summarized in an *Editorial Note* (pp. 832–833).

## (2) Comments on Chapters 2 to 5

### Chapter 2: Open-Ocean Ecosystems

To a large extent based on information collected during numerous Sowjet–russian expeditions, this chapter opens up new vistas for many western marine ecologists and elaborates important insights into the structures, functions, and dynamics of pelagic ecosystems of the open ocean. Over decades a leading expert in the field of plankton research, the reviewer synthesizes, evaluates, and condenses a vast body of information, stressing basic phenomena, principles, and problems.

The largest biotope on earth, the oceanic pelagial, comprises a vast and unique ecological supersystem which appears to have maintained its basic properties over hundreds of millions of years. Within the different regions of this supersystem, a hierarchy of ecosystems can be distinguished, from large, rather stable systems, down to the short-lived systems in small water lenses only several tens of metres in dimension. The major denominators of the functions and structures of discrete ecosystems in the open ocean are the specific biological and physico-chemical histories and properties of the life-supporting water masses concerned, as well as their mobility, mixing dynamics, and size. The water masses differ from each other and succeed each other both horizontally (geographically) and vertically. While their vertical structure may be subject to change in different climatic zones, as a rule, surface, intermediate, deep, and bottom waters can be distinguished. The boundaries between adjacent water masses vary as a function of time, water movement, and geography, thus forming—both horizontally and vertically—zones of mixing and transition.

The basic, evolutionarily old, ecosystems of the open ocean are large in size, but small in number: two cold-water ecosystems (Central Arctic and Coastal Antarctic Circumpolar); three subpolar (North Atlantic, North Pacific, Subantarctic); five central tropic (two in the Pacific, two in the Atlantic, one in the Indian Ocean); and three equatorial ecosystems. In addition, distant-neritic ecosystems of 'neutral' regions and coastal ecosystems (Chapter 3) can be distinguished.

While it may be difficult to determine the borderline between neighbouring ecosystems objectively, functions, structures, and seasonal dynamics are generally more homogeneous within each system than between different systems. Differences between the ecosystems distinguished are especially related to species composition, cyclic processes of reproduction, and differences in rates of biological activity. The borders between related oceanic ecosystems are enforced and maintained by wind, climate, earth rotation, and the dimensions and morphology of the ocean basins.

All marine organisms ultimately depend on solar energy entering through the ocean's surface and on plant nutrients, including those transported upwards into the euphotic zone by turbulence. The intensity and duration of solar radiation and the amount of nutrient replenishment decisively determines the overall productivity of pelagic ecosystems. While nutritionally dependent on plant material, animals in the pelagic ecosystems are in direct contact with primary producers only in the upper, euphotic water layer. Deeper down in the water column they utilize the remains of dead, sinking surface organisms and the products of their metabolism; in addition, they depend on vertical migrators. In open-ocean ecosystems, intensity and pattern of vertical fluxes of energy and matter are of basic significance. Biosynthesis by deep-sea bacteria is only of rather limited, local importance.

Compared to their terrestrial counterparts, oceanic ecosystems feature: (i) A neritic annual productivity which is some four orders of magnitude higher (phytoplankton vs. land plants). (ii) Much less long-term accumulation of energy and matter in the bodies of living organisms. (iii) Fast growth of plankton algae combined with short generation times (up to 11 generations $d^{-1}$). (iv) Intensive exploitation (grazing): sometimes almost the total plant production is consumed immediately over large ocean areas, while on land—even on meadows—less than one-seventh is usually consumed directly. (v) High growth rates, rapid succession of generations, and higher organismic mobilities result in much quicker material recycling than, for example, forests with their enormous reserves of living mass and long generation times of trees (10 to several 100 yr); while in the pelagial a community may attain a high degree of maturity within months, this may take decades or even centuries in terrestrial communities.

The development of an ecosystem as a function of time is a fundamental concept in ecology (see also Volume IV: MARGALEF, 1978; Volume V: PERES, 1982). Ecosystem development in general involves a progressive reduction of entropy as entities of transmitted and retained information accumulate. Confined to a drifting lens of ageing sea water, pelagic ecosystems of the open ocean undergo a predictable, successional pattern of growth, maturation, and death, beginning with the ascent of the supporting water lens to the euphotic layer and ending with its return to the ocean's depths. While the ecosystem requires, of course, continual energy for development, differentiation, and sustenance, its overall energy budget is characterized by rapid net accumulation of biologically utilizable energy during early system development and by progressive energy degradation with increasing system maturity. At early stages of community development, the biomass of autotrophic plants is high. Hence, the animal populations present find abundant food and can develop at maximum rates. As time passes, the energy and matter initially accumulated are increasingly utilized; food availability decreases and competition increases. Rates of production and consumption pass through a more balanced phase until, finally, consumption dominates.

In other words, following its birth in the euphotic zone, the pelagic ecosystem grows and differentiates as it drifts along near the ocean's suface, initially accumulating, then transforming, and later degrading its energy reserves as it sinks back to the dark and dies. After serving as a cradle of life and as a vehicle for the transport of energy and matter, the water body concerned rejuvenates during its journey through the lightless depths of the ocean, replenishing its nutrients and thus preparing to continue the cycle upon re-ascending to the surface.

Season, climate, and atmospheric plus hydrographic circulation can significantly modify or even control dynamic processes in open-ocean ecosystems. The effects of these basic denominators of variability still tend to exceed the rapidly increasing additional impact due to human activities. However, this may change in the future. Lack of 'buffers' (such as soil or microclimates due to plant growth in terrestrial ecosystems) tend to accentuate the consequences of external impacts and thus may render pelagic ecosystems of the ocean more vulnerable to anthropogenic influences than terrestrial ecosystems.

After dealing with organismic distributions in the pelagial and the phenomena influencing and controlling them, Chapter 2 considers different types of ecosystems, differentiating between high-latitude regions, northern temperate regions, northern regions of the Pacific and Atlantic Oceans, Southern Ocean, anticyclonic tropical gyres, and equatorial upwellings.

In view of the high degree of complexity of oceanic ecosystems and the fact that they are largely inaccessible for experimentation, simulation and modelling of major processes and properties have become primary tools for analytical and interpretative ecosystem research. Present models concentrate on basic parameters and relationships at the expense of (often interesting) detail. Proper experimentation with simulation models can gradually lead to a basic concept, regarding the development and dynamics of the system in time and space. Vertical stratification of plankton is caused and maintained by physico-chemical and biological factors. The latter appear to be associated primarily with competition for food which becomes increasingly scarce, energy deficient, and uniform as a function of water depth. The vertical nutrition gradient modifies the composition and reduces the individual density of communities in the deep sea; it selects for increasing size ('deep-sea gigantism'), for passively food-collecting predators characterized by elongated appendages designed to exploit a large water volume, and for specific adaptations to a unique but stable biotope. Reduced mortality and metabolic rates, neutral buoyancy, omission of all 'unnecessary' body structures, simplification of circulatory and excretory systems, and long-term physiological (enzymatic) adjustments to life at low temperatures—all this tends to maximize the efficiency of energy turnover and to minimize energy expenditure.

In the mesopelagial (200 to 1000 m), species dominate which still feed in the surface zone during diel or ontogenetic migrations. The bathypelagial (1000 to 3000 m), is characterized by low abundances and biomasses of animals, mainly subsisting on organic animal matter from the surface zone. Most predators are passive lurers. The abyssopelagial (>3000 m) features further pauperization of diversity and abundance. There do not seem to be sufficient nutrients for passive predators and even carnivores are very rare (e.g. gammarids confined to the bottom). Among the few plankters, small-sized detritus feeders and euryphages dominate.

Management of pelagic ecosystems, while ultimately considered necessary, is still a goal rather than an immediately available possibility—even after many decades of increasing commercial exploitation and scientific assessment. Maximization of benefits obtained from the open ocean, protection of life in the high seas, and resource maintenance require concerted international co-operation for developing and applying sound measures of long-term management.

The basic prerequisites for managing ecosystems in the open ocean are: thorough knowledge of the structures and functions of these living systems, of their flow patterns of energy and matter, the environmental requirements of their components, and of the external factors which affect and control the dynamics of these systems. In addition to a much-needed re-emphasis of basic ecological research, modern methods of cybernetics, system analysis, and mathematical simulation are essential if we want to understand system dynamics and to develop capacities for predicting system behaviour as a function of time.

## Chapter 3: Coastal Ecosystems

Coastal waters have long been of great, immediate significance to man—economically, culturally, and scientifically. Major human population centres and industries developed near the coast, and numerous universities and marine laboratories specialized in the study of coastal communities. More accessible to general day-to-day experience and to scientific inquiry than the open ocean, coastal ecosystems have been subject to detailed and intensive exploration in several parts of the world.

Considering the tremendous amount and diversity of information available, we have chosen to focus our attention on two perspectives: the significance for coastal ecosystems of (i) salinity and related environmental factors, and (ii) the sources and flow patterns of energy and matter. The first perspective emphasizes categories and classifications of brackish waters, estuaries, and lagoons; the second uses the knowledge available on different types of coastal ecosystems for summarizing, exemplifying, and generalizing principles of energy dynamics, trophic transfer, and nutrient cycling.

### 3.1: *Brackish Waters, Estuaries, and Lagoons*

Chapter 3.1 emphasizes the historical fact that the study of ecosystems, as we conceive them today, began in the tidal flats of northern Germany. It was here that MOEBIUS (1877) discovered, during his extensive studies of oyster banks, a now well-established ecological principle: under certain, relatively uniform (in space and time) environmental circumstances, interdependent groupings of different kinds of organisms tend to organize themselves in typical, predictable patterns. MOEBIUS recognised the oyster banks of the shallow, muddy waters of Schleswig-Holstein to represent an integrated natural community of different forms of life—a 'Lebensgemeinschaft', 'Lebensgemeinde', or 'biocoenosis'.

In most coastal ecosystems, water movement dynamics (Volume I, Part 2) are essential determinants (see also Chapter 3.2). While turbidity and sedimentation tend to be ecological master factors in numerous coastal communities, life in temperate, rocky intertidal regions and in the coral reefs flourishes best without stream-borne sediments.

Brackish waters and estuaries have been studied by marine ecologists, especially in the northern hemisphere. Here, a number of attempts have been made to classify the waters investigated. Using, in essence, the types and quantities of organisms recorded in a certain brackish-water area, several investigators have proposed—mostly locality-related and hence specific and often diverging—classification schemes and a concomitant number of terms, sometimes cumbersome and conflicting (see also Volume I, p. 822).

Nevertheless, Chapter 3.1 concludes that salinity has been, and is likely to remain, 'the most useful single factor on which to base the classification of various aquatic regions' (p. 744). 'This statement should, however, not detract from the fact that, typically, functions and structures of an ecosystem depend on several factors acting in concert.

It is for this reason that we had devoted a special chapter in Volume I (ALDERDICE, 1972) to reviewing organismic responses to factor combinations.

Interface systems between oceans, rivers, and estuaries are, in essence, shaped by geomorphology and tides. Typically, they feature pronounced changes in salinity and seasonal factors, high loads of suspended matter, a complex hydrology and chemistry, and intensive organismic migrations. Increasingly, estuaries are subject to modifications due to a variety of man's activities. Variability and instability render estuaries difficult to classify in generally acceptable terms.

Important aspects of lagoon classification have been developed from studies of Mexican coastal areas. Based on historical events which form barriers separating a coastal water body from the sea, different types of lagoons have been described. As a general framework, these categories may also be applicable to other coasts of the world.

### 3.2: *Flow Patterns of Energy and Matter*

In addition to ecosystem structures (species composition; see especially Volume V, Part 1), ecosystem functions (flow patterns of energy and matter) determine the characteristics and boundaries of the system concerned. Within a given ecosystem, species composition is more consistent and species interactions are more intensive than across the border to neighbouring systems. Typically, the flow rates of energy and matter maximize within the system, with lesser exports to or imports from other systems. Depending on habitat and environmental factors, the roles of the essential system components—producers, consumers, and decomposers—may change, thus also modifying quantitative and qualitative parameters in the flow patterns of energy and matter.

Chapter 3.2 reviews, compares, and evaluates the information available on selected examples representing nine different coastal ecosystems: Narragansett Bay (Rhode Island, USA), Lynher estuary (Cornwall, UK), Askö area (Baltic Sea, Sweden), mangrove forest (South Florida, USA), Georgia salt marsh (Sapelo Island, USA), kelp bed (Nova Scotia, Canada), kelp bed (Benguela region, South Africa), sandy beach (Port Elizabeth, South Africa), and rocky shores (False Bay, South Africa). Among these examples, ecosystems can be distinguished that are largely based on different types of nutritional resources: (i) phytoplankton, (ii) macrophytes and detritus; and (iii) attached microalgae. Therefore overlaps between the system's main sources of energy and matter and other, additional, sources may modify and complicate the picture. Nevertheless, in ecosystem analysis, characterization of different systems based on their main resources of energy and matter can provide important insights into the system's dynamics.

Abiotic factors, especially water movement (Volume I, Part 2) affect structures and functions in coastal ecosystems more effectively than in open-ocean ecosystems (Chapter 2). Near the coast, water movement tends to exert great influence on sedimentation processes, and—through affecting the type of sediment—on the composition and activity of the local benthic fauna and flora. Waves, currents, and tides determine, to a considerable degree, the distribution and flow routes of energy and matter within the system. Thus, in a kelp bed, low water movement may turn the local community into a largely closed system in which the 'faeces loop' retains detritus as a main food resource; while upwelling carries the detritus away, diminishing the nutritional support, downwelling turns the phytoplankton into the main resource of the system.

In addition to abiotic factors, biotic interactions, in particular predator control, can determine the direction, intensity, and type of energy flows. A given system tends to attain stability only within certain patterns of energy flow. Deformation due to excessive environmental change—natural or man-made—may destabilize the system and force it to shift to a new steady state with different structures and functions. The biotic controls of a system may be complex and subtle; and include changes in behaviour of key organisms. The significance of chemical interactions between organisms and within an ecosystem requires more attention.

A comparison of food webs in the nine ecosystems considered reveals that filter feeders and/or deposit feeders dominate the consumers. Forming a feedback loop, faeces contribute significantly to the food supply, even in phytoplankton-based systems, such as that in Narragansett Bay. Faeces are even more important as nutrients in macrophyte-based systems, i.e. mangrove forests, salt marshes, and kelp beds, because macrophytes are less easily digestible and tend to be assimilated at reduced efficiencies. Attached microalgae such as diatoms may also represent an important nutrient source for particle feeders in such systems. Consequently, the categorization into ecosystems based on macrophytes or phytoplankton may entail appreciable areas of overlap: macrophyte-based systems can comprise significant diatom components, and phytoplankton-based systems may have considerable detrital components.

In regard to nutrient recycling, the question as to whether the micro-organisms or the macrofauna constitute the major system component responsible for remineralization is still open to debate; it requires more attention in future research. The role of these two groups of organisms for recycling may differ in different coastal ecosystems. In soft sediments, for example, the large microbial community (bacteria, fungi)—kept in check by the micro- and meiofauna—acts as dominant decomposer, while on solid substrates with less surface area for accommodating micro-organisms, remineralization may be predominantly carried out by the macrofauna.

The degree of coupling of the flows of energy and matter is another important aspect in the dynamics of coastal ecosystems that calls for more attention in future studies. Presumably, carbon and nitrogen flows are less closely coupled in coastal ecosystems than in open-ocean ecosystems (Chapter 2). As in the open sea, nitrogen seems to represent the main limiting nutrient in coastal waters, while carbon flow is more or less synonymous with energy flow. Although coastal ecosystems may appear quite self-sufficient, they can be strongly linked by a variety of transport and/or exchange phenomena. Some systems, such as the mangrove forests in Florida, appear to depend on large amounts of nutrient input transported by river runoff.

## Chapter 4: World Resources of Marine Plants

Marine plants have been of much less importance as a resource for the human population than marine animals (Chapter 5). In spite of a gigantic potential, the actual utilization of marine plants as human food and as raw materials has remained rather modest.

### 4.1: *Benthic Plants*

Chapter 4.1 assesses the world-wide resources of marine benthic plants on an ecological basis and with emphasis on climatic regions. The data available are scarce, widely scattered, and often insufficient for presenting definite quantitative statements. Only for a few species of immediate economic importance is information available on the quantities harvested.

In order to be harvestable economically, a plant resource must be present in sufficiently large, dense, and accessible populations, essentially free from impurities (epiphytes, epizoans, clay, etc.) Ideally it should grow in monospecific stands. Further important considerations for harvestability include cost of labour, energy, and repair, as well as distance to potential customers. The climate should be beneficial to harvesting operations including on-the-spot drying of the plants collected.

While still only a very minute fraction of world-wide living resource exploitation, the commercial utilization of marine seaweeds is increasing quickly. A growing number of seaweeds are utilized as raw materials for industrial products and as food for human consumption, especially in Japan, China, and Korea—countries in which seaweed consumption has a long history. Nevertheless, the present overall importance of benthic marine plants for providing human food is very limited—less than 1·1% of the total amount of food produced by agriculture, fisheries, and aquaculture. Only intensive mariculture, ocean farming, sewage reclamation, and extension of harvesting into the coldest seas could significantly increase the contribution made by benthic marine plants. However, compared to cereals and other staple food, seaweeds are likely to remain of marginal importance. Their nutritional significance rests not in quantity but in their contents of vitamins, iodine, and trace elements. Even small dietary additions of seaweeds may relieve whole populations from goitre and from other diseases due to malnutrition.

Raw materials obtained from benthic algae include phycocolloids which serve as emulsifiers and stabilizers. Phycocolloids of brown algae—alginic acid and alginates—are used in textile printing, for alginate fibres, jellies and cakes, beer and fruit drinks, pharmaceutical products, paper-making, and ice-creams. Of the phycocolloids of red algae, carrageenan is used in dairy products, for gelated desserts, jams, and fish jellies; agar, a well-known gelling agent, serves as a culture medium for a variety of micro-organisms and as a separator in macromolecular studies; it occupies a strategic position in the production of cosmetics, dentistry materials, and food stuffs. In agriculture, seaweeds are used as fodder, food additives, fertilizer, and soil conditioner.

Chapter 4.1 considers in detail the distribution, abundance, productivity, and harvestability of benthic marine plants for the major tropical, warm-temperate, cold-temperate, subarctic, and arctic coasts of the World Ocean. World-wide harvests are summarized in Table 4-5 (p. 830) and Fig. 4-6 (p. 831).

### 4.2: *Planktonic Plants*

Chapter 4.2 had to remain unwritten because of lack of information (see *Editor's Note*, p. 832). While the potential resources of marine planktonic plants are gigantic, these small inhabitants of the euphotic zone cannot be harvested economically. Commercial cultivation of marine microalgae, on the other hand, encounter problems of marketability. Many phytoplankters have tastes, textures, or odours that render it difficult to turn them into pallatable foods.

The primary use of planktonic plants is presently as food organisms in commercial cultivation (mariculture) and research cultivation (Volume III). However, they may have considerable future potential as raw materials for pharmaceutical products and for other industrial uses.

## Chapter 5: World Resources of Fisheries and their Management

One of the oldest activities of man, fishing still involves primarily harvesting of wild stocks. However, far-reaching technical advances in the sea fisheries (ship construction, catching technology, fish preservation and processing), especially between 1950 and 1970, have so enormously increased the world-wide harvesting capacity that progress in internationally-controlled resource management has become vital. The basic prerequisites for successful fisheries management are: (i) thorough knowledge of fish distribution, abundance, reproduction, and population dynamics as a function of natural environmental conditions and man-made interferences (particularly fishing); (ii) detailed information on the general ecological situation in the area concerned; and (iii) strict and reliable co-operation between all those harvesting the same resources, with the aim of transforming ecological knowledge into practical management policies and of effectively executing the measures agreed upon.

In an attempt to summarize and evaluate general trends in world-wide sea-fish harvests, the data available have been considered from five different perspectives: catches collected in different geographic regions, landings made by different nations, categorization of landed fish according to species, documentation of the gears used, and evaluation of the procedures adopted for processing and marketing.

Catches obtained in different geographic regions reveal large differences (Table 5-1; p. 841). Close to half of the world catch was taken in the northeast Atlantic and the northwest Pacific Oceans. While highest on the continental shelves and in coastal upwelling areas, the amount of harvestable fish differs considerably as a function of geographic region and local fishing activities. In general, the open ocean contains too few fish for commercial exploitation, except for a few predators near the top of the trophic pyramid (e.g. tunas, marine mammals) which act as aggregators of the productivity of large sea areas.

Comparison of the landings by different nations in 1979 yielded the following sequence among the leading fishery countries: Japan (10 million tons), USSR (9), China (4·1), Peru (3·7), USA (3·5), Norway (2·7), Chile (2·6), India (2·3), and South Korea (2·2). The 1979 world total was 71·3 million tons (Table 5-2, p. 843). Since 1950, annual landings have increased rapidly in many developing countries. Technical progress, in particular freezing at sea, has made it possible for countries with rather limited resources on their own coasts to become major fishing nations (e.g. USSR, Poland).

Among the principal groups of species caught were cods (10·6 million tons) and herrings/sardines (15·6). However, since 1948 the relative importance of these fish has decreased, and since the early 1970s many stocks of herring and similar species have collapsed, especially on the eastern side of the Atlantic Ocean (Table 5-4, p. 847). Of the other major species harvested, most play roles ecologically similar to cod (demersal forms of temperate shelves) or herring (shoaling pelagic forms of temperate shelves or upwelling areas). In 1983, the herrings and their relatives (sardines, anchovies, etc.) are still the largest group harvested and include several of the commercially most important species. Characteristically, year-to-year changes in the landing statistics of these fish can be very extreme. Examples are the collapse of the Californian sardine in the 1940s and of the Peruvian anchoveta in 1972. Commanding a high price, the contribution of tunas to the total financial value of the world catch by far exceeds their contribution in weight (approximately 4%).

Among the gears used, trawl (about 50% of the landings) and purse-seine (about 33%) now dominate the scene—a development paralleling the replacement of manpower by modern machinery. Nevertheless, over the years, fishermen have devised and developed a large array of fishing gear, taking advantage of the behaviour of different target species and adjusting their efforts to climates and different *in-situ* conditions. A world-wide comparision witnesses that similar solutions were often found to similar problems.

Procedures for processing and marketing have undergone considerable change in recent years. While some 50 yr ago more than half the catch was consumed fresh, and most of the rest was smoked or salted, freezing, canning, and conversion to fish meal and oil (feed for farmed animals) have greatly increased in importance. Canned fish from the major fishery nations has become a standard source of cheap protein in many developing countries. However, canning and freezing are also assisting developing countries in the tropics to export high-grade expensive protein (especially shrimps, but also tunas) to richer countries.

Considering the fisheries resources in the major regions of the World Ocean, northern temperate waters, upwelling areas, tropical seas, and southern temperate waters are distinguished, as well as the Southern Ocean and open-ocean areas (Techniques, areas, animals, and output of aquaculture operations have received attention in Volume III, Part 3: KINNE and ROSENTHAL, 1977; see also KINNE, 1980, 1983, in press).

Most of the richest fishing grounds are located in northern temperate waters. Here the Northwest Pacific Ocean ranks first in terms of weight caught, followed by the Northeast Atlantic Ocean. Comprising extensive areas of relatively shallow water, both areas are made up from an array of ecologically diverse subregions. The Northwest Atlantic Ocean is in many ways similar to the Northeast Atlantic Ocean. It features many of the same, or closely related, species and a similar fisheries, except that large, long-range vessels play—or used to play—a much more important role in the western Atlantic Ocean.

Upwelling areas include some of the world's richest fishing grounds. Upwelling occurs seasonally in many coastal areas, as well as in the open ocean along the equator. The reviewer focuses his attention on the four major coastal upwelling areas: Peru/Chile, California current, Benguela current, and Canary current. In these regions the prevailing wind blows along the coast towards the equator; it thus forces surface water offshore, replacing it by cool, nutrient-rich deep water. This and the intensive solar radiation, characteristic of most upwelling areas, provide ideal prerequisites for a high productivity in the surface waters concerned (Chapter 2). Significantly, the fish populations are similarly composed in all major upwelling areas (Table 5-13), some feeding directly on the abundant phytoplankters (e.g. anchovies), most on zooplankters (sardines, mackerels, jack mackerels). While larger predatory fish (e.g. tunas) are not as abundant as might be expected, marine birds and mammals tend to establish large and locally dense populations.

Pelagic fish stocks in upwelling areas may undergo extensive and, not yet fully understood, abundance fluctuations. Whole fisheries have collapsed suddenly (e.g. Californian sardine, Peruvian anchovy, Namibian pilchard), while others which collapsed some decades ago (e.g. Japanese sardine) have recovered equally dramatically. Demersal fish populations are much smaller. Apparently, most of the marine production is recycled within the surface layer (see also Chapters 2 and 3.2).

Tropical seas, while rich in fish, have been little exploited by the large fishery fleets of developed countries and have received insufficient attention from fishery scientists interested in stock assessment and management. Among the most important future resources, tropical fish stocks pose significant management problems, and—due to the large number of different, coexisting species—render the analysis of interspecific relationships difficult.

In southern temperate waters, the narrow shelf areas and low primary productivity support only a rather limited fisheries. Local human populations are small, and are among the world's greatest producers and consumers of meat. The richest fishing grounds of the southern hemisphere—except upwelling areas—are on the Patagonian shelf with large stocks of *Merluccius hubbsi* and other gadoids. The presumably largest stocks of pelagic fishes are those of *Engraulis anchovita* and *Clupea fuegensis*.

Surrounding the Antarctic continent, the Southern Ocean is the most isolated and least known part of the World Ocean. Interest in the ecology of Antarctic waters, which feature surprisingly high biological activities—even under the pack ice—and comprise abundant fishery resources, have recently stirred considerable international attention.

Sealers and whalers harvested the Southern Ocean for Centuries. Technological developments greatly increased the efficiency of marine-mammal hunting. Hence their protection and management have become issues of great concern. The food chain in Antarctic waters is very short with often only one link (mainly krill and trophically related forms) between primary producers and maximum-size consumers. Hence, turnover efficiency of the—not particularly intensive—primary productivity into harvestable resources is very high. Maximum values of overall productivity have been recorded along the coasts and the ice edge. The considerable resources of the Southern Ocean seem to be sensitive to heavy exploitation. Following concerted harvesting, marine-mammal and fish populations were soon reduced to uneconomic densities.

While potentially suitable as human food, krill poses several problems: (i) there is very little meat in each individual; (ii) the meat is extremely delicate and hence must be processed quickly; (iii) it is difficult, and has not yet been possible, to develop a reasonably priced product that could attract a large demand; and (iv) processing is technically difficult and there is a high content of fluorine in the exoskeleton. Even the use of krill as feed or fertilizer in agriculture and aquaculture is limited by the relatively high price of the final product. The potential impact of large-scale krill harvesting on Antarctic ecosystems may be considerable and requires careful attention.

Although the open oceans cover some 80 to 90% of the area of the total World Ocean, their productivity is low and scattered. Important open-ocean resources are whales (especially sperm whales), tunas, squids, lantern fish, and small mesopelagic fish. Saury and dolphins are harvested only when they are close to the shore.

Management of fishery resources comprises all decisions that affect man's influence on exploitable populations and the benefits thus received. In 1983, the direct influence of fishing is much greater than that of other human influences (e.g. pollution and land reclamation). Only the effects of fishing are considered in this chapter. Pollution and related activities of man are dealt with in Parts 3 and 4 of Volume V.

Management policies must take into account the general ecological and social situation, as well as the economic and social conditions of the fishery and fishing industry, and the biological condition of the stock harvested. Management policies must seek and find an appropriate balance point between high exploitation and avoidance of critical long-term harm to the exploited stock. Because they did not meet these requirements, most international management measures have thus far been failures. In addition to often insufficient ecological knowledge, the major shortcomings of fishery management can be summarized in six words: too local, too little, too late. In order to be successful, management measures must be ecologically sound, universally accepted, and strictly enforced by an appropriate authority. The UN draft texts for a new regime of the sea are bringing fishing in Exclusive Economic Zones (EEZs) out to 200 miles, under the jurisdiction of the coastal state. Over 99% of present catches are taken in territorial waters and these EEZs. Nevertheless, international management co-operation would still be required in respect to fish that might migrate between two or more EEZs or that are caught on the high sea.

There are two basic questions which the fishery manager asks the fishery scientist: (i) Are changes in the amount or pattern of fishing effort necessary in order to sustain stock exploitability? (ii) If so, what kind of measures should be adopted to effect these changes and what will be their effect? Interactions between species complicate the matter. For example, if it is possible that overfishing of one species may induce replacement by other equally well harvestable and marketable fish species, there would be—from an economic viewpoint—less urgent need for careful management of the first species. Chapter 5 comprehensively summarizes and evaluates the information available on methods and models of fishery management and considers the ecological, economic, and social implications of management measures.

Major shortcomings result from the fact that most management activities are restricted to a single species and to a single fishery. Interspecific dynamics, concomitant effects of other fisheries (exploitation of interacting species), fishery damage (mortality and injury of by-catch, destruction and modification of biotopes), other detrimental effects of man (e.g. pollution), epizootics, and other disease phenomena (parasitization, proliferative disorders, consequences of large-scale catch injuries) are insufficiently taken into account. In other words there is great need to pay more attention to the general ecological background of the target populations.

Even the most advanced management models do not include all factors considered essential. They focus on feeding interactions and consider only fish of commercial size (e.g. cod eats herring), neglecting developmental stages (e.g. herring eats cod eggs or larvae). A comprehensive model incorporating all potentially essential interactions has not yet been achieved.

Over the next few decades, the contribution of fisheries to feeding the growing human world population is expected to increase but slowly. Especially the harvest of the types of fish presently marketed will not be augmentable to the extent required and originally predicted. There may be some increase though in the exploitation of hitherto unfamiliar species and sea areas. In any case, the pressure on our living marine resources will increase and so will the need for their protection and management.

*Acknowledgements.* This chapter was completed while visiting the Australian Institute of Marine Science in Townsville, Australia. It is a pleasure to acknowledge support from the Director of AIMS, Dr John Bunt, as well as the benefit of interesting discussions with several of his staff members.

## Literature Cited (Chapter 1)

ALDERDICE, D. F. (1972). Factor combinations: responses of marine poikilotherms to environmental factors acting in concert. In O. Kinne (Ed.), *Marine Ecology,* Vol. I, Environmental Factors, Part 3. Wiley, Chichester. pp. 1659–1722.

KINNE, O. (1978). Introduction to Volume IV. In O. Kinne (Ed.), *Marine Ecology,* Vol. IV, Dynamics. Wiley, Chichester. pp. 1–11.

KINNE, O. (1980). Aquaculture. A critical assessment of its potential and future. *Interdisciplinary Science Reviews,* **5**, 24–32.

KINNE, O. (1983). Aquacultur: Ausweg aus der Ernährungskrise? *Spektrum der Wissenschaft,* **December 12/1982**, 46–57.

KINNE, O. (in press). Realism in aquaculture—the view of an ecologist. In M. Bilio and H. Rosenthal (Eds), *Proceedings of a World Conference on Aquaculture, Venice, 1981.*

KINNE, O. and ROSENTHAL, H. (1977). Cultivation of animals. Commercial cultivation (aquaculture). In O. Kinne (Ed.), *Marine Ecology,* Vol. III, Cultivation, Part 3. Wiley, Chichester. pp. 1321–1398.

MARGALEF, R. (1977). *Ecologia.* Omega, Barcelona.

MARGALEF, R. (1978). General concepts of population dynamics and food links. In O. Kinne (Ed.), *Marine Ecology,* Vol. IV, Dynamics. Wiley, Chichester. pp. 617–704.

MOEBIUS, K. (1877). *Die Auster und die Austernwirthschaft.* Wiegandt, Hempel und Parey, Berlin.

PERES, J. M. (1982). General features of organismic assemblages in pelagial and benthal. In O. Kinne (Ed.), *Marine Ecology,* Vol. V, Ocean Management, Part 1. Wiley, Chichester. pp. 47–66.

PETERSEN, C. G. J. (1911). Valuation of the sea. I. Animal life of the sea-bottom, its food and quantity. *Rep. Dan. biol. Stn,* **20**, 1–81.

PETERSEN, C. G. J. (1914). Valuation of the Sea. II. The animal communities of the sea-bottom and their importance for marine zoogeography. *Rep. Dan. biol. Stn,* **21**, 1–68.

TANSLEY, A. G. (1935). The use and abuse of vegetational concepts and terms. *Ecology,* **16**, 284–307.

Marine Ecology Vol. V, Part 2
Edited by Otto Kinne


# 2. OPEN-OCEAN ECOSYSTEMS

M. E. VINOGRADOV

## (1) Introduction

The pelagial of the ocean, the largest biotope of the biosphere with its immense capacity of 1·4 milliard km$^3$ of dissolved mineral and organic substances, has preserved its structure through hundreds of millions of years. The singularities which distinguish the ocean from any other part of the biosphere are reflected in the properties of the ecosystems in this unique biotope.

All oceanic ecosystems are more or less interconnected by different-scaled horizontal and vertical water circulations so that, viewed from this standpoint, they may be regarded as a single supersystem. However, the ecosystems of different regions differ from one another and it is the extent of their similarities and dissimilarities that determines the biological structure of the ocean (ZENKEVITCH, 1948). Hierarchism is distinctly expressed in all pelagic ecosystems from the all-oceanic one down to the short-lived systems of water lenses with a scale of a few tens of metres.

The basic singularities of pelagic ecosystems which determine their structure, functioning, and their distinction from the ecosystems of the ocean bottom and the land, are associated with the singularities of their biotope—its mobility, its constant mixing, and the immensity of its water column.

### (a) Mobility of the Biotope

The biotope of a plankton community consists of the water masses which the community inhabits. Each water mass is characterized by definite temperature and salinity (T–S curves) and other physico-chemical and biological relationships formed under definite physico-chemical conditions and existing more or less stably in time and space. A plankton population may also serve as a characteristic of a water mass—in fact a most conservative one. The composition of plankton is little affected by horizontal displacements and gradual transformation of waters; the plankton organisms adjust themselves to gradual changes in the environment which they inhabit and to which they are confined. Therefore, the thermal, saline, and other limits restricting the distribution of a plankton species within a certain region are usually to be considered not as the extreme boundaries of its range but as the boundaries of the water mass to which the given species is confined. This applies also to groups of species (see also Volume V: PERES, 1982).

The water masses succeed each other not only geographically but also vertically, occupying different water layers. Their assortment from surface to bottom is termed 'structure'. This structure may vary in different climatic zones, but as a rule surface, intermediate, deep, and bottom masses are recognized.

The moving water generates in the ocean vast gyres or smaller eddies with neutral zones between them. The planetary gyres usually include surface and intermediate water masses, but may sometimes penetrate to greater depths. Eddies attaining diameters of hundreds of kilometers and existing sometimes for several months (KOSHLYAKOV and MONIN, 1978) seem to play an important, although as yet little known, part in the existence of pelagic communities.

The hydrological boundaries between water masses are not constant in time but vary within geographical coordinates with ensuing temporal variations in the boundaries between the communities of these water masses. Moreover, the mixing of water masses results in the intermixing and interpenetration of different communities and the creation of vast transitory zones (zones of mixing) both in the horizontal and vertical direction (see also Volume V: PERES, 1982).

The mobility of the biotope imparts to the large-scale pelagic ecosystems some fundamental peculiarities inherent only in them, which determine their structure and functioning (these have been considered in detail by MCGOWAN, 1974, and REID and co-authors, 1978). Noteworthy among them are the following. The basic pelagic ecosystems are large in area but not numerous, as had been pointed out earlier by BOGOROV (1970). They are confined to the main cyclonic (in cold water regions) and anticyclonic (in the tropics) gyrals and to the system of equatorial zonal flows. Thus, according to VORONINA (1978), the following regions confined to the main large-scale gyrals may be recognized in the open ocean: two cold water ecosystems (a Central Arctic and a Coastal Antarctic Circumpolar), three temperate (North Atlantic, North Pacific, and Subantarctic), five central tropical (two in the Pacific, two in the Atlantic, and one in the Indian

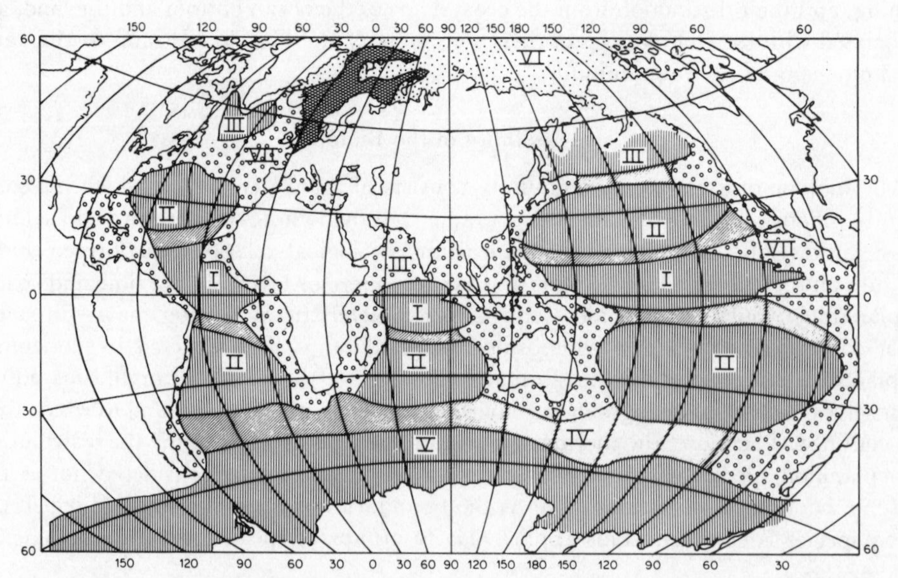


Fig. 2-1: Basic ecosystems of the oceanic pelagial. I: equatorial; II: of anticyclonic tropical gyrals; III: north subpolar; IV: subantarctic; V: antarctic; VI: arctic ice-neritic; VII: distant-neritic and neritic. Oblique hatching: transitory zones between communities. Black: North Atlantic expatriation region. (Based on information provided by BEKLEMISHEW, 1969; modified)

Ocean) and three equatorial (Fig. 2-1). To them must be added the (distant-neritic) ecosystems of neutral regions and coastal (neritic) systems.

The mixing of waters and populations at the periphery of oceanic ecosystems may be quite intensive so that the quantity of specific (endemic) species in ecosystems of the various taxonomic groups may be reduced to from 5 to 30% of their total number. Thus, for instance, in the equatorial gyral the basic ranges of about 50% of the species are situated in the northern and southern central gyres (HEINRICH, 1975). Strictly speaking in their 'pure' form these basic systems, i.e. their core, occupy a relatively small area of the ocean (Fig. 2-2), the greater part of its aquatory consists of regions transitory between ecosystems which, in fact, are ecotones.

The functional and structural processes and the course of the seasonal development of communities are relatively homogeneous within each system and their dissimilarities in the ecosystem are usually smaller than between different systems. However, the dissymmetry of the main oceanic gyres, caused by the fact that their centres are closer to the eastern coasts of continents, results in a considerable unevenness of conditions in the gyres, in the formation of additional, smaller stable circulations (Fig. 2-3) influencing the functioning of pelagic ecosystems, and the distribution of plankton.

Quantitative and qualitative differences exist between ecosystems in regard to the processes of life activity. Some of these differences are accounted for by the degree of eutrophication of surface layers and rate of life development which, in their turn, are determined by the ascending or descending water movement in cyclonically or anticyclonically oriented gyres. However, the essential attribute clearly distinguishing the

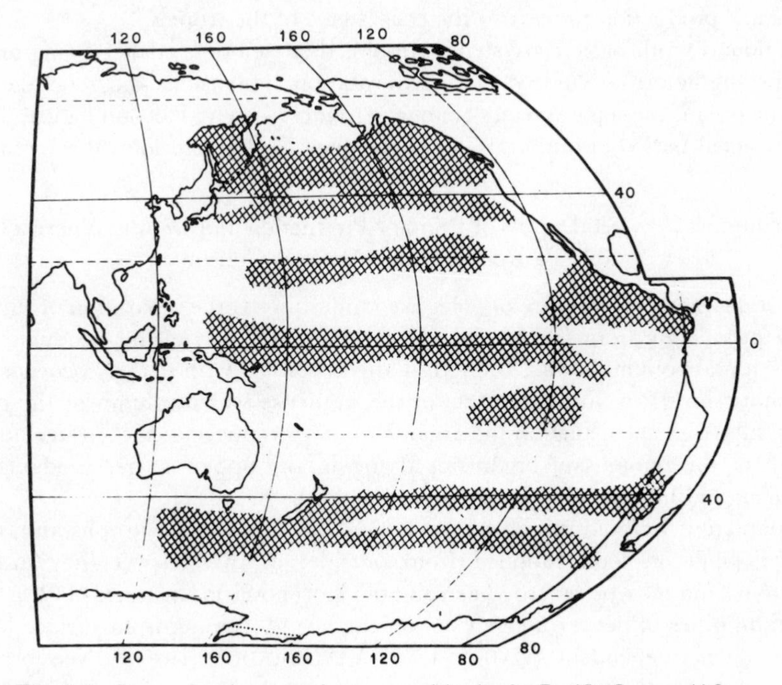


Fig. 2-2: Cores of major oceanic communities in the Pacific Ocean. (After McGOWAN, 1974; reproduced by permission of the author)

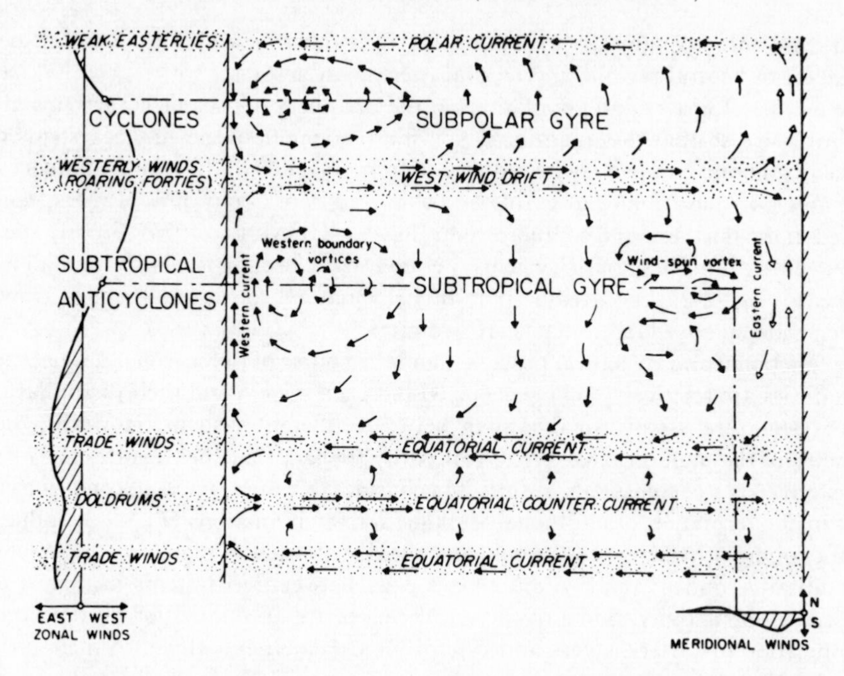


Fig. 2-3: Schematic wind-driven circulation in ocean. (After MUNK, 1950)

ecosystems is the cyclic process of production in the ecosystems of cold water regions and the constant production process in the ecosystems of the tropics.

Evolutionarily, all basic ecosystems are old, their self-dependence being assured by such constant factors as wind system, earth rotation, dimensions and general configuration of the ocean; these parameters change extremely slowly. Paleontological data point to a protracted period of immutability in the general character of water circulation.

### (b) Decoupling Between Depths of Primary Production and Depths Where a Considerable Part of Organic Matter is Consumed

A most characteristic feature of pelagic communities is the separation of the depth of primary production from the depths where the greater part of this production is consumed. Animal populations are brought in direct contact with the phytocoenosis only in the euphotic layer; in all other parts of the immense water column of the ocean the animals, although subsisting on the organic matter produced at the surface, utilize it in the form of the remains of organisms dying in the upper layers, products of their metabolism and interzonal animals migrating to the depths.

The plankton communities of the producing zone include autotrophs and, therefore, do not depend on food supplied from outside; in this respect they are energy-autonomous and may be regarded as valuable biocenoses in the sense of ODUM (1971). The communities of deeper layers exist on the energy coming from surface layers, i.e. they are energy-dependent (ZAIKA, 1967; VINOGRADOV, 1970a). Neo-formation of organic matter by deep-sea chemosynthesizing bacteria is of local importance and plays a purely subordinate part in the total balance of organic matter.

However, a constant and sometimes very intensive exchange of populations takes place between the communities of different layers. In cold water regions, the upper-interzonal phytophages *Calanus cristatus* and *C. plumchrus* in the Pacific, *C. hyperboreus* in the Arctic, *C. finmarchicus* in the North Atlantic, and *C. propinquus, Calanoides acutus, Rhincalanus gigas* in the Southern Ocean sink to depths of 1000 to 2000 m and sometimes to 4000 to 6000 m after feeding in the surface zone (VINOGRADOV, 1970a). These crustaceans account for nearly half the biomass of the total population of the meso- and bathypelagial (750–3000 m) and are thus the most important components both of surface and deep-sea communities. Viewed from this standpoint, the energy-autonomous communities of the surface layers and the energy-dependent deep-water communities may be regarded as parts of a single large community populating the whole, or nearly the whole, water column of the ocean; and the whole water column itself, with all its inhabitants, may be seen as a single ecosystem. Nevertheless, the population of the water column is stratified, which makes it possible to differentiate communities characteristic of different depths and to examine their specific features.

### (c) High Rate of Matter Turnover in Communities

Terrestrial plants develop powerful supporting systems of roots and trunks; but these are not required by marine plants except for marine phanerogams. The aquatic environment with its dissolved mineral and organic nutrients facilitates direct food assimilation by plants through cell membranes, thus keeping the living mass of phytoplankton in constant turnover. The specific annual production of phytoplankton surpasses the specific production of land plants many times. It is apparently higher than 100 and, according to some estimates, reaches 370 (BOGOROV, 1970) as compared with less than 0·1 on land (BOLIN, 1972); this amounts to a difference of four orders of magnitude.

Characteristic of the energetic exchange in terrestrial plants is the accumulation of energy in the form of carbohydrates. In plankton algae there is almost no accumulation, the energy being expended mainly on growth. In most plankton organisms intensive growth is achieved by a rapid succession of generations and only in a few zooplanktivorous animals by increased body size (some coelenterates, whale sharks, sunfish, etc.). Plankton algae may produce up to eleven generations within 24 h (FINENKO, 1977). In tropical copepods a generation lasts from 10–12 to 30 d. A most striking example of high growth rate are salpas; their weight may increase by 10% $h^{-1}$, and the numerical abundance of populations by 1·6 to 2·5 times 24 $h^{-1}$ (HERON, 1972).

The labile changes in the biomass of food organisms and of their consumers may result in grazing out of practically the total plant production over large areas of the ocean, whereas in land systems, such as pasture meadows, less than a one-seventh of the production is consumed (MACFADYEN, 1964). Due to this lability, pelagic systems depend more on environmental conditions than terrestrial ones (STEELE, 1974).

VINOGRADOV and VORONINA (1979) point out that the high growth rate and rapid succession of generations on the one hand and the mobility and variability of the biotope on the other contribute to a very rapid development of pelagic ecosystems. In the slowly developing terrestrial communities, such as the forests with their enormous reserves of living mass and long life cycles—which in trees may extend to over tens and hundreds of years—many decades and centuries are needed for the community to reach a mature state, and many years and decades are needed for herbaceous prairie areas with shorter life cycles of dominant species. In pelagic plankton communities, in which generations flash by as in movies, the developmental period is immeasurably shorter; in the tropical regions of the ocean it takes no more than 3 to 6 mo. During this period, species are succeeded by others, the number of species and trophic diversity increases, food chains extend, and a series of other successional changes take place. Under favourable hydrophysical conditions (stable water stratification) the community reaches a relatively high degree of maturity (BLACKBURN and co-authors, 1970; GUEREDRAT, 1971; VINOGRADOV and co-authors, 1973; see also Volume V: PERES, 1982).

### (d) Spatial–Temporal Aspect of the Existence of Pelagic Communities

The concept of the development of an ecosystem (or a community) over time, i.e. its succession, is one of the fundamental propositions of modern ecology (Volume V: PERES, 1982). MARGALEF (1968) writes: 'succession in ecology occupies the same place as evolution in general biology'. Succession is a process of self-organization that goes on in any system.

Any system formed by reproduction and interaction between organisms and the environment develops towards the creation of an entity in which the value of entropy per unit of retained and transmitted information is reduced to a minimum. Only those structures of a system most capable of influencing the future with least expenditure of energy are preserved over time. In other words, the process of succession is equivalent to the process of accumulation of information, the system tending towards a certain stationary asymptotic state (MARGALEF, 1968).

Such an interpretation of successional changes in an ecosystem applies also to changes occurring in developing communities of the oceanic pelagial (VINOGRADOV, 1977b), although these communities possess some singularities of their own that influence the character of succession—in the first place, a far greater degree of 'openness' as compared with terrestrial systems.

The fertility of the upper productive layer is assured by nutrients carried via turbulent processes into the euphotic zone from deep-water layers where their reserves exceed their annual expenditure by several orders of magnitude. The intensity of this inflow, i.e. a purely physical process, controls the productivity of pelagic ecosystems. It depends far less on organismic activities than the productivity of terrestrial ecosystems. Furthermore, pelagic systems are, as a rule, far more heavily exploited than land systems.* Transfer of organisms from the layer of their natural habitat to depths with unfavourable living conditions, translocation by currents away from the optimal horizontal habitat area, grazing to extinction by carnivores—all these processes are very intense in pelagic systems and involve the overwhelming part of their plankton populations. Intensified exploitation, as pointed out by MARGALEF (1968; see also Volume IV: MARGALEF, 1978), hinders or prevents the system from attaining maturity. It is interesting that plankton communities attain a relatively high degree of maturity only in stratified waters with weak mixing, where the outflow of phytoplankton cells from the euphotic layers is reduced to a minimum.

Species diversity increases with progressive maturing of the communities. Thus, the index of informative species diversity (SHANNON'S index) of marine phyto- and zooplankton varies from 1·4 to 2·0 in young systems to between 4·5 and 5·5 in the late stages

*'Exploitation' refers to any removal of living organisms from a given system (MARGALEEF, 1968).

of succession (MARGALEF, 1968; TIMONIN, 1971; ZERNOVA, 1976). In terrestrial ecosystems this index varies between the same range: from 1·5 in the earliest stages of formation (KARR, 1968) to from 3·5 to 4·5 in mature systems, such as, for instance, the tropical rain forest (LLOYD and co-authors, 1968). MARGALEF (1968) regards the value of SHANNON's index 4·5 to 5·5 as the limit attainable only by fully mature systems. The trophic diversity of the community increases simultaneously with species diversity (TIMONIN, 1971): in the early stages of succession non-specified phytophages or euryphages are most abundant, while in the late stages there is a dominance of species with more selective food habits, consumers of larger food units. Relatively mature oceanic systems are characterized by a high share of macrophages and longer food chains. Organizational development of the ecosystem creates a great number of econiches and semi-isolated subcommunities (see also Volume V: PERES, 1982).

Another determining element of succession in the oceanic pelagial is a rapid accumulation of energy during the earlier stages of the community's development and its gradual expenditure with increasing maturity. Naturally some energy is expended also at the beginning and some reserves are accumulated by 'mature' communities, but at the earlier stages the processes of energy assimilation prevail over the processes of dissimilation, while in the later stages the processes are reversed. The decrease in energy stored by the community with its maturation is reflected in its biomass. In the earlier stages of development the biomass increases rapidly and concentrates in the most labile link of the community, the phytoplankton. When the processes of destruction begin to prevail in the community the total mass of plankton decreases, but the biomass of individual groups may continue to accrue, the increase concerning first the phytophages, then the small, and eventually the larger carnivorous forms, resulting in a cardinal alteration of the trophic structure of the community (see Fig. 2-20). The pattern of this migration of biomass along the trophic chain has been clearly demonstrated in field observations (e.g. VINOGRADOV and VORONINA, 1964; GUERDERAT, 1971; TIMONIN, 1971), and in the simulation of the functioning of a community (VINOGRADOV and co-authors, 1972, 1973).

With progressing maturity not only the structural but also the functional characteristics of the community are modified. In the early stages of community development, when the biomass of autotrophs is high, the food requirements of animals are well satisfied at all trophic levels, thus assuring near-maximum increment. Later the degree of satisfaction of food requirements decreases with concomitant increases in the stress of trophic relationships between organisms of different trophic levels. There is also a higher degree of balance between the production at different trophic levels and its consumption (VINOGRADOV and co-authors, 1976).

Thus, with progressing development and with displacement in space with the moving water, pelagic communities undergo substantial changes. Therefore, when comparisons of ecosystems of different areas of a single large region are made, possible age differences responsible for many of the observed dissimilarities, should always be taken into account.

### (e) Long-term Variation of the Structure of Communities

In addition to regular seasonal or quasi–permanent changes pelagic communities are subject to more prolonged or irregular but significant variability. The high variability associated with fluctuations in climate, atmospheric circulation, and ensuing hydrophysical circulation changes is widely known (BJERKNES, 1969; MONIN and co-authors, 1974). It is natural that this variability exerts an extremely strong influence—usually explicable but not always understandable—on pelagic communities. As there are no intermediate buffers in the pelagial between organisms and environment (such as, for example, soil or microclimate created by plants in terrestrial communities), these effects may be very sharply expressed. The fluctuations induced by them in the biomasses of zooplankton and commercial fishes are quite comparable with, and often stronger than, the results of human economic activities. Several examples of such fluctuations have been examined by ZELIKMAN and KAMSHILOV (1960), LONGHURST and co-authors (1972), COLEBROOK (1978). (See also Volume V: PERES, 1982.)

Among such phenomena, perhaps the best known and most thoroughly studied is the

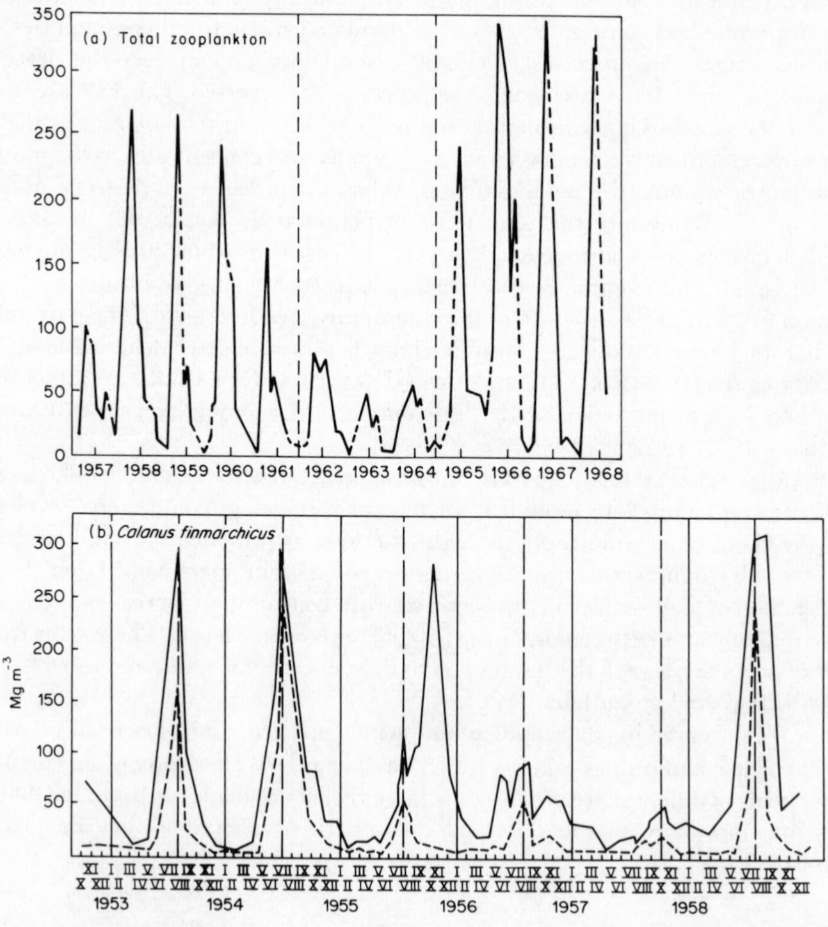


Fig. 2-4: Annual fluctuations of zooplankton biomass. (a) 0–150 layer at the weather station 'P' (50° N–150° W). (Based on LONGHURST and co-authors, 1972); (b) Barents Sea (Based on information provided by ZELIKMAN and KAMSHILOV, 1960)

regularly recurring El Niño: when the warm, strongly freshened equatorial waters penetrate far southward along the Peruvian coast they destroy the ecosystems of the Peruvian upwelling (see below), causing catastrophic losses to the Peruvian anchovy fishery and killing millions of coastal piscivorous birds, predatory fishes, and other animals. The phenomenon is caused by the variability of the trade-winds induced circulation over the Pacific Ocean (BJERKNES, 1969). The multi-year fluctuations in the intensity of the Kuroshio and the penetration of its branches into the Japan Sea may cause either the appearance of enormous shoals of *Sardinops* at the shores of the Soviet Primorye (as in 1939–1940) or their entire disappearance from these waters for several decades. Such multi-year fluctuations in the abundance of clupeid fishes can also be observed in other coastal regions.

Also in the open ocean abundance, biomass, and even composition of zooplankton and fish populations are subject to strong long-term fluctuations. They were distinctly displayed in the multi-year plankton collections at weather station 'P' in the Pacific Ocean (Fig. 2-4); in work with a plankton indicator in the North Atlantic (COLEBROOK, 1978); in long-term observations on plankton composition in the English Channel; and in plankton investigations in the Barents Sea (ZELIKMAN, 1977).

During the 1960s and 1970s directional changes in the structure and productivity of ecosystems have taken place in some inner seas and coastal regions exposed to particularly strong anthropogenic activities, but these changes bear no relation to the natural fluctuations of these ecosystems.

### (f) Structural and Functional Patterns of Communities

The functioning of marine communities is assured by an intricate gamut of relationships between populations that consist of organisms evolutionally adapted to existence under the conditions of the given system. A definite part is played also by the morphophysiological features of the organisms, their genetic characteristics, behavioural reactions, etc. Different types of relationships are formed between the components of the communities based on exchange of energy, matter, and information; but it may be assumed that the basic relationships that integrate the community as an entity and form the groundwork of its structure and productivity are trophic ones of the predator–prey type. (ELTON, 1927). That is why investigations of trophic relationships within the community, estimates of energy flow through a biological system, and its utilization by diverse trophic groups yield the most substantial information on the functioning of the communities.

In these studies the investigator is interested not only in certain isolated parts or properties of separate elements but also in their totality and interaction. Therefore, equal attention should be devoted to the integral (structural) features (such as, for instance, number of species, diversity, age and size composition, spatial distribution, abundance, and biomass) and to the processes of interaction between its components, i.e. to its differential (functional) characteristics.

The structural characteristics—far easier to analyse than the functional ones—are also far better known. The species structure of the communities and their separate taxocenes, their singularities, and the laws governing their formation, have been discussed by scores of authors. More and more works are appearing that deal with different aspects of the spatial structure of communities, the size structure of various taxocenes, etc. Intensive investigations are conducted on the trophic structure, not to mention hundreds of investigations on plankton abundance and biomass.

In contrast, few research activities have been devoted to the functional characteristics of communities, for which quantitative assessments involve great difficulties. The only exceptions are experimental mass determination of phytoplankton production (not yet quite satisfactory methodologically; see the reviews by RAYMONT, 1963; RYTHER, 1963; STRICKLAND, 1972; PARSONS and TAKAHASHI, 1973; SOROKIN, 1973; LORENZEN, 1976; KOBLENTZ-MISHKE, 1977; FINENKO, 1978) and far less intensive determinations of bacterioplankton production and destruction (reviews by SOROKIN, 1971a,b, 1973, 1977b). Determinations (or calculations) of marine zooplankton production are less frequent and until recently dealt mainly with populations rather than with trophic levels (reviews by MANN, 1969; MULLIN, 1969; GREZE, 1970; BOUGIS, 1974; TRANTER, 1976; SHUSHKINA, 1977).

An increasing number of researchers are studying production at the community level (RAYMONT, 1963; RAYMONT and CARRIE, 1964; PARSONS and co-authors, 1969; SHUSHKINA and co-authors, 1974, 1978; LEBORGNE, 1978; VINOGRADOV and SHUSHKINA, 1978). Methods of calculation have been developed (VINOGRADOV and co-authors, 1976; VINOGRADOV and SHUSHKINA, 1978) based on experimental determinations of metabolic rate, food assimilation, and conversion of assimilated food into growth. These parameters not only permit evaluation of the production of various trophic groups, trophic levels, and the community as a whole, but also the degree of satisfaction of food requirements in different trophic groups ($\delta = C_i/C_i^{\max}$),* the stress of food relationships, the degree of balance of production and consumption between elements of different trophic levels

$$\varepsilon_j = \frac{P_j}{\sum_i r_{ij}},$$

ecological efficiency (SLOBODKIN, 1962)

$$\omega_{ij} = \frac{\sum_i r_{ij}}{R_j},$$

efficiency of food assimilation

$$\eta = \frac{P + R}{R} = \frac{A}{R},$$

flow of specific energy through the community (MARGALEF, 1968) ($P_p/B_0$), rate of energy utilization

$$K_{3p} = \frac{P_p}{\sum_{i=b}^{s} R_i} = \frac{P_p}{D},$$

and many others.

*Where $C_i$ is the value of food consumption of the $i$ th element of the community; $r_{ij}$, the particular ration of $i$th element on the $j$th group of food organisms; $Ri$, metabolic rate; $P_i$ production of $i$th element; $A$, assimilated food; $D$, destruction of the heterotropic part of the community ($D = \sum_{i=b}^{s} R_i$); $U^{-1}$, food assimilability; $\Delta P$ allochthonous organic matter (detritus, DOM) assimilated by bacteria. Elements of the community: $p$, phytoplankton, $b$, bacteria; $a$, protozoans, $f$, non-carnivorous metazoans; $s$, predatory zooplankton.

Thus it becomes possible to appraise the intensity of processes in the community and their changes with progressing maturity. Finally, net and actual productions of communities have been calculated. Net production

$$P_0 = P_p - \sum_{i=b}^{s} R_i = P_p - D$$

(WINBERG, 1960)—equal to

$$P_0 = \sum_{i=p}^{s} P_i - \sum_{i=b}^{s} C_i U_i^{-1}$$

(ZAIKA, 1979; SHUSHKINA, 1977)—permits us to judge to what extent the organic matter (primary production), formed at a given moment, is utilized by all heterotrophic elements of the community. Actual production includes the utilization by heterotrophic elements (bacteria) not only of the newly formed organic matter but also of the earlier formed (during the productive phase of community development), or brought from outside ($P_{ac} = P_p + \Delta P - D$) (VINOGRADOV and co-authors, 1976, 1980; VINOGRADOV and SHUSHKINA, 1978). (For an example consult Volume V: PERES, 1982.)

## (2) General Patterns of Quantitative Life Distribution

### (a) Large-Scale Horizontal Distribution

In different latitudes varying amounts of solar energy enter the ocean which is stratified differently and whose surface waters are enriched with nutrients different from those required for the development of phytoplankton. Accordingly, distribution and intensity of the diverse abiotic and biotic factors also undergo substantial changes in a meridional direction. Although the amount of solar energy supplied to the surface of the ocean changes only gradually, the changes in the general pattern of life distribution are far less monotonous, giving rise to the formation of distinct latitudinal biological zonality. (See Volume V: PERES, 1982.)

Another important factor is the effect of coasts and coastal shallows on which depend many fundamental singularities of water circulation and the supply of nutrients to the euphotic layers. Thus, a circumcontinental zonality superimposes itself on the latitudinal one (ZENKEVITCH, 1948) and the actual picture presented by the natural zones (natural oceanic regions) is the result of the interference of both these types of zonality.

The rate of photosynthesis in the sea depends primarily on light energy and concentration of mineral salts. Another necessary condition is that turbulent mixing may not carry away too large a portion of production from the euphotic zone, i.e. that a certain relationship be preserved between the thickness of the mixed layer and the critical depth (see, for instance, PARSONS and LEBRASSEUR, 1968). The rate of grazing by zooplankton may also affect the level of phytoplankton production. It is the combined effect of all these factors that controls the level of production in oceanic regions. Since the problem of phytoplankton production in the ocean has been discussed in detail in Volume IV by FINENKO (1978), we restrict ourselves here to a very brief examination of the factors listed above. (See also Volume V: PERES, 1982.)

### *Light*

Polewards of Latitude 50° to 60° total solar radiation is less than 60 kcal $cm^{-2}$ $yr^{-1}$, but in ice-free waters, during the months with daylight, it does not limit phytoplankton development in the surface layers. However, during the winter period, illumination becomes a limiting factor substantially reducing the vegetation period. In iced seas, where ice reduces the period of the light optimum, the vegetation period is still shorter. Thus, in the Central Arctic Basin, according to USACHEV (1961) and KAWAMURA (1967), the vegetation season lasts merely 1 to 1·5 mo.

In low latitudes powerful radiation inhibits photosynthesis in the uppermost water layers and the optimum of phytoplankton development shifts to greater depths. However, the intensity of solar radiation in itself does not limit phytoplankton development neither in temperate-cold water nor in Tropical zones.

### *Mineral Nutrition*

Differences in phytoplankton production in various regions of temperate-cold water and Tropical zones depend in practice on the concentration of nutrients in the layer above the basic pycnocline, and this concentration, in turn, depends on vertical water movement through the basic pycnocline.

In temperate and cold water (Subpolar) zones, winter convection is the major factor assuring the inflow of nutrients into surface waters. The thickness of the water layer involved varies between 150 and 200 m in the northern part of the Pacific Ocean, but may reach 800 to 900 m and more in some regions of the North Atlantic Ocean. Of considerable, but apparently secondary, importance in the enrichment of surface waters are water ascents produced by hydrodynamic causes. In the tropics, on the other hand, quasi-permanent hydrodynamic phenomena are of primary importance in determining the level of production in different regions. A weakly expressed winter convection is observed only in Subtropical regions, but as it does involve the deeper waters rich in nutrients, it induces particularly intense phytoplankton development (SEMINA, 1966).

### *Stratification*

As already mentioned, the development of phytoplankton depends not only on light intensity and nutrient concentration but also on a definite thickness of the mixed layer. This must not exceed its critical value too much because otherwise too intensive an outflow of cells from the eutrophic zone may prevent the appearance of flares of phytoplankton abundance (SVERDRUP, 1953). Turbulent mixing plays the same role in the development of phytoplankton as graphite moderators do in nuclear reactors: too great an increase in their surface causes arrest of chain reactions, and their total removal—explosion.

An analogous phenomenon is observed in winter in the middle latitudes: when the thickness of the mixed layer is increased by low illumination and deep convection, phytoplankton does not develop; but in spring with increased illumination (and hence increased thickness of the euphotic layer) and the appearance of several shallow thermoclines in the heated water, an explosive bloom of phytoplankton occurs yielding a mass of surplus production. The same situation plays an important role in regions of

Tropical upwellings where intense water ascents bring the thermocline nearer to the surface.

In Antarctic regions (judging from data of VORONINA, 1977) the main factor limiting phytoplankton development even in summer is not lack in nutrients but turbulent mixing which sweeps the cells out of the euphotic zone.

### *Grazing*

Another factor limiting the burst of phytoplankton and slowing down the development is depletion by grazing (CUSHING, 1959). Recent investigations reveal that this effect is observed in some middle and polar latitudes during the relatively brief period of biological summer (VINOGRADOV and VORONINA, 1979).

### *Influence of Highly Productive Ecosystems on Adjacent Regions*

In some highly productive local oceanic aquatories (frontal zones, upwelling regions) phytoplankton production considerably exceeds the food requirements of herbivores. This 'surplus' production is usually assumed to be carried away by currents from the region of its formation into oligotrophic regions, thus enriching these (e.g. BOGOROV, 1959; MARGALEF, 1978). However, calculations of the total production of communities in the Peruvian upwelling show that the surplus of phytoplankton formed on the shelf is barely sufficient to supplement the rations of the numerous herbivores of abundant offshore communities concentrated in a narrow strip of water, some 100 to 200 miles wide (SHUSHKINA and co-authors, 1978; VINOGRADOV and SHUSHKINA, 1978). Satellite observations reveal that large eddies may be torn away from current systems in highly productive regions. Such eddies, bearing highly productive communities, may be carried into the ocean over hundreds of miles; in this way high concentrations of phytoplankton may be sporadically brought into the impoverished water of the open ocean bordering on highly productive regions.

### *Quantitative Distribution of Phytoplankton*

The combined effect of all the above-mentioned factors determines the general pattern of the distribution of primary production and the biomass (abundance) of phytoplankton (Figs 2-5, 2-6). The development of phytoplankton in the World Ocean reveals features of latitudinal and circumcontinental zonality.

Insufficient illumination during the vegetation period limits phytoplankton development only in the waters of the Arctic system covered by close pack-ice. In the high latitudes of the Central Arctic Basin, primary production proved to be minimal (ENGLISH, 1965) and the numerical abundance of phytoplankton did not exceed 2600 cells $l^{-1}$ (KAWAMURA, 1967). Beyond the continuous pack-ice cover, abundance and production of phytoplankton increase. The very low values of primary production recorded in Antarctic waters (Fig. 2-5) are apparently underestimates; they are based on data collected during the poor seasons (EL-SAYED, 1970, and others) and do not take into account phytoplankton development during the biological spring (see Fig 2-6).

In temperate-cold water subpolar ecosystems abundance and production increase, and in some regions (mainly coastal ones) they may reach maximum values recorded in

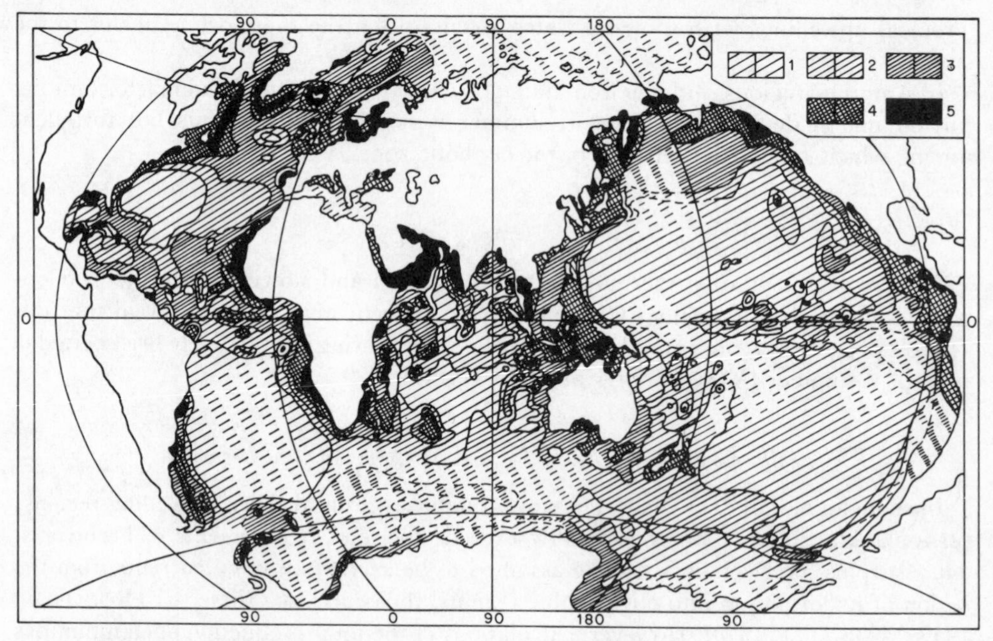


Fig. 2-5: Distribution of mean annual primary production (mg C $m^{-3}$ $d^{-1}$). 1: <100; 2: 100–150; 3: 150–250; 4: 250–500; 5: >500. (Based on information provided by KOBLENTZ-MISHKE, 1977)

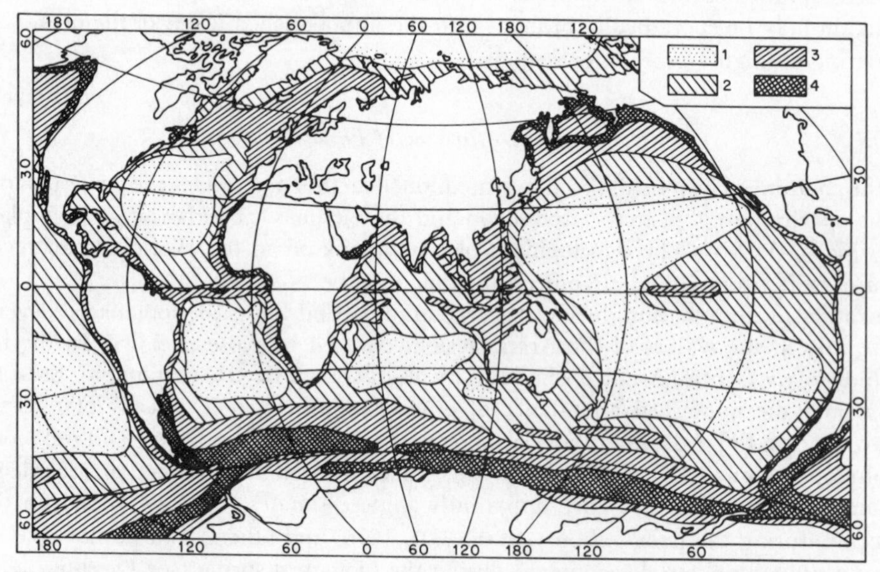


Fig. 2-6: Scheme of distribution of phytoplankton (cells $l^{-1}$) average for 0–100 m layer. 1: $<10^2$; 2: $10^2$–$10^3$; 3: $10^3$–$10^4$; 4: $>10^4$. (Based on information provided by VOLKOVINSKY and co-authors, 1972)

the oceans: $10^6$ cells $l^{-1}$ and biomass values of about 20 g $m^{-3}$ (Bering Sea, near Kamchatka; SEMINA, 1977a; Table 2-1). In Subantarctic waters the biomass of diatoms may exceed 3 g $m^{-3}$ (HASLE, 1969).

In the vast ecosystems of Central Tropical waters confined to anticyclonic gyres, where the pycnocline lies below 100 m and the ascent of deep waters is particularly difficult, phytoplankton development is pessimal. These are typical oligotrophic and ultraoligotrophic regions where daily primary production average of 20 to 50 mg C $m^{-2}$ (SOROKIN, 1977) and the biomass of water-bottle settled phytoplankton is reduced to a few milligrams, often to tens and even hundreds of a mg $m^{-3}$ (SEMINA, 1974).

In Equatorial regions, especially in their eastern part with intense upwelling, the rate of phytoplankton development sharply increases due to the complex dynamic system of currents and intense mixing. The process of mixing is most intensive in the coastal zones and particularly in the zones of coastal upwellings. Here phytoplankton development and production reach their maximum values. Thus, for instance on the shelf in the Peruvian upwelling, values of primary production of 1·2 g C $m^{-3}$ or 90 kcal $m^{-2}$ 24 $h^{-1}$ and a phytoplankton biomass of 170 g $m^{-2}$—in the layer of maximal concentration of 10 $gm^{-3}$—were recorded (SOROKIN, 1978b). Nearly as rich are the upwelling regions off the coasts of Oregon, northwestern and southwestern Africa, Somalia, and in some coastal temperate-cold water zones. However, further consideration of these most interesting regions is beyond the scope of this chapter. It must suffice to say that in oligotrophic (Arctic and Tropical) regions the numerical abundance and biomass of phytoplankton vary no more than 2 to 15 times, as compared with 5 to 100 times in productive Tropical, and 500 to 5000 times in productive temperate-cold water regions (SEMINA, 1977a).

### *Distribution of Biomass of Net Zooplankton*

The map of zooplankton biomass distribution shown in Fig. 2-7 is based on data obtained for temperate and high latitudes during the summer periods of the hemispheres, when the biomass in the 0 to 100-m layer is near maximal. For Tropical and Equatorial regions all data available were used irrespective of the season they were obtained.

As was to be expected, the pattern of zooplankton distribution follows that of phytoplankton. Some of the dissimilarities observed are accounted for by differences in the material used (location and time of collection, number of stations); the remaining dissimilarities are associated with certain peculiarities of migration, developmental cycle, seasonal abundance, and other structural characteristics of the zoocene. (See also Volume V: PERES, 1982).

A most striking feature of the map is the vast aquatories occupied by ecosystems of southern Central gyrals. The symmetrical ecosystems of northern Tropical gyrals are only weakly developed in the Pacific Ocean; in the Atlantic Ocean they occupy a small area in the Sargasso Sea region, and they are virtually absent in the monsoon Indian Ocean. In the central part of anticyclonic gyres—planetary halistases—in the presence of distinct stratification, a general slow sinking of waters takes place. The communities here are in a high stage of maturity, and the biomass of zooplankton is very low, averaging less than 25 or even less than 20 mg $m^{-3}$ in the upper 100-m layer.

In equatorial ecosystems, especially in the eastern parts of the oceans with more or

Table 2-1

Types of waters of the World Ocean differing in phytoplankton abundance and their characteristics (After SEMINA, 1977a; modified; reproduced by the permission of the author)

| Type | Region | Mean number of cells $l^{-1}$ ($\times 10^3$) | Mean biomass ($mg/m^{-3}$) | Maximum number of cells $l^{-1}$ ($\times 10^3$) | Maximum biomass ($mg/m^3$) | Vegetation period (months) | Author |
|---|---|---|---|---|---|---|---|
| 1 | 2 | 3 | 4 | 5 | 6 | 7 | 8 |
| **Oligotrophic** | | | | | | | |
| High latitudes | Central Arctic Basin | 0·8 | — | 2·6 | — | 1–1·5 | KAWAMURA (1967) |
| Low latitudes | Subtropical regions of the Pacific Ocean | 0·1 | 0·5–1 | 1·0 | 10 | 12 | SEMINA (1977) |
| **Eutrophic** | | | | | | | |
| Middle latitudes | Norwegian Sea | 40 | 120 | 216 | 1650 | 6–7 | VINOGRADOVA (1970) |
| | Coast of Kamchatka | 35 | 1000 | 1290 | 20000 | 6 | SEMINA (1974, 1977a) |
| Low latitudes | Gulf of Panama | 212 | >1000 | 4640 | >10000 | 12 | SMAYDA (1966) |
| | Peruvian upwelling | — | 500–1500 | — | 1500–10000 | 12 | SOROKIN (1977b) |

*Note*: — indicates no values available.

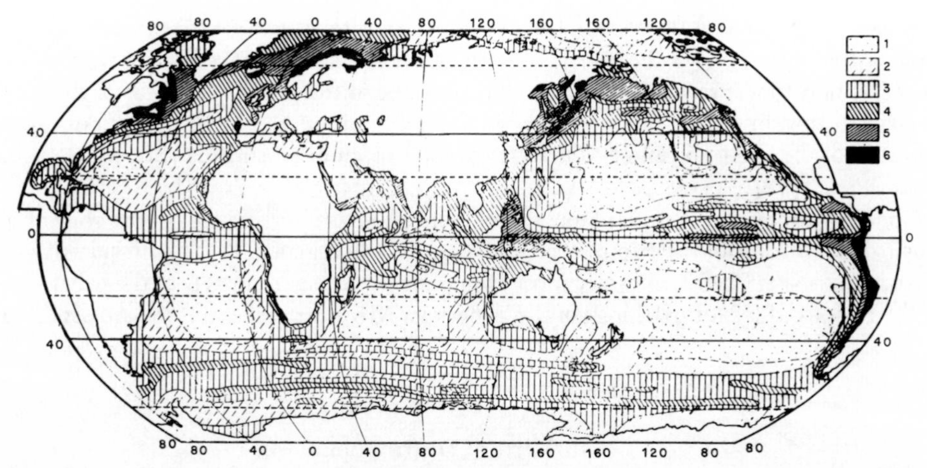


Fig. 2-7: Distribution of net zooplankton biomass (mg $m^{-3}$) in the upper 100 m layer of the ocean. 1: <25; 2: 25–50; 3: 50–100; 4: 100–200; 5: 200–500; 6: >500. (Based on information provided by BOGOROV and co-authors, 1968)

less developed upwelling, the zooplankton biomass is higher than in ecosystems of Tropical gyrals, usually exceeding 50 to 100 mg $m^{-3}$. The biomass is particularly high in eastern oceanic ecosystems along the coastal currents, associated with upwellings off the coasts of America, Africa, and Java. Here, zooplankton biomass often attains 200 to 500 mg $m^{-3}$ and may reach gigantic values of several or even tens of g $m^{-3}$.

In the northern hemisphere the ecosystems of Pacific and Atlantic Subarctic gyrals yield about equal biomasses of zooplankton measured in hundreds of mg or even in g $m^{-3}$ in the 0 to 100-m layer during the summer maximum.

The exosystems of the Subantarctic present a single formation associated with the western drift current. Plankton development is not simultaneous over the entire region; seasonal process follow a north-to-south trend with maximum concentrations of zooplankton species alternating in the surface layer. Owing to this trend, at each moment the standing crop of plankton in the Subantarctic proves to be lower than in the Subarctic (100–200 mg $m^{-3}$) although total zooplankton production is probably rather similar in both regions.

Finally, the ice-neritic ecosystems of the Central Polar Basin and near-Antarctic waters of the Eastern Winds Current are poor in plankton.

### *Macroplankton and Nekton Distribution*

The data available do not permit construction of a map showing the distribution of these groups of pelagic animals. PARIN and NESIS (1977) who devoted special attention to the peculiarities of quantitative distribution believe that distributional aspects are also governed by latitudinal and circumcontinental zonation and that 'high indices of abundance are observed only in those regions of the ocean where zooplankton is abundant' (p. 77). The greatest concentrations of macroplankton and nekton are characteristic of temperate cold water: boreal and subantarctic ecosystems and for the Equatorial zone of the oceans. It seems, however, that the 'highest biomass of cephalopod molluscs and fish is confined to the neritic oceanic zone transitory between neritic and oceanic regions and situated over the marginal parts of the shelf and immediately beyond them' (p. 77). Such concentrations are most characteristic of the peripheral areas of the zones of coastal upwellings. Apparently they exist by exploiting the highly productive communities of the shelf upwelling which have much production surplus (SHUSHKINA and co-authors, 1978).

Concentrations of macroplankton and nekton are characteristic also of the zone of the North Polar Front and the southern Subtropical Convergence. In Equatorial latitudes (PARIN and NESIS, 1977) they are most abundant in the eastern parts of the oceans and in the regions of local upwellings of the American Mediterranean and the Australiasian Seas.

## (b) Vertical Distribution

Increasing water depth and changing environmental conditions cause changes in (i) species composition of the population, (ii) morphological, physiological and biochemical singularities of organisms, (iii) organismic abundance and biomass, and (iv) structural and functional characterictics of communities.

The vertical distribution of zooplankton is far less influenced by the distribution of water masses than the horizontal one, and the pattern of vertical plankton distribution cannot be explained only by changes in water masses (e.g. BANSE, 1964; PAXTON, 1967; VINOGRADOV, 1968, 1970a, 1972). As layers with different thermohaline characteristics are often very thin and many plankton animals are capable of vertical migration, they may quickly move from one water mass to another, sometimes even passing through water layers with high density gradients or from gradient layers to homothermal ones (e.g. VUCETIĈ, 1961; VINOGRADOV 1956, 1974; VINOGRADOV and SHUSHKINA, 1980). More than that: vertical migrations through layers occupied by different water masses involve a considerable part of the population inhabiting waters of the same structure. They are evolutionally fixed and convey to the migratory species a number of biological advantages (VINOGRADOV, 1968).

Of particular importance to communities of the oceanic pelagial is their small-scale variability. The vertical plankton distribution proves to be extremely uneven, forming narrow (down to 1–5 m thick) quasi-stationary layers with increased or lowered concentrations of community components; their structural and functional characteristics differ from the average situation (e.g. SOROKIN, 1959; VINOGRADOV and co-authors, 1970, 1977; LONGHURST, 1976; TIMONIN, 1976; VINOGRADOV and SHUSHKINA, 1976). The formation of such layers depends to a considerable measure on the non-uniform hydrological microstructure of the upper oceanic waters (STOMMEL and FEDOROV, 1967; FEDOROV, 1976 (see also Volume V: PERES, 1982).

The presence of layers of above-average concentrations of plankton facilitates the existence of communities in oligotrophic regions of the ocean. If the same number of organisms were evenly distributed through the water column, the energy required by the consumers for search of food would exceed the amount of energy obtained through consumption. In layers with abundance maxima not only is the concentration of plankton above average but also its patchiness is observed more often than beyond these layers. This has been well illustrated by repeated soundings of the bioluminescent field (GITELZON, and co-authors 1971; Fig. 2-8).

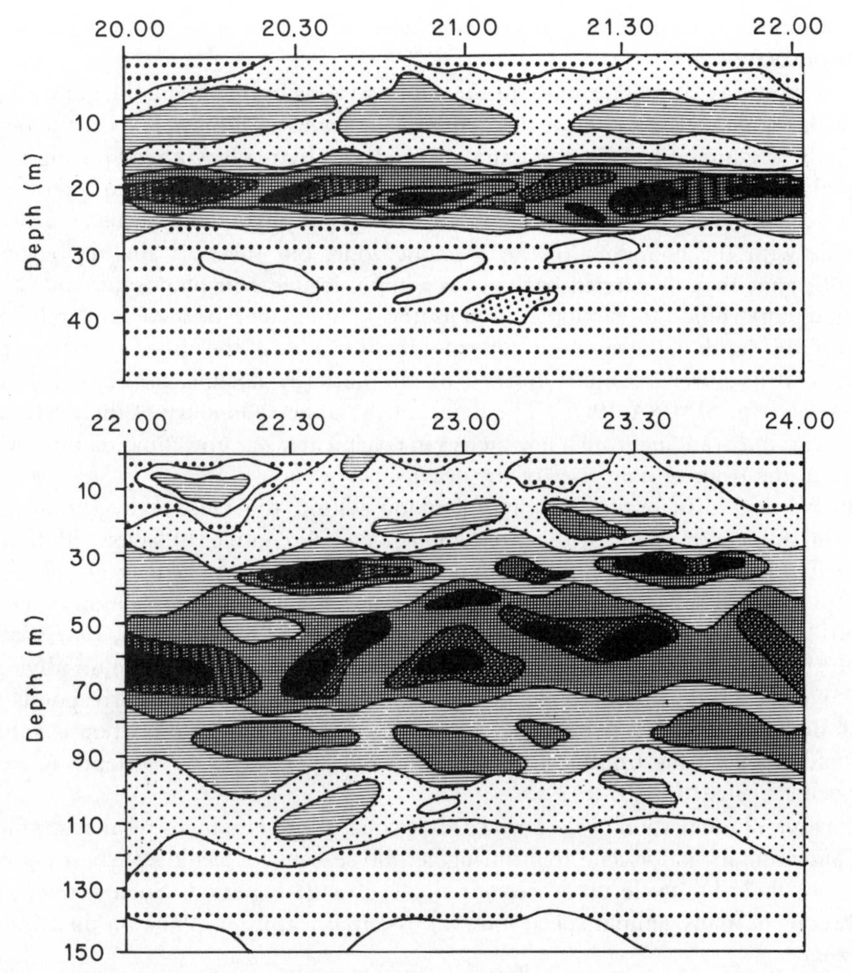


Fig. 2-8: Vertical distribution of bioluminescent field intensity according to data of continuous soundings from a drifting ship. The patchiness within the layer of bioluminescence maximum extends from 50 to 300 m. (Based on information provided by LEVIN and co-authors, 1975)

Also in the horizontal plane such aggregations (patchiness) of plankton are formed. Their size may vary from some hundreds of centimetres to hundreds of metres or kilometres. The causes of their formation, their distribution, and their role in the existence of communities have been investigated in detail by many authors (e.g. CUSHING, 1962; STEELE, 1976; WROBLEWSKI and O'BRIEN, 1976). Recent investigations on these problems are considered in a volume of collected papers edited by STEELE (1977).

### (c) Surface and Deep Zones

Continuous shifting of plankton, its concentration variations with depth (depending on season and time of the day) and changes in habitat depth of different life-cycle stages prevent the plankton population from being strictly confined to a definite water layer. However, there can be no doubt that the ocean is naturally divided into a producing surface zone and a consuming deep-water zone (e.g. BOGOROV, 1948; BRUUN, 1956; VINOGRADOV, 1968).

Photosynthesis takes place only in the euphotic zone of the ocean. The lower boundary of this zone is usually set at the depth of the compensation point, i.e. the depth at which the accumulation of organic matter produced by photosynthesis during daylight is equal to its loss due to respiration during a 24-h period. Owing to turbulent mixing, the lower boundary of the layer populated by living phytoplankton usually does not coincide with the boundary of the euphotic zone, but the cells sinking beyond the euphotic zone may be carried back again without losing their photosynthetic capacity. Of vital importance to phytoplankton is the maintenance of a definite relationship between the depth of the euphotic zone and the thickness of the upper mixed layer. This layer, with its lower boundary formed by the basic pycnocline, is the biotope of the phytocene (e.g. SEMINA, 1966). The depth of the upper boundary of the pycnocline is rather variable, ranging from a few metres in coastal areas or upwelling regions to 120 to 180 m in the tropical halistases of the oceans.

The vertical distribution of phytoplankton within the biotope is virtually never uniform, influenced as it is by the interaction of a whole complex of factors: illumination, nutrient concentration, stability of water layers (turbulence), density gradient in the basic pycnocline, presence of seasonal pycnoclines, maturity of the community, etc. (e.g. SVERDRUP, 1953; SOROKIN, 1959; SEMINA, 1966, 1977b; KAWAMURA, 1967; PARSONS and LEBRASSEUR, 1968). Generally, the thickness of the layer of maximum phytoplankton concentrations (trophic layer, according to NAUMANN, 1931) corresponds to the mean depth of the basic pycnocline. In regions where water stratification is subject to seasonal changes, the trophogenic layer is usually bounded by the depth of seasonal pycnoclines (SEMINA, 1977b).

The animal population of the surface zone consists of permanent residents (epiplankton) and animals that ascend to it during certain seasons for feeding purposes at certain stages of their life cycle or at certain times of a 24-h period. Such active, regular migrations of many animal species are an essential feature of plankton distribution in this zone.

The consuming deep-water zone is virtually unaffected by seasonal changes. It includes the intermediate, deep, and bottom waters or the lower part of the thermosphere, all the waters of the psychrosphere, and still deeper layers. Some of the species feeding primarily in the surface zone stay the rest of the time in deep waters (interzonal plankton), thus contributing to the transport of organic matter from surfaces to depths. Other species never leave the deep waters (deep-sea plankton); their connection with surface species is realized through intermediate links of the food chain or is restricted to their feeding on sinking organic remains. The populations of both vertical zones are connected by close topic and trophic relations; a continuous, rather intensive population exchange prevails, uniting the communities of surface and deep water layers.

### (d) Differences in Surface Plankton Distribution in Cold–Temperate and Tropical Regions

#### *Middle and High Latitudes*

In middle and high latitudes, the basic pycnocline often occurs at depths of about 100 m and more. According to SEMINA (1977b) in early spring, prior to the beginning of

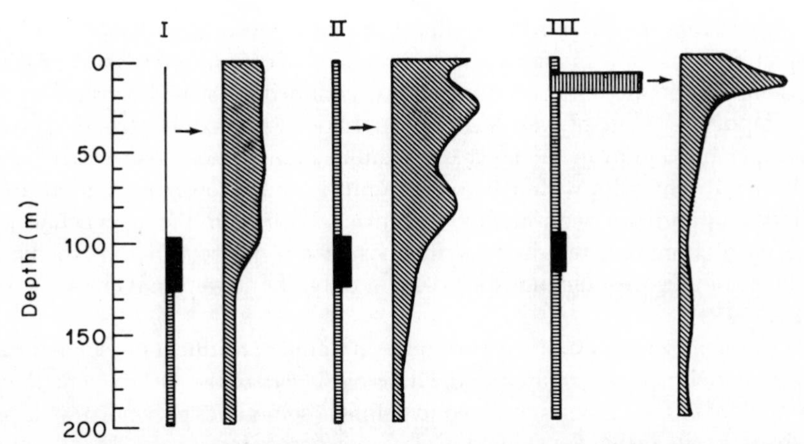


Fig. 2-9: Vertical phytoplankton distribution in the subarctic region of the Pacific Ocean. I: early spring; II: late autumn; III: summer. Right columns: phytoplankton distribution; left columns: water stability; black: permanent picnocline; vertical hatching: seasonal picnocline; arrow: depth of compensatory point. (Based on information provided by SEMINA, 1974)

stratification, phytoplankton is relatively evenly distributed in the biotope. But very soon, with the appearance of seasonal discontinuity layers, its concentration in the surface layers sharply increases. The drive to utilize most fully the basic food resources of the surface layers, the phytoplankton, causes herbivorous zooplankters to concentrate in the near-surface water layer during the spring–summer period. There animal populations fatten on the vernal bloom of phytoplankton and most of its dominant species reach their maximum annual biomass.

In summer the seasonal thermocline shifts to greater depths causing a corresponding increase in the thickness of the trophogenic layer. The vertical distribution of phytoplankton is very uneven, usually with a maximum in the discontinuity layer (Fig. 2-9). The bulk of the phytophagous animal populations now also occupies a larger water column and its dominant species with similar food spectra begin to alternate in importance at different depths (VINOGRADOV, 1956).

In autumn with the destruction of seasonal pycnoclines, phytoplankton distribution becomes more uniform. With the beginning of homothermy, the entire surface zone (down to 150–200 m, in some regions to greater depths) and with the decrease in solar radiation, the thickness of the mixed layer increases with a concomitant sharp reduction in phytoplankton. Since living conditions in the surface zone are no longer favourable to zooplankters their epipelagic populations decrease in abundance and interzonal species shift to deeper layers.

Unstable vertical stratification and long periods of complete homothermy apparently preclude the formation of groups of zooplankton species confined to definite layers of the surface zone.

### *Lower Latitudes*

Within the biotope of the phytocene, i.e. above the basic pycnocline, concentrations of phytoplankton may occur at any depth depending on the relationship between the depth of the compensation point and the depth of the basic pycnocline. Depth and position of phytoplankton maxima change with the succession of communities. At the early stages of succession, phytoplankton concentrations are confined to the wide surface layer; later, with the impoverishment of water, a characteristic two-maxima structure is formed, with an upper maximum in the layer of optimum illumination—supported mainly by excretion of nutrient salts within the community—and a lower maximum above the pycnocline—supported by an inflow of nutrients through the pycnocline and the minimum of illumination required for the existence of phytoplankton. In the mature communities of oligotrophic halistases often only the lower maximum is retained (SOROKIN, 1959).

Owing to the absence of marked variations in annual stratification, an assortment of relatively isolated niches is created in the different layers of the surface zone that favour the formation of animal groups confined to definite (sometimes very narrow) layers and, among them, an intensive development of neuston organisms.

The relatively uniform development of phytoplankton all year round, a higher annual stability of ecosystems than in cold-water regions, and constant hydrological conditions assure a more or less uniform circum-annual vertical distribution of zooplankton. Since there are no unfavourable periods, the plankton animals are not forced to leave the surface layers, so that seasonal migrations are either absent or only weakly expressed, and upper-interzonal species play no important role.

However, lack of small upper-interzonal mass species—the staple food of mid-depth macroplankton in Subpolar regions—forces most of the tropical mid-depth zooplankton species to ascend regularly to the surface zone. That is why the lower-interzonal species, mostly macroplankters, play a more important role in the communities of low latitudes than in the productive regions of high latitudes.

It must be noted that the interzonal species of cold water regions are distributed along a very extensive vertical range. A large part of their population inhabits depths down 2000 to 3000 m. In the Tropical zone, upper-interzonal species do not descend to such depths, only single individuals are encountered below 1000 m. This difference between cold water and Tropical regions seems to be accounted for by the fact that in the former seasonal migrations prevail, while in the tropics far shorter diurnal migrations are performed. Also, the effect of thermal stratification is important. In cold water regions, occupied by waters with Polar and Subpolar structures, temperature differences between surface and deep layers are relatively small— considerably smaller than seasonal temperature fluctuations in the surface layers. Therefore, there are no temperature barriers that could prevent interzonal species feeding and developing at the surface from sinking to greater depths at some stages of their life cycles. In contrast, in Tropical waters surface temperatures may differ from those of deep layers by as much as 20 ° to 25 °C.

### (e) Quantitative Plankton Distribution in the Whole Water Column

#### *Microzooplankton (Protozoa)*

No reliable concept has emerged yet regarding the vertical distribution of unicellular heterotrophs. Proper quantitative evaluation has been hindered by methodological difficulties. Representative data can be obtained only by analysing living material, but no sufficiently reliable and precise methods are as yet available, not even for assessing the microplankton of surface layers (DODSON and THOMAS, 1964; HOLM-HANSEN and co-authors, 1970; TUMANTSEVA and SOROKIN, 1975).

Observations carried out in the Mediterranean Sea, at the western coasts of Africa, and in the northwestern part of the Indian Ocean have shown that deep waters down to the 1000- to 3000-m layer are rich in unicellular heterotrophs—coccoliths and dinoflagellates (LECAL, 1952; BERNARD and LECAL, 1960; BERNARD, 1961, 1963). Their abundance varies considerably from 300 thousand to 28 million cells per one multicellular animal (BERNARD, 1958). However, repeated investigations in the Mediterranean Sea yielded 50 to 200 times lower values for different depths than obtained by BERNARD (GREZE, 1963). Heterotrophs were also far less numerous in the collection of R. V. *Vityaz* in the Pacific and Indian Oceans.

Another mass group of unicellular organisms is that of olive-green cells (see also Volume V: PERES, 1982; pp. 40–41) which apparently may be flagellates (FOURNIER, 1970). These cells, according to many authors (e.g. HASLE, 1959; HAMILTON and co-authors, 1968; MELNIKOV, 1975), occur mainly below the euphotic zone with a maximum of 500 to 1000 m; here their numbers may reach 100 to 200 thousand cells $l^{-1}$. Farther down they are less abundant, but even at depths of 3000 to 4000 m they usually remain above 25 to 50 thousand cells $l^{-1}$. These cells were discovered in great quantities in the guts of deep-sea isopods (MENZIES, 1962), some prawns (WHEELER, 1970), tunicates (FOURNIER, 1970), and copepods (FOURNIER, 1973; HARDING, 1974). Surprising is the lack of correlation between the vertical distribution of unicellular heterotrophs and net zooplankton which might feed on them. Unexpected also is the fact that the dominant mass of mesoplankton at depths of 1000 to 3000 m consists of carnivorous forms rather than of consumers of unicellular heterotrophs. The quantitative distribution of unicellular heterotrophs in the entire water column of the ocean and its changes with depth requires further special and thorough investigation.

### *Net Plankton (Mesoplankton)*

The biomass of net plankton in the deep layers of the ocean is directly dependent on its biomass (production) in the surface zone. The relationship between the abundance of plankton in the upper (0–500 m) and deeper (500–4000 m) layers remains virtually constant over the whole aquatory of the ocean: two-thirds or 65% in the upper 500-m layer of the total amount in the 0 to 4000-m layer. This relationship is disturbed only in areas with an inflow of deep waters from neighbouring regions substantially different in their productivity of surface plankton. If these regions are more productive in surface plankton, the amount of deep-sea (500–4000 m) plankton increases, possibly reaching 70% of the total mass in the 0 to 4000-m column; if the regions are less productive, the percentage decreases.

The changes in total biomass of zooplankton with depth (Tables 2-2 and 2-3) are rather monotypic: in depths over 500 to 1000 m, both in the tropics and in temperate-cold water regions, they are governed by the exponential dependence $y = ae^{-kx}$; where $y$ = biomass of plankton; $x$ = depth; $a$ = a coefficient showing the influence of the quantity of biomass in the overlying layers; and $k$ = a coefficient of the rate of decrease in biomass with depth (VINOGRADOV, 1958, 1960, 1968; JOHNSTON, 1962; BANSE, 1964). Depending on the changes in biomass in the surface layers, the coefficient $a$ may vary considerably, while $k$ remains virtually constant differing only in the temperate cold

Table 2-2

Mean biomass (mg $m^{-3}$) of meso- and macroplankton in the mesotrophic equatorial (12° N to 12° S) and oligotrophic tropical (40° N to 12° S and 12° S to 40° S) regions of the Pacific and Indian Oceans. Average of 32 series of hauls with BR nets (After VINOGRADOV, 1977a; reproduced by permission of the author)

| | Mesoplankton | | Macroplankton | | | |
|---|---|---|---|---|---|---|
| Depth (m) | 12° N–12° S | 40° N–12° N 12° S–40° S | 12° N–12° S | % of total zooplankton biomass | 40° N–12° N 12° S–40° S | % of total zooplankton biomass |
| 1 | 2 | 3 | 4 | 5 | 6 | 7 |
| 0–50 | 63·5 | 27·6 | 0 | 0 | 0 | 0 |
| 50–100 | 52·3 | 25·9 | 15·2 | 22·5 | 0 | 0 |
| 100–200 | 18·8 | 14·1 | 0·3 | 1·6 | 0 | 0 |
| 200–500 | 7·8 | 6·8 | 2·2 | 22·0 | 0·8 | 10·5 |
| 500–1000 | 5·2 | 4·8 | 2·7 | 34·2 | 1·6 | 25·0 |
| 1000–2000 | 1·2 | 1·8 | 5·9 | 83·1 | 0·3 | 14·3 |
| 2000–4000 | 0·23 | 0·40 | 0·02 | 8·0 | 0 | 0 |

Table 2-3

Mean biomass (mg $m^{-3}$) of meso- and macroplankton in the northwestern part of the Pacific Ocean. Average of 9 series of hauls with BR nets* (After VINOGRADOV, 1977b; reproduced by permission of the author)

| Depth (m) | Mesoplankton biomass | Macroplankton biomass | Macroplankton (% of total zooplankton biomass) |
|---|---|---|---|
| 0–50 | 626 ± 60 | 0 | 0 |
| 50–100 | 109 ± 34 | 0 | 0 |
| 100–200 | 108 ± 13 | 0 | 0 |
| 200–300 | 272 ± 51 | 0·3 (200–500) | 0·12 (200–500) |
| 300–500 | 234 ± 19 | | |
| 500–750 | 103 ± 15 | 3·4 (500–1000) | 4·2 (500–1000) |
| 750–1000 | 51 ± 10·5 | | |
| 1000–1500 | 27 ± 2·4 | 1·3 (1000–2000) | 5·3 (1000–2000) |
| 1500–2000 | 19 ± 1·7 | | |
| 2000–2500 | 17 ± 2·8 | 1·3 (2000–3000) | 10·6 (2000–3000) |
| 2500–3000 | 5·0 ± 0·8 | | |
| 3000–4000 | 1·6 ± 0·17 | 0·11 | 6·9 |
| 4000–5000 | 0·91 ± 0·08 | 0·03 | 3·3 |
| 5000–6000 | 0·86 ± 0·05 | 0 | 0 |
| 6000–7000 | 0·60 ± 0·04 | 0 | 0 |
| 7000–8000 | 0·32 | 0 | 0 |

*Stations of R.V. *Vityaz* occupied in summer 1966 in the Kuril-Kamchatka Trench.

water and Tropical regions where the decrease in mesoplankton with depth is less intensive (Fig. 2-10). Thus, for instance, in the Kuril-Kamchatka Trench $y = 56{\cdot}2e^{-6{\cdot}5\times10^{-4}x}$, in the Kermadek Trench $y = 7{\cdot}75e^{-6{\cdot}5\times10^{-4}x}$, in the Mariana Trench $y = 1{\cdot}82e^{-8{\cdot}5\times10^{-4}x}$, and in the Bougainville Trench $y = 5{\cdot}74e^{-8{\cdot}5\times10^{-4}x}$. Here $[y]$ = mg m$^{-3}$, $[x]$ = m. However, a more detailed analysis of changes in biomass with depth shows that the decrease is actually not so uniform and that there are layers with higher and lower gradients. In the mesoplankton distribution of the Kuril-Kamchatka Trench (Fig. 2-11) three distinct intervals may be distinguished (Fig. 2-12). The first interval encompasses the layer extending from the surface to 500 to 750 m; it is occupied by surface waters and by cold-intermediate and the upper part of warm-intermediate waters. Here plankton distribution is very uneven and subject to cardinal seasonal changes. However, generally the biomass in this interval amounts to hundreds of mg m$^{-3}$. This is the surface type of plankton distribution.

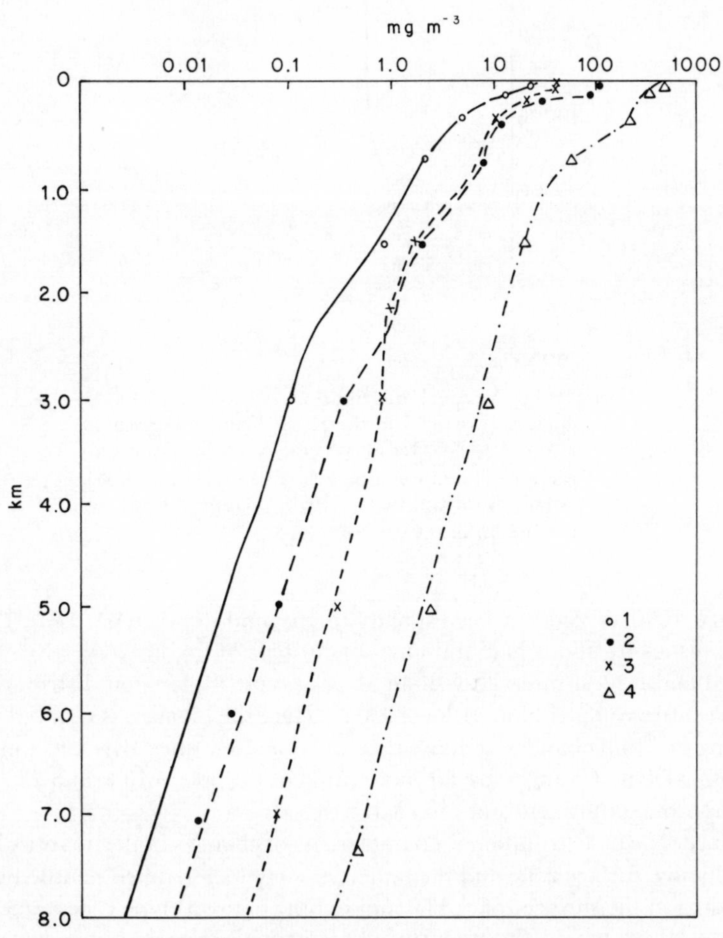


Fig. 2-10: Vertical distribution of net zooplankton biomass (mg m$^{-3}$) in different regions. 1: Mariana Trench; 2: Bougainville Trench; 3: Kermadec Trench; 4: Kuril-Kamchatka Trench, May 1953. (After VINOGRADOV, 1968; reproduced by permission of the author)

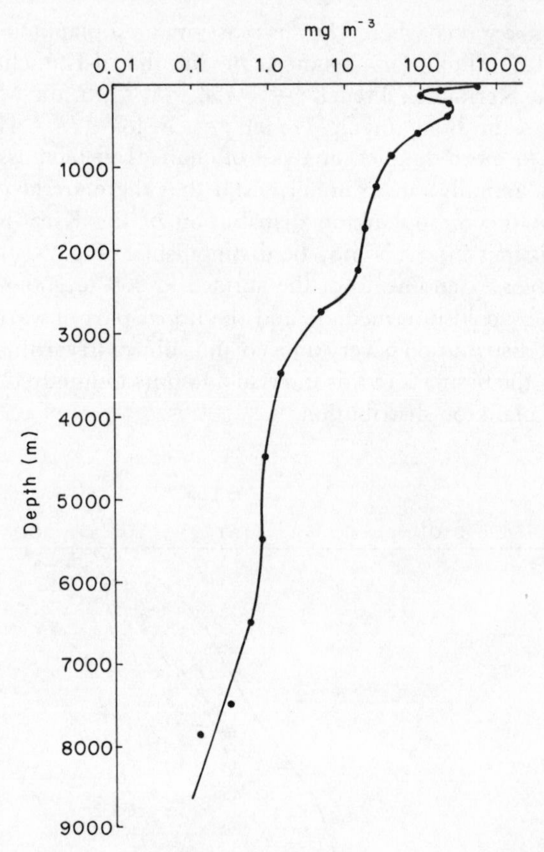


Fig. 2-11: Vertical distribution of net zooplankton biomass (mg $m^{-3}$) in the Kuril-Kamchatka region of the Pacific Ocean (averages of 9 stations in fractioned sampling layers; July–August, 1966). (After VINOGRADOV, 1970b; reproduced by permission of the author)

The range 1000 to 2500 m corresponds to the mid-depth type. Here the biomass slowly diminishes by about half and amounts to tens of mg $m^{-3}$.

The third and largest interval with an abyssal type of plankton distribution extends through the entire water column below 3000 m. Here the biomass is reduced to a few mg or tens of mg $m^{-3}$ and changes no more than by about ten times over the immense range from 2000 to 9000 m. Changes are far more rapid in the 500 to 1000 and 2500 to 3000 m intervals, approximating ten times in each case.

The surface type of distribution is characteristic of changes in the quantity of plankton in the productive surface zone and the underlying layers inhabited mainly by interzonal species feeding in the surface zone. The mid-depth (bathyal) type is characteristic of the intermediate layers where the quantity of interzonal animals diminishes but is still relatively high, and where autochthonous carnivores, both meso- and macroplankton, are dominant. The abyssal type is characteristic of waters poorest in plankton. Here euryphagous–detritophagous species dominate. The role played by carnivorous species

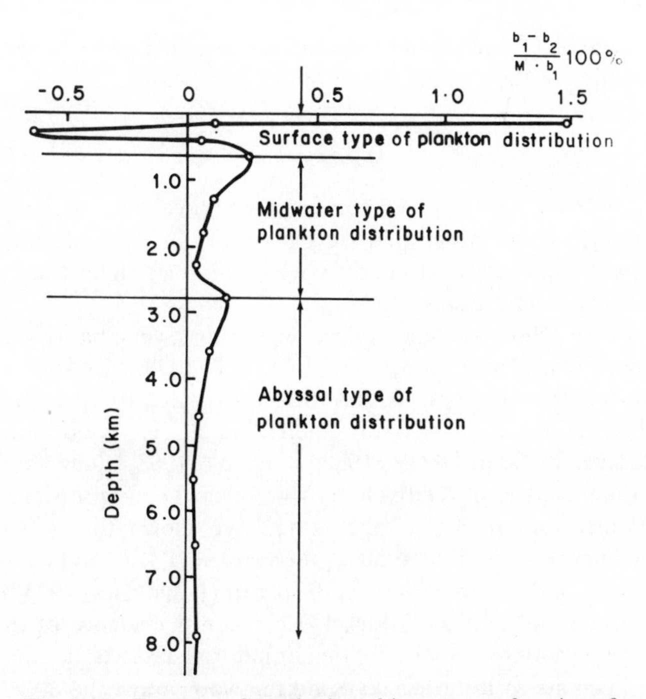


Fig. 2-12: Decrease of zooplankton biomass as a function of water depth in the Kuril-Kamchatka region of the Pacific Ocean (averages of 9 stations; July–August, 1966). M: distance (m) between the centres of adjacent sampled layers; $b_1$: biomass in the overlying sampled layer; $b_2$: biomass in the underlying sampled layer. (After VINOGRADOV, 1977a; reproduced by permission of the author)

is insignificant and comes to naught with increasing depth. These species virtually disappear from the plankton when its concentration drops below 0·5 to 0·25 mg $m^{-3}$.

The biomass distribution in the Tropical zone more or less distinctly reveals the same intervals with surface, mid-depth, and abyssal types of distribution. However, the boundaries usually lie in somewhat different depths than in boreal regions. The surface type of distribution is succeeded by the mid-depth type in the 100- to 200-m layer, and the mid-depth type by the abyssal in the 1500- to 2500-m layer. These differences are accounted for by the predominance of upper-interzonal species in the surface plankton of Subpolar regions. A considerable part of the populations inhabit also the waters beneath the surface zone. In the tropics, on the contrary, upper-interzonal species play an insignificant role. In Subpolar regions a considerable mass of interzonal species perform large-scale seasonal migrations with ranges up to 2000 to 2500 m, while in the Tropical regions diurnal migrations prevail with a far lesser depth range. Therefore, in Subpolar regions the mid-depth type of distribution—associated with the presence of great quantities of migrants from the surface zone—extends into far greater depths than in the tropics.

The uneven (stepwise) change in the rate of decrease of mesoplankton biomass with depth has a definite biological significance and is accounted for by the zonal sequence of trophic relationships in the pelagic communities (see below).

*Macroplankton*

Substantial differences are observed between the vertical distributions of macroplankton and net mesoplankton. Macroplankters perform diurnal and ontogenetic vertical migrations of great amplitude and intensity which produce significant daily changes in the depths of their concentrations.

Observations with echo-sounders on deep-scattering layers, hauls with large-meshed horizontally towed nets (JESPERSEN, 1935; LEAVITT, 1938), deep-sea trawlings in the pelagial (LEGAND and co-authors, 1972; VINOGRADOV and PARIN, 1973; PARIN, 1975) and, finally, observations from the bathysphere (BEEBE, 1934) and bathyscaphes (BERNARD, 1958; PERES, 1958a,b; DIETZ, 1962; BARHAM, 1963) allow the conclusion that mesoplankton organisms form relatively narrow vertical concentrations. Investigations on the vertical distribution of macroplankton have shown that its basic taxonomic groups are very unevenly distributed along the vertical (PARIN and co-auohors, 1977).

Collections taken in different regions of the ocean (JESPERSEN, 1935; LEAVITT, 1938; VINOGRADOV and PARIN, 1973; PARIN, 1975) confirm evidence of the variability of general vertical distribution. Thus, in the Equatorial regions of the Pacific Ocean, according to collections with the Isaaks-Kidd midwater trawl the maximum of macroplankton biomass (5·5 mg $m^{-3}$) during the day occurs in depths from 450 to 950 m. During the night the depth distribution becomes more even. But deeper down (950–1200 m) the quantity of macroplankton noticeably diminishes to about 2 mg $m^{-3}$ and here no diurnal biomass variations prevail. In other regions at mid-depths (from 500 to 1500–2000m) no distinct pattern was discovered in the rate of macroplankton decrease with depth. However, below 1800 to 2000 m its quantity was always very low.

Ecologically, it is very important to assess, even if approximately, the role of macroplankton in the total mass of plankton, at different depths and in different regions of the ocean. Tables 2-2 and 2-3 (p. 680) list the mean quantities of meso- and macroplankton* in the Tropical (40° N–12° N and 12° S–40° S) and Equatorial (12° N–12° S) regions of the Pacific and Indian Oceans, obtained from vertical hauls with large closing plankton nets of the BR 113/140 type. In oligotrophic Tropical regions the maximum of macroplankton biomass lies in the 500 to 100-m layer. In lesser and greater depths (200–500 m and 1000–2000 m) its quantity decreases. However, it is generally not so high in the 500 to 1000 m layer not more than one-third of the total mesoplankton biomass and one-quarter of the total plankton. A different picture exists in the productive Equatorial region. In the 200- to 500-m layer the macroplankton biomass accounts for nearly 30% of the biomass of mesoplankton, increasing to 52% in the 500- to 1000-m layer and in the 1000- to 2000-m layer it is five times that of the mesoplankton. Here larger animals prevail more than in the 500- to 1000-m layer, their main concentrations being confined to depths of 1000- to 1500 m.

Thus, over the entire aquatory of the Tropical ocean the depths of 500- to 2000 m are inhabited by many large carnivorous animals (prawns, fishes, cephalopods). Their concentrations form a kind of living filter, or 'live net' as termed by BARHAM (1963), which

*Macroplankton comprises animals more than 3 cm long.

extends beneath the productive zone of the ocean and consumes a considerable part of organic matter of animal origin sinking from the surface layers. Owing to this filter only very small amounts of food reach the deeper water layers which prove to be the poorest in plankton. In Tropical regions where diurnal migrations prevail with a range of no more than 500 to 800 m, the filter of carnivorous macroplankton is concentrated and pressed to the lower boundaries of the producing layers. In the filter itself migratory forms play an important role.

In low-productivity regions the layer of concentrations of carnivorous macroplankton is relatively narrow and the biomass not abundant, while in the productive regions, such as the Equatorial ones, this layer is extremely thick and encompasses a large water column.

In the productive Subpolar regions the main mass of upper-interzonal mesoplankton species perform seasonal ontogenetic migrations of immense amplitude. Not only in winter but also in summer a considerable part of their populations is dispersed in the deep waters where it may serve as food reserve for the deep-sea plankton. Here the main consumers of the upper interzonal species (mostly copepods) sinking to the deep layers are non-migratory deep-sea mesoplankton predators, including especially the large mesoplankton chaetognath (*Eukrohnia fowleri*).

But the relative quantity of macroplankton, especially of prawns and small fishes, is smaller in mid-depths of Subpolar regions than in Tropical regions and more evenly distributed over a wide range of depths down to 3000- to 4000 m.

Both in Subpolar and in Tropical regions the main concentrations of macroplankton are generally confined to layers characterized by a mid-depth type of mesoplankton distribution. However, in the tropics the share of carnivorous macroplankters in the total pelagic population of these layers is noticeably higher than in Subpolar regions (compare Tables 2-2 and 2-3); this indicates that a relatively greater portion of meso-plankters is consumed by macroplankters in the Tropical zone. It is probably this factor that accounts for the higher rate of decrease in mesoplankton biomass with depth in the Tropical regions of the ocean (p. 681).

Probably, mid-depth concentrations are formed not only by carnivorous macroplank-ters but also by large nekton forms that feed on them. Hence, commerical aggregations of mid-depth fishes may occur in the intermediate water layers. In the final analysis, these fishes subsist on organic matter formed in the producing zone; since their life cycles are significantly longer than the life cycles of the preceding links of the food chain, they may yield a relatively greater total biomass.

### *Biological Zonality of the Pelagial*

The irregular changes in the faunistic composition of plankton with increasing depth, the presence of layers with particularly sharp faunal changes caused both by physical and biological factors, the existence of zones with faunas possessing some common morphological features, the changes with depth of dominant trophic groups, and other ecological singularities of pelagic populations—all these factors justify the division of the pelagial into biological zones inhabited by specific communities (see Volume V: PERES, 1982).

Since the close of the past century many authors have suggested scores of schemes relating to different regions of the ocean or different groups of animals. These schemes have been thoroughly examined (VINOGRADOVA and co-authors, 1959) and found to be rather similar, which once more confirms the existence of general laws governing the vertical distribution of the pelagic fauna. The differences between these schemes consist mainly in the placing of the boundaries of vertical zones and in a greater or lesser fractionation of some of these zones.

By studying the distribution of diverse taxonomic groups in different regions of the ocean and by generalizing earlier schemes it became possible to devise a scheme of vertical plankton zonality for the whole World Ocean. This has been achieved by HEDGPETH (1957) and later by members of the biological laboratories of the Institute of Oceanology (BELYAEV and co-authors, 1959).

The scheme of the vertical zonality of the pelagic fauna, based on all contemporary data, may be presented in the following form (VINOGRADOV, 1968):

Surface zone or<br>Epipelagial<br>0–100 (200) m

Transitory layer or<br>Mesopelagial<br>100 (200)–750 (1000) m

| | | |
|---|---|---|
| Deep-sea zone<br>below 750 (1000) m | Upper Subzone or<br>Bathypelagial<br>750 (1000)–2500 (3500) m | |
| | Lower subzone or<br>Abyssopelagial<br>below 2500 (3500) m | Oceanic<br>depth<br>2500 (3500)–6000 m |
| | | Hadal or<br>ultraabyssal<br>depths: below<br>6000 m |

The types of vertical distribution of plankton biomass considered above (Fig. 2-12, p. 683) correlate well with this scheme and corroborate the validity of the gradations employed. The surface type of distribution corresponds to the surface zone and the transition layer (epi- and meso-pelagial), the mid-depth type corresponds to the bathypelagial, and the abyssal type to the abyssopelagial.

It should be especially noted that—in contrast to the bottom fauna—no special hadal zone is recognized for the pelagial. The endemism of the pelagic fauna extends to only a few groups and is very restricted. With the passage of trenches to the waters, in depths of about 6000 m, no significant changes occur either in the intensity of the decrease in total plankton biomass or in the relationships between trophic groups. The structure of the communities changes, but only gradually by developing tendencies that arise in the upper layers of the abyssopelagial (decreasing biomass; reduction of predatory forms down to extinction; increased oligomixedness of the communities).

### (3) Ecosystems of the Surface Productive Zone

The pelagic ecosystems of cold water (Polar and Subpolar) regions differ fundamentally from Tropical ecosystems, first of all by the relative evenness or clearly expressed cyclicity of phytoplankton vegetation (VINOGRADOV, 1968; VORONINA, 1978). The existence and development of all other plankton groups are adapted to the cycle of primary production. The utilization of nutrients brought into the surface zone, mainly by winter convection, begins only in spring when solar radiation increases and surface waters become more stable (appearance of the seasonal pycnocline). In contrast, in Tropical regions solar radiation is sufficient all year round; critical depth is nearly continously below the boundary of the upper mixed layer and stratification remains relatively stable. The development of phytoplankton is virtually circum-annual.

#### (a) Ecosystems of High Latitude Regions (Central Arctic Basin)

The ecosystem of the Arctic Basin (see Volume V: PERES, 1982, Chapter 6) includes a community that lives under drift ice beyond the shelf zone. Since access to this region is difficult, our knowledge of it is rather limited.

The greatest contributions to the study of arctic plankton have been made by large-scale expeditions (FARRAN, 1936; BOGOROV, 1946) and, especially, by permanent drifting stations (VIRKETIS, 1957, 1959; ENGLISH, 1965; DUNBAR, 1968; HUGHES, 1968; VINOGRADOV and MELNIKOV, 1980).

In character the ecosystem of the Arctic Basin is derived from the circumpolar system of the arctic pelagial, generated by the climatic perturbations of the Pleistocene; it functions in an extremely rigid oscillatory regime. Since primary production is very low, none of its species is able to realize its spawning potential which in turn makes the ecosystem a pessimal one. The system is characterized by a levelling of population waves greater than the ecosystems of ice-free waters. A seasonal variability of zooplankton of two orders of magnitude is observed only in 3 to 4 species; in the remainder it is reduced to one order or even less. Changes in biomass production and destruction are recognizable only in the upper 200-m layer of arctic waters (about -1·8 °C), where the productions of the thin euphotic pelagic zone and the cryopelagic flora are utilized.

The communities of the Arctic Basin are characterized by the following specific features: (i) short (about 1·5 mo) phytoplankton vegetation (USACHEV, 1961; KAWAMURA, 1967) with only few monocyclic mass forms HORNER, 1976; BELYAEVA, 1980). (ii) The important role of cryophilic populations as providers of organic matter (ANDRIASHEV, 1970; MELNIKOV, 1980). (iii) Absence of species with narrow food specialization; prevalence of predation; importance of microzooplankton and detritus as food sources. (iv) Capacity of storing energy in the form of lipids and ether waxes as energetic reserves (LAWRENCE, 1976); this enables the euryphages not only to survive periods of food absence but also to continue to reproduce all year round at a very low rate. (v) Circum-annual mortality. (vi) A peculiar alteration of diapause and activity inherent in both plankters and in cryophiles (GEORGE and ALLEN, 1970; KOSOBOKOVA, 1978). (vii) Marked time lag in the appearance of maxima of zooplankton abundance associated with a low level of metabolism and a correspondingly longer life cycle of species, i.e. 3- to 4 times longer than in arctico-boreal waters. (viii) Large body-size, inherent in dominant species, associated with a low metabolic rate which is evolutionally favourable to long-functioning structures. (ix) Low fecundity; spawning of eggs rich in yolk. (x) Monotony of ecological niches in the pelagial under the ice, and, hence, low species diversity in inhabitants of the upper 200 m.

Spatial heterogeneity facilitating species coexistence is achieved in several ways. The infrastructure of submerged ice surfaces, with its diversified configurations of planes that favour a patchy development of flora and fauna, gives shelter to cryophiles, and seasonal changes in the microrelief of the underwater ice surface enhance the diversity of the biotope for cryopelagic organisms. The spatial heterogeneity of the system is supported by a permanent aggregation of dominant species (HUGHES, 1968; HEINRICH and co-authors, 1980) as well as by their vertical ontogenetic migrations in which amplitude and time are specific for dominant species.

Evidence confirming the heterogeneity of the biotope is provided by the development of *Calanus glacialis* in big clearings where the number of young individuals in the 0 to 5-cm layer may reach 1 to 2 thousand $m^{-3}$ (SHUVALOV and PAVSHTIKS, 1977). Gregariousness is inherent also in the arctic cod *Boreogadus saida* and black cod *Arctogadus glacialis,* which occur under the ice cover. Probably the frail zoocene of the Arctic Basin would not withstand the stationary pressure of planktophageous vertebrates were it not for the fact that the cod which feed mainly on cryophiles ascend in winter to the surface, while the plankters sink to the depths.

Striking evidence of the ecological polyvalency of the mass species of the arctic ecosystems is seen by DUNBAR (1970) in their 'taxonomic indiscipline', i.e. the high variability of the northern representatives of the genera *Calanus*, *Pseudocalanus*, *Parathemisto*, *Hyperia*, *Thysanoessa, Sagitta, Aglantha,* and others. The existence of cryptic species in some of the genera has already been confirmed (FROST, 1974; PAVSHTIKS and VYSHKVARTSEVA, 1977).

The seasonal aspect of community existence may be assumed to be the following. Biological spring sets in with an increase in phytoplankton abundance both in and under the ice. The algal population under the ice is assumed to be potentially heterotrophic (KALFF, 1967; HORNER, 1976). Primary production is low; in summer maximum values in clearings do not exceed 5 to 6 mg $^{14}C\ m^{-2}\ d^{-1}$ (ENGLISH, 1965). The onset of photosynthesis depends on ice conditions which change from place to place and from year to year. To the ice flora proper which enriches the surface waters belong the diatoms *Pinnularia, Navicula, Melosira arctica;* the main mass of the spring bloom consists of *Thalassiosira bioculata, T. gravida, T. hyalina, Coscinosira polychorda,* the pennate diatoms *Fragillaria oceanica*, *Nitzschia frigida*, and *Licmophora* sp. The production of algae overgrowing the lower ice surfaces may be relatively high.

Summer phytoplankton is characterized by an increase in the number of diatoms *Chaetoceros*, *Rhizosolenia* which attain their maximum value of 180 thousand cells $l^{-1}$ during the two-week bloom period. Dinoflagellates and peridinians are few. A reserve for phytoplankton blooms instantaneously flaring up in a clearing during spring and summer is provided by the resting spores of diatoms which may remain in an anabiotic state for more than 500 d. These spores preserve their productive capacity even after having passed through the guts of copepods (HARGRAVES and FRENCH, 1977; DURBIN, 1978). By the middle of August almost nothing remains of the autotrophic production, and toward September the number of cells becomes reduced to a few hundred cells $l^{-1}$ (BELYAEVA, 1980).

Seasonal changes in zooplankton are characterized by an ascent in July of the copepods *Calanus hyperboreus, C. glacialis, Microcalanus pygmaeus*, and *Metridia longa* into the upper 50-m layer, where the young copepodites feed and breeding begins in the bulk of the population (parts of the populations of these species breed all year round). The abundance of these dominants increases toward July about ten times more than during the dark period, and in August copepods make up 50% to 90% of the total zooplankton population in the 0 to 25-m layer. Analogous data have been obtained from a region north of Greenland in collections taken at the ice station Arlis-II: 86% of the total biomass consisted of copepods, half of them represented by *Calanus*. On the average, during the period of its maximum in the Arctic Basin the zooplankton biomass reached 290 mg $m^{-2}$ (dry weight) in the 0 to 500-m layer (HOPKINS, 1969).

Submarine ridges and banks favour vertical mixing; thus in late July to August over the Lomonosov and Mendeleev ridges isolated patches of seston may yield biomasses of 200 to 500 mg $m^{-3}$. In the frontal zone north of the New Siberian Islands and the Wrangel Island patches of 1000 mg $m^{-3}$ were recorded (PAVSHTIKS, 1971a,b). Unfortunately, no data are available to the extent of such high-biomass patches. Important in this seston are the biomasses of the ctenophore *Beroë cucumis*, the jelly fishes *Aglantha digitale, Aeginopsis laurentiae,* and the siphonophore *Diphyes arctica.* This group has to be considered destroyers of copepod biomass; to it should also be added the chaetognaths. The greatest biomass among the copepods is that of *Calanus hyperboreus, C. glacialis, Oithona similis*, and, temporarily, *Microcalanus pygmaeus*; among the amphipods, *Parathemisto*. In August *Oicopleura labradoriensis* and *Fritillaria borealis* increase in number.

In winter under the ice juvenile amphipods *Apherusa glacialis, Lagunogammarus wilkitzkii, Pseudolibrotus nanseni* and *P. glacialis* are abundant. In winter in the surface layer the relative abundance of predatory *Metridia longa* and of two species of chaetognaths increases.

The ecosystem of the pelagial of the Arctic Basin is, in many parameters, similar to the arctic ecosystem of the North European Basin of the Atlantic Ocean (see below); it also is restricted from above by the ice cover and bounded below by cold oligotrophic water.

The functional connection of the pelagial with the shelf is realized through the ice with its cryophilic community. This latter circumstance evidences the existence of features of functional symmetry of the ice systems of the Antarctic shelf and the Arctic Basin (ANDRIASHEV, 1970).

### (b) Ecosystems of Temperate Regions

In the northern hemisphere temperate Subpolar water regions occupy the vast spaces of cyclonic gyrals in the Pacific and Atlantic Oceans north of 40° N and the marginal seas. In the southern hemisphere they extend over the immense region of the Antarctic Circumpolar Current (Westwind Drift) to the south of the Subantarctic (or Subtropical) Convergence. The structure and variability of the plankton communities of these regions have been thoroughly studied. YASHNOV (1940), KIELHORN (1952), WILBORG (1955), GRUZOV (1963), PAVSHTIKS (1969), ZELIKMAN (1977) and COLEBROOK (1978), investigated the North Atlantic Ocean; BOGOROV and VINOGRADOV (1955), VINOGRADOV (1956, 1968), BRODSKY (1957), MCALLISTER (1961), HEINRICH (1962), PARSONS and LEBRASSEUR (1968), PARSONS and ANDERSON (1970), MOTODA and MINODA (1974) and SEMINA (1974) studied the North Pacific Ocean; HARDY and GUNTHER (1935), HART (1942), FOXTON (1956), MARR (1962), ANDREWS (1966),

HASLE (1969), EL-SAYED (1970), VORONINA (1972, 1974, 1977) and many others worked in Antarctic waters.

Highly characteristic of the communities of cold water regions is a relatively low species diversity (sharp dominance of a few species) and a similarity between life cycles of leading species. The absence of stable stratification in the surface zone, long periods of winter homothermy, brevity, and temporal fluctuations of phytoplankton vegetation to which the dominant species are adjusted—all these factors preclude narrow food specialization of phytophages. Substantial temporal changes in habitat conditions and low predator pressures contribute to low species diversity (MARGALEF, 1968; PAINE, 1969; SLOBODKIN and SANDERS, 1969). The necessity of utilizing the short bursts of phytoplankton development forced the phytophages to evolve a high plasticity of life cycles. Examples are: mixed age composition of populations, seasonal changes in sex ratio, changes in fecundity, varying rates of development in different individuals, and even different ways of morphogenesis (VORONINA, 1977).

The oscillatory type of population dynamics inherent in the mass species of cold water communities enables them to adjust their development to changes in their habitats and to exist in an environment with sharply varying conditions. The communities of these regions do not reach high degrees of maturity. This, however, does not imply that they are unstable or evolutionally primitive, as suggested by DUNBAR (1960, 1972). These simpler systems prove to be most efficient and adequate to biological progress under conditions of cyclic fluctuations in their environment and under-utilization of food resources (NESIS, 1965; VINOGRADOV, 1968; ZELIKMAN, 1977; VINOGRADOV and VORONINA, 1979).

### (c) Communities of the Northern Regions of the Pacific and Atlantic Oceans

#### *Structure*

The Subarctic region of the Pacific Ocean (see also Volume V: PERES, 1982; Chapter 6) comprises all aquatories extending from the Aleutian Islands to the zone of mixing, including the waters of the Kuril-Kamchatka region, the Western cyclonic, and the Alaskan gyrals. To this region also belong the greater parts of the Bering and Okhotsk Seas. Most of these waters are dichothermal, i.e. during the summer a cold-intermediate layer of 'winter water' is preserved in depths of 70 to 200 m. The extension of these waters may vary from year to year. They may be absent eastwards of 180° to 170° W (UDA, 1963), but a sharp halocline is present all year round at a depth of 100 to 200 m also in the eastern parts (DODIMEAD and co-authors, 1963). The depth of the basic pycnocline and the presence of seasonal pycnoclines have a substantial effect on the productivity of the different regions of this aquatory (PARSONS and LEBRASSEUR, 1968; SEMINA, 1977b).

In the open regions of the ocean, the basic elements of the zoocene in Subarctic waters are the copepods *Calanus plumchrus*, *C. cristatus*, *Eucalanus bungii*, *Metridia pacifica*, *Oithona similis*, and the chaetognaths *Sagitta elegans* s.l. and *Eukrohnia hamata*. The three first-named species yield 80 to 95% of the total mesoplankton biomass in the upper 200-m layer during the spring–summer season and determine the general aspect of the community. At this time in the surface layers the main mass of the zoocene (90% of the biomass and more) is represented by upper-interzonal species. East of 160° W the surface layers are occupied by transformed waters of the Pacific Current, with a somewhat different distribution of the mass species *Calanus cristatus, C. plumchrus* and *Eucalanus bungii* which occur here mainly in depths below 100 m.

There is no dichothermy in the waters of the North Atlantic Ocean. Here the saline surface waters sink in winter to considerable depths, so that not only is the 0 to 150-m (200 m) layer involved in winter homothermy, as in the Pacific, but far deeper ones down to 400 to 1000 m and more.

The basic element of the communities' zoocene is *Calanus finmarchicus* s.l. Less important are usually *Calanus hyperboreus, Metridia longa, M. lucens, Pseudocalanus elongatus, Oithona similis, O. atlantica, Pareuchaeta norvegica, Sagitta elegans* and appendicularians. However, among these only *Calanus finmarchicus* s.l. occupies a dominant position everywhere (except, partly, in the North Sea).

### *Seasonal Changes*

The urge toward maximum possible utilization of phytoplankton, the main food resource of the surface layers, causes all the phytophages of the surface and upper-interzonal zooplankton to concentrate during the spring–summer season in the sub-surface water layer. The vernal bloom of phytoplankton represents the main feeding and fattening period for these populations. If at this time a surplus of phytoplankton production is available, herbivorous species with similar food spectra do not compete for food. It is also during the vernal bloom that the annual maxima of biomasses of dominant species are formed. Naturally, as there is not perfect balance between the production and consumption of phyto- and zooplankters, their abundance maxima diverge somewhat in time. For fattening, which lasts only a short time, an extremely rich source of food is used. This results in the strong dominance of a few species best adapted to the exploitation of this resource and in possession of the highest biotic potential.

The sinking of the seasonal thermocline during the summer is accompanied by a corresponding increase in the thickness of the layer of main phytoplankton concentrations, while in the open sea regions the total quantity of phytoplankton diminishes. Now the greater part of phytophage populations occupy a larger water column and a tendency becomes apparent in the dominant species toward alternating an importance at different depths.

In winter with homothermy of the entire surface zone (down to 150–200 m, in some regions even below that) and diminished solar radiation, the thickness of the mixed layer increases beyond the critical depth. This results in a nearly complete disappearance of phytoplankton. When the living conditions in the surface zone are no longer favourable to the zooplankton, its populations either greatly decrease in abundance or sink to deeper layers, the process starting in early autumn. The first alternative takes place among smaller epiplankton animals which produce several generations per year (*Oithona similis* etc.); the second one is available only to animals large enough to have a long life cycle (like *Calanus*) and capable of performing vertical migrations of great amplitude. The second alternative seems to be more profitable judging from the fact that the upper-interzonal herbivores substantially surpass epiplankton animals in biomass.

Three basic types of life cycles may be recognized in boreal communities of herbivorous species (HEINRICH, 1961). To the first type belong species in which the shedding of eggs begins only during the vernal burst of phytoplankton development. The biomass of species of the first type reaches it annual maximum in summer with the older copepodite stages becoming dominant. To this type belong *Calanus finmarchicus* s.l., the most dominant of North Atlantic communities (e.g. REES, 1949; MARSHALL and ORR, 1955) and *Eucalanus bungii* in the Pacific Ocean, a species in which not only breeding but also the passage of older copepodites into the adult stage takes place with the onset of vernal phytoplankton vegetation. In the dominant species of the North Pacific zoocene—*Calanus plumchrus* and *C. cristatus*—on the contrary, reproduction does not depend on phytoplankton abundance and may or may not coincide with its vegetation period. They breed in deep water layers and their first copepodite stages rise to the surface into the zone of 'bloom'. The adult females have reduced mandibles and do not feed. *Calanus hyperboreus* also has a life cycle of the second type: its main breeding takes place not long before the beginning of phytoplankton vegetation below the euphotic zone (WIBORG, 1940; DIGBY, 1954). The third type of life cycles is characteristic of species capable of circum-annual reproduction, but with generation abundances strongly affected by the amount of available phytoplankton. Most epipelagic species belong to this type.

The breeding periods of carnivorous species are very protracted and may extend over the whole year. In any case, their seasonal fluctuations are slight as compared with the corresponding fluctuations of the herbivorous zooplankters (GRAINGER, 1959).

In different regions of the ocean the extent of divergence in the time of phyto- and zooplankton biomass maxima is determined by the type of life cycles of the dominant species. In the North Atlantic Ocean during the vernal phytoplankton maximum, *Calanus* species are represented by relatively few females and early young. After a 'period of delay' their effect on phytoplankton increases and reaches its maximum in summer with the passage of the main mass of the population into elder copepodite stages. This, according to CUSHING (1959), accounts for the lack of balance between the cycles of phytoplankton production and consumption (Fig. 2-13). In the northwestern Pacific Ocean the picture is substantially different due to a different type of life cycle of the dominant species *Calanus plumchrus* and *C. cristatus*. From the very onset of phytoplankton

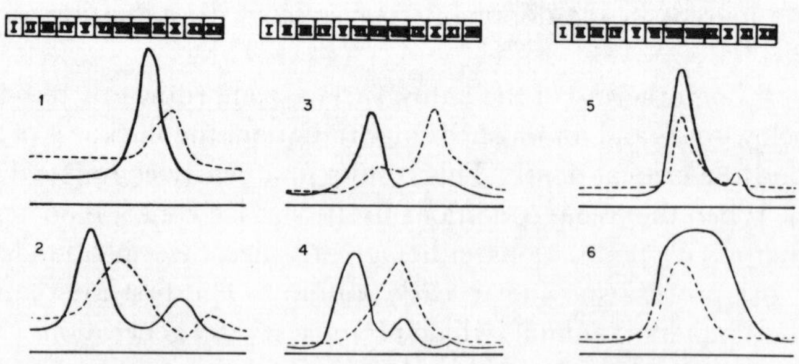


Fig. 2-13: Annual cycles of plankton communities in temperate and high latitudes. 1: Arctic-Siberian Seas; 2: Norwegian coast; 3: Baltic Sea; 4: southwestern part of Japan Sea; 5: neritic zone of Bering Sea; 6: Georgia Strait. Solid line: phytoplankton biomass; broken line: zooplankton biomass. Same scale for absolute quantities of zooplankton during the annual maxima in all regions. (Based on information provided by HEINRICH, 1961)

development, the early copepodites of these species are already in the surface layers, rapidly growing and actively feeding on phytoplankton. Consequently, there is no 'period of delay' in grazing and phytoplankton production is better balanced with consumption.

Actually, however, the trophic relationships in the communities are not restricted to that of phytoplankton versus herbivorous zooplankton, so that no dependable concept of the succession of productive and destructive phases in community development (its balance) can be obtained by mere evaluation of the biomasses of phytoplankton and herbivorous net zooplankton. The links bacterioplankton—protozoans—heterotrophic naked flagellates—infusorians greatly complicate the situation. With the exhaustion of nutrients above the thermocline, phytoplankters cease to develop and their main mass dies; but the organic matter of the dead biomass provides energy for a violent development of bacteria and heterotrophic protozoans feeding on them. Their biomass exceeds 10 g $m^{-3}$ in the 0 to 150-m layer (SOROKIN, 1974). This phase in the succession of the community seems to play an important role in assuring a high production of minute animal food, which is actively consumed by the young of dominant mesoplankters.

*Productive Characteristics of Mesoplankton*

Estimates of herbivorous mesoplankton production, or rather of the production of dominant copepod species, have been made by many authors. According to YASHNOV (1940) and KAMSHILOV (1958), the annual production of *Calanus finmarchicus* in the Barents Sea—where it accounts for about 80% of the total mesoplankton biomass–averages 55 to 65 g $m^{-2}$. In the northwestern part of the Pacific Ocean the annual production of the dominant *Calanus* species is estimated at 60 to 70 g $m^{-2}$ (MEDNIKOV, 1960), in the western part of the Bering Sea even at 115 g $m^{-2}$ (HEINRICH, 1956). However, the total production of mesoplankton is much lower than the values given above owing to elimination of herbivorous species by mesoplankton carnivores (chaetognaths, coelenterates, ctenophores, copepods, etc.).

Naturally, the amount of production varies from one year to another, but despite the substantial climatic changes of our century and increasing anthropogenic interferences, the annual production (biomass) of zooplankton in the open regions of the seas and the ocean apparently does not undergo substantial directional changes. Thus, for instance, in the Norwegian Sea the mean annual fluctuations of biomass and production of zooplankton during 1959 to 1963 were not more than twofold (TIMOCHINA, 1968), in the southeastern part of the Barents Sea not more than threefold (ZELIKMAN, 1977). Relatively small mean annual fluctuations were found also in multi-year analyses (1948–1975) taken with the Longhurst/Hardy plankton sampler in the North Atlantic Ocean (COLEBROOK, 1972, 1978). Thus, as pointed out by ZELIKMAN (1977), zooplankton

> 'despite the considerable climatic perturbations and a century long pressure of anthropogenic factors preserves a stable enough mean level of production which is evidence of a well expressed homeostasis of the system' (p. 55).

Of course changes in productivity may be more strongly expressed in heavily polluted shelf regions and inner seas. Data are available on changes in the productivity and even species structure of communities—for instance in the Baltic, Black, and Mediterranean Seas, as well as Georges Bank and northeastern coastal waters of the United States.

More important changes involving open sea regions are wrought in the structure of nekton communities. The formerly dominant, commercially valuable fish species (herring, cod, saithe, haddock) have been greatly reduced by fishery and their niches fully or partially occupied by other, commercially less valuable plankton consumers, e.g. blue whiting, arctic code, and capelin (PONOMARENKO, 1968; SONINA, 1973).

### (d) Ecosystems of the Southern Ocean*

The global ecosystems of the Antarctic and Subantarctic (see also Volume V: PERES, 1982; Chapter 6) occupy, respectively, the waters extending from the Antarctic Convergence (South Polar Front) to the Antarctic Continent, and from the Antarctic Convergence in the south to the Subtropical Convergence (Subantarctic Convergence) in the north. The antarctic ecosystem is confined to waters of Antarctic structure comprising antarctic surface, deep, and near-bottom water. The subantarctic water structure consists of Subantarctic Surface, Antarctic intermediate, deep, and near-bottom waters.

Light intensity limits phytoplankton development at the surface from April to September in 60° S and from March to August in 70° S. Nutrient concentrations in the euphotic zone remain high throughout the year (HART, 1934; HARDY and GÜNTHER, 1935; CLOWES, 1938). During the months of daylight the rate of phytoplankton development depends on the stability of water stratification (GRAN, 1931), i.e. on the formation of the thermocline at depths less than critical. (The formation of summer stratification is associated with ice thawing and heating of the water.) In the northern Antarctic summer stratification begins in mid-October (CURRIE, 1964) and gradually progresses southward. Correspondingly, the phytoplankton bloom begins in the north in October and attains its maximum in December.

Seasonal variations in phytoplankton distribution in the Antarctic are associated mainly with the formation of a belt of bloom and its advance to the south. The shifting from the Antarctic Convergence to the boundary of pack-ice takes about 2 months. (HART, 1942). During the vegetation season a succession of groups takes place differing in size and taxonomic status (HART, 1942; STEYAERT, 1974). This schedule may be disturbed by various factors. Extreme variations in the abundance of algae extend over five orders of magnitude.

#### *Structure and Functioning of Communities*

The main mass of the zoocene consists of herbivorous copepods which supply more than 70% of the total zooplankton biomass (VORONINA, 1966a.) Most important among them are three species: *Calanoides acutus*, *Calanus propinquus*, and *Rhincalanus gigas*. Their life cycles, like those of the dominant copepods of boreal plankton, are adjusted to the utilization of brief periods of food abundance and all of them follow virtually the same course. Development and changes in the age composition of populations are accompanied by changes in vertical distribution, location depth of maximum concentrations, and sequence of dominance of development stages along the vertical (Fig. 2-14).

In the early period of the community development, i.e. during the biological spring, the overwintered individuals concentrate within a narrow near-surface layer where,

*This section is based on data of VORONINA (1977, 1978).

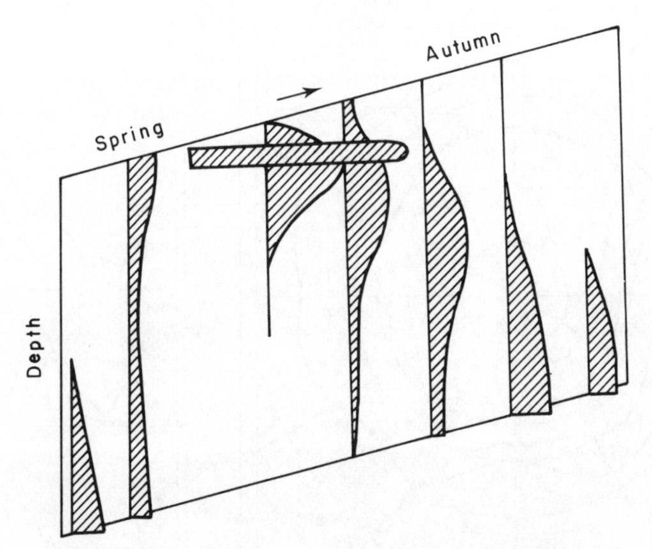


Fig. 2-14: Seasonal changes in the vertical distribution of interzonal phytophage abundance. (Based on information provided by VORONINA, 1972)

feeding intensely, they soon reach sexual maturity and spawn. The young, likewise, are confined to the narrow surface layer, but with progressing development the new generation disperses as the older individuals move to greater depths. For a certain time the population occupies the entire euphotic layer, but later the animals begin to sink below it. During its vernal ascent the population extends widely along the vertical but the migration ends at the surface in a dense concentration of all individuals.

Seasonal changes in total plankton abundance are determined by the combined effects of the life cycles of different populations. In the upper 100-m layer two maxima occur during the year (Fig. 2-15). The first is associated with the vernal ascent of mass filter-feeding copepods, the second with the development of their new generations. The relatively poor overwintered populations do not fully utilize the primary production and thus do not impede the accretion of phytoplankton biomass which reaches its maximum after the first peak of zooplankton. Later, growing-up new generations begin to utilize phytoplankton more intensively. This probably accounts for the seasonal reduction of its abundance.

An important feature in the organization of a plankton community is the divergence in time of certain stages of life cycles in closely-related species. This may be demonstrated, for instance, by comparing the age composition of copepod populations collected at random stations. The population of *Calanoides acutus* is everywhere in a more advanced state than that of *Calanus propinquus*, while *Rhincalanus gigas* lags behind both of them (Fig. 2-16). These differences are caused by different timing of such events as ascent to the surface, breeding and sinking to the depth. This results in partial divergence of maxima of different populations along the vertical (VORONINA, 1966b,1972).

Another important effect of the temporal divergence of life cycles in closely-related species is the spatial divergence of their biomass maxima. The asynchronism of plant and animal maxima is reflected in the spatial distribution of their biomass. The cir-

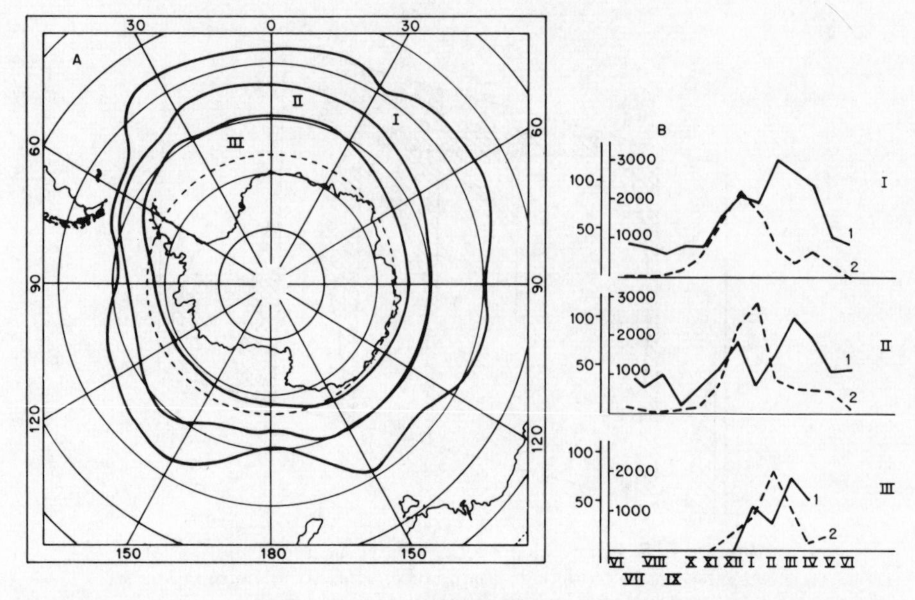


Fig. 2-15: Seasonal changes in the quantity of phyto- and zooplankton in the Antarctic Ocean. A: Latitudinal zones; B: seasonal changes in biomass of phyto- (broken line) and zooplankton (solid line) in the indicated zones. (Based on information provided by VORONINA, 1977)

cumantarctic ring of phytoplankton bloom always occurs farther south than the ring of zooplankton summer maximum (VORONINA, 1970a,b). During the period preceding the zooplankton peak the highest concentrations of both groups of organisms usually coincide in the upper layer of the pelagial. Later they become vertically decoupled; first, when—due to the overgrazing of phytoplankton—its absolute maximum is located at a greater depth than the abundance maximum of its consumers, and then—at the end of the season—when the main mass of animals shift into the subsurface layer so that the algal maximum is above them (see also Volume V: PERES, 1982).

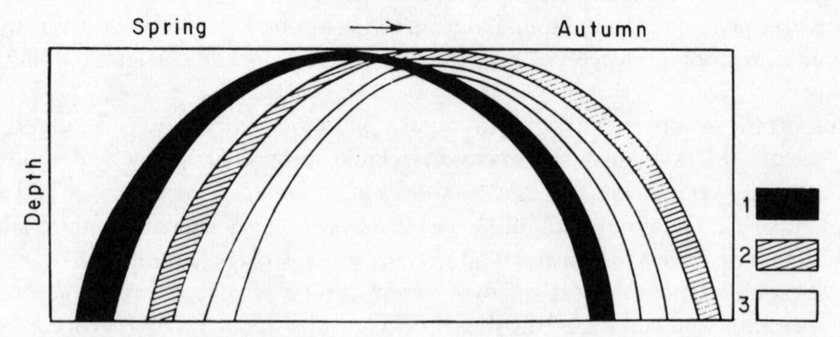


Fig. 2-16: Seasonal changes in vertical structure of the zoocene in the Antarctic Ocean. Position and extension of population cores (portions of populations between 25% and 75% levels of abundance). 1: *Calanoides acutus*; 2: *Calanus propinquus*; 3: *Rhincalanus gigas*. (Based on information provided by VORONINA, 1972)

In the Antarctic the wide amplitude of fluctuations in the abundance of phytoplankton and the protracted period of its underexploitation are symptoms of unbalance in the cycles of phyto- and zooplankton. However, the degree of this unbalance seems to be subject to significant local variations depending on hydrological conditions and the composition of zooplankton. In regions where the thermocline is sharply expressed so that the copepods of the wintering stock are unable in spring to penetrate through it into the productive layer, primary production is underexploited more than anywhere else (VORONINA, 1970a, 1974). At the Antarctic Convergence, where primary production is reduced by sinking water and zooplankton accrues on allochthonic individuals, the relationship between phytoplankton and the biomass of herbivores is considerably lower than in other regions, and phytoplankton consumption is correspondingly considerably higher.

A mathematical simulation of seasonal distributions of *Calanus propinquus* and *Calanoides acutus* populations in the Southern Ocean has been developed, based on the scheme of VORONINA and co-authors (1980a), which permits assessment of the annual production of these species and estimation of the production of all mesoplanktonic herbivores. In copepods this amounts to about 65 g $m^{-2}$, or including small euphausiids, to 70 g $m^{-2}$ with average annual P/B coefficients of 4·5 for *Calanoides acutus* and 3·8 for *Calanus propinquus*, i.e. to values very close to those obtained for mesoplankton production in temperate regions of the northern hemisphere.

### *Trophic Web*

In the Southern Ocean the trophic-web structure is relatively simple. Phytoplankters are consumed by filter feeders: copepods, euphausiids, and tunicates; copepods and euphausiids, by hyperiids, chaetognaths, predatory copepods, squids, fishes, and whalebone whales. Fishes and squids also feed on amphipods and are in turn eaten by sperm whales and small-tooth whales. Thus, the trophic net comprises four basic levels: producers, phytophages, and predators of first and second order. Their quantitative characteristics are probably most reliably represented by the following estimates: in the Antarctic Ocean the mean annual biomass of net plankton (mesoplankton) amounts to 26 g $m^{-2}$ in the 0 to 100-m layer (FOXTON, 1956), and in the Indian and Pacific sectors of the Southern Ocean, to 10 to 20 g $m^{-2}$ during the summer maximum in the upper 100-m layer (VORONINA, 1966b). Most important in biomass are copepods (73%), followed by chaetognaths (10%) and euphausiids (8%). The biomass of euphausiids, according to trawling data, is about 0·7 g $m^{-2}$, increasing in regions of maximum abundance to 30 g $m^{-2}$ (MARR, 1962). The biomass of whales during the period of high abundance was 0·56 g $m^{-2}$, and the quantity of plankton consumed by the whales during the summer season 10 g $m^{-2}$ (NEMOTO, 1968; MACKINTOSH, 1973).

Recently, a radical reorganization is taking place in the quantitative relationships between the separate elements of this trophic net. The initial stock of whales during the last 40 years was reduced by 85 to 90% by rapacious extermination (MACKINTOSH, 1970) and a corresponding decrease is observed in the consumption of their food objects (VORONINA, 1977, p. 86).

Thus, the communities of temperate regions are characterized by: low species diversity, relatively simple trophic net, ordered spatial structure which regularly changes in time, maintenance of a mean annual level of productivity, i.e. well-expressed homeostasis, greater or lesser gap between phyto- and zooplankton maxima (depending on the type of seasonal cycles in the dominant species), and significant local differences in the balancing of phytoplankton production and consumption.

### (e) Ecosystems of Tropical Regions

The ecosystems of Tropical regions occupy a great part of the ocean aquatory, approximately from 40° N to 40° S. Most important reorganizations of the structural and functional singularities of their communities take place at the northern and southern boundaries, where the tropical ecosystem is replaced by the arctic-boreal and the subantarctic ones.

Within the Tropical region the composition and mode of existence of plankton communities are rather similar. This suggests a structural unity of the tropical community as a whole. However, there are hierarchical differences. The most substantial differences are observed between the communities of Central anticyclonic gyres, the community of Equatorial zonal flows, as well as distant-neritic and neritic communities. We shall restrict ourselves here to a brief consideration of the two first-mentioned types inhabiting the water of the open ocean. Coastal and estuarine ecosystems receive attention in Chapter 4. Many features of oceanic tropical ecosystems have been discussed in detail in a number of recently published papers and reviews (PARIN, 1968; VINOGRADOV, 1971,1975; BE, 1977; HEINRICH, 1977a,b; REID and co-authors, 1978; VORONINA, 1978; and many others). (See also Volume V: PERES, 1982; Chapter 6.)

### (f) Ecosystems of Anticyclonic Tropical Gyres

The anticyclonic gyres that form the halistases of central (southern and northern) waters are zones of planetary convergences homologous in all oceans (Fig. 2-1). They are generally characterized by a prevalence of descending water motion, a thick, heated mixed water layer, and a deep pycnocline. Nutrients are brought into surface layers mainly at places of quasi-stationary water ascents induced by current divergences. At the northern and southern peripheries of the gyrals some seasonal mixing takes place, but the areal-size of such regions is small as compared with the total aquatory of the gyre. Over the greater part of a gyre access of nutrients to the surface is impeded; they reach it through the pycnocline very slowly and in very small quantities, though still sufficient to ensure a maximum of phytoplankton development at depths of 70 to 100 m (SOROKIN, 1959; VINOGRADOV and co-authors, 1970). Above these depths the community subsists mainly on regenerated nutrients (see below).

Of substantial importance in the penetration of nutrients into surface layers from under the pycnocline seems to be local cyclonic synoptic eddies moving over the aquatory of the gyre (KOSHLYAKOV and MONIN, 1978), which contribute to a greater concentration of nutrients in the euphotic layer and to the appearance of short bursts of phytoplankton development. Nevertheless, it would be no great error to assume that the system is balanced at every given moment, i.e. that the inflow of energy into the system is equal to its dispersion. Therefore, it becomes possible to obtain a concept of the communities' production cycles from the results of short-term observations. This greatly facilitates the study of tropical ecosystems and the modelling of the processes occurring within them. As the ecosystems of Central gyres and of the Equatorial region have many common structural and functional features, it is convenient to consider them together, but some of the singularities inherent in equatorial ecosystems must be given special consideration.

*Structural Characteristics*

The main mass of species of the Tropical zone are members of communities of both anticyclonic gyres and the Equatorial current system. HEINRICH (1975,1977a), who has closely examined this, points out that—in the Pacific Ocean—widely tropical species of euphausiids, chaetognaths, and rhizopods account for 60% of the total species number of these groups. A far lesser percentage have their basic distributional range in the gyre. Thus, for instance, in the northern cyclonic gyre they account for only 19%, and in some groups for 5% to 30% of the species. Certain species may be absent at the periphery of the gyre and occur together only in one of its parts, as a rule a rather small one (Fig. 2-2). Many species are characterized by extended areas of expatriation encroaching upon the biotopes of neighbouring communities (Fig. 2-17). Therefore, in the Tropical zone the boundaries between communities with different species structure

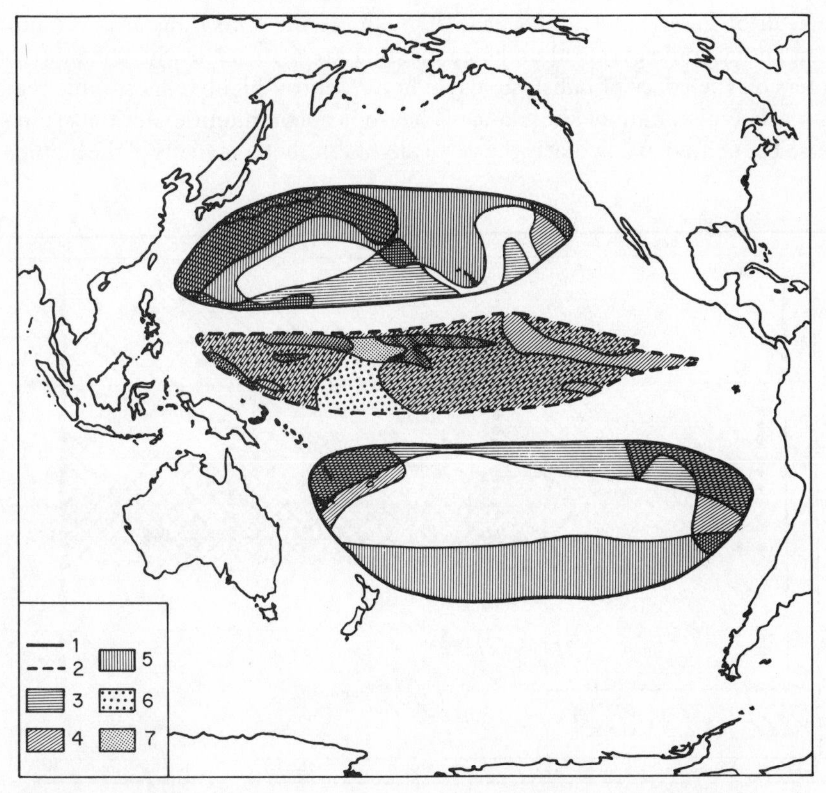


Fig. 2-17: Distribution of expatriated species. Boundaries of communities. 1: central; 2: equatorial. Regions of expatriation of species with ranges based on the communities; 3: equatorial; 4: distant-neritic; 5: transitory zones; 6: central. (Based on information provided by HEINRICH, 1975)

are not sharply outlined, but their reality has been confirmed by numerical methods (HEINRICH, 1977b).

Most abundant in tropical communites are, as a rule, tropical species and species with ranges limited by a given community. Expatriating species are few (HEINRICH, 1977a). Depending on ecological and biogeographical singularities of distribution, maximum abundances of some species may be confined to the highly productive waters of eastern Equatorial upwellings; of others, to oligotrophic Central waters (VINOGRADOV and VORONINA, 1964; GUEREDRAT, 1971). For instance, the copepod *Candacia pachydactyla* and the pteropod *Creseis virgula virgula* in the Equatorial Pacific are most numerous in a narrow productive belt between 8° N and 8° S. *Candacia aethiopica* is confined to eutrophic and mesotrophic waters, *C. bispinosa* to oligotrophic waters (Fig. 2-18). In such species, belts of increased abundance are associated with divergences and convergences (VINOGRADOV and co-authors, 1961; VINOGRADOV and VORONINA, 1962,1963). Lastly, there are species with little or no differences in abundance in Central and Equatorial waters, or species with maximum abundance in some distant-neritic areas (HEINRICH, 1977a).

Small-scale changes in species abundance within a single gyre are caused by heterogeneity of the biotope. REID and co-authors (1978) relate, for instance, a decrease in species diversity of euphausiids in Tropical regions with a well-expressed oxygen deficit (Gulf of Panama; Arabian Sea). This limits the spreading of many migratory forms. In anticyclonic gyrals abundance maxima are associated with the more productive waters on the edges of halistase, while in its Central highly oligotrophic regions no species reach a maximum of abundance. The tropical communities are richest in species numbers. The spatial, particularly the vertically stable, heterogeneity of the biotope seems

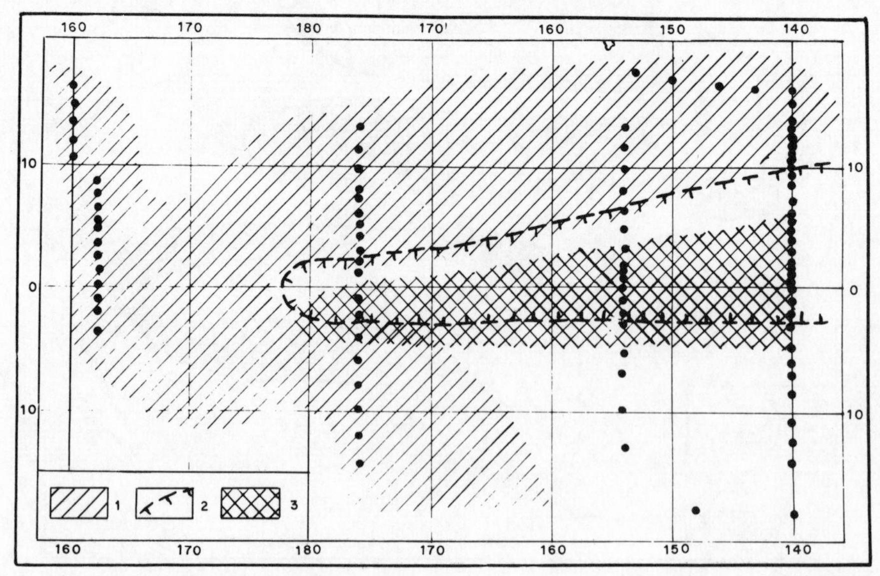


Fig. 2-18: Zone of distribution of *Candacia bispinosa* and *C. aethiopica* in the equatorial zone of the Pacific Ocean. 1: *C. bispinosa*; 2: *C. aethiopica*; 3: region with phytoplankton abundance above 1000 cells $m^{-3}$ in the 0–100 m layer. Dots: location of stations with observations. (Original)

to be a major factor ensuring high faunistic diversity. Indeed, in Tropical regions, in most taxonomic groups species diversity changes little, showing a small increase only in the coastal regions of western currents (REID and co-author, 1978). HEINRICH (1977a), however, points out a more varied picture: abundance increase of some groups in waters of certain ecosystems (e.g. Foraminifera in Equatorial waters), of other groups in other gyres (e.g. Euphausiacea in Central waters). In communities transitory between high-latitudinal and tropical, the number of species decreases in all groups.

### *Development of Communities*

Investigations of the distribution of diverse groups of plankton (phyto-, bacterio-, and zooplankton), as related to the variability of physical, hydrochemical, and biophysical fields, permitted the advance of a theoretical scheme of the turnover of nutrient salts and organic matter which determine the pattern of development of pelagic communities in the tropical waters of the ocean (Fig. 2-19).

In zones of ascending water (quasi-stationary divergence of currents or another type of upwelling) where the thermocline is situated near the surface, the nutrients intensely penetrate into the surface layer and phytoplankton begins to develop. Simultaneously, water withdraws from the zone of ascent, the temperature (density) stratification of water layers becomes more marked, mixing slackens, and the upper boundary of the

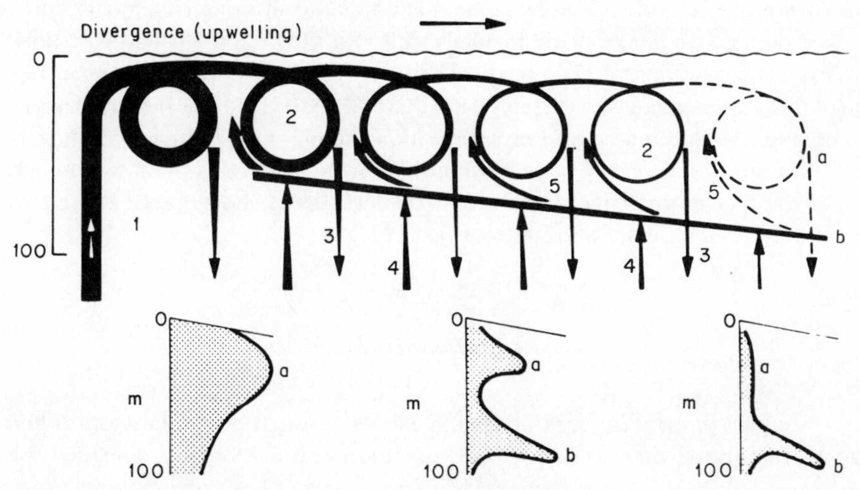


Fig. 2-19: Turnover of nutrient salts and organic matter in the succession of a tropical pelagic community. 1: Ascent of nutrient salts and dissolved organic matter in the divergence zone; 2: their repeated use in the productive–destructive cycles of the community below the thermocline; 3: loss with sinking organic remains and migrating organisms; 4: turbulent ascent of nutrient salts through the thermocline; 5: their incorporation into the productive cycles of the surface community, which does not compensate the losses with a descending flow; 6: thermocline. Lower: Change in the pattern of phytoplankton distribution at different distances from the divergence zone. a: Upper concentration maximum; b: lower concentration maximum. (Based on information provided by VINOGRADOV and co-authors, 1970)

thermocline shifts to a greater depth. Nevertheless, the nutrients under the effect of turbulent mixing continue to penetrate into the upper mixing layer, although now in lesser quantity. The combined activities of phytoplankton and bacteria consume the nutrient salts supplied from below, forming a maximum abundance of life above the thermocline. Nutrient salts hardly penetrate the overlying water layers and hence this water cuts off the overlying parts of the community from its biogenous resource. However, this isolation is not complete.

Organisms inhabiting the layers above the lower maximum subsist mainly on nutrients transported by water currents from zones of ascending to zones of descending water motions. These nutrients, being involved in the production–destruction cycle of the community, are virtually totally incorporated in the bodies of organisms and their concentration in the surrounding water is near to analytical zero. With the 'ageing' of waters and the constant removal of nutrients by the migration of organisms into deeper layers and precipitation of phytoplankton and detritus, the biogenous environment of the surface layers becomes gradually depleted; daily primary production reduces its value from several tens of mg C $m^{-3}$ to some hundreds of mg C $m^{-3}$.

Maturing communities are carried by the water withdrawing from the zone of water ascent so that the temporal pattern of their development becomes extended in space. Since the process of water ascent is quasi-stationary, the communities near the upwelling zone in the early stages of development are characterized by a maximum of phytoplankton, which farther downstream is succeeded by a maximum of herbivorous zooplankton with a more protracted development, and still farther from the zone of water ascent by a maximum of carnivores. The maxima of groups differing in the duration of their development do not coincide in space (VINOGRADOV and VORONINA, 1964; VINOGRADOV and co-authors, 1973; see also Volume V: PERES, 1982). A clear-cut representation of this phenomenon was obtained by TIMONIN (1971) in the Indian Ocean. In the zones of water ascent and community formation the plankton biomass reaches its maximum, and species diversity its minimum (SHANNON's index $<2$); with increasing maturity the species diversity of communities increases in convergence zones to a maximum SHANNON's index of about 4·5 to 4·5.

*Trophic Structure and Production*

A certain part of primary organic matter passes through phytoplankton to herbivorous zooplankton and then to carnivorous plankton and nekton (the so-called 'pasture' type of food chain). In the early stages, primary production is not completely consumed by heterotrophs (including bacterioplankton) and a considerable portion of it (sometimes up to 50% and more) passes out of the community. In highly productive regions such 'surplus' organic matter may occur in very large quantities.

On the other hand, as shown by calculations, in the vast aquatories of the oligotrophic Tropical regions primary production is not sufficient to satisfy the energy requirements of the communities (e.g. SOROKIN, 1971a,1977b; SOROKIN and co-authors, 1975; VINOGRADOV and co-authors, 1976). However, there is another route by which energy is supplied to the community, namely through detritus and dissolved organic matter—first to bacteria and protozoans and then to filter feeders and predatory invertebrates.

Although in this 'detritus type of food chain' much energy is lost through the appearance of additional trophic links, this route plays an important role in the food supply of pelagic communities (for details see Volume IV: CONOVER, 1978).

SOROKIN (1971a; see also Volume IV: SOROKIN, 1978) assumes that the dissolved organic matter formed in surplus in the highly productive cold water regions sinks into the deep layers with the descending flow of cooled water; it is then transported by meridional currents and brought to the active layer of the ocean in zones of quasi-stationary water ascent. Here, owing to high temperatures, it becomes susceptible to bacterial decomposition. Thus, in these tropical communities bacteria play the role of local producers. The amount of allochthonous organic matter brought into the communities of oligotrophic Tropical regions through the detritus food chain may surpass several times (2–5) the quantity of organic matter generated by primary production (SOROKIN, 1977b, 1978a). Some authors consider these values over-estimated (e.g. STEEMANN-NIELSEN, 1972; BANSE,1974; see also SIEBURTH, 1977). In the succession of pelagic communities the period of organic matter accumulation (burst of phytoplankton development) is very short as compared with the time during which energy expenditure prevails over energy accumulation. In tropical communities this relationship may be very roughly estimated as 1 : 30 or even 1 : 50. Therefore episodic, instantaneous recordings will almost always fall in the destructive period of community metabolism which is supported to a considerable extent by the earlier accumulated organic matter. If this were true, the role of allochthonous organic matter may prove to be less important than assumed by SOROKIN. Nevertheless, there can be no doubt that the energy supplied to the community through dissolved organic matter plays a significant role in its existence.

The food chain in the highly mature communities of anticyclonic gyrals is extremely long and may reach seven trophic levels (PARIN, 1968,1970). Owing to the highly mixed composition of tropical communities, zooplankton production cannot be evaluated by the production of a few dominant species. The data available on the production of its diverse trophic components are very scanty and do not allow average evaluations for tropical ecosystems. However, some assessments may be mentioned: GREZE (1973a) estimates the production of phytophages in some points of the Tropical Atlantic Ocean at 150 g $m^{-2}$, predators at 84 g $m^{-2}$, and detritus feeders at 44 g $m^{-2}$ (see Volume V: PERES, 1982; Chapter 6). In the mesotrophic Tropical regions of the Pacific Ocean at the periphery of gyres, the annual production of mesoplankton herbivores, according to data of SHUSHKINA (1971), SHUSHKINA and KISLYAKOV (1975), and VINOGRADOV and co-authors (1976), has an approximately similar value of 120 to 180 g $m^{-2}$ in ca. 0 to 150 m. However, in waters with different trophic conditions—depending on the state of maturity of the community—the production of separate trophic groups of zooplankton may vary within a very wide range and the diurnal net production of its entire zoocene as a whole community may be equal to or below zero during the destructive phase of community development (SHUSHKINA and KISLYAKOV, 1975; SHUSHKINA, 1977; VINOGRADOV and SHUSHKINA, 1978).

### (g) Ecosystems of Equatorial Upwellings

In the immense aquatory of the oceanic Tropical low-productivity zone regions of quasi-stationary ascents of intermediate waters stand out which carry nutrients into the euphotic zone and ensure a high productivity of their plankton communities. In open-ocean waters ascents are most intense in Equatorial regions.

The region of the Equatorial upwelling *sensu lato*—fairly narrow and only slightly expressed in the western part of the ocean—widens in its eastern part and the phenomenon itself is substantially intensified (WYRTKI, 1966). In the Atlantic Ocean upwelling is constantly observed east of 30° to 20° W, from 6° to 7° N to 7° to 8° S. In the eastern Pacific Ocean it is more constant and clearly traceable to the east of 180° to 160° E. The region of its manifestation, gradually widening in an eastern direction, extends from 8° to 12° N to 6° to 8° S. In the Indian Ocean, owing to the monsoon circulation, coastal upwellings are most clearly expressed, but occasionally a strong upwelling on the equator is observed in the eastern part of the basin.

The rate of water ascent to the surface directly on the equator, the Equatorial upwelling *sensu stricto* so-to-speak, or the Equatorial Divergence also increases from west to east, reaching its maximum at 93° to 100° W in the Pacific Ocean and at 5° to 10° W in the Atlantic Ocean. Concomitantly, the upper boundary of the thermocline rises, contributing to the enrichment of nutrients in the surface layers.

The speed of water ascent at the equator may attain $10^{-2}$ and even $10^{-1}$ cm $s^{-1}$ (POLOSIN, 1967; ROTSCHI and JARRIGE, 1968; CHEKOTILLO, 1969). In the zone of upwelling it is rather variable at different points. This leads to the formation of lenses of water rich in nutrients and in a 'cloudy' distribution of plankton (patchiness). Judging from data of soundings of the luminescent field, the horizontal extension of such 'clouds' is of the order of hundreds of metres (LEVIN and co-authors, 1975).

### *Formation of Plankton Communities*

The composition, distribution, and productivity of equatorial communities are assured by the rise to the euphotic zone of water rich in nutrients, the transport of developing communities by zonal flows, and the transverse meridional transport of communities from zones of rising to zones of sinking water. The narrow zones of ascending water extending along the equator and alternating with zones of descending movement impart a 'banded' pattern to the distribution of regions with high and low plankton biomasses. Thus, in the eastern Equatorial part of the Pacific Ocean the upward movement is most intense on the equator and the divergence at the northern boundary of the Equatorial Countercurrent (about 10° N). Along this line extend well-outlined maxima of plankton biomass (Fig. 2-6) observed by many authors (e.g. BRANDHORST, 1958; KING and HIDA, 1957; KING and IVERSEN, 1962; VINOGRADOV and VORONINA, 1963). A similar 'banded' pattern of distribution in the zone of Equatorial upwelling is observed in the Atlantic Ocean (GRUZOV, 1971) and in the Indian Ocean (VINOGRADOV and VORONINA, 1962).

The time required for the development of algae is too short for the upwelled waters to shift away from the place of their ascent, but the burst of zooplankton following the phytoplankton bloom occurs at a considerable distance from the zone of ascending movement. According to a rough estimate by SETTE (1955) about 50 to 150 d are required in Equatorial regions for the development of all trophic links of a community and, according to BLACKBURN and co-authors (1970) the maxima of small fishes and cephalopods in these regions lag behind the chlorophyll maximum by about 4 mo. The simulations performed by VINOGRADOV and co-authors (1973) yield analogous values:

Table 2-4

Biomasses (g $m^{-2}$) of the basic elements of the plankton community at the equator in the Pacific Ocean (0 to 120 m layer; January to February, 1974) (Original)

| Group of organisms | 97° W | 122° W | 140° W | 155° W |
|---|---|---|---|---|
| Phytoplankton, *p* | 46·5 | 5·1 | 4·5 | 4·8 |
| Bacterial, *b* | 16·5 | 2·8 | 5·2 | 2·9 |
| Protozoans, *a* | 2·0 | 2·8 | 1·6 | 2·7 |
| Mainly herbivorous metazoans, *f* | 6·0 | 2·0 | 3·1 | 2·8 |
| Mainly predatory metazoans, *s* | 4·6 | 2·1 | 1·9 | 2·5 |
| Total zooplankton (including protozoans), *z* | 12·6 | 6·9 | 6·6 | 7·9 |
| Total plankton | 75·6 | 14·8 | 16·3 | 15·6 |

70 to 100 d. During this interval the community may be carried along the equator over a distance of 1000 to 1500 miles and be shifted by the meridional component 150 to 260 miles to one side of the equator. The shift will be greater the more prolonged the development of the species or the farther the place of the organism from the producer in the food chain. Therefore, as a rule, concentrations of macroplankton and of large fishes feeding on it occur far down the zonal flow or to the side of the zone of divergence.

The communities of the euphotic zone (vertical mixed layer) formed in the zone of the Equatorial upwelling are carried with the surface water of the South Equatorial Current westwards along the equator, undergoing on the way certain successive changes. With advance to the west the biomasses of all their elements including zooplankton, macroplankton, and fishes also undergo certain changes (e.g. KING and DEMOND, 1953; VINOGRADOV and VORONINA, 1963; VORONINA, 1964; BLACKBURN, 1968; BLACKBURN and co-author, 1970; Table 2-4).

*Seasonal Changes in Plankton Composition and Abundance*

Seasonal changes in plankton communities in the tropics are determined by the singularities of wind regime and water circulation. In the zone of upwelling they depend mainly on the rate of water ascent and location of divergence zones which, in turn, depend on seasonal strengthening or weakening of the trades. Special investigations of seasonal plankton biomass variability in the vast Equatorial zone of the open ocean (100° 30' W–121° 30' W; 16° N–10° 30' S) were carried out by the 'Estropac' expedition. This expedition disclosed substantial seasonal fluctuations of diurnal phytoplankton production—from 127 to 318 mg C $m^{-2}$ (OWEN and ZEITZSCHEL, 1970). The zooplankton biomass was found to change in the same phase as chlorophyll but with a very small amplitude. Seasonal fluctuations of very small amplitude were observed in the biomass of small fishes and cephalopods, which lagged 4 mo behind the fluctuations of chlorophyll (BLACKBURN and co-authors, 1970).

In the eastern shore area of the Equatorial Atlantic Ocean two periods may be distinctly recognized in the life cycle of plankton communities (GRUZOV, 1971; VINOGRADOV, 1971; see also Volume V: PERES, 1982). The first (February–May) is a period of depression with relatively stable stratification and heating of surface layers. At this time, and particularly toward the close of the period, the Equatorial zone (1° N–5° S) differs but little in plankton abundance from oligotrophic Tropical regions and only directly at the equator does its biomass show a certain increase.

The second period (June–December) is a period of increased intensity of water ascent. The biomass and production of algae reach their annual maximum, thus contributing to an intensive growth of phytophagous copepods and a 'rejuvenation' of their populations. The favourable food conditions of this period result in increased fertility and the appearance of new abundant generations of phytophages consisting of very large individuals. The zooplankton biomass accrues rapidly. Towards the middle of this period (August), phytoplankton development begins to decrease, and its species composition undergoes a cardinal change: peridinians and coccoliths become dominant. In October the rate of water ascent increases again, and diatoms again become dominant. There is a wave of mass development of tunicates (*Thalia democratica*) followed by a wave of filter-feeding crustaceans. Zooplankton biomass reaches its annual maximum during the period of attenuation of water ascent. At this time, the reduced rate of photosynthesis can no longer satisfy the food requirements of the augmented mass of herbivorous zooplankters. Food conditions become unfavourable and, gradually, spring depression sets in (GRUZOV, 1971).

### *Functional Characteristics*

Let us consider these characteristics on the basis of an example of the community of the Equatorial upwelling in the eastern part of the Pacific Ocean. This will allow some conclusions regarding the development of the community during its displacement by the stream of the South Equatorial Current from east to west (VINOGRADOV and co-authors, 1976; VINOGRADOV, 1978a; VINOGRADOV and SHUSHKINA, 1978. (See also Volume V: PERES, 1982.)

The sharp decrease in biomass (Table 2-4) and production (Table 2-5) of all elements of the community west of the region of most intense water ascent (97° W) substantially changes the intensity of many processes in the community, although the

Table 2-5

Functional characteristics of the plankton community at the equator in the Pacific Ocean (0 to 120 m layer; January to February, 1974) (kcal $m^{-2}$ 24 $h^{-1}$) (Original)

| Longitude | $^*B_0$ | $P_p$ | $P_b$ | $P_a$ | $P_p$ | $P_s$ | $P_p/B_0$ | $K_{sp}$ | $D_0$ | $P_0$ | $P_{ac}$ | $P_{ac}/B_0$ |
|---|---|---|---|---|---|---|---|---|---|---|---|---|
| 97° | 50·3 | 24·1 | 6·6 | 1–9 | 2·8 | 0·98 | 0·47 | 1·33 | 18·2 | 5·9 | 19·9 | 0·39 |
| 122° | 10·7 | 5·3 | 2·9 | 0·94 | 0·61 | 0·25 | 0·50 | 0·58 | 9·1 | −3·8 | 3·0 | 0·28 |
| 140° | 12·7 | 4·5 | 2·3 | 0·89 | 1·0 | 0·33 | 0·35 | 0·58 | 7·7 | −3·8 | 1·9 | 0·15 |
| 155° | 11·7 | 5·7 | 5·0 | 0·83 | 0·83 | 0·31 | 0·49 | 0·43 | 13·6 | −7·8 | 4·9 | 0·42 |

$^*B_0$ = Biomass of the community.
$P_i$ = Production of different groups of the community (Table 2-4).
$P_p/B_0$ = Specific energy flow.
$K_{3p}$ = Efficiency of primary production.
$D_0$ = Total heterotrophic destruction of the community.
$P_0$ = Net production.
$P_{ac}$ = Actual production.

value of the specific flow of energy $P_p/B_0$) remains virtually unchanged (Table 2-5). All groups are most productive in the zone of most intense upwelling (97° W). Here, too, positive and net production of the whole community are high, indicating that the community is now in the productive phase of development.

With the transport of the surface-layer community to the west by the South Equatorial Current upwelling becomes less intense and net production negative, i.e. the organic matter produced by phytoplankton within the community is now sufficient to compensate for the energy expenditure of bacterial flora and zoocene. This means that the community in this region subsists not only on the organic matter that it produces but also on energy accumulated earlier (organic matter) or on allochthonous organic matter brought into the community from outside, for instance, from eastern regions of more intensive upwelling or in the process of the ascent of Antarctic Intermediate waters, according to SOROKIN's (1971a, 1973a) suggestion. The production deficit increases from east to west and, concurrently, decreases from east to west the efficiency of primary production of organic matter ($K_{3p}$ (p. 706) from 1·3 at 97° W to 0·4 at 155° W). These values of $K_{3p}$, together with the values of primary production given above, permit us to consider the waters of the region of most intense upwelling as eutrophic and the waters of the rest of the region as mesotrophic (VINOGRADOV and SHUSHKINA, 1978).

The decrease in the biomass of phyto- and bacterioplankton from west to east reduces the degree of satisfaction of food requirements ($\delta = c/c^{max}$)* by 1·5 to 2 times in nearly all trophic groups. The value of specific grazing of different groups increases considerably, while the efficiency of food assimilation diminishes ($\eta$). There is also a concomitant increase in the stress of trophic relationships, balance of production, and consumption between some trophic levels ($\varepsilon$) and efficiency of energy transfer (ecological efficiency) through the system ($\omega$). Thus, the ecological efficiency increased in flagellates from 3% at 97° W to 17% at 150° W; in infusorians, from 14% to 21–26%; and in fine-filter feeders, from 5% to 11%. Owing to the high concentration of carnivores in the whole upwelling zone (44–55% of zooplankton biomass) 50% to 80% of the mesoplankton production is consumed by its own member organisms, and in mesotrophic waters of reduced upwelling the production intensity of mesozooplankton, after a sharp decrease becomes negative. However, the actual production of the community developing on primary production and allochthonous organic matter, remains positive over the whole aquatory of the Equatorial upwelling investigated east of 155° W. This confirms the values calculated by SOROKIN and co-authors (1975) which point to the prevailing importance of the 'detritus food chain' in the community of the Equatorial upwelling. According to SOROKIN and co-authors (1975), in the productive regions of intensive upwelling (97° W) up to 80% of the energy of primary production is brought into the community through the 'detritus food chain'.

Distinctly expressed in the Equatorial upwelling is one of the basic singularities of developing communities: increase in biomass and accumulation of energy ($K_{3p} > 1$) in the early stages of development (97° W) and the expenditure of energy in more mature communities (west of 120° W). This general process of energy accumulation by the community and the subsequent expenditure of the stored reserves permits us to speak of the entirety of the succession process in the plankton community on the equator.

*See p. 666.

### (h) Models of the Functioning of Pelagic Ecosystems

The ecosystems of the ocean are extremely complex and variable. Investigations of their functioning are beset with many difficulties and their development or reorganization under the effect of changing abiotic and biotic factors entirely unpredictable. Investigation by experimentation is largely inapplicable to natural ecosystems. However, there is another way—that of simulating major processes in the ecosystems; this allows the prediction of certain aspects of system behaviour. Moreover, the construction of a model is a valuable tool for checking the co-ordination of separate experimental facts and observations. Therefore, modelling should be regarded as a specific method for the study and description of ecosystems.

However, in the construction of such a model only basic parameters and relations can be used and many interesting details must be omitted—a circumstance which inevitably incites protest from scientists who have devoted much time and effort in studying these interesting details (e.g. HEDGPETH, 1977; see also KINNE and BULNHEIM, 1977; and Chapter 4). This critical attitude towards the modelling of biological systems is supported by the fact that mathematicians are often inclined to create and develop ecological models which are obviously inadequate.

In modelling the behaviour and development of populations it is convenient to base the evaluations on the numerical abundance of individuals, but this criterion is not applicable in the analysis of multi-species communities. As early as 1925 LOTKA suggested using as criterion the energy flow through the system or the turnover of substances within the system (e.g. fixed nitrogen). The energy approach has the advantage that the potential energy of primary production can be utilized only once—the energy flow in the system is undirectional. This permits evaluation of the production processes in a given block of the system by the difference in energy input and output. However, with the energy approach, processes of mineralization and reutilization of nutrients, influences on system development of complex catalytic substances, aminoacids, vitamins, etc. escape the attention of the investigator because the role of all these components in the community is unrelated to their energy content. Recently, much attention has been given to the modelling of turnover of matter (primarily nitrogen), but the obtaining of initial data (parameters) for such models is rather complicated and cumbersome so that as yet relatively few data are available.

The construction and investigation of mathematical models of the functioning of marine pelagic ecosystems has a fairly long history (PATTEN, 1968; STEELE, 1975; DUGDALE, 1975). RILEY (1946,1947a) seems to have been the first to formulate problems of modelling ergocenes of a community and to suggest a practical solution. RILEY's model of seasonal phyto- and zooplankton development on Georges Bank and on the coasts of New England takes into account the basic abiotic and biotic parameters, many of which had been determined in expeditional work. The model, like most of the subsequent ones, is based on trophic relationships on the interpretation of VOLTERRA (1936), while the processes of biosynthesis are limited by the principle of minimum. In elaboration of the ideas of RILEY a classical model of the system nutrients–phytoplankton–zooplankton (RILEY and co-authors, 1949) was developed where the first attempt was made at a conjoint modelling of biotic and abiotic processes in the ecosystems and at the evaluation of their role in the functioning of the system. As noted by PATTEN (1968) this model is 'a classical contribution to quantitative plankton ecology'.

STEELE (1958), using the ideas of RILEY's model, developed a model of the dynamics of plankton in the North Sea, which describes the seasonal dynamics of the community in a two-layer sea with a thermocline. A model of the production cycle of the entire community (including fishes) was proposed by CUSHING (1959). Noteworthy are the interesting work of STEELE (1974) on a model of the 'plankton ecosystem' and the model of VORONINA and co-authors (1980a,b) for assessing the production of the populations of dominant species and of the entire herbivorous zooplankton in the Southern Ocean. Detailed analysis of the methods of modelling the phyto- and zooplankton dynamics in natural communities is given by PLATT and co-authors (1977) and STEELE and MULLIN (1977).

Among models of nutrient turnover in ecosystems we mention those of nitrogen turnover in the sea by MIYAKE and WADA (1968) and of the turnover of dissolved silicon in the ocean by GRILL (1970). The most detailed models of this trend are those of the dynamics of ecosystems and their components in zones of upwellings (DUGDALE and MACISAAC, 1971; WALSH and DUGDALE, 1971; WALSH, 1975, 1976; WHITLEGE, 1978). Recently, a comprehensive model of a multi-component physico-chemico-biological marine system has been proposed by SERGEEV and co-authors (1977). On the whole, modelling is an extensively developing trend of modern scientific research.

With the use of the energy principle, some progress has been achieved in the simulation of marine ecosystems. WINBERG and ANISIMOV (1966,1969) proved that it is possible to create a detailed model of the development of a pelagic community by strictly adhering to the energy principle in the description of the processes. MENSHUTKIN and UMNOV (1970) widened their model by introducing nutrient elements into it and factually transforming it into a model of an aquatic ecosystem. The most comprehensive formulation of a model of the balanced relationship of matter and energy in marine pelagic ecosystems was presented by LYAPUNOV (1971).

The approach of LYAPUNOV (1971) to the modelling of complicated systems (LYAPUNOV and YABLONSKY, 1963) is in its general features similar to the macroscopic method of ODUM (1971). The system is presented in an assemblage of relatively autonomously functioning elements linked by connecting channels. The role of signals may be played by portions of matter or information and, accordingly, material and informational connections may be recognized between the elements of the system.

A mathematical model of balance relationships in an ecological system can be constructed only on the condition that a certain degree of completeness has been achieved in the study of the simulated object. In other words, a concept must be formed of the energy distribution between the elements, of the laws governing the intensity of energy flow between the elements, of what enters the system, and of what (and in what quantities) leaves the system or is extracted from it.

Later on, in accordance with the results of investigations carried out during the specialized cruises of the R.V.V. *Vitjaz* and *Akademik Kurchatov*, LYAPUNOV's model was substantially supplemented and refined. Thus, in particular, horizontal advection of the developing community by currents was introduced, the three-layered hydrological structure of the euphotic zone (water under, in, and above the thermocline) was taken into account, and trophic relationships were calculated—not according to the scheme of VOLTERRA (1936) but to the equation proposed by IVLEV (1955). In addition, the various groups of phyto- and mesoplankton, bacteria, and protozoans were recognized as substantial elements of the system. Most important of all, coefficients of equations could be quantitatively evaluated for defined regions (Central Tropical part of the Pacific Ocean). As a result, a model of the development of the pelagic community of the Tropical epipelagial—from the moment of its formation in the zone of divergence to its reaching a quasi-stationary state in the halostase of the anticyclonic gyral—was developed ready for processing in an electronic computer (VINOGRADOV and co-authors, 1972, 1973).

Simulation models of seasonal changes in the areal distribution of elements (for the Sea of Japan) were constructed on the same principles, beginning with phytoplankton and bacteria and ending with squids and fishes (MENSHUTKIN and co-authors, 1974), a volumetric model of the variability of areal and vertical distribution of plankton elements in the Equatorial zone of the Pacific Ocean (VINOGRADOV and MENSHUTKIN, 1977), and, finally, an as yet preliminary model of the pelagic ecosystem of the entire Pacific Ocean (MENSHUTKIN, 1979).

Let us consider in greater detail a simulation of the development of the community of the Tropical epipelagial; its relevance to the actually observed pattern of plankton distribution has been clearly demonstrated (VINOGRADOV and co-authors, 1972, 1973). This made it possible to carry out model experiments on the development of the system (VINOGRADOV and co-authors, 1973,1975).

An analysis was made of the community inhabiting the surface layer down to a depth of 200 m. The layer was divided into an upper mixed layer, the thermocline layer, and the layer below the thermocline. The changes in the system were investigated by 24-h periods and intervals of 5 to 10 m along the vertical. The vertical connection between these layers was realized by light penetration, turbulent diffusion, phytoplankton and detritus sinking, and diurnal migrations of zooplankton. The principles on which the equations of the model are worked out, the schemes of its spatial structure, and determinations of its parameters have been described by VINOGRADOV and co-authors (1973) and VINOGRADOV and MENSHUTKIN (1977).

The simulated changes in the biomass of the system elements with time and, consequently, with distance from the zone of water ascent show the highest rate of accretion (Fig. 2-20) in the biomass of phytoplankton and bacteria. Phytoplankton reaches its maximum development on the 3rd to 7th day, bacteria on the 5th to 8th day. Small filter feeders lag somewhat behind, and large filter feeders have a still slower rate of development; the biomasses of various groups of carnivores reach their maximum values only on the 35th to 50th day.

By the 50th to 80th day the system attains a near-stationary state. This is characterized by a low biomass of all living elements and a nearly balanced relationship between photosynthesis and nutrient inflow. In this quasi-stationary state the biomass of the basic elements of the system is subject to autofluctuations. The model also permits prediction of changes in vertical distributions of all elements as the system develops; the results obtained prove to be close to those observed in nature.

The results of a comparison of absolute biomass values derived from the model with observations in nature are given in Table 2-6. Considering the relative roughness of the model, its agreement with nature may be regarded as acceptable and sufficiently reliable to be used for predictions of the behaviour of a real system under varying parameters. Let us consider the results of some model experiments.

The possibility of 'fertilizing' oceanic waters by introducing mineral salts of nutrient elements into the euphotic zone has been repeatedly suggested. The model permits

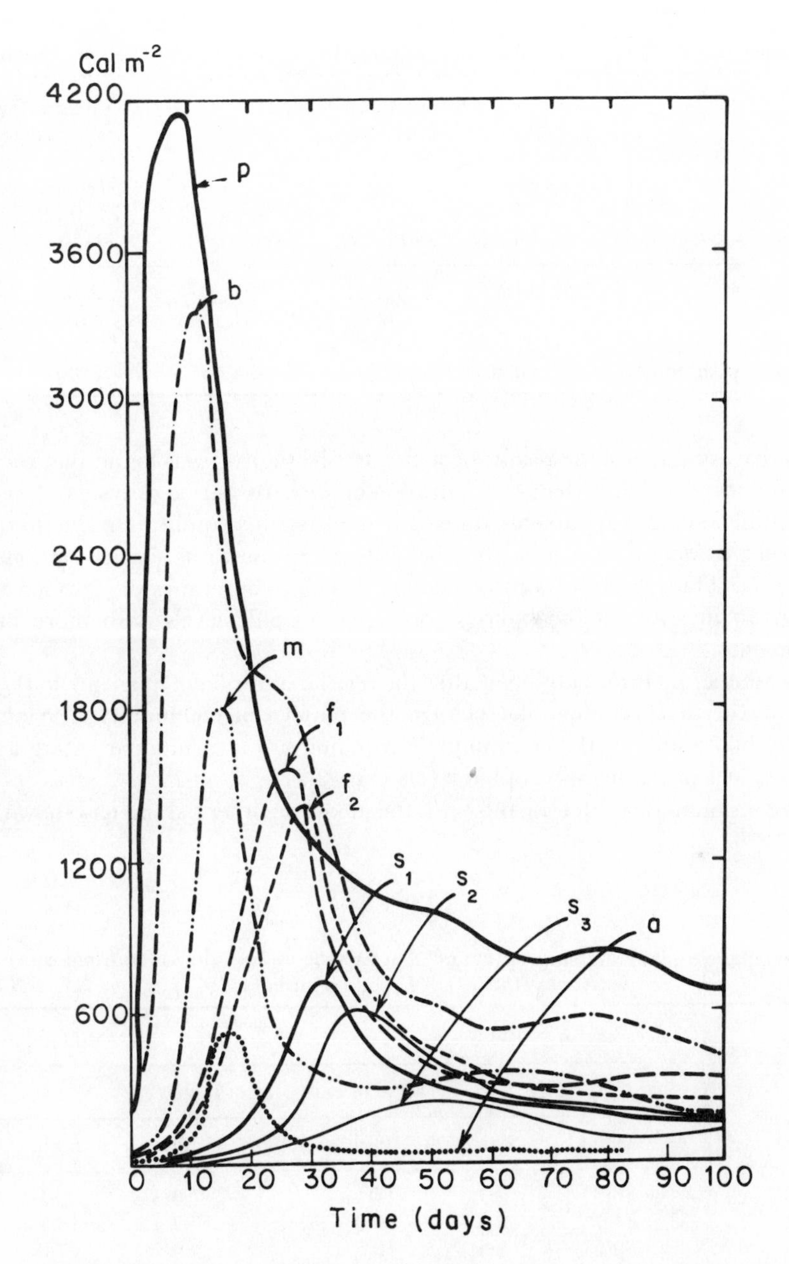


Fig. 2-20: Simulation of biomass changes as a function of time. Elements of a tropical community in the 0–150 m layer. p: phytoplankton; b: bacterioplankton; m: micro-zooplankton; $f_1$ and $f_2$: small (<1 mm and large (>1 mm) filter-feeders; $s_1$: cyclopopoids; $s_2$: carnivorous calanoids; $s_3$: chaetognaths; a: protozoans. (Based on information provided by VINOGRADOV and MENSHUTKIN, 1977)

Table 2-6

Biomass (cal $m^{-2}$) of the elements of a community in the 0 to 150-m layer (Original)

| Elements of the community | Observations in the Model output | | | Observations in the ocean | |
|---|---|---|---|---|---|
| | 30th day | 40th day | 80th day | *Vityaz* St. 6429 30–40 d | *Vityaz* St. 6493 60–80 d |
| Phytoplankton | 1319 | 1092 | 827 | 2000 | 900 |
| Bacteria | 1673 | 864 | 564 | 4100 | 2180 |
| Filter-feeding copepods | 2754 | 1338 | 542 | 945 | 238 |
| Carnivorous plankton | 896 | 1274 | 496 | 1100 | 462 |

quantitative estimates of the results of such a fertilization. It was found that the ensuing substantial increase in nutrient concentration of the early period of system development is perceptible only during a short time as it soon regains equilibrium due to increased grazing of phytoplankton; thereafter the system returns to its former stationery level (Table 2-7). Thus, 'fertilization' of regions where communities are formed does not result in an increase in the biomass of higher trophic levels with more prolonged development.

Quite another picture is obtained after the regular addition of nutrients to the system. Their short-term effect does not change the pattern of community development in principle, but results in the community's reaching a quasi-stationary state at greater biomass values of the higher trophic levels (Fig. 2-21).

Analogous numerical experiments with the model on other parameters show that even

Table 2-7

Simulation of phytoplankton biomass (cal $m^{-3}$) at varying initial values of nutrient concentrations (After VINOGRADOV and co-authors, 1973)

| Layer(m) | n = 100 mg $m^{-3}$ | | | n = 400 mg $m^{-3}$ | | |
|---|---|---|---|---|---|---|
| | Time of existence of system (d) | | | | | |
| | 10th | 50th | 100th | 10th | 50th | 100th |
| 0–10 | 0·3 | 0·1 | 0 | 0·3 | 0·1 | 0 |
| 10–20 | 31·7 | 8·7 | 10·9 | 37·4 | 16·1 | 15·4 |
| 20–30 | 39·6 | 9·4 | 10·7 | 289·7 | 7·8 | 10·4 |
| 30–40 | 39·4 | 9·2 | 10·6 | 207·7 | 7·8 | 10·4 |
| 40–50 | 36·6 | 9·7 | 10·9 | 295·1 | 7·9 | 10·7 |
| 50–60 | 36·7 | 9·5 | 10·9 | 153·7 | 7·1 | 10·5 |
| 60–70 | 41·7 | 11·4 | 10·9 | 62·1 | 18·9 | 8·8 |
| 70–80 | 13·4 | 15·7 | 11·2 | 15·6 | 10·3 | 17·9 |
| 80–90 | 1·9 | 31·4 | 17·0 | 2·1 | 39·0 | 17·9 |
| 90–100 | 0·6 | 3·6 | 12·8 | 0·6 | 4·2 | 1·4 |
| 100–150 | 3·1 | 2·1 | 2·1 | 3·1 | 2·2 | 2·4 |

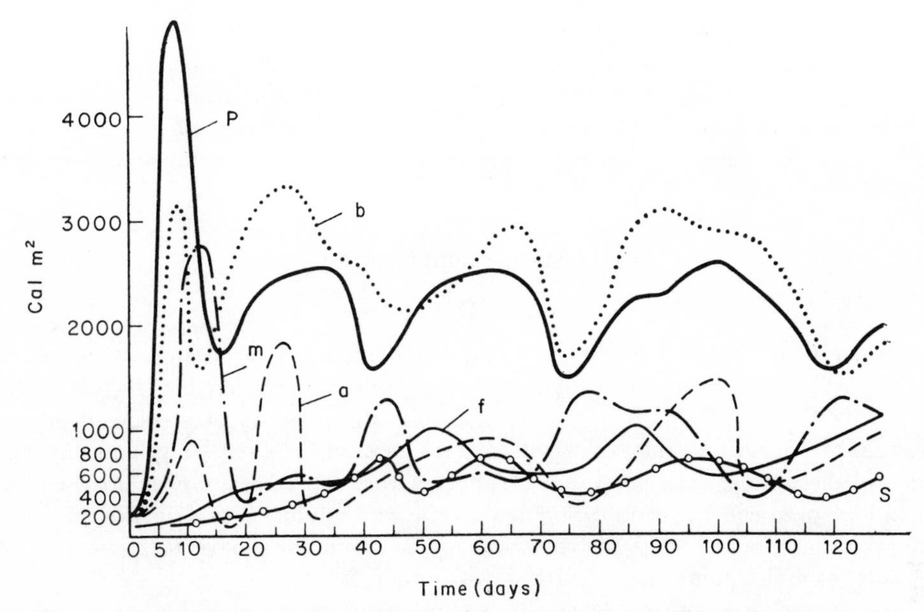


Fig. 2-21: Biomass changes as a function of time. Elements of a tropical pelagic community in the 0–150 m layer at regular (every 5 d) intrusions of nutrient salts (nitrogen) into the upper mixed layer. p: phytoplankton; b: bacterioplankton; a: protozoans; m: nauplii; $f_1$: small filter-feeders; $s_1$: carnivorous mesoplankton. (Based on information provided by VINOGRADOV and co-authors, 1975)

very sharp changes in the initial conditions—such as for instance a hundredfold or thousandfold increase in the concentration of carnivores—will affect the population only for a short time, after which the system returns to its standard level.

Far more important is the influence of such constantly acting factors as rate of vertical water motion, intensity of turbulent mixing, or solar radiation. Thus, for example, the pattern of vertical distribution of diverse elements has been proven to undergo a sharp change if the rate of water ascent is reduced from $10^{-3}$ cm $s^{-1}$, as assumed in the model, to less than $10^{-4}$ cm $s^{-1}$, conversely, elevated to $10^{-2}$ to $10^{-1}$ cm $s^{-1}$.

It was found that the functioning of the ecosystem depends substantially on the value of the specific production of community elements and, first of all, on the specific production of phytoplankton. Thus, if we assume that a maximum diurnal specific phytoplankton production exceeding a unit in the initial phase of system development reaches a constant level of less than 0·6 on the 15th day (THOMAS,1970), then the community will virtually die off on the 30th to 50th day. On the other hand, a very high specific phytoplankton production in oligotrophic waters (KOBLENTZ-MISHKE and VEDERNIKOV, 1976) will also lead to situations not observed in nature.

Thus, experiments with the simulation model allow us to form a concept of the development of a system in different situations and to derive sometimes trivial but sometimes also quite unexpected conclusions.

Models of pelagic ecosystems are important not only from the standpoint of the actual results obtained. No less important is the fact that the application of modelling in oceanographic research has greatly modified our approach to the collection and treatment of samples and to the organization of fieldwork. If a simulation of processes going on in the ocean is to be recognized as the resulting and generalizing stage of the investigations, then the need for a complex and comprehensive study of the object being modelled becomes imperative. Omissions or insufficient knowledge of processes referring to one of the elements may depreciate efforts directed toward the study of other elements of the system.

## (4) Deep-Sea Communities

### (a) Vertical Distribution

The vertical stratification of plankton in the water column of the ocean is too extensive to be caused only by the gradients of physical and hydrochemical factors (BANSE, 1964, VINOGRADOV, 1968,1972). It is preserved despite the presence of vertical turbulent eddies that contribute to the mixing of separate parts of the community and biotope. Evidently there must exist some inner factors within the community which help to uphold high gradients of stratification against the external forces that operate to lower them. It seems possible that the members of the community have to spend energy especially for maintaining their stratified distribution.

The biological mechanisms responsible for plankton stratification within the water column are not quite clear. They may include, for example, competitive exclusion of dominant ecologically similar species according to the principle of GAUSE (1934).

In marine pelagic communities which occupy rather similar biotopes, ecological differences comprise mainly differences in food habits of their component species. In situations of 'overensured' food supply, the coexistence of ecologically similar species with the same food habits is possible, but with the diminution of food resources competition increases (see also Volume V: PERES, 1982). At first a partial overlapping of the trophic projections of the ecological niches of closely-related species takes place, preventing coexistence; at this period general competition between trophic groups may still be low. But with a further decrease in food supply more and more species and, finally, entire trophic groups, become involved in competition.

In the western part of the Pacific Ocean VINOGRADOV and PARIN (1973) discovered a vertical sequence in the importance of mass groups of carnivorous macroplankton: in the 80- to 120-m and 300- to 400-m layers, Myctophidae were dominant in biomass; in depths of about 200 m, prawns; in 120 to 170 m, cephalopods. Below 500 m, both predators and planktophagous fishes were dominant. The pattern of stratified distribution and alternation of dominant species is most evident in direct observations. All researchers who have explored great depth, beginning with BEEBE (1934), mention the striking alternation of layers with dominance of different species (e.g. BERNARD,1958; PERES,1958a,1965; DIETZ,1962).

Owing to vertical water mobility, the extremely small extent of vertical habitat zones compared with regions of horizontal distribution and the capacity of most plankton animals for active vertical replacement, there is no total exclusion of one species by others, nor complete divergence with depth. In vertical plankton distributions, the principle of competitive exclusion is manifested primarily in a vertical divergence of layers with a dominance of ecologically similar species with the same food spectra. Nevertheless, even a gradual change in the amount of food with increasing depth—when reaching a certain limit—will produce sharp changes in the structure of the community.

### (b) Reduction of Energy Expenditures

Energy is brought into the communities of oceanic depths with the organic matter formed in the surface layers in the euphotic zone. The ways by which the organic matter penetrates from the surface zone into different depths, its quantity or rate of supply, the form in which it occurs there, are deeply rooted in the singularities of the existence of the pelagic fauna and the character of its vertical distribution. The fact that the pattern of vertical quantitative plankton distribution is generally similar throughout the entire ocean (e.g. VINOGRADOV, 1960,1962b,1968; YASHNOV, 1961; LONGHURST and WILLIAMS, 1979) is evidence that in each case this distribution depends not only on the complex of local conditions (abiotic and biotic), but also on some general cause. This determinant factor proves to be the quantity of available food, i.e. the value of food resources (see also Volume V; PERES, 1982).

The ways in which the organic nutrient matter penetrates into the depths, the form in which it is present there, have been discussed repeatedly and in great detail (MENZIES, 1962; VINOGRADOV, 1962a,1968; FOURNEIR, 1972; MACDONALD, 1975; MCCAVE, 1975). Whatever the route and source—whether in the form of metabolic products of higher living animals, their remains, or in the bodies of animals performing vertical migrations or entrained from the coasts and the shelf by bottom current—in every case the amount of this organic matter will be gradually reduced with its sinking to deeper layers due to partial consumption by the inhabitants of intermediate layers and mineralization, so that the deficit of food available to the inhabitants of the pelagial will augment with increasing depth.

Apparently the quantity of deep-sea plankton depends not only on the value of its production in the euphotic zone but also on the balance of the life cycles of surface communities which determines the extent of the utilization of this production in the euphotic zone within the communities themselves. BANSE (1964) was the first to note this possibility. However, since there are no experimental determinations of the ecologico-physiological characteristics of deep-sea plankton, the suggestion could not as yet be verified. Where aquatories, as well as the structure of deep-sea communities themselves, differ in the balance of production and destruction of surface plankton such a relationship would never arise (VINOGRADOV, 1968). At least, in temperate regions of the ocean the flow of organic matter from the surface down to the depths has a certain seasonal cycling. This cycling apparently should bring seasonal patterns in the life cycles of the deep-sea animals (SCHOENER, 1968; MAUCHLINE, 1972). In fact, in the Kuril-Kamchatka Trench seasonal changes were found in the age of the populations of some pelagic animals at the depths more than 6000 m (VINOGRADOV, 1970b).

In the upper nutrient-enriched layers the animals may easily compensate for even considerable losses of energy associated with active movement by more intensive feeding; this is not always possible in the impoverished deep waters. JØRGENSEN (1955) is of the opinion that the energetic expenditures of plankton filter feeders are refunded only, provided that the quantity of assimilable detritus in the water exceeds 25 mg $l^{-1}$. Actually, the content of organic detritus in deep waters may be significantly lower. SHMALGAUZEN (1968, p. 52) points out that:

> 'abundance of means of subsistence prove beneficial [for evolutionary development] also for perhaps less adapted, but in return active and fertile individuals. Con-

trariwise in the absence of the required materials the selective advantages will be on the side of most strictly adapted and even narrowly specialized species with the most economical metabolism, even if at a decrease in the general level of their activity".

Adaptations to existence under conditions of extremely scarce food resources in various animal groups with different food habits may manifest themselves in different ways and to different degrees. However, the tendency towards a reduction of energy expenditure must be inherent in the entire population of great depths. Even the upper-interzonal animals that spend only a part of their lives in deep waters greatly reduce their energy exchange during this period. Thus, *Calanus hyperboreus* which winters in depths of 500 to 2000 m reduces its energy expenditure for respiratory exchange by three times. This change in breathing rate is purely adaptive and does not depend on temperature (CONOVER, 1962).

With increasing depth and poverty of food resources the animals become subject to changes in mean body size. In bathypelagic depths, carnivores and some other forms increase noticeably in size, the increase being accounted for not only by a predominance in representatives of large-sized genera but also by 'deep-sea gigantism', i.e. increase in size with depth in species of a genus or closely-related genera. Thus, in *Sergestes* the mean body size increases in different species from 28 mm (*S. vigilax*, habitat depth: 110–650 m) to 94 mm (*S. robustus*, habitat depth: 550–800 m); in *Meningodora* species, from 36 mm (*M. myccila*, mean habitat depth: 1280 m) to 74 mm (*M. millis*, mean habitat depth: 1550 m).

To explain the phenomenon of deep-sea gigantism some authors (WOLFF, 1960, 1962; BIRSHTEIN, 1963; BELYAEV, 1966) ascribed it to the abiotic factors, first of all to pressure and low temperature. MAUCHLINE (1972) suggested that the more important factors are different aspects of ecological pressure and, especially, increased longevity combined with low growth rate. But TSEITLIN (1977) proved that the main factor determining the increase in size of carnivorous species is the impoverishment of food resources. A definite role is played also by the low temperature. Proceeding from the known formulae of food rations and their dependence on temperature, TSEITLIN suggested a method for quantitative size evaluation in planktophages of different groups with increasing depth and changes in food concentration (mesoplankton biomass). The calculated result obtained by him coincide with the observed data.

The principle of economy in energy expenditure is expressed first of all in a decrease of activity. Deep-sea animals are predators in the overwhelming majority, not actively pursuing their prey but passively waiting for it. This is evidenced in particular by their (sometimes significantly) elongated appendages which ensure control over a considerable volume of water. A number of deep-sea cephalopods have greatly elongated arms which may be 15 times longer than the animal's body (*Chirhoteuthis veranyi*). In some predatory deep-sea copepods the buccal appendages are noticeably increased by elongated articles of the maxilliped basipodite and bear rare but strong and long setae. WICKSTEAD (1962) observed in a live copepod such setae extending downward and forward, forming a trap which closes instantly when the prey enters the field of its action, while the predatory copepod itself makes no attempt to pursue the prey that swim past it. PERES (in MARSHALL, 1960) observed from the bathyscaphe the behaviour of *Chauliodus*; he writes that

> 'these fishes hover in the water with the long axis of the body at an acute angle to the horizontal plane, the head being above the tail. At the same time the long second dorsal ray, which is tipped with luminescent tissue, is curved forward over the head so that the extremity of the ray lies in front of the mouth.'

And about *Cyclothone* and *Gonostoma*:

> 'Ces animaux sont toujours immobiles, paraissant flotter; ils semblent attendre la proie sans la poursuivre, et il m'a été impossible de voir quelles sont les nageoires dont le mouvement aide à leur equilibre' (cited after MARSHALL, 1960, pp. 107 and 110).

How many prey organisms are needed by such a predator for its subsistence? GORELOVA and TSEITLIN (1979), who studied the changes occurring with increasing depth in the rations of the very same *Cyclothone* observed by PERES, found that its specific diel rations decrease with depth with a concomitant increase in the body size of the fish (Table 2-8). The authors, having estimated the size of the prey of *Cyclothone*, note that *C. alba* on an average captures a single prey once in 2 d, *C. pseudopallida* once every 3 d, *C. pallida* once every 10 d. Proceeding from entirely different premises, MACDONALD (1975) calculated that the ostracod *Giganthocyris* requires for its subsistence on the average of 1 prey $mo^{-1}$.

The drastic decrease in mobility is accompanied by a reorganization of the entire organism resulting in its pronounced specialization (e.g. DENTON and MARSHALL, 1958; MARSHALL, 1960, 1971; WALTERS, 1961; VINOGRADOV, 1968).

The basic adaptations tend towards neutral buoyancy at which the animal may remain in a suspended state without any muscular effort and thus reduce energy expenditure. A decrease in specific weight is achieved mainly by the reduction of heavy tissues. The bony skeleton is replaced by a cartilaginous one, the muscular fibres weaken. The subdermal region and spaces between muscular tissues are filled with jelly-like mesenchyma, and a part of the subcutaneous cavities are filled with fat. Most of the soft tissues

Table 2-8

Specific diel rations and body sizes of *Cyclothone* species living in different depths (Based on information provided by GORELOVA and TSEITLIN, 1979)

| Fish species | Mean depth of habitat (m) | Weight (g) | Length (cm) | Specific diel ration (C/W) |
|---|---|---|---|---|
| *Cyclothone alba* | 550 | 0·056 | 25 | 0·016 |
| *C. pseudopallida* | 750 | 0·145 | 36 | 0·012 |
| *C. pallida* | 800 | 0·289 | 46 | 0·004 |

become jelly-like in consistency so that the average water content in the tissues of some deep-sea fishes increases by 80 to 90% and their protein content proves 2 to 4 times lower than that in fishes living in shallower waters (DENTON and MARSHALL, 1958; WALTERS, 1961; CHILDRESS and NYGAARD, 1973). The swim bladder loses it basic purpose (preservation of neutral buoyancy at changing pressure) and either disappears entirely or becomes reduced and filled with fat (MARSHALL, 1960).

The integuments of deep-sea pelagic decapods are far thinner than in shallower-water living species, thus *Hymenodora frontalis* has a relatively hard exoskeleton, but the deeper-living *H. glacialis* has an extremely soft one. Moreover their tissues have high lipid contents—up to 72% of dryweight (VINOGRADOV and co-authors; 1970; CHILDRESS and NYGAARD, 1973). The same phenomenon is characteristic of other plankton crustaceans. In the deep-sea chaetognaths *Eukrohnia fowleri* and *Sagitta macrocephala* the gut is encircled by parenchymatous tissue filled with fat. Deep-sea cephalopods (*Vampyroteuthis*, *Vitreledonella*, *Amphitretes*, *Japetella*, *Eledonella*) have lost the greater part of their supporting and muscular tissues, and their muscles, thus partly degenerated, are replaced by jelly-like tissue so that in body consistency they resemble jelly fish.

The reduction of musculature results in a simplification of the circulatory and excretory systems. In pelagic deep-sea fishes the surface of gill plates is greatly decreased, pointing to a less intensive breathing rate, while the low breathing rate, the secondarily simplified excretory system, and the reduction of sense organs are indications of reduced energy expenditure. Indeed, measurements of metabolic rates in deep-sea species carried out at habitat depths show that they are several times lower in mesopelagic plankton animals than in surface species (CHILDRESS, 1971; MEEK and CHILDRESS, 1973); and at a depth of 1800 m the rates are two orders of magnitude lower than in shallow-water fishes of the same size (SMITH and HESSLER, 1974). The calculations of the intensity of the metabolic rates in surface and bathypelagic fishes by protein equivalent make apparent differences of one order of magnitude (CHILDRESS and NYGAARD, 1973).

The adaptations associated with the tendency to reduce energy expenditure and food habits are extremely profound and varied. They alter the appearance of pelagic deep-sea animals far more than adaptations to changes in all other environmental factors. In other words, it is the change in food availability—not the change in physical parameters (light, pressure, or temperature)—that primarily determines the structure and mode of life of pelagic animals in the deep sea.

The principle of economy in energy expenditure is not restricted to the lowering of the level of basic and active metabolism. According to WALTERS (1961) neoteny, widely occurring in many groups of deep-sea animals, is one of the main evolutionary adjustments observed in waters of great depths poor in food resources. Owing to their neotenic development, deep-sea animals seem to require less energy for reaching sexual maturity.

The amount of food resources also determines the structure of communities at different depths. The number of species decreased with depth to a lesser degree than plankton abundance. As a result, each species is far more dispersed in deep waters than in surface layers. The species diversity of communities increases, as well as their stability which, according to MACARTHUR (1955), rises with increasing polymixedness and the number of links in the food web. It is further known that the efficiency of ultilization of food resources by a community increases with the latter's stability, while higher stability enables it to subsist on a relatively smaller supply of energy to the system (MARGALEF, 1963; see also Volume IV: MARGALEF, 1978).

### (c) Structural Characteristics of Communities

*Mesopelagial (200–1000m)*

The main mass of the population consists of species which are directly associated with the surface zone and feed there during their ontogenetic or diel migrations. Here, especially in temperate-cold water regions, zooplankton concentrations are in some cases nearly identical with those in the surface layers. An important part is played by bacteriophages (radiolarians: phaeodarians, etc.), which utilize the high concentrations of bacteria and unicellular heterotrophs usually confined to depths of 500 to 600 m (SOROKIN, 1977a). In addition to migrating animals (carnivorous and non-carnivorous) carnivorous forms that do not ascend to the surface layers are of substantial importance. The highly polymixed communities of the mesopelagial have a somewhat lower species diversity than the communities of the surface zone in the tropics and exceed surface cold-water communities in this parameter.

*Bathypelagial (1000–3000 m)*

The communities of the bathypelagial are characterized by a relatively low numerical abundance and biomass which even in the most productive regions of the ocean does not exceed 20 to 30 mg $m^{-3}$, and by a considerable predominance of zoophages over detrito- and phytophages among which in cold water regions upper interzonal filter feeders also play an important role. In the upper layers of the bathypelagial they may account for 30 to 40% of the total mesoplankton biomass.

The bulk of the population consists of carnivorous forms subsisting on animal organic matter coming from the surface zone (Table 2-9). However, most of the predators are not mobile; they do not pursue their prey but wait for it. The polymixedness of the communities is high: many groups of pelagic animals reach a high degree of species diversity here. The scantiness of food resources and necessity of reducing the energy expenditure for basal and active metabolism are accompanied in many groups by substantial morphological reorganization, which results in a taxonomical detachment of the bathypelagic fauna from that of the surface. This is true first of all for the groups which in the surface layers are represented by active predators, but in the bathypelagial restrict themselves to luring prey (BIRSHTEIN and VINOGRADOV, 1971).

Low population abundance combined with a relatively high species diversity results in a very high dispersion in the distribution of individual species, far higher than in lesser depths. The increase in the index of community diversity evidences its greater stability which, according to MARGALEF (1963, 1978), enables it to exist with a lesser inflow of energy into the system, i.e. on more restricted food resources.

*Abyssopelagial (below 3000 m)*

Community structures at depths below 3000 m have been very little studied. Here, numerical abundance and plankton biomass are extremely low. There are only a few species. No group exhibits a maximum of species diversity below 3000 m. In contrast to the bathypelagial, abyssopelagic communities may be characterized as being oligomixed. Owing to the extreme degree of plankton rarefaction, even passive predation

Table 2-9

Changes with depth in the trophic structure of the communities in the northwestern part of the Pacific Ocean (summer 1966). Biomass of major trophic groups in percentage of total net plankton in each layer, not taking into account interzonal filter feeders which do not feed at the depth of capture. Averages from 9 stations (Original)

| Depth (m) | Filter feeders (phyto- and detritophages) | Snappers (predators and necrophages) | Euryphages (mixed type of feeders) |
|---|---|---|---|
| 1 | 2 | 3 | 4 |
| 0–50 | 85 | 11 | 4 |
| 50–100 | 58 | 34 | 8 |
| 100–200 | 42 | 53 | 5 |
| 200–300 | 55 | 28 | 17 |
| 300–500 | 55 | 27 | 18 |
| 500–750 | 10 | 54 | 36 |
| 750–1000 | 13 | 54 | 33 |
| 1000–1500 | 5 | 77 | 18 |
| 1500–2000 | 2 | 83 | 15 |
| 2000–2500 | 3 | 89 | 8 |
| 2500–3000 | 5 | 72 | 23 |
| 3000–4000 | 4 | 39 | 57 |
| 4000–5000 | 4 | 22 | 74 |
| 5000–6000 | 2 | 15 | 83 |
| 6000–7000 | 2 | 21 | 77 |
| >7000 | 8 | 29 | 63 |

seems to be untenable as a source of energy, and carnivorous species lose their importance. They are represented mainly by gammarids confined to the bottom. Plankton in these depths consists primarily of small-sized detritus feeders or euryphages.

The species impoverishment in the abyssopelagial is undoubtedly also caused by the fact that—due to the high degree of rarefaction—the dispersion of some species becomes so high that the meeting of sexes for reproduction is no longer assured. Because of this the inhabitation of abyssal depths, where total numerical abundance is extremely low, can prevail only under the condition of a relatively monotonous species composition.

Evidence from the community of the oceanic water column completely confirms the statement by ELTON (1927) that the essential factor maintaining the integrity of communities, i.e. the essential integrating force, is of trophic nature (see also KINNE, 1977, p. 724). The organic matter photosynthesized by phytoplankton and penetrating into the greatest depths of the ocean is the main trophic link connecting the populations of all layers of the water column. The wide eurybathycity of many species, the vertical migrations during which the animals cover hundreds and thousands of metres and in which most plankton animals are involved, assure this connection (see Table 2-9).

### (5) Conclusions and Possibilities of Management

At present, the exploitation of biological resources of the ocean is directed nearly exclusively towards the removal of the production of natural communities in the form of the final links of the trophic chains—fishes, cephalopods, molluscs, whales. Today, commercial fishery encompasses practically the whole aquatory of the ocean.

The need for further increase in catches calls for a sound scientific fundament on which the most rational trends of commercial fishery can be based. The attention of researchers must focus first on the elucidation of quantitative patterns of distribution of forage organisms, determination of the productivity of the ocean, study of relationships between the feeding areas of fishes, and hydrophysical and hydrochemical factors, i.e. on what today would be termed biological ocean assessment. These investigations have indicated ways of rational fishery management and helped to maximize fishing efficiency. Fundamental questions of the general structure of the ocean have been elucidated, its productivity assessed, and a wealth of material on species composition and biology of its inhabitants has been accumulated and systematized.

At the same time, the scantiness of the commercial resources of the ocean was pointed out (for details consult Volume V, part 3), and an upper limit of 80 to 100 million t was set for commercial catches. This limit only slightly exceeds the present level, which in 1974 reached 60 millions (GULLAND, 1976; see also Chapter 5). The greatest increase was observed between 1958 and 1961, when the world catch rose every year by 3·4 million t or 9%. Between 1964 and 1967 this increase dropped to 2·6 millions and between 1967 and 1971 to merely 1·8 million t or 3·7%. In 1972 there was even a decrease in total world catch which, by 1974, still failed to return to the 1970 level.

It is obvious that a further, even very profound intensification of the sea fishery and the inclusion of the few remaining untapped regions and fish stocks could not help much to attain the goal of 100 to 150 million t of fishery products set for the year 2000 (MOISEEV, 1969). This goal cannot be achieved by the methods used today in the exploitation of fishery resources. Catches may be increased but only for a short time, until the 'free fishery' reaches its ceiling. Any further increase may undermine the fish resources of the ocean, a process which may prove irreversible for many valuable marine species.

There is an urgent need for turning from 'free hunting' to the management of the biological productivity of the ocean and to the type of husbandry practiced on land (Volume III: KINNE and ROSENTHAL, 1977; KINNE, 1983). To achieve this task—new in the very principle of its concept—of exerting purposeful influence on marine communities new propositions have to be evolved. The development of rational methods for influencing marine communities and effective marine husbandry has to be based on a thorough knowledge of the structure and functioning of these communities, clearcut concepts of production processes, matter conversion, and energy flow at different trophic levels (Volume IV). A theory must be evolved for the management of biological systems in coastal waters and in the open ocean. The complexity of marine ecosystems and the difficulties which beset their investigation makes absolutely indispensable—as pointed out by MONIN and co-authors (1974)—the cybernetization of marine ecology, i.e. the use of methods of mathematical simulation for predicting the behaviour of systems subject to directed changes of one or other of their parameters.

In the international program Global Oceanic Research (1969) developed by the UNESCO, special emphasis is placed on the necessity of studying the production process and the functioning of oceanic ecosystems as groundwork for their directed modification.

Sufficiently new and as yet imperfectly developed methods for investigation of the functioning of ecosystems should be applied at first in studies of systems simpler and more easily accessible, but at the same time highly important in the general economies of the ocean. Such are the communities of the pelagial where the uniformity of environmental conditions contributes to a strong dominance of trophic relationships. In contrast to benthic systems, any relationships not directly trophic are of secondary importance in pelagic communities. At the same time, abiotic conditions which directly affect the functioning of communities are more easily determined and assessed in the pelagial.

In temperate and cold water regions, for assessing the energy balance of a system during its entire production cycle, circum-annual characteristics of the given community are required; technically, these are not easy to obtain. In this respect communities of Tropical waters which are subject to far less seasonal variations offer great advantages—as at each given moment the system may without great error, be regarded as being balanced. The complicated species structure and high species diversity of tropical communities present no special difficulties when the energy principle is used to estimate their functioning.

Having methods for investigating pelagic communities at our disposal would greatly facilitate our turning to the study of predicting the behaviour of systems, including the populations not only of the water column but also of the ocean floor, i.e. of systems which may be factually used in cultivation and husbandry on the shelf.

Similarly, the problem of management also comprises the creation of models designed to study the developmental dynamics of commercial fish populations in different aquatories; this will permit their more rational and efficient exploitation. Today such models already exist or are being created for several fish species (e.g. for *Scomber japonicus* in California, *Cololabis saira* in the northwestern Pacific Ocean, and yellow fin tuna in the eastern Pacific Ocean) (FRANCIS, 1974; PARRISH, 1978).

It may be expected that elaboration of the problem of managing biological systems will provide a theoretical basis for (i) rationally directed changes in oceanic communities, (ii) increased output of economically valuable products, and (iii) effective farming in the ocean. All this could increase the production of animal protein obtainable from the ocean.

Another task of increasing importance is the assessment of the effect of various anthropogenic activities (pollution, fishery) on the biosystems of the ocean (Volume V, Parts 3 and 4). Viewed from this perspective the essential fact is that anthropogenic pollution primarily affects the surface layers of the ocean, i.e. the most productive ecosystems in the euphotic waters which ensure the supply of energy for populations in the entire water column. Naturally, the study of the effect of pollution on the surface community is of primary importance.

Abundant information has already been accumulated on the biology of marine organisms, their interrelationships with one another and with the abiotic environment (Volumes I, II, III, IV), including anthropogenic influences (Volume V, Parts 3 and 4). Actually, an approach has become possible to analyse the functioning of communities exposed to various external influences.

Any consideration of the pelagic ecosystems of the open ocean should take into account that a system developing in time is shifted with the flow of 'ageing' water through space. Therefore, it is the succession of the community along the whole journey of the water—from the moment of its ascent to the euphotic layer to the moment of its sinking—that should be considered as the elementary unit. Of course, the ecosystem will not be in equilibrium (energy balanced) at any given moment of time. Near the zone of its formation the system accumulates energy, while 'farther downstream' it spends it. It is during the 'accumulation period' that the system is particularly vulnerable to depressing factors. Therefore, pollution exerts maximum damage where the systems form in regions of upwellings and current divergences in the tropics or during the periods of plankton development in temperate latitudes.

Consequently, the problem of assessing the effect of anthropogenic pollution on the ecosystems of the ocean should be approached in two ways: (i) By obtaining definite data on the character and extent of influences of the most common pollutants on mass forms (or groups) of hydrobionts; this would enable us to account for their effect in simulations of ecosystems and to predict their development. (ii) By developing mathematical models which take into account the stochastic character of pollution and the possibilities of controlling its extent.

A model of the functioning of ecosystems under pollution represents only a particular case of a more general model which accounts for the presence of management, where pollution may be understood as any kind of anthropogenic interference and management taken as a substantial optimal use of natural resources in the broadest sense of these concepts.

Thus, generally formulated, the problem exceeds the limits of separate domains of natural sciences and becomes a subject of engineering ecology in which not just simulations but analytical optimizational models are used, models of theoretical cybernetics (systemology) with their components of information and management theory. The main premise of these investigations is based on the assumption that the probabilistic characteristics of the level of pollution of an ecosystem depends of the amount of resources allotted to combat it.

The task, as pointed out by BRUSILOVSKY (1975), is reduced to the distribution of available resources in such a way that during a certain period of time the expectation of the level of pollution should prove to be a minimal constant value and the dispersion of the pollution level to be lower than a certain pre-assigned value. The converse problem is also interesting.

The resolution of the direct task—the distribution of resources—should facilitate a 'freezing' of the level of ecosystem pollution at the minimal value possible for a given value of expenditure, and the resolution of the converse task will allow us to find out what quantity of resources (expenditure) is necessary for the 'freezing value' to correspond to the pre-assigned one.

Thus, large ecosystems may be considered as subsystems of a single economic system. This perspective will permit us to develop a strategy of optimal marine management considering the effect of commercial and non-commercial (pollution) human activities on marine ecosystems, and to suggest the most rational method for their directed changes. The transition of man's activities from hunting to agriculture required thousands of years; in the ocean it is expected to be realized within the next few decades.

## Literature Cited (Chapter 2)

ANDREWS, K. J. H. (1966). The distribution and life-history of *Calanoides acutus* (Giesbrecht). *'Discovery' Rep.*, **34**, 119–161.

ANDRIASHEV, A. P. (1970). Cryopelagic fishes of the Arctic and Antarctic and their significance in polar ecosystems. In M. W. Holdgate (Ed.), *Antarctic Ecology*. Academic Press, London. pp. 297–305.

BANSE, K. (1964). On the vertical distribution of zooplankton in the sea. In M. Sears (Ed.), *Progress in Oceanography*, Vol. II. Pergamon Press, Oxford. pp. 53–125.

BANSE, K. (1974). On the role of bacterioplankton in tropical ocean. *Mar. Biol.*, **24**, 1–5.

BARHAM, E. G. (1963). Siphonophores and the deep-scattering layer. *Science, N.Y.*, **140** (3568), 826–827.

BE, A. W. H. (1977). An ecological, zoogeographic and taxonomic review of recent planktonic Foraminifera. In A. T. S. Ramsey (Ed.), *Oceanic Micropaleontology*, Vol. I. Academic Press, London. pp. 1–100.

BEEBE, W. (1934). *Half Mile Down*. Harcourt, Brace and Co., New York.

BEKLEMISHEV, C. W. (1969). *Ecology and Biogeography of the Open Ocean* (Russ., Engl. summary). Nauka, Moscow.

BELYAEV, G. M. (1966). *Bottom Fauna of the Ultra-abyssal Depths of the World Ocean* (Russ., Engl. summary). Nauka, Moscow.

BELYAEV, G. M., BIRSHTEIN, Ja. A., BOGOROV, B. G., VINOGRADOV, M. E., VINOGRADOVA, N. G. and ZENKEVITCH, L. A. (1959). On vertical zonality of the ocean (Russ.). *Dokl. (Proc.) Acad. Sci. U.S.S.R.*, **129**, 658–661.

BELYAEVA, T. V. (1980). Phytoplankton in the region of drifting station 'North Pole-22'. In M. E. Vinogradov and I. A. Melnikov (Eds), *Biology of Central Arctic Basin*. Nauka, Moscow, 133–142.

BERNARD, F. (1958). Données récents sur la fertilité elémentaire en Méditerranée. *Rapp. P.-v. Réun. Cons. int. Explor. Mer*, **144**, 103–108.

BERNARD, F. (1961). Problems de fertilité elémentaire en Méditerranée de 0 a 3000 metre de profondeur. Resúltats scientifiques des campagnes de la 'Calypso', fasc. V. *Annls Inst. océanogr., Monaco*, **39**, 5–160.

BERNARD, F. (1963). Vitesses de chute chez *Cyclococcolithus fragilis*. Conséquences pour le cycle vital des mers chaudes. *Pelagos*, **1**, 5–34.

BERNARD, F. and LECAL, J. (1960). Plancton unicellulaire récolté dans l'Océan Indien par le 'Charcot' (1950) et le 'Narsel' (1955–1956). *Bull. Inst. océanogr. Monaco*, **1166**, 1–59.

BIRSHTEIN, Ja. A. (1963). *Deep-Sea Isopods (Crustacea, Isopoda) in the Northern-Western Part of the Pacific Ocean* (Russ., Engl. summary). Izd. Academii Nauk SSSR, Moscow.

BIRSHTEIN, Ja. A. and VINOGRADOV, M. E. (1971). Role of the trophic factor in the taxonomic discreteness of marine deep-sea fauna (Russ., Engl. summary). *Bull. Mosc. obshchestvo ispitatelei prirodi (Series Biol.)*, **76**, 59–92.

BJERKNES, J. (1969). Large-scale ocean-atmosphere interaction. In *Morning Review Lectures of the Second International Oceanographic Congress*. UNESCO. pp. 11–20.

BLACKBURN, M. (1968). Micronekton of the eastern tropical Pacific Ocean: family composition, distribution, abundance, and relations to tuna. *Fish. Bull. Fish Wildl. Serv. U.S.*, **67**, 71–115.

BLACKBURN, M., LAURUS, R. M., OWEN, R. W. and ZEITZSCHEL, B. (1970). Seasonal and areal changes in standing stocks of phytoplankton, zooplankton and micronekton in the eastern tropical Pacific. *Mar. Biol.*, **7**, 14–31.

BOGOROV, B. G. (1946). Zooplankton (according to collections of expedition on ice-breaker 'G. Sedov', 1937–1940). (Russ.). *Trudy drejfuiuschej expeditsii Glavsevmorputi na ledokolnom parokhode 'G. Sedov'*, **3**, 336–370.

BOGOROV, B. G. (1948). Vertical distribution of zooplankton and vertical zonation of waters in the ocean (Russ.). *Trudy Inst. Okeanol.*, **2**, 43–57.

BOGOROV, B. G. (1959). Biological structure of the ocean (Russ.). *Dokl. (Proc.) Acad. Sci. U.S.S.R.*, **128**, 819–822.

BOGOROV, B. G. (1970). Biogeocoenoses of the ocean's pelagial (Russ.). In L. A. Zenkevitch (Ed.), *Program and Method Investigation of Biogeocoenological Water's Surroundings*. Nauka, Moscow. pp. 28–46.

BOGOROV, B. G. and VINOGRADOV, M. E. (1955). Some essential features of zooplankton distribution in the north-western Pacific (Russ., Engl. summary). *Trudy Inst. Okeanol.*, **18**, 113–123.

BOGOROV, B. G., VINOGRADOV, M. E., VORONINA, N. M., KANAEVA, I. P. and SUETOVA, I. A. (1968). Distribution of zooplankton biomass within the superficial layer of the World Ocean (Russ.). *Dokl. (Proc.) Acad. Sci. U.S.S.R.*, **182**, 1205–1208.

BOLIN, B. (1972). Carbon circulation. In M. S. Ghilarov (Ed.), *Biosphere*, Izdat, 'Mir', Moscow. pp. 91–104.

BOUGIS, P. (1974). *Ecologie du Plancton Marin*, II. Le Zooplancton, Massonetcie, Paris.

BRANDHORST, W. (1958). Thermocline topography, zooplankton standing crop, and mechanisms of fertilization in the eastern tropical Pacific. *J. Cons. perm. int. Explor. Mer*, **24**, 16–31.

BRODSKY, K. A. (1957). *Fauna of the Copepods (Calanoida) and Zoogeographical Regionalization of the Northern Pacific and Adjacent Water* (Russ.). Izd. Akademii Nauk SSSR, Moscow-Leningrad.

BRUSILOVSKI, P. M. (1975). Model of pollution in an ecosystem (Russ.). *Uchen. Zapiski Bashkir. Gos. Universitet, Ser. mathemat. Sci.*, Ufa.

BRUUN, A. F. (1956). The abyssal fauna: its ecology, distribution and origin. *Nature, Lond.*, **177**, 1105–1108.

CHEKOTILLO, K. A. (1969). On the vertical circulation in the western part of the Pacific Ocean (Russ.). In *Scientific Conference on the Tropical Zone of the World Ocean* (Abstracts). Nauka, Moscow.

CHILDRESS, J. J. (1971). Respiratory rate and depth of occurrence of midwater animals. *Limnol. Oceanogr.*, **16**, 104–106.

CHILDRESS, J. J. and NYGAARD, M. H. (1973). The chemical composition of midwater fishes as a function of depth of occurrence off southern California. *Deep Sea Res.*, **20**, 1093–1109.

CLOWES, A. (1938). Phosphate and silicate in the Southern Ocean. *'Discovery' Rep.*, **19**, 1–120.

COLEBROOK, J. M. (1972). Variability in the distribution and abundance of the plankton. *Spec. Publs. int. Commn NW. Atlant. Fish.*, **8**, 167–186.

COLEBROOK, J. M. (1978). Continuous plankton records: zooplankton and environment north-east Atlantic and North Sea, 1948–1975. *Oceanol. Acta*, **1**, 9–23.

CONOVER, R. J. (1962). Metabolism and growth in *Calanus hyperboreus* in relation to its life cycle. *Rapp. P.-v. Réun. Cons. int. Explor. Mer*, **153**, 190–197.

CONOVER, R. J. (1978). Transformation of organic matter. In O. Kinne (Ed.), *Marine Ecology*, Vol. IV, Dynamics. Wiley, Chichester. pp. 221–499.

CURRIE, R. J. (1964). Environmental features in the ecology of Antarctic seas. *Biologie Antarctique*, Paris.

CUSHING, D. H. (1959). The seasonal variation in oceanic production as a problem in population dynamics. *J. Cons. perm. int. Explor. Mer*, **24**, 455–464.

CUSHING, D. H. (1962). Patchiness. *Rapp. P.-v. Réun. Cons. int. Explor. Mer*, **153**, 152–164.

DIETZ, R. S. (1962). Deep-sea scattering layers. *Scient. Am.*, **207**, 44–50.

DENTON, E. J. and MARSHALL, N. B. (1958). The buoyancy of bathypelagic fishes without a gas-filled swimbladder. *J. mar. biol. Ass. U.K.*, **37**, 753–768.

DIGBY, P. S. B. (1954). The biology of the marine planktonic copepods of Scoresby Sound, East Greenland. *J. Anim. Ecol.*, **23**, 298–338.

DODIMEAD, A. J., FAVORITE, F. and HIRANO, T. (1963). Review of oceanography of the Subarctic Pacific Region. *Bull. int. N. Pacif. Fish. Commn*, **13**, 1–195.

DODSON, A. N. and THOMAS, W. H. (1964). Concentrating of plankton in a gentle fashion. *Limnol. Oceanogr.*, **9**, 455–456.

DUGDALE, R. C. (1975). Biological modelling, I. In J. Nihoul (Ed.), *Modelling of Marine Systems*, Series 10. Elsevier, Amsterdam. pp. 187–205.

DUGDALE, R. C. and MACISAAC, I. I. (1971). A computation model for the uptake of nitrate in the Peru upwelling region. *Investigación pesq.*, **35**, 299–308.

DUNBAR, M. J. (1960). The evolution of stability in marine environments. Natural selection at the level of the ecosystem. *Am. Nat.*, **94**, 129–136.

DUNBAR, M. J. (1968). *Ecological Development in Polar Regions: A Study in Evolution*. Prentice-Hall, New Jersey.

DUNBAR, M. J. (1970). Ecosystem adaptation in marine polar environments. In M. W. Holdgate (Ed.), *Antarctic Ecology*. Academic Press, London. pp. 105–111.

DUNBAR, M. J. (1972). The ecosystem as a unit of natural selection (Ecology essays in honour of C. E. Hutchinson). *Trans. Conn. Acad. Arts Sci.*, **44**, 113–130.

DURBIN, E. G. (1978). Aspects of the biology of resting spores of *Thalassiosira nordenskiöldii* and *Detonula confervacea*. *Mar. Biol.*, **45**, 31–37.

EL-SAYED, S. Z. (1970). On the productivity of the Southern Ocean. In M. W. Holdgate (Ed.), *Antarctic Ecology*. Academic Press, London. pp. 119–135.

ELTON, C. (1927). *Animal Ecology*. Sidgwick and Jackson, London.

ENGLISH, T. S. (1965). Some biological oceanographic observations in the central north Polar Sea drift station 'Alpha 1957–58'. *Special Report* (Air Force Cambr. Res. Lab.), **38**, 195–232.

FARRAN, G. P. (1936). The arctic plankton collected by the Nautilus-Expedition, 1931, II. Report on the Copepoda. *J. Linn. Soc.*, **39**, 404–410.

FINENKO, Z. Z. (1977). Phytoplankton adaptation to main factors of oceanic environment, 1. Marine organisms and factors of external environment (Russ.). In M. E. Vinogradov (Ed.), *Oceanology*, Biology of the Ocean, Vol. I. Nauka, Moscow. pp. 9–18.

FINENKO, Z. Z. (1978). Production in plant populations. In O. Kinne (Ed.), *Marine Ecology*, Vol. IV, Dynamics. Wiley, Chichester. pp. 13–88.

FOURNIER, R. O. (1970). Studies on pigmented micro-organisms from aphotic marine environments. *Limnol. Oceanogr.*, **15**, 675–682.

FOURNIER, R. O. (1972). The transport of organic carbon to organisms living in the deep oceans. *Proc. R. Soc. Edinb. (B)*, **73**, 203–212.

FOXTON, P. (1956). The distribution of the standing crop of zooplankton in the Southern Ocean. *'Discovery' Rep.*, **28**, 191–236.

FROST, B. W. (1974). *Calanus marshallae*, a new species of Calanoid copepod closely allied to the sibling species *C. finmarchicus* and *C. glacialis*. *Mar. Biol.*, **26**, 77–99.

FRANCIS, R. C. (1974). TUNPØP. A computer simulation model of the yellow fin tuna population and the surface tuna fishery of the eastern Pacific Ocean. *Bull. inter. Am. Trop. Tuna Commn.*, **16**, 235–279.

GAUSE, G. F. (1934). *The Struggle for Existence*. Williams and Wilkins, Baltimore.

GEORGE, R. J. and ALLEN, Z. P. (1970) USC-FSU biological investigations from the Fletcher's Ice Island T-3 on deep-sea and under-ice benthos of the Arctic Ocean. *Department of Biological Sciences Technical Report*, No. 1. University of Southern California.

GITELZON, J. J., LEVIN, L. A., SHEVYRNOGOV, A. P., UTYUSHEV, R. N. and ARTEMKIN, A. (1971). Bathyphotometric sounding of the pelagic zone of the ocean and its use for the studies of spatial structure of a community (Russ.). In M. E. Vinogradov (Ed.), *Functioning of Pelagic Communities in the Tropical Regions of the Ocean*. Nauka, Moscow. pp. 50–64.

GORELOVA, T. A. and TSEITLIN, V. B. (1979). Feeding of the mesopelagic fishes of the genus *Cyclothone* (Russ., Engl. summary). *Okeanologija*, **19**, 1110–1115.

GRAINGER, E. H. (1959). The annual oceanographic cycle at Igloolik in the Canadian Arctic. *J. Fish. Res. Bd Can.*, **16**, 453–501.

GRAN, H. H. (1931). On the conditions for the production of plankton in the sea. *Rapp. P.-v. Réun. Cons. int. Explor. Mer*, **75**, 37–46.

GREZE, V. N. (1963). Specific traits noted in the structure of the pelagial in the Ionian Sea. *Oceanology*, **3**, 100–109.

GREZE, V. N. (1970). The biomass and production of different trophic levels in the pelagic communities of the South Seas. In J. H. Steele (Ed.), *Marine Food Chains*. Oliver and Boyd, Edinburgh. pp. 458–467.

GREZE, V. N. (1973a). Secondary production of the seas and oceans (Russ.). In *General Ecology, Biogeozenology and Hydrobiology*, Vol. I, Results, Science and Technique. VINITI, Moscow. pp. 102–136.

GREZE, V. N. (1973b). Secondary production in Southern Seas. *Oceanology*, **13**, 80–85.

GREZE, V. N. (1978). Production in animal populations. In O. Kinne (Ed.), *Marine Ecology*, Vol. IV, Dynamics. Wiley, Chichester. pp. 89–114.

GRILL, E. V. (1970). Mathematical model for the marine dissolved silicate cycle. *Deep Sea Res.*, **17**, 245–266.

GRUZOV, L. M. (1963). Particular features of the seasonal development of zooplankton in different regions of the Norway Sea (Russ.). In *The Norway Sea*. AtlantNIRO, Kaliningrad. pp. 5–31.

GRUZOV, L. N. (1971). On the balance between the reproduction and the consumption processes in a plankton community of the equatorial Atlantic (Russ., Engl. summary). In *Trudy Atlant-NIRO*, **37**, 429–449.

GUEREDRAT, I. A. (1971). Evolution d'une population de copépodes dans le sistéme des courants equatoriaux de L'Ocean Pacifique. Zoogéographic écologie et diversité spècifique. *Mar. Biol.*, **9**, 300–314.

GULLAND, J. A. (1976). Production and catches of fish in the sea. In D. Cushing and J. Walsh (Eds), *Ecology of the Seas*. Blackwell, Oxford. pp. 283–314.

HAMILTON, O. D., HOLM-HANSEN, O. and STRICKLAND, J. P. (1968). Notes on the occurrence of living micro-organisms in deep water. *Deep Sea Res.*, **15**, 651–656.

HARDING, G. C. H. (1974). The food of deep-sea copepods. *J. mar. biol. Ass. U.K.*, **54**, 141–155.

HARDY, C. and GUNTHER, M. A. (1935). The plankton of the South Georgia whaling grounds and adjacent waters, 1926–1927. *'Discovery' Rep.*, **11**, 1–156.

HARGRAVES, P. E., and FRENCH, S. (1977). Resistance of diatom resting spores to grazing. *J. Phycol.*, **13** (2).

HART, T. J. (1934). On the phytoplankton of the south-west Atlantic and Bellingshausen Sea. *'Discovery' Rep.*, **8**, 1–268.

HART, T. J. (1942). Phytoplankton periodicity in Antarctic surface waters. *'Discovery' Rep.*, **21**, 261–356.

HASLE, G. R. (1959). A quantitative study of phytoplankton from the Equatorial Pacific. *Deep Sea Res.*, **6**, 38–59.

HASLE, G. R. (1969). An analysis of the phytoplankton of the Pacific Southern Ocean: abundance, composition, and distribution during the Brategg expedition 1947–1948. *Hvalråd, Skr.*, **52**, 1–168.

HEDGPETH, J. W. (1957). Classification of marine environments. In *Treatise on Marine Ecology and Paleoecology*, 1. Geol. Soc. Am., Mem., **67**, 17–27.

HEDGPETH, J. W. (1977). Models and muddels. Some philosophical observations. *Helgoländer wiss. Meeresunters.*, **30**, 92–104.

HEINRICH, A. K. (1956). Production of copepods in the Bering Sea. *Dokl. (Proc.) Acad. Sci. U.S.S.R.*, **111**, 199–201.

HEINRICH, A. K. (1961). Seasonal phenomena in plankton of the World Ocean. 1. Seasonal phenomena in the plankton of high and temperate latitudes (Russ., Engl. summary). *Trudy Inst. Okeanol.*, **51**, 57–81.

HEINRICH, A. K. (1962). The life histories of plankton animals and seasonal cycles of plankton communities in the oceans. *J. Cons. perm. int. Explor. Mer*, **27**, 15–24.

HEINRICH, A. K. (1975). The significance of the expatriated species in the structure of the Pacific tropical plankton communities (Russ., Engl. summary). *Oceanology*, **15**, 721–726.

HEINRICH, A. K. (1977a). Communities of the tropical regions of the ocean. Pelagic communities (Russ.). In M. E. Vinogradov (Ed.), *Oceanology*, Biology of the Ocean, Vol. 2. Nauka, Moscow. pp. 91–104.

HEINRICH, A. K. (1977b). A quantitative estimate of the similarity of population, hierarchy and boundaries of planktonic communities in the Pacific (Russ., Engl. summary). *Zool. zh.*, **56**, 181–187.

HEINRICH, A. K., KOSOBOKOVA, K. N. and RUDJIAKOV YU, A. (1980). Seasonal changes in depth distribution of some common copepod species in the Arctic Ocean (Russ.). In M. E. Vinogradov and I. A. Melnikov (Eds), *Biology of the Central Arctic Ocean*. Nauka, Moscow. pp. 155–166.

HERON, A. C. (1972). Population ecology of a colonizing species: the pelagic tunicate *Thalia democratica*. *Oecologia*, **10**, 294–312.

HOLM-HANSEN, O., PACKARD, T. T. and POMEROY, L. R. (1970). Efficiency of the reverse-flow filter technique for concentration of particulate matter. *Limnol. Oceanogr.*, **15**, 832–835.

HORNER, R. A. (1976). Sea-ice organisms. *Oceanogr. mar. Biol. A. Rev.*, **14**, 167–182.

HOPKINS, T. L. (1969). Zooplankton biomass related to hydrography along the drift-truck of Arlis-II in the Arctic Basin and the East Greenland Current. *J. Fish. Res. Bd Can.*, **26**, 305–310.

HUGHES, K. H. (1968). *Seasonal Vertical Distributions of Copepods in the Arctic Water in the Canadian Basin of the North Polar Sea*. M.S. Thesis, University of Washington, Seattle.

IVLEV, V. A. (1955). *Experimental Ecology of Nutrition of Fishes* (Russ.). Pischepromisdat, Moscow. (Engl. transl. by D. Scott, Yale Univ. Press, New Haven, 1961).

JESPERSEN, P. (1935). Quantitative investigations on the distribution of macroplankton in the different oceanic regions. *Dana Rep.*, **7**, 44.

JOHNSON, M. V. (1963). Zooplankton collections from the high Polar Basin with special reference to the Copepoda. *Limnol. Oceanogr.*, **8**, 89–102.

JOHNSTON, R. (1962). An equation for the depth distribution of deep-sea zooplankton and fishes. *Rapp. P.-v Réun. Cons. Int. Explor. Mer*, **153**, 38.

JØRGENSEN, C. B. (1955). Quantitative aspects of filter feeding in invertebrates. *Biol. Rev.*, **30**, 391–454.

JØRGENSEN, C. B. (1962). The food of filter feeding organisms. *Rapp. P.-v. Réun. Cons. int. Explor. Mer*, 153, 99–107.

KALFF, J. (1967). Phytoplankton dynamics in Arctic larvae. *J. Fish. Res. Bd Can.*, **24**, 1861–1871.

KAMSHILOV, M. M. (1958). Production of *Calanus finmarchicus* Gunner in littoral zone of the eastern Murman (Russ.). *Trudy murmansk. biol. Sta.*, **4**, 45–55.

KARR, J. R. (1968). Habitat and avian diversity on strip-mined land in the east-central Illinois. *Condor*, **71**, 348–357.

KAWAMURA, A. (1967). Observations of phytoplankton in the Arctic ocean in 1964. *Inf. Bull. Plankol. Japan*, December, 71–90.

KEILHORN, W. V. (1952). The biology of the surface zone zooplankton of Boreo-Arctic Atlantic Oceanarea. *J. Fish. Res. Bd Can.*, **9**, 223–264.

KING, J. E. and DEMOND, J. (1953). Zooplankton abundance in the central Pacific. *Fish. Bull. Fish. Wildl. Serv. U.S.*, **54**, 111–144.

KING, J. E. and HIDA, T. S. (1957). Zooplankton abundance in the central Pacific, 2. *Fish. Bull. Fish. Wildl. Serv. U.S.*, **57**, 365–395.

KING, J. E. and IVERSEN, R. T. B. (1962). Midwater trawling for forage organisms in the central Pacific, 1951–1956. *Fish. Bull. Fish Wildl. Serv. U.S.*, **62**, 271–321.

KINNE, O. (1977). International Helgoland Symposium on ecosystem research : summary, conclusions and closing. *Helgoländer wiss. Meeresunters.*, **30**, 709–727.

KINNE, O. (in press). Realism in aquaculture—the view of an ecologist. In M. Bilio and H. Rosenthal (Eds), *Proceedings of a World Conference on Aquaculture, Venice 1981.*

KINNE, O. (1983). Aquacultur: Ausweg aus der Ernährungskrise? *Spektrum der Wissenschaft*, **December 12/1982**, 46–57.

KINNE, O. and BULNHEIM, H.-P. (Eds) (1977). International Helgoland Symposium on ecosystem research. *Helgoländer wiss. Meeresunters.*, **30**, 1–735.

KINNE, O. and ROSENTHAL, H. (1977). Commercial cultivation (Aquaculture). In O. Kinne (Ed.), *Marine Ecology*, Vol. III, Cultivation, Part 3. Wiley, Chichester. pp. 1321–1398.

KOBLENTZ-MISHKE, O. J. (1977). Primary production. Quantitative horizontal distribution of plants and animals in the Ocean. In M. E. Vinogradov (Ed.), *Oceanology*, Biology of the Ocean, Vol. I. Nauka, Moscow. pp. 62–64.

KOBLENTZ-MISHKE, O. J. and VEDERNIKOV, V. I. (1976). A tentative comparison of primary production and phytoplankton quantities at the ocean surface. *Mar. Sci. Commun.*, **2**, 357–374.

KOSHLYAKOV, M. N. and MONIN, A. S. (1978). Synoptic eddies in the ocean. *A. Rev. Earth Planet. Sci.*, **6**, 495–523.

KOSOBOKOVA, K. N. (1978). Diurnal vertical distribution of *Calanus hyperboreus* Krøyer and *Calanus glacialis* Jashnov in the central Polar Basin. *Oceanology*, **18**, 722–728.

LAWRENCE, J. M. (1976). Patterns of lipid storage in post metamorphic marine invertebrates. *Am. Zool.*, **16**, 747–762.

LEAVITT, B. B. (1938). The quantitative vertical distribution of macrozooplankton in the Atlantic Ocean Basin. *Biol. Bull. mar. biol. Lab., Woods Hole*, **74**, 376–394.

LE BORGNE, R. (1978). Évaluation de la production secondaire planctonique en milieu océanique par la méthode des rapports C/N/P. *Oceanol. Acta*, **1**, 107–118.

LEGAL, J. (1952). Répartition en profondeurs des Coccolithophorids en quelques stations Méditerraneennes occidentales. *Bull. Inst. océanogr. Monaco*, **1018**, 1–14.

LEGAND, M., BOURRET, P., FOURMANOIR, P., GRANDPERRIN, P., GUEREDRAT, J. A., MICHEL, A., RANCUREL, P., REPELIN, R. and ROGER, C. (1972). Relations trophiques et distributions verticales en milieu pélagique dans l'Océan Pacifique intertropical. *Cah. ORSTOM, Océanogr.*, **10**, 303–393.

LEVIN, L. A., UTYUSHEV, R. N. and ARTEMKIN, A. S. (1975). Intensity of bioluminescence in the equatorial part of the Pacific Ocean (Russ., Engl. summary). *Trudy Inst. Okeanol.*, **102**, 94–101.

LLOYD, M., INGER, R. F. and KING, F. W. (1968). On the diversity of reptile and amphibian species in a Bornean rain-forest. *Am. Nat.*, **102**, 497–515.

LONGHURST, A. R. (1976). Interactions between zooplankton and phytoplankton profiles in the eastern tropical Pacific Ocean. *Deep Sea Res.*, **23**, 729–754.

LONGHURST, A. R., COLEBROOK, M., GULLAND, J. A., LEBRASSEUR, R. J., LORENZEN, C. and SMITH, P. (1972). The instability of ocean populations. *New Scient.*, **54** (798), 500–502.

LONGHURST, A. R. and WILLIAMS, R. (1979). Materials for plankton modelling: vertical distribution of Atlantic zooplankton in summer. *J. Plankt. Res.*, **1**, 1–28.

LORENZEN, C. J. (1976). Primary production in the sea. In D. H. Cushing and J. J. Walsh (Eds), *The Ecology of the Seas*. Blackwell, Oxford. pp. 173–185.

LOTKA, A. I. (1925). *Elements of Physical Biology*. Williams and Wilkins, Baltimore.

LYAPUNOV, A. A. (1971). On the construction of a mathematical model of balance correlations in the ecosystem of the tropical ocean. In M. E. Vinogradov (Ed.), *Functioning of Pelagic Communities in the Tropical Regions of the Ocean*. Nauka, Moscow. pp. 13–24.

LYAPUNOV, A. A. and YABLONSKY, S. V. (1963). Theoretical problems of cybernetics (Russ.). *Problemȳ Cybernetiki*, **9**, Nauka, Novosibirsk.

MCALLISTER, C. D. (1961). Zooplankton studies at Ocean Weather Station 'P' in the north-east Pacific Ocean. *J. Fish. Res. Bd Can.*, **18**, 1–29.

MACARTHUR, R. H. (1955). Fluctuations of animal populations and measure of community stability. *Ecology*, **36**, 533–536.

MCCAVE, I. N. (1975). Vertical flux of particles in the ocean. *Deep Sea Res.*, **22**, 491–502.

MACDONALD, A. G. (1975). *Physiology Aspects of Deep-Sea Biology*. University Press, Cambridge.

MACFADYEN, A. (1964). Energy flow in ecosystems and its exploitation by grazing. In D. J. Crisp (Ed.), *Grazing in Terrestrial and Marine Environments*. Blackwell, Oxford. pp. 3–24.

MCGOWAN I. A. (1974). The nature of oceanic ecosystems. In C. Miller (Ed.), *The Biology of the Oceanic Pacific*. Oregon State University Press, Corvallis. pp. 9–28.

MACKINTOSH, N. A. (1970). Whales and krill in the twentieth century. In M. W. Holdgate (Ed.), *Antarctic Ecology*. Academic Press, London. pp. 195–212.

MACKINTOSH, N. A. (1973). Distribution of postlarval krill in the Antarctic. *'Discovery' Rep.*, **36**, 95–156.

MANN, K. H. (1969). The dynamics of aquatic ecosystems. In I. B. Crogg (Ed.), *Advances in Ecological Research*. Academic Press, New York. pp. 1–81.

MARGALEF, R. (1963). Succession in marine populations. In Raghuvira (Ed.), *Advancing Frontiers of Plant Science*, Vol. II. New Delhi, India. pp. 137–188.

MARGALEF, R. (1968). *Perspectives in Ecological Theory*. University of Chicago Press, Chicago.

MARGALEF, R. (1978a). What is an upwelling ecosystem? In R. Boje and M. Tomczak (Eds), *Upwelling Ecosystems*. Springer-Verlag, Heidelberg. pp. 12–14.

MARGALEF, R. (1978b). General concepts of population dynamics and food links. In O. Kinne (Ed.), *Marine Ecology*, Vol. IV, Dynamics. Wiley, Chichester. pp. 617–704.

MARR, J. W. S. (1962). The natural history and geography of the Antarctic krill *Euphausia superba* Dana. *'Discovery' Rep.*, **32**, 33–464.

MARSHALL, N. B. (1960). Swimbladder structure of deep-sea fishes in relation to their systematics and biology. *'Discovery' Rep.*, **31**, 1–121.

MARSHALL, N. B. (1971). *Explorations in the Life of Fishes*. Harvard University Press, Cambridge, Mass.

MARSHALL, S. M. and ORR, A. P. (1955). *The Biology of a Marine Copepod Calanus finmarchicus Gunner*. Oliver and Boyd, Edinburgh.

MAUCHLINE, I. (1972). The biology of bathypelagic organisms, especially Crustacea. *Deep Sea Res.*, **19**, 753–780.

MEDNIKOV, B. M. (1960). Production of calanids in north-western part of the Pacific (Russ.). *Dokl. (Proc.) Acad. Sci. U.S.S.R.*, **134**, 1208–1210.

MEEK, R. P. and CHILDRESS, J. J. (1973). Respiration and the effect of pressure in the mesopelagic fish *Anoplogaster cornuta* (Beryciformes). *Deep Sea Res.*, **20**, 1111–1118.

MELNIKOV, I. A. (1976). Microplankton and organic detritus in the Southeastern Pacific. *Oceanology*, **15**, 103–110.

MELNIKOV, I. A. (1980). Biology of a drifting Arctic ice (Russ.). In M. E. Vinogradov and I. A. Melnikov (Eds), *Biology of the Central Arctic Basin.* Nauka, Moscow. pp. 61–97.

MENSHUTKIN, V. V. (1979). Model of the pelagic ecosystem of the Pacific Ocean. *Oceanology*, **19** (2), 205–209.

MENSHUTKIN, V. V. and UMNOV, A. A. (1970). Mathematical simulation of the simplest water ecosystem. (Russ.). *Gidrobiol. Zh.*, **6**, 28–35.

MENSHUTKIN, V. V., VINOGRADOV, M. E. and SHUSHKINA, E. A. (1974). Mathematical model of the pelagic ecosystem in the Sea of Japan. *Oceanology*, **14** (5), 717–723.

MENZIES, R. I. (1962). On the food and feeding habits of abyssal organisms as exemplified by the Isopoda. *Int. Revue ges. Hydrobiol.*, **47**, 339–358.

MIYAKE, Y. and WADA, E. (1968). The nitrogen cycle in the sea. *Rec. oceanogr. Wks Japan*, **9**, 197–208.

MOISEEV, P. A. (1969). *Biological Resources of the Ocean* (Russ.). Izd. Pishevaja Promyshl, Moscow.

MONIN, A. S., KAMENKOVITCH, V. M. and KORT, V. G. (1974). *The Variability of the Ocean* (Russ.). Gidrometeoizdat., Leningrad.

MOTODA, S. and MINODA, T. (1974). Plankton of the Bering Sea. In D. W. Hood and E. I. Kelley (Eds), *Oceanography of the Bering Sea.* Institute of Marine Science, University Alaska, Fairbanks. pp. 207–241.

MULLIN, M. M. (1969). Production of zooplankton in the oceans: the present status and problems. In H. Barnes (Ed.), *Oceanography and Marine Biology*, Vol. VII. Allen and Unwin, London. pp. 293–314.

MUNK, W. H. (1950). On the wind-driven ocean circulation. *J. Met.*, **7**, 79–93.

NAUMANN, E. (1931). Limnologische Terminologie. *Handbuch der Biologischen Arbeitsmethoden*, Abt., 9, 8, Hf. 3, 321–776.

NEMOTO, T. (1968). Feeding of baleen whales and krill, and the value of krill as a marine resource in the Antarctic. In *Proceedings of a Symposium on Antarctic Oceanography*. Scientific Council of Agricultural Research, Scientific Council of Oceanic Research, Santiago, Chile.

NESIS, K. N. (1965). Some problems involved with the food structure of marine biocoenosis. *Oceanology*, **5**, 701–714.

ODUM, E. P. (1971). *Fundamentals of Ecology*. Saunders, Philadelphia.

OWEN, R. W. and ZEITZSCHEL, B. (1970). Phytoplankton production: seasonal change in the oceanic, eastern tropical Pacific. *Mar. Biol.*, **7**, 32–36.

PAINE, P. T. (1969). Food web complexity and species diversity. *Am. Nat.*, **100**, 65–75.

PARIN, N. V. (1968). *Ichthyofauna of the Epipelagic Zone* (Russ., Engl. summary). Nauka, Moscow.

PARIN, N. V. (1970). Large pelagic predatory fishes in the trophic system of the Tropical Ocean. In *Sovremennoe Sostojanie Biologicheskoj Productivnosti, Sirjevikh Biologicheskikh Resursov Mirovogo Okeana* (Russ.). Kaliningrad.

PARIN, N. V. (1975). Change of pelagic ichthyocoenoses along the section at the equator in the Pacific Ocean between 97° and 155° W. *Trudy Inst. Okeanol.*, **102**, 311–334.

PARIN, N. V. and NESIS, K. N. (1977). Macroplankton and nekton. Quantitative distribution of plants and animals in the ocean. In M. E. Vinogradov (Ed.), *Oceanology*, Biology of the Ocean, Vol. I. Nauka, Moscow. pp. 69–77.

PARIN, N. V., NESIS, K. N. and KASHKIN, N. I. (1977). Macroplankton and nekton. Vertical distribution of plants and animals in the ocean. In M. E. Vinogradov (Ed.), *Oceanology*, Biology of the Ocean, Vol. I. Nauka, Moscow. pp. 159–173.

PARISH, R. H. (1978). Climatic variation and exploitation in the Pacific mackerel fishery. *Fish. Bull. Calif.*, **167**, 1–110.

PARSONS, T. R. and ANDERSON, G. C. (1970). Large-scale studies of primary production in the north Pacific Ocean. *Deep Sea Res.*, **17**, 765–776.

PARSONS, T. R. and LEBRASSEUR, R. J. (1968). A discussion of some critical indices of primary and secondary production for large-scale ocean surveys. *CALCOFI Rep.* (California Marine Research Commission), **12**, 54–63.

PARSONS, T. R., LEBRASSEUR, R. J., FULTON, J. D. and KENNEDY, O. D. (1969). Production studies in the Strait of Georgia. 2. Secondary production under the Fraser River plume, February to May, 1967. *J. exp. mar. Biol. Ecol.*, **3**, 39–50.

PARSONS, T. R. and TAKAHASHI, M. (1973). *Biological Oceanographic Processes*. Pergamon Press, Oxford.

PARRISH, R. H. (1978). Climatic variation and exploitation in the Pacific mackerel fishery. *Fish. Bull. Calif.*, **167**, 1–110.

PATTEN, B. S. (1968). Mathematical models of plankton production. *Int. Revue ges. Hydrobiol.*, **53**, 357–408.

PAVSHTIKS, E. A. (1969). Effect of streams on seasonal changes of zooplankton in Devis Strait. *Gidrobiol. Zh.*, **5**, 85–92.

PAVSHTIKS, E. A. (1971a). Hydrobiological characteristics of Arctic Basin water masses in the region of drifting station 'North Pole-17' *Trudy arkt. antarkt. nauchno-issled. Inst.*, **302**, 63–69.

PAVSHTIKS, E. A. (1971b). On seasonal variations of zooplankton quantity in the North Pole region. *Dokl. (Proc.) Acad. Sci. U.S.S.R.*, **196**, 441–444.

PAVSHTIKS, E. A. and VYSHKVARTZEVA, N. V. (1977). Age composition and morphological characters of the population *Calanus finmarchicus s.I.* near Spitzbergen and Franz Josef Land (80°n.z.) in September 1970. *Issled. Fauny Morej.*, **14**, 219–236.

PAXTON, I. R. (1967). A distributional analysis for the lantern fishes (family Myctophidae) of the San Pedro Basin, California. *Copeia*, (2), 422–443.

PEARCY, W. G. (1964). Some distribution features of mesopelagic fishes off Oregon. *J. mar. Res.*, **22**, 83–102.

PERES, J. M. (1958a). Remarques générales sur un ensemble de quinze plongées effectuées avec le bathyscaphe FNRS III. *Annls. Inst. océanogr., Monaco*, **35**, 259–285.

PERES, J. M. (1958b). Trois plongées dans la canyon du Cap Sicié effectuées avec le bathyscaphe FNRS III, de la Marine Nationale. *Bull. Inst. océanogr. Monaco*, **1115**, 1–21.

PERES, J. M. (1965). Apercu sur les résultats de deux plongeés effectuées dans le ravin de Puerto-Rico par le bathyscaphe 'Archimede'. *Deep Sea Res.*, **12**, 883–891.

PERES, J. M. (1982). Zonations and organismic assemblages. In O. Kinne (Ed.), *Marine Ecology*, Vol. V, Ocean Management, Part 1. Wiley, Chichester. pp. 9–576.

PLATT, T. DENMAN, K. L. and JASSBY, A. D. (1977). Modeling the productivity of phytoplankton. In E. D. Goldberg, I. N. McCave, J. J. O'Brien and J. H. Steele (Eds), *The Sea*, Vol. VI. Wiley, New York. pp. 807–856.

POLOSIN, A. S. (1967). On the zero surface in the equatorial zone of the Atlantic Ocean. *Oceanology*, **7**, 89–97.

PONOMARENKO, V. P. (1968). The polar cod migrations in the soviet sector of the Arctic. *Trudӯ polyar. nauchno- issled. Inst. morsk. rӯb. Khoz. Okeanogr.*, **23**, 500–512.

RAYMONT, J. E. G. (1963). *Plankton and Productivity in the Oceans.* Pergamon Press, Oxford.

RAYMONT, J. E. G. and CARRIE, B. G. A. (1964). The production of zooplankton in Southampton water. *Int. Revue ges. Hydrobiol.*, **49**, 185–232.

REES, C. B. (1949). The distribution of *Calanus finmarchicus* and its two forms in the North Sea, 1938–1939. *Hull Bull. mar. Ecol.*, **2**, 215–275.

REID, J. L., BRINTON, E., FLEMINGER, A., VENRICK, E. L. and MCGOWAN, J. A. (1978). Ocean circulation and marine life. In H. Charnock and G. Deacon (Eds), *Advances in Oceanography.* Plenum Press, New York. pp. 65–130.

RILEY, G. A. (1946). Factors controlling phytoplankton populations on Georges Bank. *J. mar. Res.*, **6**, 54–73.

RILEY, G. A. (1947a). Seasonal fluctuations of the phytoplankton population in New England coastal waters. *J. mar. Res.*, **6**, 114–125.

RILEY, G. A. (1947b). A theoretical analysis of the zooplankton population of Georges Bank. *J. mar. Res.*, **6**, 104–113.

RILEY, G. A., STOMMEL, H. and BUMPUS, D. F. (1949). Quantitative ecology of the plankton of the western north Atlantic. *Bull. Bingham oceanogr. Coll.* **12**, 1–169.

ROTSCHI, J. and JARRIGE, F. (1968). Sur le renforcement d'un upwelling équatorial. *Cah. ORSTOM*, Océanogr., **34**, 87–90.

RYTHER, J. H. (1963). Geographic variations in productivity. In M. N. Hill (Ed.), *The Sea*, Vol. III. Wiley, New York. pp. 347–380.

SCHOENER, A. (1968). Evidence for reproductive periodicity in the deep-sea. *Ecology*, **49**, 81–87.

SEMINA, H. J. (1966). Biotope and quantity of phytoplankton in the oceans (Russ., Engl. summary). *Usp. sovrem. Biol.*, **62** (2), 289–306.

SEMINA, H. J. (1974). *Phytoplankton of the Pacific* (Russ., Engl. summary). Nauka, Moscow.

SEMINA, H. J. (1977a). Phytoplankton. Quantitative horizontal distribution of plants and animals in the Ocean (Russ.). In M. E. Vinogradov (Ed.), *Oceanology*, Biology of the Ocean, Vol. I. Nauka, Moscow. pp. 58–62.

SEMINA, H. J. (1977b). Phytoplankton. Vertical distribution of plants and animals in the ocean (Russ.). In M. E. Vinogradov (Ed.), *Oceanology,* Biology of the Ocean, Vol. I. Nauka, Moscow. pp. 117–124.

SERGEEV, YU. N., SAVCHUK, O. P., KULESH, V. P. and KOMAROVA, T. S. (1977). *Mathematical Simulation of the Marine Ecosystems*. University of Leningrad, Leningrad.

SETTE, O. E. (1955). Consideration of mid-ocean fish production as related to oceanic circulatory systems. *J. mar. Res.,* **14**, 398–414.

SHMALGAUZEN, I. I. (1968). Control and regulation in evolution (Russ.). In R. L. Berg and A. A. Lyapunov (Eds), *Kiberneticheskie Voprost Biologn.* Nauka, Novosibirsk, pp. 31–73.

SHUSHKINA, E. A. (1971). Estimates of the production intensities of the tropical zooplankton (Russ., Engl. summary). In M. E. Vinogradov (Ed.), *Functioning of Pelagic Communities in the Tropical Regions of the Ocean.* Nauka, Moscow. pp. 157–166.

SHUSHKINA, E. A. (1977). Zooplankton production. Production of marine community. In M. E. Vinogradov (Ed.). *Oceanology,* Biology of the Ocean, Vol. I. Nauka, Moscow. pp. 233–247.

SHUSHKINA, E. A. and KISLYAKOV, Yu. Ya. (1975). An estimation of the zooplankton productivity in the equatorial part of the Pacific Ocean and Peruvian upwelling (Russ., Engl. summary). *Trudŷ Inst. Okeanol.*, **102**, 384–395.

SHUSHKINA, E. A., KISLYAKOV, Yu. Ya. and PASTERNAK, A. F. (1974). Combination of the radiocarbon method with mathematical simulation for the estimation of productivity of the marine zooplankton. *Oceanology,* **14**, 319–326.

SHUSHKINA, E. A., VINOGRADOV, M. E., SOROKIN, Yu. I., LEBEDEVA, L. P. and MIKHEEV, V. N. (1978). The peculiarities of functioning of plankton communities in the Peruvian upwelling. *Oceanology,* **18**, 886–902.

SHUVALOV, V. S. and PAVSHTIKS, E. A. (1977). Composition and distribution of the undersurface zooplankton (hyponeuston) off Franz-Josef Land (Russ., Engl. summary). *Issled. Fauny Morej.,* **14**, 55–79.

SIEBURTH, J. McN. (1977). International Helgoland Symposium: convener's report on the informal session on biomass and productivitiy of micro–organisms in planktonic ecosystems. *Helgoländer wiss. Meeresunters.,* **30**, 697–704.

SLOBODKIN, L. B. (1962). *Growth and Regulation of Animal Populations.* Holt, Reinhart and Winston, New York.

SLOBODKIN, L. D. and SANDERS, H. L. (1969). On the contribution of environmental predictability to species diversity. *Brookhaven Symp. Biol.,* **22**, 82–93.

SMAYDA, T. J. (1966). A quantitative analysis of the phytoplankton of the Gulf of Panama. III. *Bull. inter-Am. trop. Tuna Commn.,* **11**, 355–612.

SMITH, K. L. and HESSLER, R. R. (1974). Respiration of bentho-pelagic fishes: *in situ* measurements at 1230 meters. *Science, N. Y.,* **184**, 72–73.

SONINA, M. A. (1973). Arcto-norwegian Haddock fertility in connection with population dynamic (Russ.). *Trudŷ polyar. nauchno-issled. Inst. morsk. rŷb. Khoz. Okeanogr.,* **33**.

SOROKIN, Yu. I. (1959). On the effect of water stratification on the primary production of photosynthesis in the sea (Russ. Engl. summary). *Zh. obshch. Biol.,* **20**, 455–463.

SOROKIN, Yu. I. (1971a). On the role of bacterioplankton in the biological productivity of the tropical waters of the Pacific Ocean. In M. E. Vinogradov (Ed.), *Functioning of Pelagic Communities in the Tropical Regions of the Ocean.* Nauka, Moscow. pp. 92–122.

SOROKIN, Yu. I. (1971b). On bacterial numbers and production in the water column in the Central Pacific. *Oceanology,* **11**, 105–116.

SOROKIN, Yu. I. (1973). Primary production of the seas and oceans. In *General Ecology, Biogeozenology and Hydrobiology*, Vol. I, Results, Science and Technique (Russ.). VINITI, Moscow. pp. 7–46.

SOROKIN, Yu. I. (1974). Vertical structure and production of a microplankton community in the sea of Japan in summer. *Okeanologija*, **14**, 327–333.

SOROKIN, Yu. I. (1977a). Bacterioplankton. Vertical distribution of plants and animals in the ocean. In M. E. Vinogradov (Ed.), *Oceanology,* Biology of the Ocean, Vol. I. Nauka, Moscow. pp. 124–132.

SOROKIN, Yu. I. (1977b). Bacterial production. Production of marine community. In M. E. Vinogradov (Ed.). *Oceanology*, Biology of the Ocean, Vol. I. Nauka, Moscow. pp. 209–233.

SOROKIN, Yu. I. (1978a). Decomposition of organic matter and nutrient regeneration. In O. Kinne (Ed.), *Marine Ecology,* Vol. IV, Dynamics. Wiley, Chichester. pp. 501–616.

SOROKIN, Yu. I. (1978b). Characteristics of primary production and heterotrophic microplankton in the Peruvian upwelling. *Oceanology,* **18**, 97–110.

SOROKIN, Yu. I., PAVELIEVA, E. B. and VASILYEVA, M. I. (1975). Productivity and trophic role of bacterioplankton in the area of equatorial divergence (Russ., Engl. summary). *Trudӯ Inst. Okeanol.,* **102**, 184–198.

STEELE, J. H. (1958). Plant production in the northern North Sea. *Mar. Res.,* **7**, 1–36.

STEELE, J. H. (1974). *The Structure of Marine Ecosystems.* Harvard University Press, Cambridge, Mass.

STEELE, J. H. (1975). Biological modelling, II. In J. Nihoul (Ed.), *Modelling of Marine Systems,* Series 10. Elsevier, Amsterdam. pp. 207–216.

STEELE, J. H. (1976). Patchiness. In D. H. Cushing and I. I. Walsh (Eds), *Ecology of the Sea.* Blackwell, Oxford. pp. 98–115.

STEELE, J. H. (Ed.) (1977). *Spatial Pattern in Plankton Communities.* NATO Conference Series IV. Marine Science, Vol. III. Plenum Press, New York.

STEELE, J. H. and MULLIN, M. M. (1977). Zooplankton dynamics. In E. D. Goldberg, I. N. McCave, J. J. O'Brien and J. H. Steele (Eds), *The Sea,* Vol. VI. Wiley, New York. pp. 857–890.

STEEMANN-NIELSEN, E. (1972). The rate of primary production and the size of the standing stock of zooplankton in the oceans. *Int. Revue ges. Hydrobiol.,* **57**, 513–516.

STEYAERT, J. (1974). Distribution of some selected diatom species during the Belgo-Dutch Antarctic expedition of 1964–65 and 1966–67. *Investigación pesq.,* **38**, 259–288.

STOMMEL, H. and FEDOROV, K. N. (1967). Small-scale structure in temperature and salinity near Timor and Mindanao. *Tellus,* **19** (2), 306–325.

STRICKLAND, J. D. H. (1972). Research on the marine planktonic food web at the Institute of Marine Research: a review of the past seven years of work. *Oceanogr. mar. Biol. A. Rev.,* **10**, 349–414.

SVERDRUP, H. V. (1953). On conditions for the vernal blooming of phytoplankton. *J. Cons. perm. int. Explor. Mer,* **18**, 287–295.

THOMAS, W. H. (1970). Effect of ammonium and nitrate concentration on chlorophyll increase in natural tropical Pacific phytoplankton populations. *Limnol. Oceanogr.,* **15**, 386–394.

TIMOCHINA, A. F. (1968). Production of mass species of zooplankton in Norwegian Sea (Russ.). *Trudy polyar. nauchno-issled. Inst. morsk. rӯb. Khoz. Okeanogr.,* **23**, 173–192.

TIMONIN, A. G. (1971). The structure of plankton communities of the Indian Ocean. *Mar. Biol.,* **9**, 281–289.

TIMONIN, A. G. (1976). Study of the vertical microdistribution of oceanic zooplankton. *Oceanology,* **16**, 79–82.

TRANTER, D. I. (1976). Herbivore production. In D. H. Cushing and J. J. Walsh (Eds), *The Ecology of the Seas.* Blackwell, Oxford. pp. 186–224.

TSEITLIN, V. B. (1977). On the causes of size changes in the pelagic zooplankton feeders with changing depth of a habitat. *Oceanology,* **17**, 132–138.

TUMANTSEVA, N. T. and SOROKIN, Yu. I. (1975). Microzooplankton of the area of equatorial divergence in the eastern part of the Pacific Ocean (Russ., Engl. summary). *Trudӯ Inst. Okeanol.,* **102**, 200–212.

UUDA, M. (1963). Oceanography of the subarctic Pacific Ocean. *J. Fish. Res. Bd Can.,* **20**, 119–179.

USACHEV, P. I. (1961). Phytoplankton of the North Pole (Based on the collections of P. P. Shirshov. First drifting station. 'The North Pole' 1937–38) (Russ.). *Trudӯ vses. gidrobiol. Obsheh.,* **11**, 189–208.

VINOGRADOV, M. E. (1956). Distribution of zooplankton in the western regions of the Bering Sea (Russ.). *Trudӯ vses. gidrobiol. Obshch.,* **7**, 173–203.

VINOGRADOV, M. E. (1958). On the vertical distribution of deep-sea plankton in the west part of the Pacific ocean. *Fifteenth International Congr. Zool.,* sect. III, pap. 31, London.

VINOGRADOV, M. E. (1960). Quantitative distribution of deep-sea plankton in the western and central Pacific (Russ., Engl. summary). *Trudӯ Inst. Okeanol.,* **41**, 55–84.

VINOGRADOV, M. E. (1962a). Feeding of the deep-sea zooplankton. *Rapp. P.-v. Réun. Cons. Int. Explor. Mer,* **153**, 114–120.

VINOGRADOV, M. E. (1962b). Quantitative distribution of deep-sea plankton in the western Pacific and its relation to deep-water circulation. *Deep Sea Res.*, **8**, 251–258.

VINOGRADOV, M. E. (1968). *Vertical Distribution of the Oceanic Zooplankton* (Russ., Engl. summary). Nauka, Moscow.

VINOGRADOV, M. E. (1970a). Some peculiarity of the change of the oceans' pelagial communities with the change of depth. In L. A. Zenkevitch (Ed.), *Program and Method Investigation of Biogeocoenological Water's Surroundings.* Nauka, Moscow. pp. 84–96.

VINOGRADOV, M. E. (1970b). The vertical distribution of zooplankton in the Kuril-Kamchatka region of the Pacific Ocean (Russ., Engl. summary). *Trudy̆ Inst. Okeanol.*, **86**, 99–116.

VINOGRADOV, M. E. (1971). Studies of the functioning of oceanic biological systems (Russ., Engl. summary). In M. E. Vinogradov (Ed.), *Functioning of Pelagic Communities in the Tropical Regions of the Ocean.* Nauka, Moscow. pp. 5–12.

VINOGRADOV, M. E. (1972). Vertical stratification of zooplankton in the Kuril-Kamchatka trench. In A. Y. Takenouti (Ed.), *Biological Oceanography of the Northern North Pacific Ocean.* Idenitsu shoten, Tokyo. pp. 333–340.

VINOGRADOV, M. E. (1974). Depth of the night-time rise of deep-scattering layers in the central Pacific. *Oceanology,* **14** (6), 891–895.

VINOGRADOV, M. E. (1975). The study of ecosystems in the pelagic zone of the eastern Pacific upwellings during the 17th cruise of the R. V. *Academic Kurchatov* (Russ., Engl. summary). *Trudy̆ Inst. Okeanol.,* **102**, 7–17.

VINOGRADOV, M. E. (1975). Ecosystems of the pelagic zone of the Pacific (Russ., Engl. summary). *Trudy̆ Inst. Okeanol.,* **102**, 408.

VINOGRADOV, M. E. (1977a). Zooplankton. Vertical distribution of plants and animals in the ocean. In M. E. Vinogradov (Ed.), *Oceanology*, Biology of the Ocean, Vol. 1. Nauka, Moscow. pp. 132–151.

VINOGRADOV, M. E. (1977b). Spatial-dynamic aspect of the pelagic community's existence. Some principles of structure and development of marine communities. In M. E. Vinogradov (Ed.), *Oceanology,* Biology of the Ocean, Vol. 2. Nauka, Moscow. pp. 14–23.

VINOGRADOV, M. E. (1978a). Some physical and biological features of equatorial upwellings (Russ., Engl. summary). *Bull. Mosc. obshchestvo ispitatelei prirodi (Series Biol.),* **83**, 5–16.

VINOGRADOV, M. E. (1978b). Problems in modelling the effect of pollution on biological systems of the ocean. In *First American-Soviet Symposium on the Biological Effects of Pollution on Marine Organisms.* U.S. Environmental Protection Agency, Florida. pp. 10–21.

VINOGRADOV, M. E., GITELZON, I. I. and SOROKIN, Yu. I. (1970. The vertical structure of a pelagic community in the Tropical Ocean. *Mar. Biol.*, **6**, 187–194.

VINOGRADOV, M. E., KRAPIVIN, V. F., FLEYSHMAN, B. S. and SHUSHKINA, E. A. (1975). The use of a pelagic ecosystem in the mathematical model to analyze the behaviour of the ocean. *Oceanology,* **15** (2), 215–220.

VINOGRADOV, M. E., KRAPIVIN, V. F., MENSHUTKIN, V. V., FLEYSHMAN, B.S. and SHUSHKINA, E. A. (1973). Mathematical model of the functions of the pelagic ecosystem in tropical regions from the 50th voyage of the R. V. *Vityaz. Oceanology*, **13** (5), 704–717.

VINOGRADOV, M. E., KULINA, I. V., LEBEDEVA, L. P. and SHUSHKINA, E. A. (1977). Variation with depth of the functional characteristics of a plankton community in the equatorial upwelling region of the Pacific Ocean. *Oceanology,* **17** (3), 345–350.

VINOGRADOV, M. E. and MELNIKOV, I. A. (Eds) (1980). *Biology of the Central Arctic Basin*. Nauka, Moscow.

VINOGRADOV, M. E. and MENSHUTKIN, V. V. (1977) . The modelling of open-sea ecosystems. In E. D. Goldbergy, I. N. McCave, J. I. O'Brien and J. H. Steele (Eds), *The Sea.* Wiley, New York. pp. 891–921.

VINOGRADOV, M. E., MENSHUTKIN, V. V. and SHUSHKINA, E. A. (1972). On mathematical simulation of a pelagic ecosystem in tropical waters of the ocean. *Mar. Biol.,* **16**, 261–268.

VINOGRADOV, M. E. and PARIN, N. V. (1973). On the vertical distribution of macroplankton in the tropical Pacific. *Oceanology,* **13** (1), 104–113.

VINOGRADOV, M. E. and SHUSHKINA, E. A. (1976). Some characteristics of the vertical structure of a planktonic community in the equatorial Pacific upwelling region. *Oceanology*, **16** (4), 389–393.

VINOGRADOV, M. E. and SHUSHKINA, E. A. (1978). Some development patterns of plankton communities in the upwelling areas of the Pacific Ocean. *Mar. Biol.*, **48** 357–366.

VINOGRADOV, M. E. and SHUSHKINA, E. A. (1980). Particular features of vertical distribution of zooplankton in the Black Sea. In M. E. Vinogradov (Ed.), *Pelagic Ecosystems of the Black Sea.* Nauka, Moscow. pp. 179–191.

VINOGRADOV, M. E., SHUSHKINA, E. A. and KUKINA, I. N. (1976). Functional characteristics of a planktonic community in an equatorial upwelling region. *Oceanology*, **16** (1), 67–76.

VINOGRADOV, M. E. and VORONINA, N. M. (1979). The development of pelagic communities. In L. M. Breckhovsekih (Ed.), *Progress in Soviet Oceanology.* Nauka, Moscow. pp. 50–63.

VINOGRADOV, M. E. and VORONINA, N. M. (1962). The distribution of different groups of plankton in accordance with their tropic level in the Indian Equatorial currents area. *Rapp. P.-v. Réun. Int. Explor. Mer,* **153**, 200–204.

VINOGRADOV, M. E. and VORONINA, N. M. (1963). Quantitative distribution of plankton in the upper layers of the Pacific Equatorial currents. 1 (Russ., Engl. summary). *Trudy̆ Inst. Okeanol,* **71**, 22–59.

VINOGRADOV, M. E. and VORONINA, N. M. (1964). Some peculiarities of the plankton distribution in the Pacific and Indian Ocean's Equatorial currents. (Russ., Engl. summary). *Okeanol. Issled., 10 Sect. ICJ Programm,* **13**, 128–136.

VINOGRADOV, M. E., VORONINA, N. M. and SUKHANOVA, I. N. (1961). The horizontal distribution of the tropical plankton and its relation to some peculiarities of the structure of water in the open sea areas. *Oceanology,* **1**, 283–293.

VINOGRADOVA, L. A. (1970). Seasonal cycles in phytoplankton development in different water masses of the Norway Sea (Russ.). *Trudy̆ Atlant. vses. nauchno-issled. Inst. morsk. ry̆b. Khoz. Okeanogr.,* **27**, 77–96.

VINOGRADOVA, L. A. (1971). Seasonal development of phytoplankton in the Gulf of Guinea (Russ.). *Trudy̆ Atlant. nauchno-issled. Inst. morsk. ry̆b. Khoz. Okeanogr.,* **37**, 17–159.

VINOGRADOVA, N. G., BIRSHTEIN, Ja. A. and VINOGRANDOV, M. E. (1959). Vertical zonation in the distribution of the deep-sea fauna (Russ.). In L. A. Zenkevitch (Ed.), *Itogi Nauki*, Vol. I, Dostijenia Okeanologii. Publications of the Academy of Science, Moscow. pp. 166–187.

VIRKETIS, M. A. (1957). Some data of zooplankton of the central part of the Arctic Basin (Russ.). *Results of Scientific Investigations of the Drifting Station 'SP-3' and 'SP-4', 1954–1955,* Vol. I. Izdat, Morskoj Transport, Leningrad. pp. 238–342.

VIRKETIS, M. A. (1959). Materials of zooplankton of the central part of the Arctic Basin (Russ.). *Results of Scientific Investigations of the Drifting Station 'NP-4' and 'NP-5', 1955–1956,* Vol. II. Izdat, Morskoj Transport, Leningrad. pp. 133–138.

VOLKOVINSKY, V. V., ZERNOVA, V. V., SEMINA, H. J., SUKHANOVA, I. N., MOVCHAN, O. A., SANINA, L. V. and TARKHOVA, I. A. (1972). Distribution of phytoplankton in World Ocean (Russ.). *ZNIITEIRH. Min. Pyb. Choz. USSR. Expressinformation (Series 9),* **3**, 1–13.

VOLTERRA, V. (1936). Le principe de la moindre action en biologic. *C.r. hebd. Séanc. Acad. Sci., Paris,* **203**, 417–421.

VORONINA, N. M. (1964). The distribution of macroplankton in the waters of equatorial currents of the Pacific Ocean. *Oceanology,* **4**, 884–895.

VORONINA, N. M. (1966a). Some results of studying the Southern Ocean zooplankton. *Oceanology,* **6**, 681–689.

VORONINA, N. M. (1966b). On the distribution of zooplankton biomass in the Southern Ocean. *Oceanology,* **6**, 1041–1054.

VORONINA, N. M. (1970a). Annual cycle of Antarctic plankton (Russ., Engl. summary). In C. V. Beklemishev (Ed.), *Fundamentals of the Biological Productivity of the Ocean and its Exploration.* Nauka, Moscow. pp. 64–7.

VORONINA, N. M. (1970b). Seasonal cycles of some common Antarctic copepod species. In M. W. Holdgate (Ed.), *Antarctic Ecology,* Vol. I. Academic Press, New York. pp. 162–172.

VORONINA, N. M. (1972). The spatial structure of interzonal copepod populations in the Southern Ocean. *Mar. Biol.,* **15**, 336–343.

VORONINA, N. M. (1974). An attempt at a functional analysis of the distributional range of *Euphausia superba*. *Mar. Biol.*, **24**, 347–352.

VORONINA, N. M. (1977). Communities of temperate and cold waters of the southern hemisphere. Pelagic communities (Russ.). In M. E. Vinogradov (Ed.), *Oceanology*, Biology of the Ocean, Vol. 2. Nauka, Moscow. pp. 68–90.

VORONINA, N. M. (1978). Variability of ecosystems. In H. Charnock and G. Deacon (Eds), *Advances in Oceanography*. Plenum Press, New York. pp. 221–243.

VORONINA, N. M., MENSHUTKIN, V. V. and TSEYTLIN, V. B. (1980a). Production of the mass species of the Antarctic copepods *Calanoides acutus* (Russ., Engl. summary). *Okeanologija*, **20**, 137–141.

VORONINA, N. M., MENSHUTKIN, V. V. and TSEYTLIN, V. B. (1980b). The secondary mesoplankton production of the Antarctic. In *Production Primaire et Secondaire*, Franco-Soviet Colloquium, Jan. 1979. Centre Nat. pour l'Exploitation des Oceans (10), Station Marine D'Endoume. pp. 77–90.

VUCETIĈ, T. (1961). Vertical distribution of zooplankton in the Bay Veliko Jezero on the island of Mljet. *Acta adriat.*, **6**, 3–20.

WALSH, I. I. (1975). A spatial simulation model of the Peru upwelling ecosystem. *Deep Sea Res.*, **22**, 201–236.

WALSH, I. I. (1976). Models of the sea. In D. H. Cushing and I. I. Walsh (Eds), *The Ecology of the Seas*. Blackwell, Oxford. pp. 388–407.

WALSH, I. I. and DUGDALE, R. C. (1971). A simulation model of the nitrogen flow in the Peruvian upwelling system. *Investigación pesq.*, **35**, 309–330.

WALTERS, V. (1961). A contribution to the biology of the Giganturidae, with description of a new genus and species. *Bull. Mus. comp. Zool. Harv.*, **125**, 297–319.

WHEELER, E. H. (1970). Atlantic deep-sea Copepoda. *Smithson. Contrib. Zool.*, **55**, 1–16.

WHITLEDGE, T. E. (1978). Regeneration of nitrogen by zooplankton and fish in the north-west Africa and Peru upwelling ecosystems. In R. Boje and M. Tomczak (Eds), *Upwelling Ecosystems*. Springer-Verlag, Heidelberg. pp. 90–100.

WIBORG, K. F. (1940). The production of zooplankton in the Oslo fjord in 1933–34. *Hvalråd. Skr.*, **21**, 1–87.

WIBORG, K. F. (1955). Zooplankton in relation to hydrography in the Norwegian Sea. *FiskDir. Skr. (Series Havunders.)*, **11**, 66.

WICKSTEAD, J. H. (1962). Food and feeding in pelagic copepods. *Proc. zool. Soc. Lond.*, **139**, 545–555.

WINBERG, G. G. (1960). *Primary Production of Bodies of Water* (Russ.). Ed. Academy of Sciences, Misnk, USSR.

WINBERG, G. G. and ANISIMOV, S. I. (1966). A mathematical model of an aquatic ecosystem (Russ.). In *Photosynthesizing System of High Productivity*. Nauka, Moscow. pp. 213–223.

WINBERG, G. G. and ANISIMOV, S. I. (1969). An attempt to investigate the mathematical model of aquatic ecosystems (Russ.). *Trudÿ. vses. nauchno-issled. Inst. morsk. rÿb. Khoz. Okeanogr.*, **67**, 49–76.

WOLFF, T. (1960). The hadal community. An introduction. *Deep Sea Res.*, **6**, 95–124.

WOLFF, T. (1962). The systematics and biology of bathyal and abyssal *Isopoda asellota*. *Galathea Rep.*, **6**, 1–320.

WROBLEWSKI, I. S. and O'BRIEN, I. I. (1976). A spatial model of phytoplankton patchiness. *Mar. Biol.*, **35**, 161–175.

WYRTKI, K. (1966). Oceanography of the eastern equatorial Pacific Ocean. *Oceanogr. mar. Biol. A. Rev.*, **4**, 33–68.

YASHNOV, V. A. (1940). Plankton productivity of northern seas of the USSR (Russ.). *Bull. Mosk. obshchestvo ispitatelei prirodi (Series Biol.)*, Moscow, 1–85.

YASHNOV, V. A. (1961). Vertical distribution of the mass of zooplankton throughout the tropical zone of the Atlantic (Russ.). *Dokl. (Proc.) Acad. Sci. U.S.S.R.*, **136**, 705–709.

ZAIKA, V. E. (1967). The objects of studies and the boundaries of application of some conceptions in Synecology (Russ.). In *Struktura I Dinamika Vodnikh Soobshchestv I Populjatsii*. Naukova Dumba, Kiev. pp. 5–15.

ZELIKMAN, E. A. (1977). Arctic pelagic community. Pelagic communities. In M. E. Vinogradov (Ed.), *Oceanology*. Biology of the Ocean, Vol. 2. Nauka, Moscow pp. 43–58.

ZELIKMAN, E. A. and KAMSHILOV, M. M. (1960). Long-term dynamics of zooplankton biomass in the southern part of the Barents Sea and some factors affecting it (Russ.). *Trudy̆ murmansk. biol. Inst.*, **2**, 68–114.

ZENKEVITCH, L. A. (1948). The biological structure of the ocean (Russ.). *Zool. Zh.*, **27**, 113–124.

ZERNOVA, V. V. (1976) Peculiarities of seasonal variations of tropical phytoplankton in the Atlantic Ocean in a standard hydrological area at 16° N 32° W (Russ., Engl. summary). *Trudy̆ Inst. Okeanol.*, **105**, 97–105.

Marine Ecology Vol. V, Part 2
Edited by Otto Kinne


# 3. COASTAL ECOSYSTEMS

## 3.1 BRACKISH WATERS, ESTUARIES, AND LAGOONS

J. W. HEDGPETH

### (1) Introduction

The ecosystems of concern in this chapter are those of the coastal regions of the oceans—brackish waters, including brackish seas (which are part of the estuarine continuum of environments), estuaries, and lagoons. Ecological subdivisions of these regions which have been treated as ecosystems include the shallow near-shore bottoms, the sheltered tidal flats, open sandy beaches, subtidal seagrass meadows, the marshlands of estuarine regions of the temperate zone and the corresponding mangrove complexes of tropical seas, the upwelling regions of the western sides of the continents, and the coral reefs of tropical seas. In most near-shore and estuarine systems turbidity and sediment supply are necessary components, but the rocky intertidal regions of temperate seas and the massive coral reefs of the tropics flourish best remote from the influence of stream-borne sediments.

It is inevitable that marine ecologists should know much more about these regions and their ecosystems because they are not only more accessible, but are of utmost economic as well as scientific interest, and because many of them are near population centres with established universities and marine laboratories dedicated to the study of the marine and estuarine environment. Although studies of the Elbe Estuary and the Baltic Sea date from the 1860s, intensive work on most coastal and estuarine environments did not begin until the 1920s, when the Dutch began to dam off their waters to convert the drained regions to agriculture, and industrial pollution stimulated the study of estuaries in England. Since the end of the Second World War studies of all sorts of coastal and estuarine environments increased at a rapid rate. The first stock-taking of this research on an international basis was made at an International Symposium held in the United States, at Jekyll Island, Georgia (LAUFF, 1967).

Since that time, the flood gates have been opened, stimulated by substantial research grants from public agencies in the United States, Canada, and many parts of Europe, and intensive activities by the research staffs of the various academies in the USSR. In the United States, the Estuarine Research Federation was established in 1971, and has since carried on the International Symposium with as many as 700 participating members every other year. In 1978, the Federation assumed publication of *Chesapeake Science* as its official journal and 'Estuaries' as a new serial. In England, the Estuarine and Brackish Water Sciences Association was organized in 1971 and established as its journal '*Estuarine and Coastal Marine Science*' (new title: '*Estuarine Coastal and Shelf Science*'), with an international board of editors representing most of the research centres in this field. The association also publishes separate monographs on estuarine and coastal topics. Important contributions to estuarine ecology are published in the international journal *Marine Ecology—Progress Series*, and extensive papers on estuarine and coastal topics appear in the several 'annual review' and 'advanced series', as well as in journals devoted to marine biology, oceanography, and limnology.

In addition to all this material there is the voluminous 'grey literature' of reports and documents by government agencies, UNESCO, and private concerns involved with projects in estuaries and coastal regions. Most of this material is reproduced on office equipment without editorial control and duplicated in limited quantities. Many of these documents are ephemeral or based on extant published literature, but some of the reports contain original information and data. Unfortunately, many of these documents are generally unavailable or obscurely catalogued by libraries, and extensive citation of them in the formal literature can lead to frustrating and futile searches. Eventually some of these items are published in available form.

It is impossible now for any one person to absorb all the material published about estuaries and coastal zones in the last two decades, and the cost of keeping up with the expensive separate volumes, of varying scientific standards, would bankrupt all but the greatest libraries.

In spite of all this activity no special term has been proposed for the study of estuaries, an indication, perhaps, of the complexities of the subject. Students of estuarine and brackish waters seem content to call themselves marine ecologists, oceanographers, engineers, or limnologists who happen to be interested in estuaries, lagoons, and brackish waters according to their specialized approaches. Since no need has been expressed for some special word to describe this activity, it seems best not to suggest one. Such organizations as the Baltic Marine Biologists and Baltic Oceanographers indicate common geographical affinities of research interests, but propose no new words to describe themselves.

The study of ecosystems began in the tidal flats of northern Germany, with the recognition by the fisheries biologist KARL MOEBIUS that there were natural groupings of organisms, interrelated and associated with certain environmental factors or conditions. He suggested that the oyster banks of the 'Waddensee' in the shallow waters of Schleswig-Holstein were a natural community or biocoenosis (MOEBIUS, 1877). His term, a hellenization of the German 'Lebensgemeinde' or 'Lebensgemeinschaft' endures to this day. The now commonly used term 'ecosystem' is a broader, more flexible concept. In the words of its creator, the English botanist, A. G. TANSLEY (1935, pp. 299–300), ecosystems

> 'are of the most various kinds and sizes. They form one category of the multitudinous physical systems of the universe, which range from the universe as a whole down to the atom . . . Some of the systems are more isolated in nature, more autonomous, than others. They all show organisation, which is the inevitable result of the interactions and consequent mutual adjustment of their components.'

TANSLEY (1935) considered the ecosystem, as he defined it, to comprise groups of plants or units of vegetation, rather than single species. In the sea, especially the shallow coastal regions and in estuaries, plants, in many cases, especially of the 'level bottoms' and rocky intertidal regions, are the short-lived, often seasonal members of the natural complex in the marine environment, whereas animals are the long-lived, stable members of the characteristic marine community. The most characteristic organisms of the sea are the great variety and numbers of animals that subsist on microscopic floating plants, eggs, and larvae and material suspended in the water, especially bivalve molluscs, but also the hosts of such smaller specialists as barnacles, bryozoans, hydroids, and corals (JØRGENSEN, 1966). Of course, it must be pointed out that the oyster banks of MOEBIUS (1877) represented a relict, dying system even in his time, for already the beds of the Danish region of Schleswig were depleted, and by 1895 they were gone from the southern part of Schleswig-Holstein (HAGMEIER and KÄNDLER, 1927). Today there is hardly a trace of oyster shells (CASPERS, 1950) to demonstrate that this was the region in which a fundamental idea of ecology was developed. This does not, of course, invalidate the idea, for the oyster banks of the northern Wadden regions were a survival of warmer climates and warmer seas, and their disappearance was inevitable with climatic changes and reduced reproductive potential.

MOEBIUS (1877) did not attempt to count the members of his biocoenosis. The introduction of numerical methods in marine ecology was made by VICTOR HENSEN in 1884 with the first attempts at quantitative plankton sampling in the North Sea and North Atlantic Ocean (HENSEN, 1887). Inspired by HENSEN's example, DAHL began the first counts of bottom organisms per unit area in 1893 (DAHL, 1921). C. G. JOHANNES PETERSEN introduced these methods in the shallow estuarine seas and bays of Denmark to ascertain the amount of food available to the important commercial fish of Denmark. He carried these studies on for many years, culminating in a series of papers on the valuation of the sea (PETERSEN, 1918). In the course of these studies, PETERSEN noticed that his samples of worms, molluscs, echinoderms, etc. from the bottom of Danish waters had recurrent similarities of species composition characteristic of the various types of bottom in the different areas. He recognized these groupings as 'communities', with less well-defined groupings as 'associations'. PETERSEN considered his communities to be of a lower level of complexity than the biocoenosis, and eventually summarized his researches in his famous diagram of the trophic structure of the sea.

In recent years the distinction between biocoenosis and ecosystem has become blurred (Volume V: PERES, 1982). In the few references to 'ecosystem' in the First Estuary Symposium (LAUFF, 1967) there was no attempt to define or restrict the term. MARGALEF (1967) equated the biocoenosis in its biotope with the ecosystem (see also Volume IV). In Volume IV of the present treatise on Marine Ecology, 'ecosystem' has been defined as

> 'an integrated spatial entity of interacting and interdependent biotic (living matter) and abiotic (non-living matter) components which are linked by energy flow and material cycling, by exchange between biotic and abiotic and among biotic constituents, as well as by homeostatic control mechanisms. The boundaries of an ecosystem, or those between adjacent ecosystems, depend on the specificity of their structures (e.g. species composition, habitat properties) and functions, but often also on subjective criteria such as convenience, conceptual perspective, or choice of methods. Most of the ecosystems studied are themselves components of larger ecological systems' (KINNE, 1978, p. 1).

LEWIS (1980) accepted ecosystems as a broad, usefully vague term which indicates that we must look at the shore primarily as ecologists. But there are many kinds of ecologists, as any casual examination of the abundant literature on coastal and estuarine ecology demonstrates. The ecologists interested in ecosystems work in the tradition of PETERSEN, who was the first to attempt a reckoning, imperfect as it was, of nature's budget. The essence of the ecosystem approach is the understanding of the processes, not the production of balance sheets. This has been well stated by MANN (1980).

## (2) Categories and Classifications

Brackish water and estuarine systems are best known from the northern hemisphere. The majority of the continental land mass is north of the equator, and there are more rivers and embayments, despite the fact that two of the world's largest river systems are in South America and another one in Africa, for the most part below the equator. There are three great low-salinity seas in the northern hemisphere, relics of glaciation and pluvial stages of the past: Hudson Bay (520 000 km$^2$), Black Sea (430 000 km$^2$; or 461 000 km$^2$ including the Sea of Azov), and Baltic Sea (384 000 km$^2$). Of these, Hudson Bay has an average summer surface salinity of 23‰ S and is ice-bound for months every year. We know much more about the Baltic and Black Seas because they are more accessible and more scientists live along their shores. Another large relict sea is the completely isolated Caspian Sea (436 000 km$^2$), with an average surface salinity (as of the early 1970s) of 12 to 13‰ S. This sea falls outside the brackish–estuarine continuum, and cannot be considered in detail here. Since the summaries by ZENKEVITCH (1959, 1963), the Caspian Sea is increasing in salinity and its biota is undergoing changes because of diversions from the Volga River, and it may be some time before an adequate summary of the state of affairs is available.

## (3) Brackish Waters

According to the *Oxford English Dictionary*, the word 'brackish' comes from an Old or Middle Dutch word, 'brak' meaning perhaps fresh water 'broken' or spoiled by salt, therefore worthless. This became 'brakwasser' and 'brackwasser' in modern Dutch and German. The French 'saumâtre' (from Latin 'salmacidus') implies water of salt concentrations up to that of sea water. In Russian, the word is 'solonovatii' ( содоноватый ). None of these words is precise with respect to the degree of saltiness, and in English the word has often been used for waters of greater salt concentration than that of sea water. In the ecological sense, especially in Europe, brackish waters are relatively stable environments of salinities lower than the adjacent ocean waters. Two entire seas, the Baltic and the Black Sea, are considered brackish waters, although the Black Sea is often considered marine in Russian literature. In addition to these large seas whose lower salinities are maintained by narrow connections with the ocean, Europe has many embayments and backwaters that form series of brackish waters, such as the 'fjords', bays, and broads of Denmark, the 'limans' (actually, the word диман means estuary) along the northwestern edge of the Black Sea, and the 'étangs' of southern France. Holland, once one of the richest brackish water regions, has diked off most of its brackish water environment and converted it to farmlands. In North America, the greatest development of brackish or low salinity waters is along the northern coast of the Gulf of Mexico, where they are known as back or inner bays, or 'lakes'. In general, brackish water regions are best developed in regions of subsiding shores with small or moderate tidal ranges, including deltaic coasts (e.g. The Netherlands), where rainfall or drainage patterns produce an excess of fresh water increment over evaporation. In xerophytic tropical regions such as Mexico, where evaporation predominates during much of the year, embayments of the sea tend to develop salinities higher than those of the adjacent ocean.

In the years before closing off the 'Zuider Zee', the Dutch, under the leadership of H. C. REDEKE, studied their brackish waters intensively, and continued to study the changes in the 'Zuider Zee' after its separation from the North Sea in 1932. The name 'Zuider Zee' was changed to the 'Ijsselmeer,' and the changes in the flora and fauna of its waters were studied as the waters became less saline. Some large areas have been diked off and converted to polders.

SEGERSTRÅLE (1959) provided a historical review of the classification of brackish waters up to that date. In this review he considers that brackish water biology, 'as a science', began with REDEKE's significant paper, 'Zur Biologie der niederländischen Brackwasser typen. Ein Beitrag zur regionalen Limnologie' (REDEKE, 1922). In this paper, REDEKE put forward a classification of brackish waters according to chlorinity content, based on his intimate knowledge of faunal distributions in the brackish waters of Holland. The terminology of this classification was based on an earlier classification of fresh waters in which the terms 'oligohaline', 'mesohaline', and 'polyhaline' were used. REDEKE moved these terms further seaward, and they have since been widely used. REDEKE's system was modified by the Finn, VÄLIKANGÅS (1933), who expressed the concentrations in salinity rather than chlorinity and made some modifications. This joint classification is usually referred to as the 'Redeke–Välikangas system'. In the context of limnologists looking toward the sea, 'marine water' was regarded as 'ultrahaline'. Researchers in environments where evaporation exceeded fresh water input, either by reduced stream flow or precipitation, applied such terms as ultrahaline and hyperhaline to salinity conditions higher than those of sea water, that is, above 38‰ S, and usually above 40‰ S.

Various workers have attempted their own classification; many of these are summarized by SEGERSTRÅLE (1959). It is obvious from these various schemes of salinity classification that most of these attempts reflect the environments and distributions of plants and animals familiar to the individual researchers. In this context, the information presented mainly constitutes a guide to the types of environments concerned. For example, ZERNOV (1913), who wrote one of the great monographs of marine ecology, considered the marine region to extend from 15 to 47‰ S in his 'General Hydrobiology', and classified it as 'polyhaline', a term applied in most other characterizations to waters of suboceanic salinity up to about 18 or 20‰ S.

Both REDEKE and REMANE derived their classifications from the numbers of animals in the various regions from fresh water to the sea. REMANE (1933) summarized his observations of animal distribution according to salinity in a famous curve, often criticized as being a subjective diagram without objective data, but reproduced in almost every marine biology and ecology text. Later REMANE (1971) has been careful to emphasize that salinity is not the only factor governing the distribution of organisms, since temperature, substrate, depth, and other factors also influence occurrence (KINNE, 1964; Volume I: GESSNER and SCHRAMM, 1971; KINNE, 1971; ALDERDICE, 1972).

Originally, REMANE (1933) did not include any indication in his curve that there is a group of completely euryhaline organisms (mostly ciliates and diatoms) that are found in every range from fresh water to near saturation levels. HEDGPETH proposed inclusion of these by a black line at the base of the curve (HEDGPETH, 1967). In the English edition of his book on brackish waters, REMANE included this in an odd way that suggested some change in abundance through the gradient rather than an almost constant, or very low, number of species. Although including 'brine species', KINNE (1971: Volume I, Fig. 4-73, p. 824), while considering this holo-euryhaline group in his text (p. 823), omitted it from his revised graph. It seems appropriate therefore to resubmit this revised diagram herewith (Fig. 3-1).

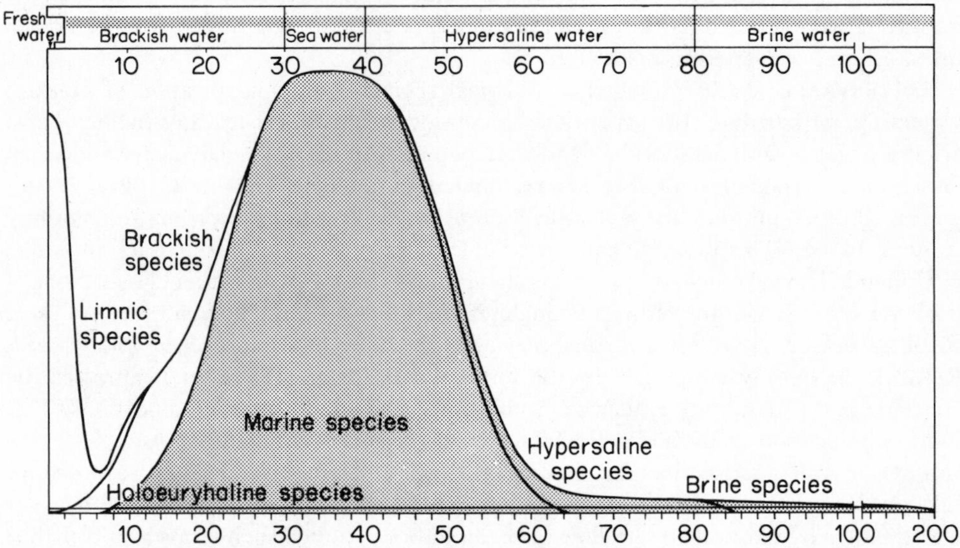


Fig. 3-1: Quantitative relations between aquatic invertebrate species occupying fresh, brackish, sea, hypersaline or brine waters. For each salinity (0–200‰ S) the relative number of species is indicated by the vertical extension of the respective areas. Rough estimations. (Based on REMANE, 1934; HEDGPETH, 1959; KINNE, 1971; after KINNE, 1971; modified; reproduced by permission of Wiley, Chichester)

Salinity has been, and will probably continue to be, the most useful single factor on which to base the classification of various aquatic regions, from those containing hardly any salt at all to those of saturated brine. In the 25 yr after REDEKE's scheme was advanced, several varying schemes of classification were proposed, including extensions of the system into hypersaline coastal lagoons such as those of southern France and the Gulf Coast of the United States.

In his review of 25 yr of the study of European brackish waters, SEGERSTRALE (1958) concluded with the comment that 'the time seems ripe for the creation of a system which covers the whole salinity range of the sea', and urged that an international committee be nominated to revise and integrate the system of classification. A committee given this charge did meet, under the auspices of the International Union of Biological Sciences, in Venice, during the Easter week of 1958. The result of this symposium was the so-called 'Venice System' (CASPERS, 1959a,b). The full proceedings of the symposium were published as a supplement to Volume 11 of *Archivio di Oceanographia e Limnologia* (1959). As could have been predicted, the classification recommended was a compromise that recognized each participant's particular area of concern, so that subdivisions, especially in the lower salinity ranges, were included.

This double-layered system seems to have met with no particular resistance, and the terminology, at least of the major divisions, with the possible exception of 'mixohaline', has been accepted. As one of the members of the symposium, I was not pleased with this word, and have found no use for it myself, although it could be used as a category for estuaries as opposed to more stable brackish water regions.

Almost immediately, the classification was criticized as inadequate for universal application, since it did not take temperature and biological factors, especially in tropical regions, into consideration (HARTOG, 1960). Later, HARTOG (1964) declared the Venice System 'not fit for a biological classification' and proposed eight categories in a typological system:

(1) Brackish seas
(2) Stream mouths in a tideless sea
(3) Estuaries
(4) Small streams in tidal regions subject to rapid and drastic salinity changes ('Schockbiotopen')
(5) Supralittoral pools
(6) Lagoons and isolated backwaters
(7) Tidal zones
(8) Coastal groundwaters

While his paper provoked considerable discussion, enthusiasm was restrained, and there seem to have been no converts.

A scheme has been developed for Australian waters by BAYLY (1967) on the basis of permanency and stability of the aquatic environments, and the degree of salinity variation within each category. This scheme includes both coastal and inland or 'athalassic' waters. Two categories of salinity variation within these typological groups are recognized: homoiohaline (narrow range) and poikilohaline (broad range). It is probable that application of this somewhat restricted classification on a 'universal' basis would produce subdivisions as complicated as HARTOG's system.

As REMANE remarked in his comments on HARTOG's (1960, 1964) presentation, there were two groups at the Venice Symposium, one concerned with salinity changes and the other with stable salinities; but it was possible to be on both sides since the two aspects are combined in nature. The Venice System was a compromise, born of the inevitable biopolitical nature of the meeting. It was not emphasized that a classification scheme that attempted to meet all possible examples known is not really a classification that among other uses provides acceptable adjectives to deal with the complexities of the continuum of nature. We need categories to handle knowledge, and our need is increasing as our computer technology, with its digital processing of information, increases.

A different scheme has been proposed by CARPELAN (1978) which is an expansion of KOLBE's (1927) halobien system, based on the distribution of diatoms according to the abundance of species. CARPELAN bases his system on the distribution of 86 species of diatoms found in salinities from 2 to 96‰ S in coastal lagoons in southern California. He recognizes seven categories of distribution, including a holo-euryhaline group found in equal abundance throughout the salinity range, and a hyperhaline group (of two species) in salinities from 52 to 100‰ S with a lower tolerance to 30 to 37‰ S. This proposed change in the 'concept of salinity boundaries' is intended 'to be consistent with current genetic, ecological and evolutionary theory'. This places the emphasis on classification of the environment on the tolerances and occurrences of species, rather than on groups of species comprising a characteristic biota. Such a scheme based on distributions and tolerances of groups of species in a single taxonomic category also invites specialists in other groups to attempt classifications that reflect evolutionary trends and genetic theory.

However, while ecosystems and communities are the products of interaction with their environment, the concept of an ecosystem as an evolving complex under natural selection on the analogy of a species implies a more coherent gene pool and possession of hereditary characteristics than can be demonstrated to exist. Obviously, communities and ecosystems change with time, but the action of the changing factors is effected on the separate species rather than on the system as a whole. The mechanism of change in ecosystems is different, reflecting the response of individual species to the modification, loss, and possible restoration of niches or functional roles within the system.

The salinity classification developed for stable brackish waters on the basis of the observed occurrence of biota is difficult to apply to estuarine conditions, and many estuarine workers ignore the terms or apply them only in the most general way. CASPERS (1959a) identified several regions in the lower Elbe with brackish water categories, indicating their seasonal changes in the river (Volume I, p. 960: Fig. 4–138).

In his exhaustive review of the distribution of species in the estuarine environment, WOLFF (1973, p. 176) concludes that 'a biological subdivision of the estuarine environment based on salinity is impossible'.

One solution to this problem is to ignore the biological characteristics and prepare diagrams combining the two most significant factors involved—temperature and salinity—according to their monthly means (Figs. 3-2, 3-3). These 'hydroclimagraphs' (HEDGPETH, 1951, 1953) present the seasonal variations of means in an easily understandable manner, but require a considerable body of data that is not available for many parts of the world. They have been used for Baltic localities and for Danish waters by RASMUSSEN (1973).

Presentation of averaged data for each month, or even the extremes, does not express a real condition, but simply provides an impression of the type of environment concerned. The monthly means or extremes of temperature and salinity do not always fall in the same month. Presentation of data as actually measured *in situ* may produce very different polygons, that give a better impression of the variables actually experienced by organisms, especially in near-shore temperature localities. PETER GLYNN and DONALD P. ABBOTT (pers. comm.) prepared a diagram (Fig. 3-4) of temperature and salinity conditions at Pacific Grove, California (USA), a locality of strong upwelling, showing the conventional polygon of means lost within a cloud of data points (excluded from the centre of Polygon A for clarity) and the maximum and minimum conditions (B,C) well beyond the limits of the mean. The minimum temperature extremes are typical of the coasts of Oregon and California during upwelling periods. Unfortunately, the kind of data necessary for such a presentation are not generally obtained. For earlier years of records before recording instruments became generally available, many of the temperatures were taken by immersing a thermometer in a bucket at noon, and determining salinity with a hydrometer at the same time.

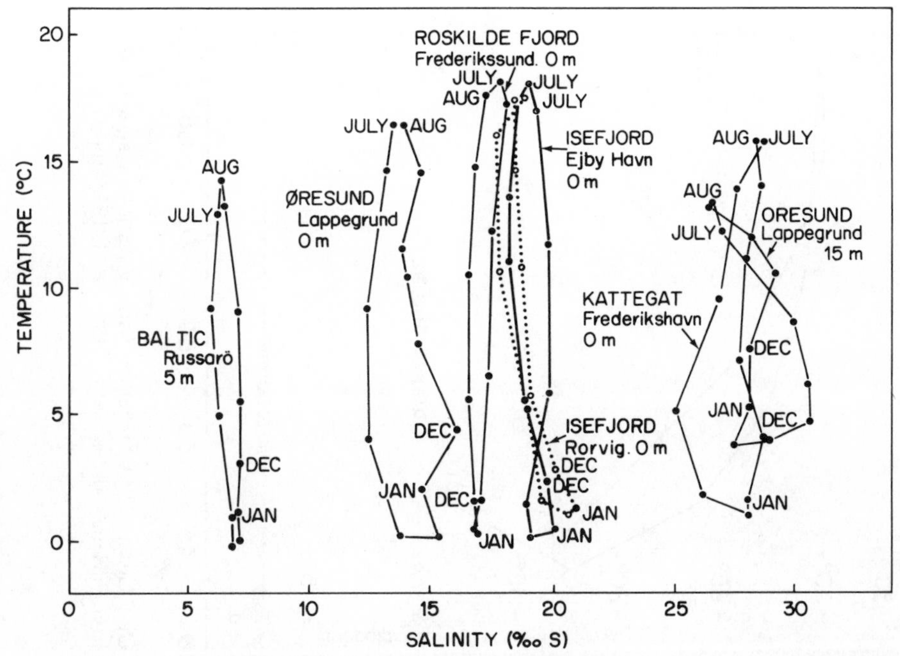


Fig. 3-2: Hydroclimagraphs showing the annual cycle of the mean monthly salinity and temperature (averages of 1957–1966), from Kattegat, Øresund, and Isefjord complex, calculated from measurements published in *Nautical Meterological Annual*. Values from Baltic Sea are from 1932–36. (After RASMUSSEN, 1963; reproduced by permission of the Marine Biological Laboratory (University of Copenhagen), Helsingør, Denmark)

## (4) Estuaries and Lagoons

Estuaries are strongly influenced by geography and tidal flows. They constitute interface systems between rivers and ocean, often characterized by complex water masses, changing salinities, large seasonal variations and migrating organisms, by high silt loads, complex chemistry, and, increasingly, by system-foreign deformations due to human activities. Estuaries are difficult to classify, and there is no common ground where those interested in the various aspects can agree upon the kinds of environments that are estuaries. A definition like that proposed by KETCHUM (1951, p. 199), 'an estuary may be defined as a body of water in which the river water mixes with and measurably dilutes sea water', could without much straining refer to the lenses of diluted sea water that drift for hundreds of miles away from the mouth of the Amazon or to the measurable dilution of waters on the continental shelf of North America between Cape Cod and the Chesapeake Bay. Nevertheless, in his last paper, 30 yr later, KETCHUM (1983, p. 1) still favoured his 1951 definition. However, the northeast Pacific coastal region, especially the Gulf of Alaska, is a region of precipitation in excess of evaporation, and the resulting salinity structure of the coastal waters had been suggested as an estuarine analogy (TULLY and BARBER, 1960). Recent studies by ROYER (1981, 1982) indicate that the volume of fresh water added to the area may exceed $23 \times 10^3$ m$^3$ s$^{-1}$. If this water, known as the Alaska Coastal Current, were from a point source it would constitute one of the largest rivers in the world, ranking between the La Plata System and the Lena. A similar condition on a smaller scale occurs in Faxa Bay, Iceland, where

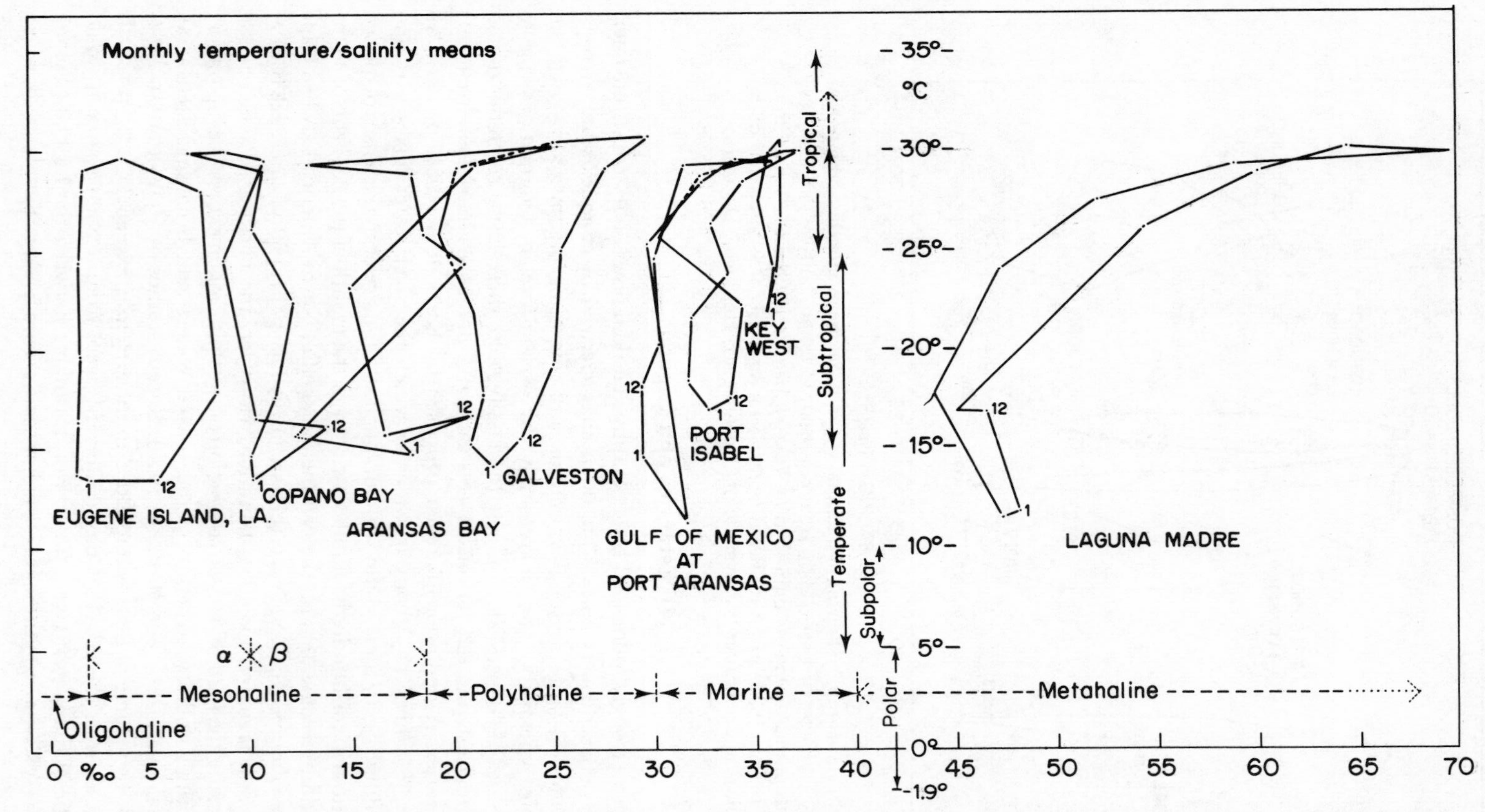


Fig. 3-3: Mean hydrographic climates for various stations and areas on the Gulf Coast (USA) together with VAUGHAN's system of temperature zonation and REDEKE's salinity spectrum classification. Compiled from various sources. (After HEDGPETH, 1953; reproduced by permission of the author)

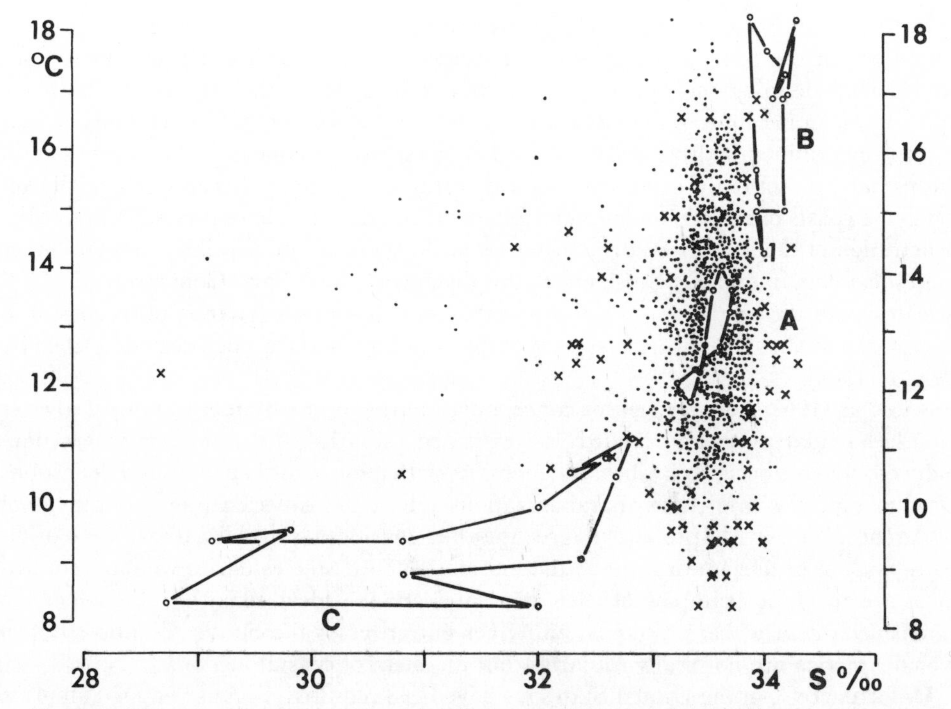


Fig. 3-4: Diagram of temperature and salinity conditions at Pacific Grove, California, USA. (Based on data provided by GLYNN and ABBOT; Original)

several small streams and coastal current from the south produce estuarine characteristics in the bay.

Not everyone is satisfied with the concept of an estuary as the tidal reach of a river as it meets the sea, bounded by the geomorphological features of the environment. This was obvious at the time of the First International Symposium on Estuaries held on Jekyll Island, Georgia, in 1964 (LAUFF, 1967).

The Symposium was opened by PRITCHARD (1967) who cited five different dictionary definitions of 'estuary' and then proceeded to offer his own: 'An estuary is a semi-enclosed coastal body of water which has a free connection with the open sea and within which sea water is measurably diluted with fresh water derived from land drainage'. This same definition was already published in a paper by CAMERON and PRITCHARD (1963, p. 306). It attempts to define an estuary as an environment in which the fresh water dilutes the sea water and 'provides the density gradients which drive the characteristic estuarine circulation patterns'. As restricted, this definition withdraws from considering PRITCHARD's (1955) earlier terms, 'negative' and 'neutral' estuaries, as more applicable to lagoons, since they are environments in which there is no appreciable fresh water influx as contrasted with sea water or in which circulation is so restricted and influx lacking or reduced by evaporation that salinities are higher than those of the ocean water. Nevertheless, these basic patterns of salinity distribution are useful in demonstrating the essential difference between estuaries *sensu stricto* and lagoons. PRITCHARD's definitions are an attempt to 'lead to a practical limitation of the term'. One is tempted to suggest that the preferred definition would be that which includes only conditions that conform most nearly to the equations or mathematical models of the definer.

In the second paper of the Jekyll Island Symposium, CASPERS (1967, p. 7) reviewed definitions in the context of biological characteristics of estuaries. He was more concerned with defining the limits of an estuary than defining the estuary itself. In his opinion, the upper limit of the estuary is not defined by salinity, but by tidal forces, that is, 'it is determined hydrodynamically rather than hydrochemically'. The lower end of the estuary is 'fixed by geomorphological features'. In many streams, especially on low-lying coasts or regions of flatlands and low relief, tidal action occurs miles above the penetration of the salinity wedge. In the Amazon there are perceptible tides 600 miles from the ocean, but no one considers all this distance as an estuary. Conversely, the Gulf Stream is not thought of as an estuary, although it has hydrodynamic properties of 'a river in the ocean'. The riverine aspects of the Gulf Stream have since been described by PRATT (1966).

CASPERS (1967) does restrict his concept of estuaries as environments 'limited to river mouths in tidal seas', and regards as 'extended estuaries' the situations where tide-induced currents occur in fresh water zones. As with most of the concepts and definitions of what estuaries are, these characterizations reflect the experience of the individual researcher. We can all, probably, agree that like considering the climate we were originally born in or acclimatized to as the 'normal climate', the estuaries we have studied most are the best examples of their kind and are the ideal estuaries. To some, the extensive system of bays, sounds, and river entrances of the South Atlantic coast of North America are not really estuaries, but a system of coastal lagoons.

Unfortunately, in the United States we have been required, because of our Estuarine Sanctuary Act, to select for consideration estuaries 'representative' or 'characteristic' of each physiographic region of the country, and for obvious political reasons must also consider the mouths of streams flowing into the Great Lakes as 'estuaries' for Sanctuary status. Such political considerations aside, estuaries are environments where estuarine conditions can be recognized on the basis of salinity, sedimentation, and nutrient conditions, and the presence of certain types of organisms, such as salt marsh plants, sea grasses, mangrove swamps, and various animal forms, especially massive populations of burrowing and sedentary bivalves, migratory crustacea, and fishes, many present only at certain stages of their lives, and so on. Very few if any of the animals that predominate in these low-lying sheltered situations can live permanently in fresh water, although most of them can live in the sea, but thrive best in the sheltered conditions behind barrier islands, offshore bars, bays, and river mouths of various shapes and sizes. We can recognize estuarine conditions even if we cannot always agree on how to define an estuary, or, for that matter, a lagoon.

Lagoons, like estuaries, are often associated with streams, but are more characteristic of regions where stream flow and sedimentation are intermittent and wave action builds up sand bars that isolate their waters from the sea. Under such conditions, tidal action is effected in the upper levels and a salinity wedge does not occur. These conditions are more characteristic of dry climates such as southern Texas, western Mexico, and western Australia, but coastal lagoons are not restricted to semi-arid or desert regions. They also occur along the Arctic Ocean, the Black Sea and the northern Adriatic, and South Africa.

On the northeast Pacific coast, especially northern California, wave action may prevail over the drainage of small streams to build up substantial bars; in some cases, these bars are permanent and the result is a fresh water lagoon. In many small streams along the California coast, summer flows of streams are inadequate to prevent formation of bars across the mouth; these bars are usually broken by the first heavy rains of the winter or 'rainy season'. In such environments, fresh water may seep through the bar and form substantial fresh water lenses in subtidal sands. Because of the lack of large streams, lagoons are shallow, and usually the sediment is predominantly sandy, instead of having the fine silts characteristic of estuarine environments. The origin, dynamics, and productivity of coastal lagoons were reviewed and to some extent summarized in the symposium held in Mexico City in 1967 (AYALA CASTAÑARES and PHLEGER, 1969).

As FAIRBRIDGE (1980) emphasizes, PRITCHARD's (1967) definition of an estuary as 'a semi-enclosed coastal body of water which is in free connection with the open sea and within which sea water is measurably diluted with fresh water derived from land drainage' describes estuaries familiar to PRITCHARD; 'but has totally lost the original and critical tidal and river qualifications' of the classic definitions. To put it another way, the estuarine bodies of water along the coastal plain of the Mid- and South Atlantic regions of North America do not fit the traditional definition of estuaries, which according to FAIRBRIDGE requires relationship to a stream valley with 'tidal and river qualifications'. To meet the requirements of a classification applicable on a world-wide scale, FAIRBRIDGE proposes a scheme of 'physiographic' types:

(i) High relief estuary, U-shaped valley profile: *fjord*; related, but more subdued relief: *fjärd, firth, sea-loch.*

(ii) Moderate relief estuary, V-shaped profile, winding valley: *ria; aber** (western Britain); special limestone karst type: *calanque, cala*.

(iii) Low relief estuary, branching valleys, funnel-shaped plan: *coastal plain estuary* (open); flask-shaped, partly blocked by bar or barrier island (with lagoons or sounds): *coastal plain estuary* (with barrier).

(iv) Low relief estuary, L-shaped plan, lower course parallel to coast: *bar-built estuary*.

(v) Low relief estuary, seasonally blocked by longshore drift and/or dunes, with/or without eolianite bar: *blind estuary*.

(vi) Delta-front estuary, ephemeral distributary: *deltaic estuary*; in interlobate embayment: *interdeltaic estuary*.

(vii) Compound estuary, flask-shaped ria backed by low plain: *tectonic estuary*.

FAIRBRIDGE's (1980) diagram of these basic physiographic types is reproduced here as Fig. 3-5. In a tabular summary of selected rivers of the world, FAIRBRIDGE designates the estuaries according to three categories—delta, funnel, and barrel.

With the exception of the tidal river requirement, FAIRBRIDGE's (1980) classification includes lagoons as well as estuaries. His categories 4 and 5 are identical with those for lagoons, especially no. 5 for 'blind estuary' with stagnation during dry seasons. Many lagoons, however, are not closed off so completely, and stagnation (which implies reduced oxygen content) does not occur. The essential difference between a lagoon and an estuary is that processes of aggradation, especially the buildup of sand to form a barrier beach or sand spit, are dominant in a lagoon, whereas in an estuary stream flow is the dominant force that prevents the buildup of a barrier system. Most lagoons are

*According to Welsh dictionaries, *aber* is estuary; in place names it is combined with the name of the stream, e.g. Aberystwyth, Aberdovey. In Breton, however, the same word is also applied to rocky bays without tributary streams.

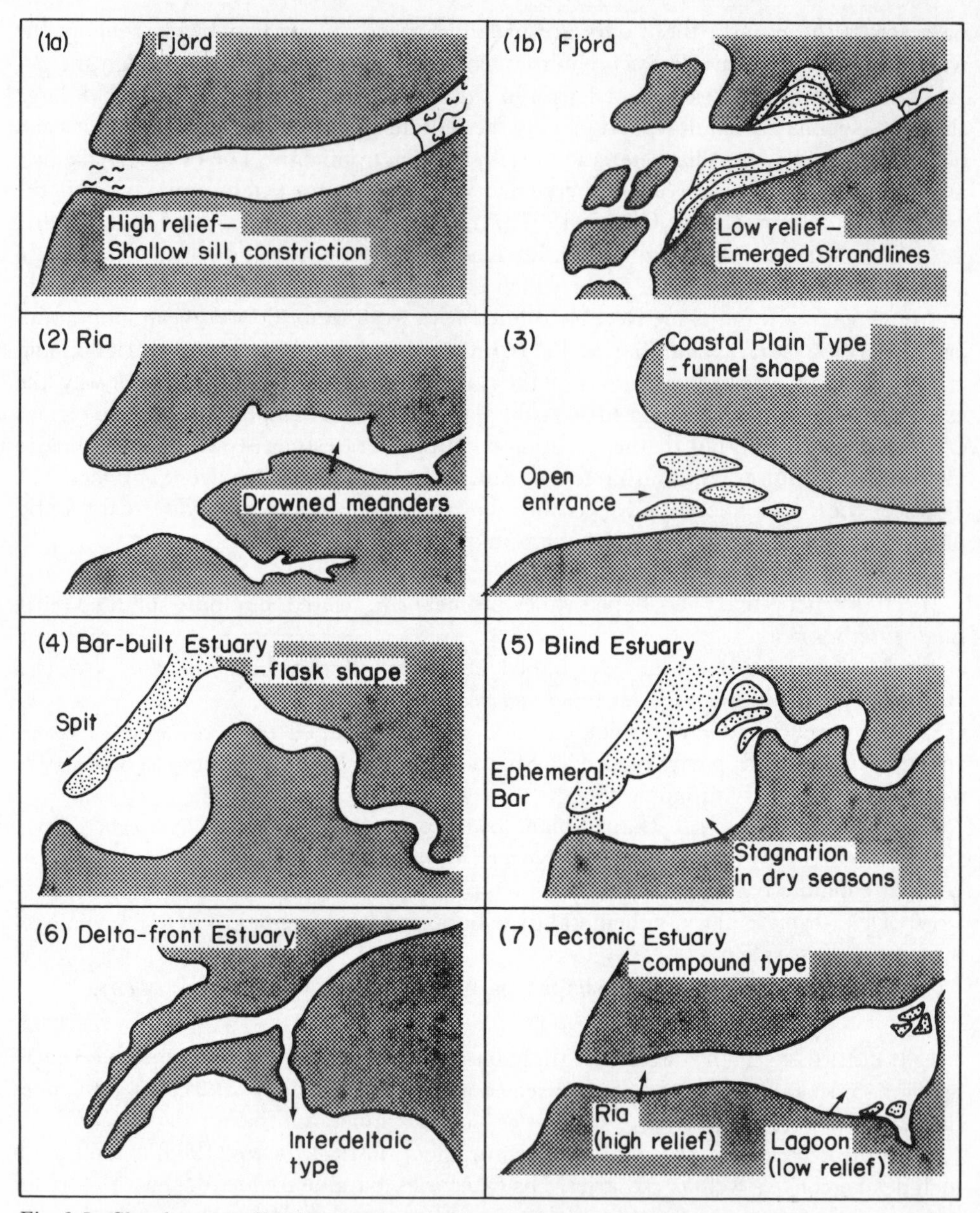


Fig. 3-5: Sketch maps of basic estuarine physiographic types. Hydrodynamic characteristics not considered; discharge, tidal range, latitude (climate), and exposure all play important roles in modifying these examples, besides long-term secular processes such as tectonics and eustasy. (After FAIRBRIDGE, 1980; reproduced by permission of Wiley, Chichester)

associated in one way or another with rivers, but the flow may be intermittent, or reduced almost to non-existence.

A very detailed classification of coastal lagoons is offered by LANKFORD (1976) as a result of his study of Mexican coastal lagoons. The primary basis of this classification is the historical events leading to the formation of barriers that separate the coastal body of water from the sea. Five types are recognized (Fig. 3-6): (i) Differential erosion—depressions formed by non-marine processes during lowered sea level, i.e. drowned valleys and river mouths, canyons, or karat depressions. (ii) Differential terrigenous sedimentation—lagoons associated with fluvial deltaic systems that occur most frequently along deltaic plains and some may be very young (only a few hundred years). (iii) Barred

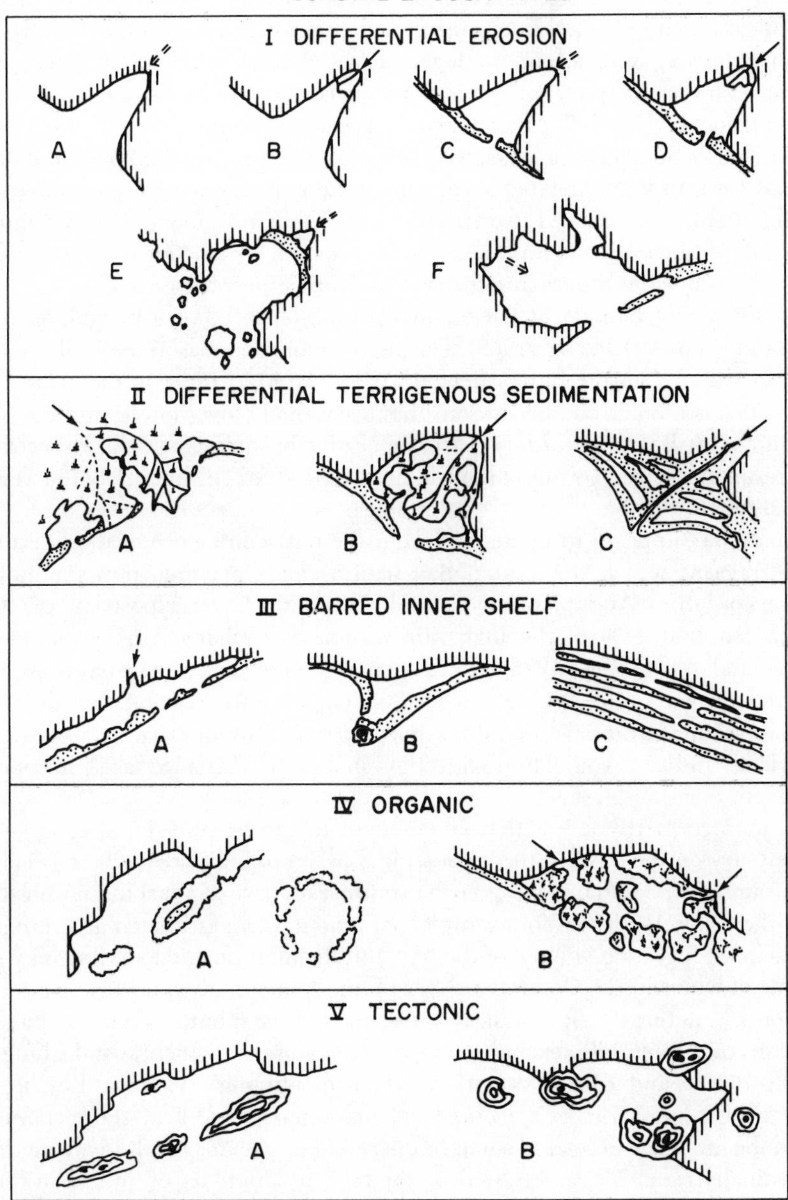


Fig. 3-6: Coastal lagoon types and subtypes in Mexico; numerals and letters refer to classification in text. (After LANKFORD, 1976; reproduced by permission of Academic Press, Inc., New York)

inner shelf—the inundated inner margin of the continental shelf, protected by wave/current produced sand barriers. Usually the major orientation of the lagoon is with the long axis parallel to the coast. (iv) Organic—the depressions are formed by growth of algae, corals, or mangroves. (v) Tectonic—formed by faulting, folding, or volcanic activities; these are divided into structural and volcanic categories (LANKFORD, 1977, pp. 196–197).

While LANKFORD (1976) gives examples of the various types occurring in Mexico, this classification can also apply to other coastal areas. Physiographically, the coast of northern California, for example, is a lagoon coast. There are several large lagoons, one of them now almost completely fresh and stabilized by a highway causeway. The most interesting lagoon is Humboldt Bay, formerly associated with two rivers which no longer flow into it. It lies on a tectonic coast with a history of recent earthquakes, and diversion of the Eel River is probably related to earthquake activity, although the very low summer flow of this stream has reduced its capacity to maintain an estuary. Heavy winter flows extirpate marine or estuarine organisms from its tidal reach, but during the summer period the coastal water moves into the river. In this respect the estuary of the Eel River resembles the Vellar Estuary at Porto Novo (India) on the Bay of Bengal, as described by DYER and RAMAMOORTHY (1969). During the monsoon season the Vellar Estuary is occupied by water from the Bay of Bengal. As for the Mad River to the north of Humboldt Bay, this is a much smaller stream that occasionally overflowed into the northern arm of Humboldt Bay (Arcata Bay) A levee or dike has prevented this in recent years, and as a result of the heavy rains of the winter of 1981–1982, the mouth of the Mad River has shifted north.

Humboldt Bay appears to be derived from the past confluence of the Eel and Mad Rivers; at present it is a Compound Bay with a single opening, now maintained by breakwaters or jetties. About half the high tide area of the bay is exposed at low tide, and it is estimated that 44% of the high tide volume is exchanged on each tidal cycle (HARDING and co-authors, 1978). Its principal source of fresh water is from small streams and surface runoff. The bay is well mixed, without seasonal stratification, and the phosphate concentrations 'resemble values for rich, upwelled waters' (*op. cit.*). These characteristics indicate that Humboldt Bay should be classified as a marine dominated lagoon.

It does not necessarily follow that an embayment can be considered a lagoon rather than an estuary on the basis of the ratio of tidal prism to high tide volume because that ratio is around 50%. Willapa Bay, on the southwest coast of Washington north of the mouth of the Columbia River, for example, has a tidal prism estimated at approximately 45% of the total high tide volume of the bay, but flushing time, depending on wind and the volume of water in the Columbia River plume (which moves northward during the winter months), acting as a low-salinity water mass at the mouth of Willapa Bay, results in conditions that can inhibit exchange between sea and embayment to a flushing time of 20 d (HEDGPETH and OBREBSKI, 1981). Hence, although Willapa Bay is formed primarily by a sand barrier or spit, the local characteristics of near-shore conditions in the sea on the influence of several small but permanent streams result in an environment that is distinctly estuarine, as attested by the high productivity of oysters in the bay.

The influence of the Columbia River plume on Willapa Bay was dramatically demonstrated by the detection of unexpectedly high levels of $^{65}Zn$ in a worker at the Hanford nuclear installation more than 300 miles upstream from the mouth of the river. This person had eaten oysters from Willapa Bay, in which the isotope, released from the Hanford Works, was concentrated by a factor of more than 200 000 times that of sea water (PERKINS and co-authors, 1960).

During summer, when stream flows are low, estuary flushing may be enhanced by the process of coastal upwelling, as demonstrated for Grays Harbor, just north of Willapa Bay (DUXBURY, 1979). This is made possible by the upwelling process, in which subsurface water rises to the surface and is moved seaward by the driving force of the wind. Where this process impinges on an estuarine system, downstream flow may be entrained and flushing action maintained in spite of reduced stream volume. This interaction between oceanic and estuarine processes may occur in those parts of the world where upwelling is a regular phenomenon.

A strong case of coastal upwelling as a delimited ecosystem has been made by recent students of the phenomena, as reviewed by various contributions in BOJE and TOMCZAK (1978) and by BARBER and SMITH (1981), although CUSHING (1971) considers upwelling to be essentially the same process as the seasonal overturn and mixing in shallow neritic seas, e.g. the North Sea. The most significant difference is that in a seasonally mixed system there is 'one major productivity event per year, while in coastal upwelling systems the duration of conditions favourable for maximum productivity is significantly longer' (BARBER and SMITH, 1981, p. 34).

Along the northwestern and northern shore of the Black Sea, from the mouth of the Danube to the Sea of Azov, there is a series of shallow embayments and lagoons, usually associated with river drainages of the present or geologically recent past. The term 'liman' is applied specifically to these bodies of water. Several of them are estuarine in nature as the word liman implies, others are restricted or completely closed off from the sea. ROZENGURT (1967) has recognized three types of limans: (i) periodically closed off from the sea; (ii) completely separated from the sea; and (iii) open to the sea with continuous fresh water inflow. Tides are almost negligible in the Black Sea (about 8-cm range), and the most noticeable changes in sea level are the result of wind action and stream runoff. This action results in well mixed, unstratified systems.

On coastal-plain shorelines of seas with well-developed tidal characteristics, sedimentation patterns correspond to the tidal ranges of the various localities. HAYES (1975) has summarized these conditions in three categories according to tidal ranges as 'microtidal, mesotidal, and macrotidal'. As examples of the microtidal category, he includes most of the estuaries of the Gulf of Mexico, and his mesotidal type includes the situations encountered on the northeastern coast of the United States and in the Wadden Sea. Although HAYES does not give an example of the macrotidal model, the estuary of the Plate is an obvious example of this general condition.

HANSEN and RATTRAY (1966) proposed a classification of estuary types computed from salinity distribution, and river runoff flow velocity. OFFICER (1976) considers this scheme the best system so far developed for estuary classification. It is based on stratification–circulation diagrams.

While all of this may satisfy many researchers concerned with salinity exchange problems in estuaries, it is somewhat cumbersome for ecologists, and we are reminded by FISCHER (1976) that these idealized models based on vertical profiles in midstream do not adequately consider conditions where there is pronounced lateral flow into the shallows, especially of such an estuarine system as San Francisco Bay. The field experience of the estuarine ecologists suggests that the more interesting, complex, and productive ecosystems are those where tidal fluxes, mixing of stream and oceanic waters, are complicated by lag effects related to physiographic as well as hydrographic conditions. Where the water flows out in an overwhelming mass, as from the Amazon River, there is no estuary within the boundaries of the river, and it falls to oceanographers to trace the mixing processes and to study the organisms involved far at sea (GIBBS, 1970, 1972). A more characteristic estuary is developed in the lower reaches of the world's second largest river, the Zaire (formerly the Congo). The salt wedge penetrates about 40 km upstream; beyond this there is a flood plain region with divided channels, 'the braided area' of about 50 km. The estuarine reach is actually at the head of a submarine canyon, and from it cold saline water is upwelled into the river. This is a different process from the coastal upwelling associated with wind patterns. The influence of the river has been detected for at least 400 miles at sea (for a description of the physical features of the lower Zaire and its estuary, see EISMA and van BENNEKOM (1978) and other papers in the same issue of the *Netherlands Journal of Sea Research*). General information on the biological characteristics of this region is not yet available.

The ecosystems of estuaries and lagoons are interface systems between the sea and fresh waters. Tidal action is a significant component of typical estuaries although in some cases it may be subordinated to wind action. In estuaries of the northwestern Gulf of Mexico, for example, where the tidal range is about 60 cm, the water level in the bays is often controlled by the wind, and there is also a pronounced seasonal variation in sea level, with the high levels occurring in spring and autumn (COLLIER and HEDGPETH, 1950). In the Sea of Azov, where tidal range is greatly reduced from that of the Black Sea, wind tides, associated with a pattern of daily onshore and offshore winds, may cause sea-level fluctuations in the order of 4 m (ZENKEVITCH, 1963).

The physical properties of a tidal estuarine system were studied in detail by PRITCHARD (1952) in the James River, a tributary of Chesapeake Bay. His observations demonstrated that there was an upper water layer with a net downstream movement, and a deeper water layer with a net upstream motion. These essential features of estuarine systems were studied by LORENZ (1863) who identified and measured the salinity wedge ('Keil') in the Elbe River. PRITCHARD designated the boundary between these layers, where there is a negative vertical velocity, as the surface of 'no net motion'. This region is of considerable ecological significance, since it enables organisms with limited swimming powers, such as crab and oyster larvae, to maintain their place in the estuary and to colonize localities upstream by moving into the upstream flow, and evading the downstream flow by modifying their vertical migration. Further studies of this transport mechanism, with special reference to larvae of decapod crustacea, have been carried out by SANDIFER (1975), who suggests two types of mechanism, one involving recruitment by retention of larvae, the other recruitment by immigration of juveniles and adults. Various combinations may occur within different populations in the same estuary.

It is at this interaction between outflowing fresh water and inflowing sea water that sediment is 'entrapped', very small particles flocculated and precipitated by the change of charge, with a resulting increase in nutrient concentrations. This is the characteristic phenomenon of the estuarine system. The region where this occurs has been termed the 'null zone' (HANSEN, 1965; PETERSON and co-authors, 1975), or the 'entrapment zone' (ARTHUR and BALL, 1979). The best characterization is that of ODUM (1970), who refers to estuaries as 'nutrient traps'.

Other prominent interface zones associated with tidal or meteorological sea level action are those between the soft sediments of the estuary margins, the exposure of sediments of tidal flats to air, fresh water runoff and rain, and the vertical or steep gradient exposure of rocky intertidal surfaces. On rocky or hard surfaces wave action is often a significant component, especially on the outer shores, but wave effects may be damped on sheltered surfaces inside estuaries.

All of these interface or ecotone areas support rich and varied ecosystems or subsystems, and may be considered the aquatic analogue of the 'edge effect' of terrestrial ecologists (e.g. LEOPOLD, 1933). Classification of these systems presents as many difficulties as that of environments according to their salt content, although such salinity classifications have the advantage of being based on one objective criterion. A scheme for systems classification based on energy flow characteristics has been proposed by ODUM and COPELAND (1972). This scheme attempts to take into consideration the major factors of environmental support and perturbation, such as light, wave force, tidal action, pollution, thermal alteration (in addition to seasonal variation), harvesting, fishing, and the influence of the organisms themselves. Unfortunately, this system includes such a mix of criteria, some of them not easily quantifiable, that it does not result in a simple or co-ordinated scheme, although it has served adequately as the organization scheme for a 4-volume catalogue of Coastal Ecological Systems of the United States (ODUM and co-authors, 1969). The classification recognizes six major divisions including 46 'categories' of coastal ecosystems. Some ideas, over-emphasized by sanitary engineers, have been treated with restraint. For example, the 'health' of a system is not recognized as an entity, and 'indicator species' are, on the whole, not considered useful criteria in this context. Diversity is considered in a general way, but no magic index or numerical expression is attempted.

Obviously the ODUM–COPELAND scheme does not meet MANN's (1982) need for a cogent theory of ecosystems, and in his discussions of the various kinds of coastal ecosystems he uses the traditional categories of systems according to the dominant or characteristic organisms, e.g. sea or marsh grasses, mangroves, coral reefs, etc., and sediment communities. MICHAELIS (1981), working in the Wadden Sea, classifies the benthic communities in the traditional way according to the characteristic species. DAY (1981) in the comprehensive summary of many years study of estuaries of southern Africa, avoids categories of systems, and uses a simplified scheme for estuary classification: normal estuaries, divided into 'saltwedge', highly stratified, partically mixed, and vertically homogenous; hypersaline estuaries; closed or blind estuaries, and lagoons (of various types). In the summary treatment, of the estuarine ecosystem and environmental constraints, the emphasis is on general properties and processes (without a single ecosystem diagram or model!) (DAY and GRINDLEY, 1981).

Eventually it may be possible to agree upon a classification of ecosystems according to a hierarchical scheme as MANN (1982a) suggests, but estuaries and coastal waters, as intrinsically non-hierarchical situations, can best be recognized according to their descriptive and dynamic properties. Such circumstances do not provide concepts for classification scheme that would meet all possible conditions encountered in nature.

## 3.2 FLOW PATTERNS OF ENERGY AND MATTER

J. G. FIELD

### (1) Introduction

This chapter reviews the structures and functions in coastal ecosystems by means of few selected examples of coastal regions which have been studied as ecosystems. The essential structures of primary producers, consumers, and decomposers may vary according to the habitat and physical forces involved, and these determine the quality and quantity of the flows of energy and matter and of the ecological processes that occur.

Table 3-1 lists estimates of primary production from seven studies of coastal ecosystems. Three different sources of primary producers are distinguished: phytoplankton, microalgae attached to sediment particles, and macrophytes (comprised of both large seaweeds and higher plants). Most of the systems receive primary products from more than one of these sources. Only Narragansett Bay (USA) is based solely on phytoplankton production with negligible inputs from attached algae or macrophytes (KREMER

Table 3-1

Primary production in some selected coastal ecosystems. Values are expressed in g C $m^{-2}$ $yr^{-1}$. (Compiled from the sources indicated)

| System (Reference) | Source of primary production: Phyto-plankton | Attached microalgae | Macrophytes | Total |
|---|---|---|---|---|
| Narragansett Bay (KREMER and NIXON, 1978) | 98 (100%) | — | — | 98 |
| Lynher mud flat Cornwall (WARWICK and co-authors, 1979) | 82 (36%) | 143 (64%) | — | 225 |
| Baltic, Askö (JANSSON and co-authors, 1982) | 191 (72%) | — | 75 (28%) | 266 |
| Mangrove Forest, Florida (LUGO and co-authors, 1978) | — | — | 890 (100%) | 890 |
| Sapelo Isl. Marsh/Estuary (POMEROY and WIEGERT, 1981) | 79 (6%) | 150 (10%) | 1216 (84%) | 1445 |
| Nova Scotian Kelp Bed (MILLER and co-authors, 1971) | 226 (11%) | — | 1750 (89%) | 1976 |
| Benguela Kelp Bed (NEWELL and co-authors, 1982) | 502 (40%) | — | 767 (60%) | 1269 |

and NIXON, 1978). This example is therefore closest to the classical open-sea food chain. At the other extreme, the Florida mangrove forest example (LUGO and co-authors, 1976) is based solely on mangrove production and other sources are regarded as negligible. Comparison of the estimates of total production should be made with caution because of the differing methods of measurement used and since the estimates are based on different assumptions. For example, the Nova Scotia kelp production can be averaged over the seaweed belt only or expressed as an average for the whole of St Margaret's Bay (Canada), while the phytoplankton production estimate is based on the whole of St Margarets Bay (MANN, 1972). There is a general trend of increasing primary production from cool and temperate coastal waters to the tropics although the highest estimates are from kelp beds in cool waters. We shall follow this trend in our series of examples, adding complexity with successive examples.

## (2) Basic Ecosystem Processes

### (a) Narragansett Bay: A Phytoplankton-based System

KREMER and NIXON (1978) have described the development of a numerical simulation model which they use to gain insight into the functioning of a coastal ecosystem. Narragansett Bay is located at 41° N on the Rhode Island coast of New England (USA), has an area of 265 $km^2$, a mean depth of 9 m (75% is less than 12 m deep), and sediments ranging from sand to silt/clay. The oceanography of Narragansett Bay has been studied for at least 20 yr and the model developed after several years of intensive biological study. The model is divided into eight geographic compartments and is mainly concerned with the dynamics and interactions of phytoplankton, zooplankton, and the nutrients nitrogen, phosphorus, and silicon. The system is based on primary production by phytoplankton, which is grazed by zooplankton, principally copepods (*Acartia*) and filter-feeding benthos, primarily the hard clam (*Mercenaria*). Migratory Atlantic menhaden (*Brevoortia*) enter the Bay in summer, the juveniles feeding on phytoplankton and adults on zooplankton. Unused phytoplankton forms detritus which is also contributed to by nearby saltmarsh grass and fringing seaweeds, both of which are considered to be too minor to be included in the model. Detrital particles, including the attached bacteria, are fed on by deposit- and suspension-feeding benthos: bivalves (*Nucula* and *Yoldia*), and a polychaete (*Nephtys*). Suspended detrital particles also contribute to the diet of *Mercenaria* and the zooplankton. Carnivores in the system are a ctenophore (*Mnemiopsis*) which feeds on copepods and various demersal fish feeding on the benthos. The top carnivores are butterfish (*Preprilus*), which prey on ctenophores, and abundant sport fish such as striped bass (*Morone*) and blue fish (*Pomatomus*) that feed on menhaden.

Fig. 3-7 shows the conceptual model used to simplify the basic food web described above to five biological and two abiotic components linked by lines indicating the flows of energy and materials. The biotic components include the essential ones of most coastal ecosystems: primary producers, zooplanktonic herbivores feeding on phytoplankton, benthic consumers feeding on phytoplankton, or detritus, planktonic, and benthic carnivores. The positive feedback to primary producers of nutrient regeneration is shown, as is the negative feedback of self-shading by phytoplankton. An obvious

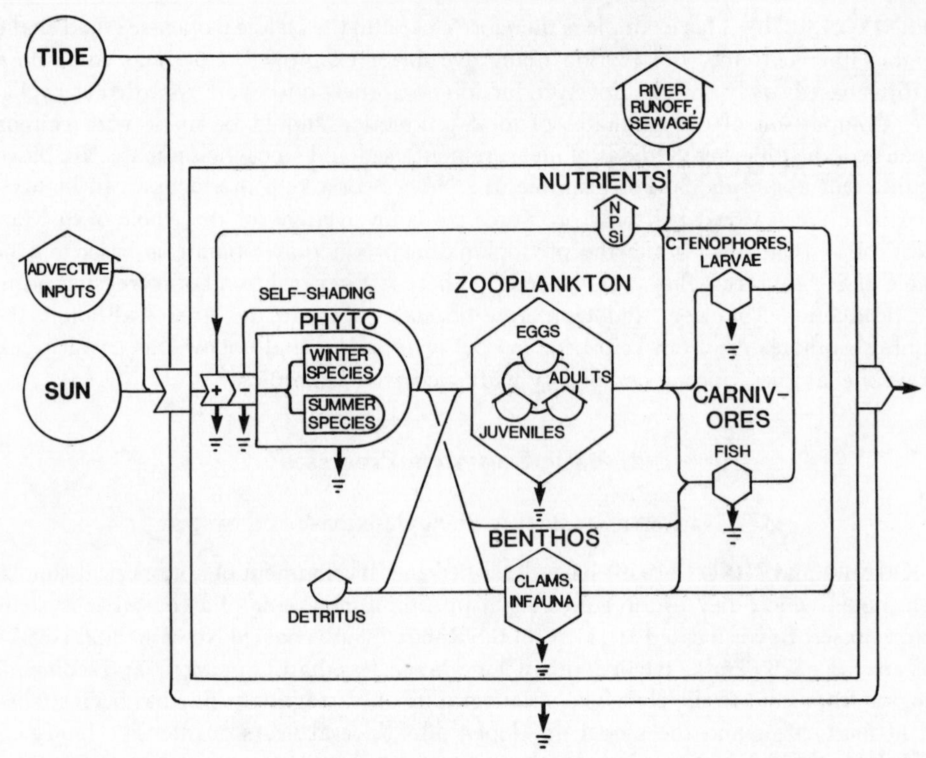


Fig. 3-7: Energy-flow diagram depicting the main processes modelled in the Narragansett Bay ecosystem, using the circuitry symbols developed by ODUM (1971). (See also PLATT and co-authors, 1981.) Main energy sources (circles) are sun and tides with arrows indicating the direction of energy flow. 'Bird houses' represent passive stores of energy or materials, such as detritus and mineral nutrients: N, P, and Si. Bullet-shaped symbol represents primary producers, here phytoplankton (Phyto). Hexagons represent consumer populations, such as zooplankton and benthos. Downward arrows: energy loss by respiration. Positive and negative feedbacks to phytoplankton production indicated by 'work-gates' which regulate the flow of energy and matter. (After KREMER and NIXON, 1978; reproduced by permission of Springer-Verlag)

omission from the diagram is a living decomposer component which is implied to be associated with the detritus.

By taking mean values of energy flows in the eight geographic compartments of the model explained by KREMER and NIXON (1978) rough estimates of the fate of phytoplankton carbon production in Narragansett Bay can be made: 59% is grazed by zooplankton, 25% by benthic filter feeders, and the remaining 16% becomes detritus directly while some 22% of primary production is voided as faeces and adds to the detritus. Thus, although the system is based on phytoplankton production, and the important planktonic food chains which are traditionally regarded as being centred around herbivores, at least one-third of the carbon available to consumers as food is detritus (Table 3-2).

Similarly, the flows of nitrogen, usually regarded as the most important mineral nutrient in the sea, can be analysed. The zooplankton excrete some 40% of the nitrogen

Table 3-2

Estimated annual carbon and nitrogen fluxes in Narragansett Bay. Averaged over 8 compartments (Based on KREMER and NIXON, 1978)

| Category | Carbon (g C $m^{-2}$ $yr^{-1}$) | Nitrogen (g N $m^{-2}$ $yr^{-1}$) |
|---|---|---|
| Primary production | | |
| Phytoplankton | 98·1 | 16·3 |
| Secondary producers | | |
| *Consumption* | | |
| Zooplankton on phytoplankton | 58·1 | (9·7)* |
| Clams, benthos on phytoplankton | 24·9 | 4·1 } (>9·4)‖ |
| Benthos on detritus | (24)* | >5·3 } |
| *Production* | | |
| Zooplankton | 7·1 | (1·7)** |
| Clams, benthos | (5†) | (1·3)†† |
| *Respiration* | | |
| Zooplankton | (39·4)‡ | — |
| Clams, benthos | (14·6)‡ | — |
| *Faeces and urine* | | |
| Zooplankton | (11·6)§ | 6·60 |
| Clams, benthos | (24·5)¶ | 8·1 |
| *Detritus formation* | | |
| (primary production − consumption + faeces) 98.1 − 83.0 + 21.6 = 36.7 gC | | ? |
| *Detritus respiration and other losses* | | |
| (Formation − consumption by benthos) = 36.7 − 24 = 12.7 gC | | |
| *Carnivores* | | |
| Ctenophores, larvae | | 0·053 |
| Fish | | ? |

*Calculated assuming C/N ratio is 6.
†Calculated assuming $K_1$ for clams = $P/C$ = 20% (STUART and co-authors, 1982a,b).
‡Calculated by difference between carbon gain and loss.
§Calculated assuming assimilation efficiency = 20% (KREMER and NIXON, 1978).
¶Calculated assuming assimilation efficiency = 50% (STUART and co-authors, 1982a,b).
‖Minimum based on production and release rates.
**Calculated assuming C/N ratio of 4.3 (KREMER and NIXON, 1978).
††Calculated based on mussel C/N ratio of 4 (A. HAWKINS and B. BAYNE, pers. comm.).

required by phytoplankton annually (16.34 g N $m^{-1}yr^{-1}$), 50% is released by the benthic community, and the remaining 10% is not accounted for, but may come from planktonic decomposition or carnivore excretion.

The importance of detritus and the benthic component of the Narragansett Bay system has subsequently been confirmed in Marine Ecosystem Research Laboratory ('MERL') mesocosm tanks on land, which simulate the natural system. These tanks do not reproduce the natural fluctuations of the bay unless the benthic sedimentary component is included (PILSON and NIXON, 1980), confirming the importance of sediments to nutrient cycling in shallow coastal systems. Water exchange in the MERL mesocosms is based on an average turnover time for the bay of about once per month (M. PILSON, pers. comm.), suggesting that it functions as a 'relatively closed' system in terms of water exchange.

The role of detritus in a phytoplankton-based coastal system is also shown by the analysis of trophic structure and fluxes in the Belgian region of the North Sea by JOIRIS and co-authors (1982). Some 50% of the annual phytoplankton production sinks to the sea floor as detritus, supplemented by an additional component from the faecal pellets of zooplankton. Thus, the benthos is heavily implicated in the nutrient cycles of this coastal system too.

### (b) The Lynher Estuary, Cornwall: A Mudflat

WARWICK and PRICE (1975) and WARWICK and co-authors (1979) have described an intertidal mudflat in the estuary of the Lynher River, Cornwall (UK), in which energy flow has been studied over a number of years. The study area consisted of a 25 × 100 m grid just below mid-tide level, with exposure times at spring tide ranging from $4\frac{3}{4}$ to $5\frac{1}{2}$ h, depending largely on the degree of fresh water inflow. The salinity ranged from 32‰ S in summer to 8·2‰ S in winter, while interstitial salinities ranged from 19·3 to 34·4‰ S.

WARWICK and co-authors (1979) summarized information on carbon flow through the mudflat community by means of a steady-state descriptive model in which each of the 13 major animal populations in the community is treated separately. The animal populations have been pooled into two main functional groups, filter feeders and deposit feeders (Fig. 3-8), the latter subdivided into macro- and meso-fauna (>45 μm sieve size), and the true meiofauna, which pass through a 45 μm sieve. The main source of primary production is the phytobenthos, mainly diatoms on the mud surface, which contribute 143 g C $m^{-2}$ $yr^{-1}$. Primary production by phytoplankton when the mudflat is covered by water at high tide is estimated to be some 82 g C $m^{-2}$ $yr^{-1}$, of which only 19.3 g C is filtered by filter feeders, the remainder either being exported from the mudflat with the receding tide or being deposited as particulate organic carbon in the sediments. About half the estimated food of filter feeders (20.7 g C) consists of bacteria suspended in the water column which presumably utilize phytoplankton exudates as their energy source. The filter feeders on the mud flat are mainly species of the bivalves *Mya* and *Cardium* and of the 'mesofaunal' polychaetes *Manayunkia* and *Fabricia*. *Manayunkia* species are the most important filter feeders in terms of energy flow, consuming some 19.5 g C $m^{-2}$ $yr^{-1}$.

The food of the mudflat deposit feeders is derived mainly from phytobenthos (143 g C $m^{-2}yr^{1}$), in strong contrast to the Narragansett Bay subtidal system where benthic diatoms are negligible. There is almost equally as large a potential contribution of sedimentary particulate organic carbon (111 g C $m^{-2}$ $yr^{-1}$), but WARWICK and co-authors (1979) estimated that only 10% of this is available as assimilable food, the remainder being refractory. A large proportion (75%) of the particulate organic carbon contributed to the sediment annually is derived from the faeces of benthic animals. Some 34% (76.7/225) of the total primary production is voided as faeces and returned to the sediments, in contrast to 22% for the phytoplankton-based Narragansett Bay community.

Amongst the deposit feeders, the true meiofauna (<45 μm sieve) account for 65% of the production (25.9 g C $m^{-2}$ $yr^{-1}$), of which 3.3 g is consumed by predators within the

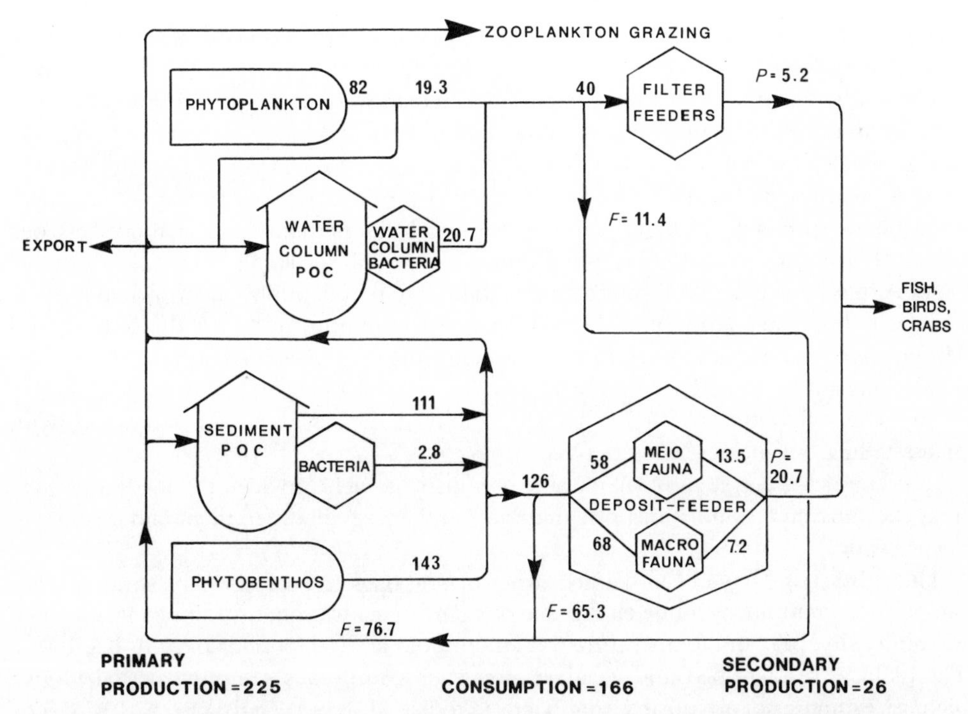


Fig. 3-8: Carbon flow on a mudflat in the Lynher River estuary, Cornwall (UK). Values are in g C $m^{-2}$ $yr^{-1}$. Primary production by phytobenthos (mainly diatoms) and phytoplankton. Storages as particulate organic carbon (POC) in sediment and water column; consumers are filter feeders and deposit feeders, both of which recycle matter through their faeces (F). Some material is exported by tides, including re-suspended sediments, and zooplankton grazing accounts for a small fraction of suspended organic carbon. Totals are given at the bottom of diagram; symbols as in Fig. 3-7. (Based on data in WARWICK and co-authors, 1979; reproduced by permission of Blackwell Scientific Publications)

infauna (*Nephtys* and *Protohydra*). The remaining 22.6 g C $m^{-2}$ $yr^{-1}$ is available as food to mobile predators such as birds, fishes, and crabs, whose impact and contribution has not been estimated.

No nitrogen budget has been calculated for the mudflat community, but it is reasonable to postulate that the excretory activities of the infauna would meet the nutrient needs of the phytobenthos (RAINE and PATCHING, 1980).

### (c) The Askö/Landsort Area of the Baltic Sea: Mixed Rock and Soft-Bottom Communities

A team of 21 scientists have collaborated in intensive studies of a shallow-water ecosystem in the Baltic Sea archipelago. During the early 1970s much effort was concentrated in a sound at Askö (Sweden), 200 m long by 50 m wide, with a maximum depth of 10 m. The location at 61° N is subject to ice-cover in winter, as well as perennial low salinities of about 6.3‰ S, resulting in a very low species diversity and communities comprised of both marine and fresh water species (JANSSON and WULFF, 1977). In spite of this, most features are marine, and interesting techniques were used to gain an insight into the functioning of the system which—combined with observations over a wide area—have allowed some generalizations for the whole Baltic Sea (e.g. JANSSON, 1978).

The study recognized four main subsystems which were each studied separately and together to estimate biomass, energy flow, and nutrient cycling (JANSSON and WULFF, 1977). The subsystems are as follows:

(i) A 50-cm wide band of the filamentous green alga *Cladophora* in the region where atmospheric pressure changes cause water-level fluctuations and winter ice-scour occurs. It is an important habitat and a nursery ground for small crustaceans.

(ii) A rock zone colonized from 0.5 m to about 2 to 6 m depth by the brown alga *Fucus vesiculosus*; this zone supports a distinct animal community in which the blue mussel *Mytilus edulis* is dominant. *M. edulis* extends below the *Fucus* zone to depths of about 25 m on suitable substrates (KAUTSKY, 1981).

(iii) The soft bottom/*Ruppia* subsystem found at depths exceeding some 4 m, with plants fading out at increasing depth.

(iv) The pelagic system of planktonic organisms which provides the medium which links the subsystems and transports materials and energy between them and offshore to deeper water.

The subsystems have been studied using plastic bags to enclose representative samples of each community while changes in oxygen concentration, nutrient concentration, light intensity, pH, and temperature were monitored for 24-h periods by SCUBA divers. The biota were identified and weighed after each continuous sampling period to give biomass estimates of organisms from bacteria to fish (JANSSON and WULFF, 1977). The results of diurnal oxygen changes showed that production of the *Cladophora* and *Fucus* subsystems is approximately balanced by the fast respiration of consumers. The pelagic subsystem was found to be an overall producer with phytoplankton primary production outstripping respiration of the planktonic community. The *Ruppia* and soft-bottom subsystems were found to be overall consumers, with production rates amounting to about 60% of respiration rates, half of which being due to bacteria. Putting together the results, the total system was found to have a high degree of self-maintenance ($P/R = 1.1$) which might mislead one into believing that the 100 m × 50 m sound system was largely self-contained. Analysis of the food web shows otherwise.

Since 1977, work has extended to cover a wider (160 $km^2$) area in the archipelago south of Stockholm. Fig. 3-9 shows the three principal subsystems: hard bottoms, pelagic, and soft-bottom communities, and indicates some of the material and energy flows between them. The diagram emphasizes the role of fish in energy transfers, since they migrate between the subsystems at different seasons (JANSSON and WULFF, 1977). Fig. 3-10 shows the same systems, this time emphasizing the role of the mussel, *Mytilus edulis* (KAUTSKY, 1981). The pelagic system has an estimated phytoplankton production of 150 g C $m^{-2}$ $yr^{-1}$ (KAUTSKY and WALLENTINUS, 1980); of this, 56 g C $m^{-2}$ $yr^{-1}$ is estimated to be filtered by the mussel population. The mussels in turn contribute copious amounts of faeces to the pool of detritus, some of which is degraded *in situ* and some of which is transported out of the subsystem to soft-bottom areas where the consumers depend on outside sources of food. Most of the production of *M. edulis* is channelled into reproduction and relatively little is taken by predators (KAUTSKY, 1981). *M. edulis* larvae, however, are an important source of food for herring larvae and other carnivorous zooplankton in the pelagic subsystem. The most important role of mussels is in recycling nutrients to the algae. KAUTSKY and WALLENTINUS (1980) have

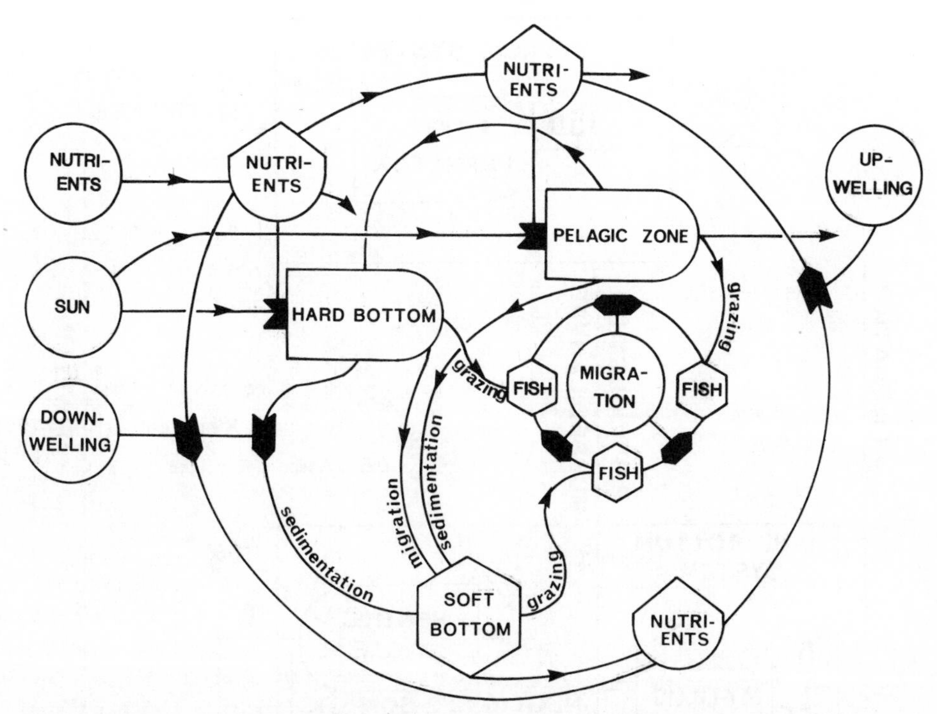


Fig. 3-9: Diagram of some of the main energy and material flows linking hard-bottom, pelagic, and soft-bottom subsystems in the Baltic Sea. Seasonal migrations of fish serve to link subsystems; advective processes of upwelling and downwelling can be seen to be important in maintaining the cycle of nutrients. Symbols as in Fig. 3-7, with work-gates allowing bidirectional or unidirectional energy flows as indicated by arrows. (After JANSSON, 1978; reproduced from *Advances in Oceanography* by permission of the Plenum Publishing Corpn)

calculated that mussels release more nitrogen and phosphorus than needed by all the benthic algae, leaving enough excess to supply 6% of the nitrogen and 17% of the phosphorus demand of the pelagic system over the entire 160 $km^2$ Askö/Landsort study area.

This recycling role is particularly important in summer, since it allows continued growth of benthic algae after the spring phytoplankton bloom has depleted the nutrients in the water column. In Fig. 3-10 broken arrows indicate nutrient flows from mussels to the pelagic subsystem and to benthic algae. Of the benthic algae, *Fucus vesiculosus* has the highest biomass, but it is not readily eaten by herbivores (RAVANKO, 1969) and most *F. vesiculosus* production is carried out of the hard-bottom area to deeper soft bottoms when plants are broken up during storms (JANSSON and co-authors, 1983). Slow decomposition occurs in the soft-bottom subsystem, and where this is below the thermocline, upwelling is required to bring nutrients back to the photic zone.

This example has shown that different coastal ecosystems may be strongly linked, although they may be apparently self-maintaining from purely energetic considerations, with oxygen production matching oxygen consumption. Analysis of the structure of the system into its main functional components (e.g. filter feeders, macrophytes, phytoplankton, decomposers) is necessary and where these are geographically separated, the

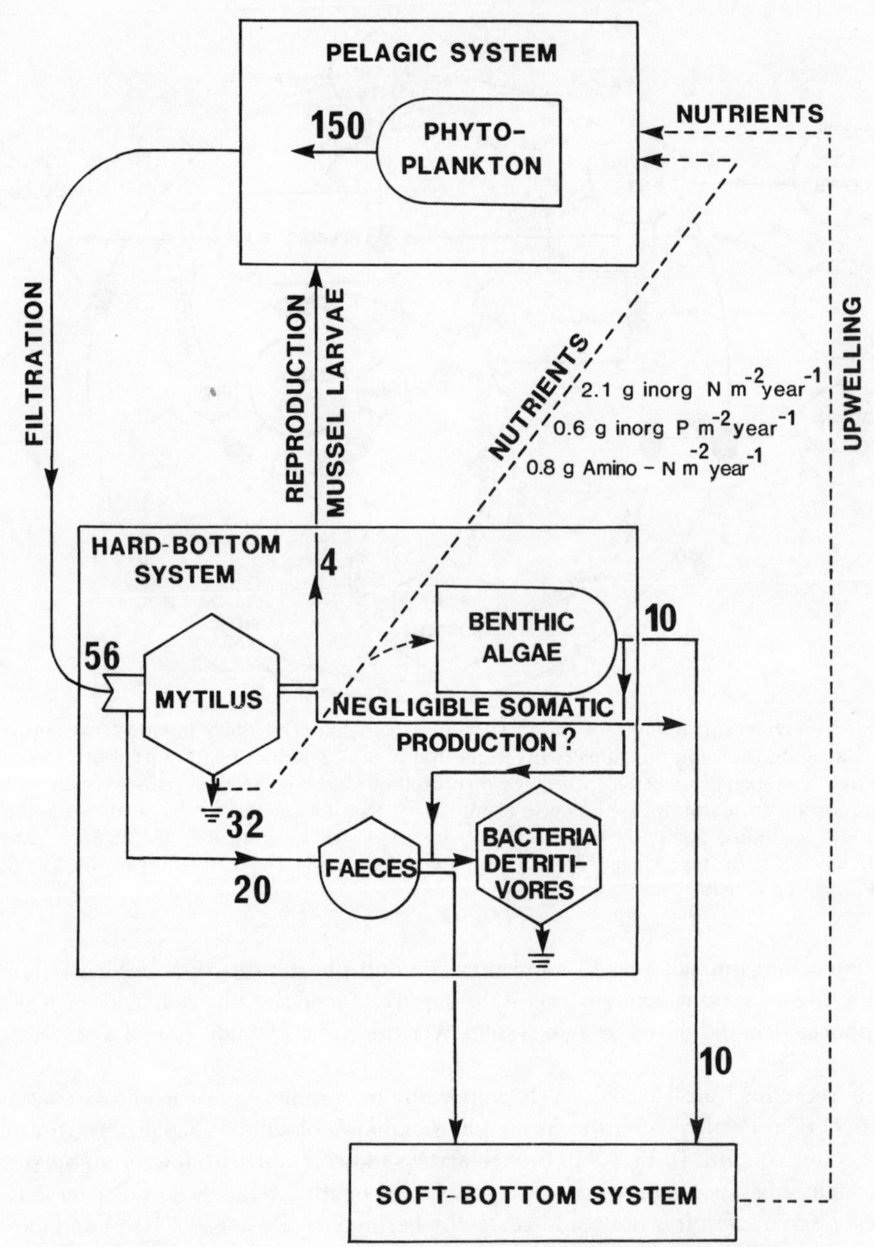


Fig. 3-10: Some of the carbon flows (solid lines) and mineral nutrient flows (broken lines) linking the pelagic, hard-bottom, and soft-bottom subsystems in the Baltic Sea. Role of *Mytilus edulis* emphasized, showing how it depends on phytoplankton advected from the pelagic system for food, returning its major production, larvae, as food for zooplankton, and nutrients needed by phytoplankton. The major fraction of minerals released by *M. edulis* is taken up by *Fucus* and other benthic algae, while a portion of the faeces and of broken algae is carried down to the soft-bottom system. Flows are in g C $m^{-2}$ $yr^{-1}$ unless otherwise indicated. Symbols as in Fig. 3-7. (After KAUTSKY, 1981; modified; reproduced by permission of the Askö Laboratory)

physical mechanisms by which materials are transported from one component to the other are also important. In both examples in which nutrients were studied (Narragansett Bay and Baltic Sea), the benthic fauna has been shown to be of major importance in recycling nutrients.

## (3) Macrophyte-Based Systems

### (a) A South Florida Mangrove Forest

Mangroves are unique coastal ecological systems in that primary production occurs in the leaves of plants which are never covered by water. The food webs are based on detritus derived from these leaves and thus provide one extreme in the continuum of coastal systems, ranging from those based on detritus derived from higher plants to those based on phytoplankton (ODUM and HEALD, 1975). This chapter does not attempt to cover all aspects of mangrove ecology; they have been reviewed elsewhere (e.g. MACNAE, 1968; LUGO and SNEDAKER, 1974). The chapter concentrates on energy flow through the food web and the little that is known about nutrient fluxes.

ODUM and HEALD (1975) studied the mangrove forests of southern Florida (USA) which are flooded seasonally by fresh-water runoff from the Everglades, while tidal flows from the Gulf of Mexico introduce saline water resulting in salinities ranging up to 30‰ S. The North River Basin formed the main study area of 21·2 $km^{-2}$, two-thirds of which is covered by red mangroves *Rhizophora mangle* and one-third by open water. Primary production was found to be 2·41 g dry organic matter $d^{-1}$ or about 350 g C $m^{-2}$ $yr^{-1}$; this agrees well with calculations by MILLER (1972), who used a sophisticated model of photosynthesis to predict net production by Florida mangroves at some 400 g C $m^{-2}$ $yr^{-1}$; 83% of the mangrove primary production was in the form of leaves. Algal production was negligible. ODUM and HEALD (1975) ascribe this fact to the high tannin content and low nutrient status of the water, and to shading by the mangrove plants.

Fig. 3-11 shows a conceptual model of the North River food web. The main flow of energy is indicated by broad arrows; it depicts a food chain of mangrove leaves—bacteria and fungi—detritivores—middle carnivores—top carnivores. Detritus decomposed very slowly in air on the forest floor, but faster in freshwater (54% remained in litter bags after 4 mo); while only 9% of leaf material remained in litter bags placed in seawater for 4 mo); mainly because of the greater variety and activity of leaf-chewing animals in sea water. The leaf material changed in composition as it decomposed through autolysis and microbial activity, with fungi playing a much larger role than reported in marine situations (see also FELL and MASTERS, 1973). The detrital particles became smaller, the proportion of protein increased from 5 to 21% (see also FENCHEL and BLACKBURN, 1979), and the calorific content increased over 6 mo, from 19·7 to 22·2 kJ $g^{-1}$ ash-free dry mass. Roughly half the detritus formed was exported from the North River Basin into surrounding bays, mainly as particles in the range 50 to 350 $\mu$m which contained bacteria, fungi, and yeasts. The remainder was consumed locally and only 2% of leaves remained permanently to form peat.

The detritivores can be subdivided into three types: the 'grinders', such as the crab *Rhithropanopeus* and the amphipod *Melita*, which take larger particles (mean 75 $\mu$m) and chew them; the deposit feeders, such as killifish *Cyprinodon* and *Adinia*, the goby *Lopho-*

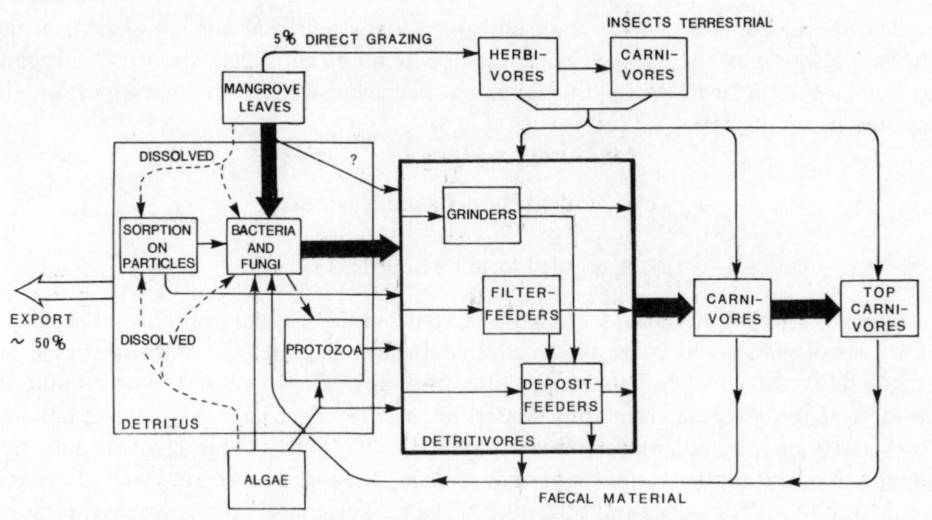


Fig. 3-11: Conceptual model of energy flows in a Florida mangrove food web, showing the major flows as heavy arrows. Subdivision of detritus and detritivores into component categories is also indicated and dissolved organic matter is indicated by broken lines. The role of faeces as a feedback to detritus is shown. (After ODUM and HEALD, 1975; modified; reproduced by permission of the Academic Press)

*gobius*, and the shrimp *Palaemonetes* (eating particles of 22 to 55 $\mu$m); and the filter-feeding mussels *Congeria* and *Brachidontes* (filtering particles of 12 $\mu$m mean size). All of these utilize food of various types, including the faeces produced by the detritivores and other animals, so that refractory detrital material may be ingested, voided, colonized by micro-organisms, and re-ingested several times over, provided it is not transported out of the system by water currents. This process was first described by NEWELL (1965); it has been referred to as the 'faeces loop' by NEWELL and co-authors (1982). The important food of detritus feeders has usually been believed to be the micro-organisms associated with the detritus, which undoubtedly enrich the protein content of detritus (ODUM and DE LA CRUZ, 1967; FENCHEL, 1970, 1977; MANN, 1972, FENCHEL and HARRISON, 1976; FENCHEL and JØRGENSEN, 1977). However, CAMMEN (1980) and STUART and co-authors (1982a) have shown that considerable amounts of the plant material itself can be assimilated by some detritivores in the absence of micro-organisms.

The final steps in the main food web in South Florida mangroves are the small carnivores (minnows and small game fish) and top carnivores (game fish, piscivorous birds, etc.). Of minor importance is the side chain to grazing terrestrial insects, which consume some 5% of mangrove-leaf production (HEALD, 1969). The importance of other side chains was not quantified by ODUM and HEALD (1975), but dissolved matter given off by mangrove leaves and algae would clearly be a source of nourishment for bacteria and fungi, which together with algae are preyed on by protozoans. As indicated in Fig. 3-11, the protozoans are available as food to the detritivore trophic group, either the filter feeders if suspended, or the deposit feeders if in the sediments. Fig. 3-11 also indicates that 50% of the detrital material is exported by the North River to neighbouring bays and inshore waters. This is almost certainly utilized in similar fashion by other detritivores, the question of 'export' being mainly a matter of how one defines the system boundaries.

Much less is known about nutrient flows than energy flows in Florida mangrove forests. What is known suggests that they are unusual in that only a small proportion of the nutrients appear to be cycled in the mangrove forests, a large proportion of the nutrients are supplied by fresh-water runoff from the Everglades (LUGO and co-authors, 1976). Their simulation model demonstrated the dependence of the mangroves on nutrients (phosphorus and nitrogen) derived terrestrially, and measurements of nutrient concentrations suggested that river runoff accounted for 92 to 96% of nutrients, while tidal flow supplied the remaining 8 to 4%. These do not take account of rapid recycling which is likely to occur amongst micro-organisms at low nutrient concentrations, especially at the warm temperatures found in mangrove forests.

### (b) A Georgia Salt Marsh

Energy and material flows have been more intensively studied in the salt marshes of Sapelo Island, Georgia (USA) than in any other marine system, indeed probably in any ecosystem. These studies have been continuing since the mid-1950s and have built up an impressive body of knowledge, drawn together in one volume edited by POMEROY and WIEGERT (1981). The ecology of salt marshes has been studied from many different angles to elucidate the physical, chemical, and biological processes involved. Some of the earlier classic salt-marsh studies included ODUM and SMALLEY (1959) and TEAL (1962). Since the early 1970s the research at Sapelo Island has included a succession of simulation models (WIEGERT and co-authors, 1975) which led to further hypotheses, research to validate them, and newer models based on the findings.

Fig. 3-12a summarizes the main carbon flows over the whole marsh and estuary. Twenty-one percent of the area is assumed to be subtidal. WIEGERT and co-authors (1981) depict carbon fluxes in the salt marsh in three phases: air, water, and sediments. The communities also differ from short *Spartina* to tall *Spartina* to subtidal areas. For simplicity these are not indicated separately in Fig. 3-12, although processes in air (involving the insects and arachnids) are clearly different from those in water (which are aerobic) and those in the sediments (which are largely anaerobic).

The major primary producer in the salt marsh is the marsh grass *Spartina*, whose growth is about equally apportioned between above-ground (shoots) and below-ground (roots) (POMEROY and co-authors, 1981). Roughly 5% of the above-ground production is assimilated by insect grazers (TEAL, 1962) and about 10% of insect assimilation is passed on to spider carnivores. Thus, in energetic or carbon terms the aerial grazer food chain is unimportant. The remainder of the *Spartina* production becomes dissolved or particulate organic matter, either in the water or in the sediments. The microbial community utilizes the organic matter as an energy source and respires some 30% (340 g C $m^{-2}$ $yr^{-1}$) of the *Spartina* production left after grazing (CHRISTIAN and co-authors, 1981). A large part of the remaining organic matter is processed anaerobically in the sediments, but its proportion has not been quantified in nature (WIEBE and co-authors, 1981). Nevertheless these authors have shown that the four main anaerobic processes—fermentation, dissimilatory nitrogenous oxide reduction, sulphate reduction, and methanogenesis—are closely coupled in the salt marsh and need to be studied simultaneously in order to explain changes in one rate from the state of the other three. The main factors affecting these processes appeared to be interstitial water flow rate and substrate availability.

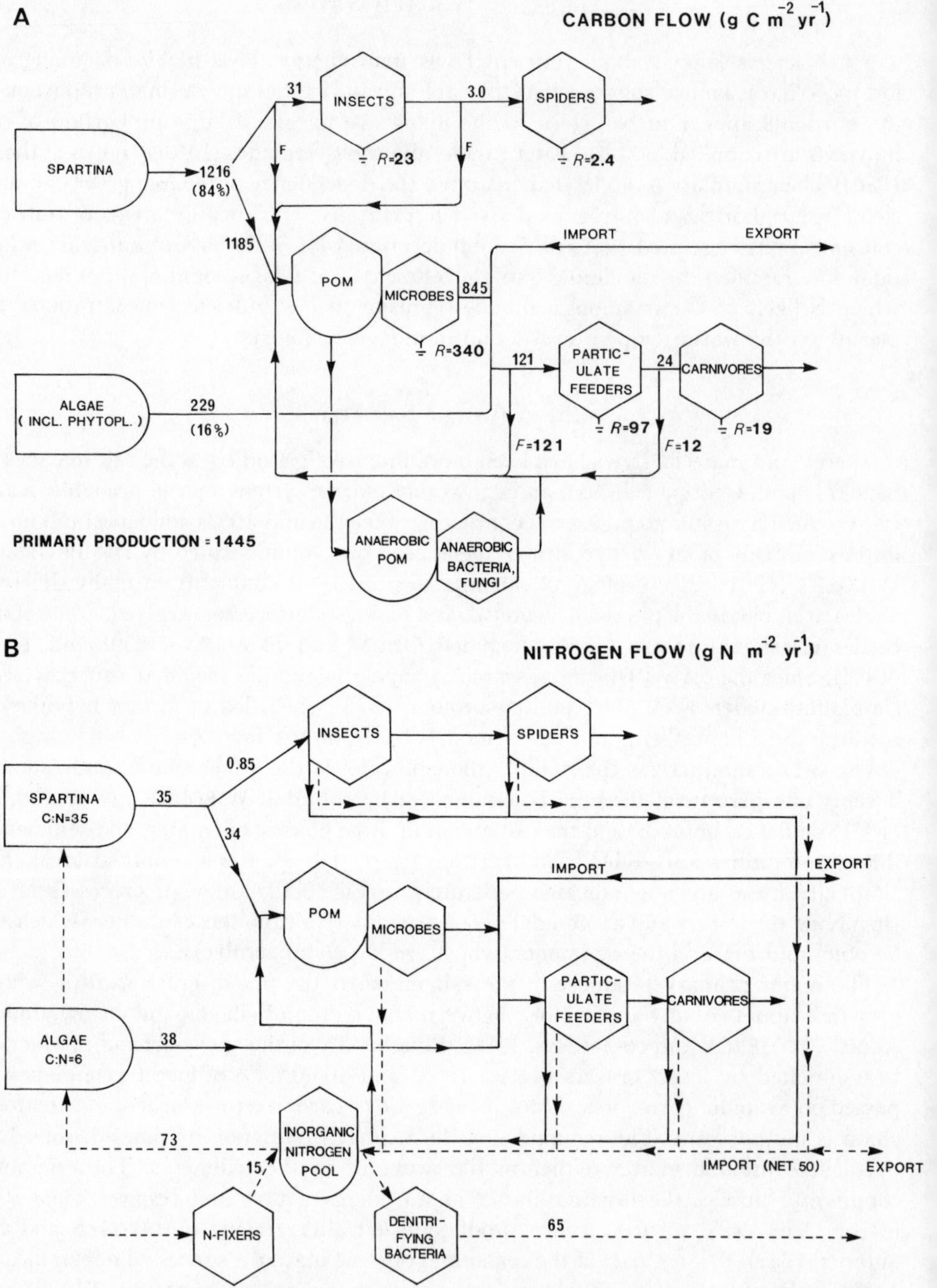


Fig. 3-12: Flows of carbon and nitrogen in a Georgia saltmarsh. (A) Estimated flows of carbon showing primary production, faeces produced (F), carbon assimilated, and carbon respired (R), all in g C m$^{-2}$ yr$^{-1}$. Particulate feeders assumed to have an assimilation efficiency of 50%. Tidal import and export of detritus (particulate organic matter, POM, plus microbes) shown as unestimated amount of carbon processed anaerobically in the sediments which feeds back to the aerobic POM. (B) shows the corresponding flows of nitrogen assuming the C : N ratios given for *Spartina* and algae. Nitrogen is recycled via organic faeces (solid lines) and inorganic ions (broken lines). Tidal import and export of detrital and inorganic nitrogen also shown. Symbols as in Fig. 3-7. (Based on POMEROY and WIEGERT, 1981; reproduced by permission of Springer-Verlag, New York)

The various particulate feeders include filter feeding mussels, clams, and oysters, deposit-feeding fiddler crabs, snails and polychaetes, and 'deposit–suspension' feeders such as mullet and shrimp (MONTAGUE and co-authors, 1981). These animals feed on phytoplankton and benthic algae, as well as on detritus derived from *Spartina*, faeces, and other sources. Most salt-marsh literature assumes that detritus is the main food source (e.g. TEAL, 1962) but recent evidence suggests that algae are more important in the diet than detritus, both from gut content analyses and from $^{12}C/^{13}C$ ratios (WETZEL, 1976; HAINES and MONTAGUE, 1979). Thus, although it appears that some 845 g C $m^{-2}$ $yr^{-1}$ might be available to particulate feeders as detritus, much of this refractory (VALIELA and TEAL, 1979) and a larger fraction of the algal production of some 229 g C $m^{-2}$ $yr^{-1}$ is incorporated in the estimated 121 g C $m^{-2}$ $yr^{-1}$ assimilated by particulate feeders. If one included all *Spartina* and algal production, the carbon transfer efficiency to particulate feeders would be 121/(1185 + 229) = 8·6%, which is misleadingly low in view of the transfer of some organic matter through several processes in the sediments or its transport out of the system. The carbon transfer efficiency from particulate feeders to scavengers and carnivores—such as blue crab, catfish, and shark—is about 20%. The diagram (Fig. 3-12a) depicts the feedback of carbon to the detritus pool via faeces and also indicates that aerobic microbial respiration accounts for some three times more carbon than all the macro-consumers together. In spite of the apparent surplus of carbon, much of it is refractory and poor in nitrogen, and it appears that food limits the population size of consumers, as evidenced by the widespread occurrence of omnivory (MONTAGUE and co-authors, 1981). The micro-organisms appear to be in equilibrium with the DOC supply with bacteria being kept in check by ciliates grazing on particles, often of faecal origin (CHRISTIAN and co-authors, 1981).

Nitrogen flows are depicted in Fig. 3-12b. The estimates are calculated from assumed C : N ratios of 35 : 1 for *Spartina* and 6 : 1 for algae. Thus, in contrast to the carbon flow figures, algae contribute over half the nitrogen available to consumers. Nitrogen fixation contributes 14·8 g N $m^{-2}$ as opposed to the estimated 65 g N lost by denitrifying bacteria. Some 3·6 g N is contributed by sedimentation and rain and the balance of about 47 g N is believed to be gained from seawater through tidal exchange. Thus, like the Great Sippewissett marsh (VALIELA and TEAL, 1979) the high rates of nitrogen fixation are exceeded by denitrification, but, unlike Sippewissett, the Sapelo Island marsh gains negligible nitrogen from groundwater and there is a net transport of nitrogen by seawater into, not out of, the marsh.

Flows of phosphorus have also been studied at Sapelo Island. Standing stocks suggest that phosphorus is never limiting (WHITNEY and co-authors, 1981), and that its recycling rate is rapid and independent of inputs from the ocean (POMEROY, 1960). The role of the mussel *Geukensia demissus* in transferring phosphorus from the water to sediments was highlighted by KUENZLER (1961), demonstrating that a population of negligible importance in energy flow could have a major role in the ecosystem.

Salt marsh studies have shown that other macro-consumers may also play an important role as controllers of energy flow, rather than as direct metabolizers of energy. As illustrated in Fig. 3-12, macroscopic animals degrade less energy than microbes (TEAL, 1962; POMEROY and co-authors, 1977) and nutrient remineralization through excretion is generally inversely proportional to body size (JOHANNES, 1964; BUECHLER and DILLON, 1974). However, many large animals masticate food, reducing the particle size and making the egested material more accessible to microbial attack by increasing its surface area. Some animals such as marsh crabs aerate the sediments by burrowing, others such as fiddler crabs and mullet turn over and aerate the top layers of sediment (EDWARDS and FREY, 1977), allowing aerobic micro-organisms to play their role.

Finally, WIEGERT and POMEROY (1981) conclude that the physical and chemical attributes of the salt-marsh soil confer stability on the system. These in turn are most affected by the tidal flows which are also responsible for maintaining nitrogen balance, and for the transfer of carbon. Long-term or permanent changes in the pattern of tidal flow affect the vegetation and the soil characteristics whereas perturbations of a more biological nature apparently have much less effect on the ecology of the marsh.

### (c) Nova Scotia Kelp Beds: Commercially Important Food Chains

MILLER and co-authors (1971) published the first synthesis of energy flow through a kelp-bed food web. Their paper addresses questions concerning the relationships between the main primary producers, kelp, and the yield to man of lobsters, the top carnivores in the community. The main food chain in terms of energy flow leads from kelps to sea urchins to lobsters to man, although predators such as rock crabs (*Cancer*) and wolffish (*Anarhicas*) also feed on sea urchins. To complicate matters, lobsters and wolffish also prey on crabs (BREEN and MANN, 1976). It is apparent that a large portion of kelp production is not grazed by urchins or periwinkles, and that there is a 'surplus' of kelp-derived detritus which is exported from St Margaret's Bay.

During this study of the kelp beds in Nova Scotia (Canada) between 1968 and 1976, it was observed that aggregations of sea urchins became more frequent and larger, leaving rock areas devoid of kelp; 90% of the former subtidal seaweed beds disappeared during this period (MANN, 1977). The urchins in healthy kelp beds showed cryptic behaviour, hiding in crevices by day and appeared to feed mainly on pieces of drift weed. Those in aggregations showed active feeding behaviour, were not cryptic, and advanced on kelp beds eating through the holdfasts and stipes of plants so that they drifted away. Large areas of 'barrens' covered by sea urchins spread down the south shore of Nova Scotia, far beyond St Margaret's Bay. Lobster catches in these barrens eventually became reduced (WHARTON and MANN, 1981), possibly due to reduced recruitment (G. HARDING, pers. comm.) or to overfishing (BREEN and MANN, 1976) and probably due to both (MANN, 1982b). Thus it appears that a feedback from the barrens further reduces lobster stocks after a time lag. This may be due to increased mortality of lobsters without kelp cover, or due to depleted primary productivity and hence lobster food, suggested by the drastically reduced urchin gonads in the barrens (MANN, 1982a,b).

It appears that the reduced densities of lobsters and possibly also other predators (MANN, 1982a,b), led to less predator control of urchins, and with a resultant increase in population size, urchin behaviour changed, causing destructive grazing on kelps. This removed the main source of primary production and also altered the habitat, increasing the exposure of lobsters and crabs to predators. This was not an isolated phenomenon but has been documented over a 400 km stretch of coast (WHARTON and MANN, 1981; MANN, 1982a,b). Fig. 3-13 suggests how this system appears to have switched in structure from a macrophyte-based system controlled by predators to a food chain in the 'barrens' based on the much lower productivity of microalgae, supplemented by settling detritus (MANN, 1977), possibly of phytoplankton origin. The latest reports are that there have been mass mortalities of urchins in some barren areas during the warmest months of 1980 and 1981, and that kelp beds are reappearing there (A. R. O. CHAPMAN and K. H. MANN, pers. comm.). Apparently the 'barrens' community is an unstable one in which

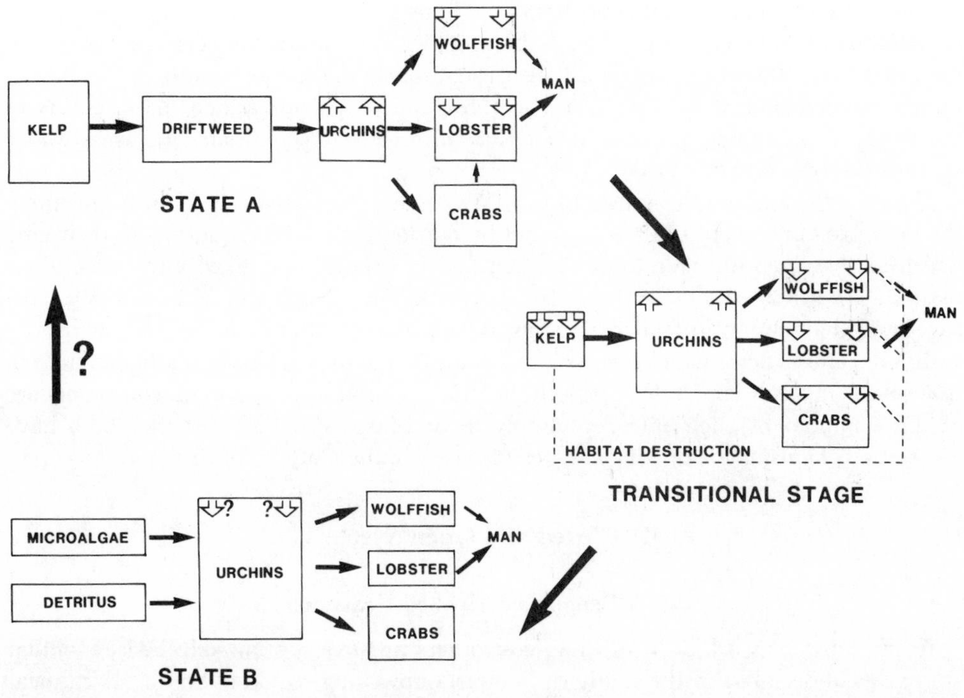


Fig. 3-13: Diagrammatic representation of the possible structural changes in a kelp—urchin—lobster food chain documented in St Margaret's Bay, Nova Scotia. Solid arrows in the food chain represent energy flow of food. State A: 'healthy' state of the community, but declining predator populations result in increasing sea-urchin populations (open arrows). The transitional stage is unstable as the urchins' behaviour changes to destructive grazing on kelps, resulting in loss of optimal habitat for predators. State B: 'barrens' with sea-urchins subsisting on the lower productivity of microalgae and detritus on rock surfaces. Indications are that epidemic disease may start to cause a crash in the urchin populations in which case the community may naturally revert to State A, based on the higher primary production of kelp (Original; based on BERNSTEIN and co-authors, 1981; CHAPMAN, 1981; MANN, 1982a)

the urchins are limited by food availability since their growth rate is greatly reduced (WHARTON and MANN, 1981), and after several years the increased metabolic demands of higher temperatures at the end of summer make them vulnerable to fungus disease. It remains to be seen whether the system will revert to the more productive condition based on kelp plants and whether lobster yields will increase again.

Aspects of nitrogen flow have been studied in Nova Scotian kelp beds, although a nutrient budget has not been attempted for the complete ecosystem. At the primary producer level, it has been shown that the availability of combined nitrogen limits kelp growth at certain times of year. During the winter months it has been shown that kelp plants in St Margarets Bay take up nitrate and store it in their tissues, enabling them to maintain fast growth rates after the phytoplankton spring bloom had depleted the nitrate concentration in seawater (CHAPMAN and CRAIGIE, 1977). It appears that high rates of photosynthesis are maintained through the summer, but growth is reduced once the nitrogen reserves are used up and excess photosynthate is released as dissolved organic matter (HATCHER and co-authors, 1977), presumably to be rapidly utilized by bacteria which can meet their nitrogen requirements from extremely low concentrations in seawater (AZAM and co-authors, 1983a,b). Thus, carbon fixation and nitrogen uptake by kelps are only loosely coupled when nitrogen becomes limiting at certain seasons. On the other hand the same species in an area of upwelling at the mouth of the Bay of Fundy, Nova Scotia (GAGNE and MANN, 1981) and these populations, like *L. pallida* in the Benguela upwelling system, do not store nitrogen in their tissues for use at other seasons (GAGNE and co-authors, 1982).

Amongst the consumers, sea urchins in Nova Scotia have been shown to supplement the poor protein content of their kelp diet by harbouring micro-organisms in their guts which fix dissolved nitrogen from seawater and incorporate the fixed nitrogen in their tissues (GUERINOT and co-authors, 1977). In addition, FONG and MANN (1980) have suggested that urchin gut micro-organisms help digest plant proteins and synthesize additional amino acids which are essential to the diet of the urchins and are not found in *Laminaria*. One may conclude that both the primary producers and main consumers are at times limited by their nitrogen supply in St Margaret's Bay and that both have evolved mechanisms for making the best use of a limited supply of nitrogen.

## (4) Closed and Open Systems

### (a) A Benguela Kelp-Bed Ecosystem

Inspired by the work of MANN and co-workers in Nova Scotian kelp beds, a similar study was undertaken in the southern Benguela upwelling region of South Africa, near Cape Town. Preliminary accounts of the food web have been described in two publications (FIELD and co-authors, 1977; VELIMIROV and co-authors, 1977). The principal difference from Nova Scotian kelp beds is that the South African spiny lobster *Jasus lalandii* feeds mainly on the mussel *Aulacomya ater*, while *Homarus americanus* in Nova Scotia feeds mainly on the most abundant prey, urchins and crabs. After analysis of the biomass composition and distribution in South African kelp beds(FIELD and co-authors, 1980a) a series of studies was aimed at the dynamics and energetics of the dominant populations: kelps (MANN and co-authors, 1979; DIECKMAN, 1980), sea urchin (GREENWOOD, 1980); mussel (GRIFFITHS and KING, 1979a,b; STUART and co-authors, 1982a); amphipod *Talorchestia* (MUIR, 1978), isopod; *Cirolana* (SHAFIR and FIELD, 1980a,b) and *Ligia* (KOOP and FIELD, 1980, 1981), and lobster (NEWMAN and POLLOCK, 1974; POLLOCK, 1979; GRIFFITHS and SEIDERER, 1980; SEIDERER and co-authors, 1982). Other studies were directed at the composition and degradation of detritus from dissolved matter (LINLEY and co-authors, 1981; LUCAS and co-authors, 1981) and from particulate kelp and mussel faeces (STUART and co-authors, 1981, 1982b). Based on these studies and assuming that export losses balanced import gains, NEWELL and co-authors (1982) drew up an energy budget which showed the annual primary production in the kelp beds was about balanced by the energy requirements of the consumers.

The assumption that imports balance exports is not a trivial one on an open coast and current measurements at different stations and depths around a kelp bed showed that under active upwelling or downwelling conditions off the Cape Peninsula the water column may turn over as much as seven times per day (FIELD and co-authors, 1980b; 1981). WULFF and FIELD (in press) set out to explore the implications of such upwelling and downwelling water transport on the food-web structure within the kelp-bed community, using a simple mathematical simulation model. The model showed that when there was no water transport, or there are rapid changes in transport direction so that the same water is transported into and out of the kelp bed, re-suspended faeces can make an important contribution to the detritus and to the main food chain which adequately supported the filter feeders and lobsters. This situation is analogous to food chains of many estuarine and salt-marsh detritus communities in which there is relatively little export of material (see pp. 763, 769). Modelling downwelling conditions which prevail in the four winter months and occur sporadically in summer, phytoplankton blooms from the offshore regions of the Benguela region were advected into the model kelp bed so that the main food chains were based on phytoplankton, which formed increasing proportions of the consumer diets with faster downwelling. The model thus depicts a food web analogous to the phytoplankton-based one in Narragansett Bay (p. 759). When the model was used to simulate upwelling conditions, increasing portions of detritus were exported from the community with faster rates of water transport. Thus, although detritus remained the major proportion of the diet available, smaller quantities were available to filter feeders with faster upwelling and only at the slowest upwelling rates was there sufficient food to meet consumer requirements. Thus, paradoxically, the consumers tended to starve under upwelling conditions because their detritus supply was advected away.

Fig. 3-14 summarizes results obtained when a realistic seasonal upwelling index was used. The model shows that the animal populations could maintain a stable biomass on the food available, but they revealed a net energy gain (equivalent to positive scope for growth and reproduction) in the four winter months when downwelling predominates, and a negative scope for growth during the upwelling season. The model prediction has been confirmed by lipid analyses which show a large proportion of neutral lipids (storage products) in winter samples of mussels, while summer samples suggest starvation with small proportions of neutral lipids (I. E. HORGAN, pers. comm.). Sensitivity analyses on the kelp-bed simulation model show that large changes in biological parameters, such as food selectivity and assimilation efficiency, have relatively little effect on the energy gain of filter feeders compared to their great sensitivity to the rate of water transport in upwelling and downwelling. Thus, one may conclude that physical advection processes may have a dominating influence on the nature of, and biomass maintained by, near-shore communities.

### (b) A South African Sandy Beach Ecosystem

Sand beaches are found on coasts exposed to water movement in the form of waves or currents which prevent the sedimentation of fine muddy material. Traditionally they are regarded as 'open' systems which filter living and non-living particulate material imported from primary producers elsewhere, and return mineral nutrients to the ocean (PEARSE and co-authors, 1942; BROWN, 1964). This may be true of intertidal sandy sediments, which are insufficiently stable for significant populations of algae or higher plants to develop on or in the sand (BROWN, 1964; MUNRO and co-authors, 1978). However, MCLACHLAN (1980) has pointed out that the beach should be considered together with its adjacent surf zone, which may then comprise the necessary producer, consumer, and decomposer components to form a functioning ecosystem. The principal remaining criterion for a viable ecosystem is that the flows of energy and materials

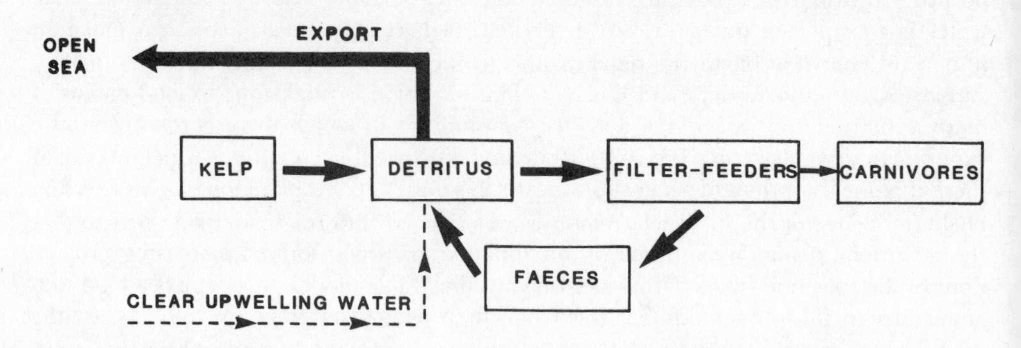
A UPWELLING FOODWEB
OPEN SEA
EXPORT
KELP
DETRITUS
FILTER-FEEDERS
CARNIVORES
FAECES
CLEAR UPWELLING WATER

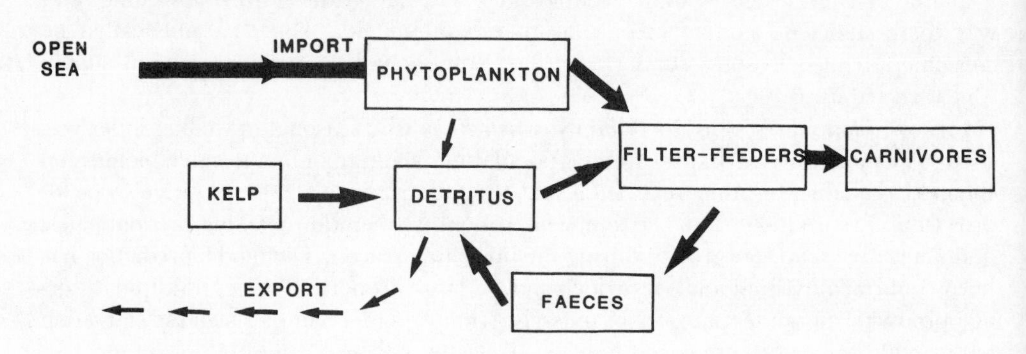
B DOWNWELLING FOODWEB
OPEN SEA
IMPORT
PHYTOPLANKTON
FILTER-FEEDERS
CARNIVORES
KELP
DETRITUS
FAECES
EXPORT

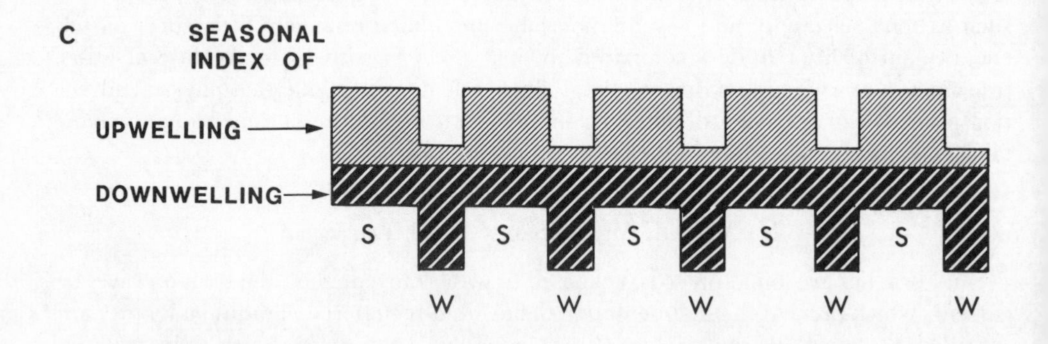
C
SEASONAL INDEX OF
UPWELLING
DOWNWELLING
S
S
S
S
S
W
W
W
W
W

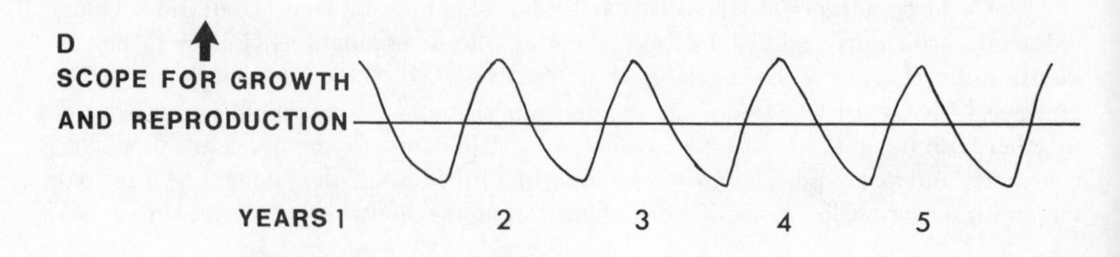
D
SCOPE FOR GROWTH AND REPRODUCTION
YEARS 1
2
3
4
5

within the system should be greater than the imports and exports to and from it. McLachlan and co-authors (1981) present evidence to demonstrate that this is true for a beach at Port Elizabeth, South Africa, in the first study of a sandy shore as a complete ecosystem.

Several workers have observed that surf zones are areas of intensive phytoplankton growth, with dense blooms of diatoms (Rapson, 1954; Lewin and Norris, 1970; Lewin and co-authors, 1975; McLachlan, 1980), by implication productivity being higher in the surf zone than outside it. This, supported also by evidence of cellular water circulation patterns on some sandy shores (Harris, 1978), suggests that flows within these surf-zones and beaches may indeed be greater than the imports and exports.

Fig. 3-15 depicts energy and nutrient flows in the sandy shore ecosystem at Port Elizabeth (South Africa). The width of sandy shore ecosystems, from the high-water-springs mark to the outer limit of circulation cells and turbidity beyond the breakers, varies according to the state of tide and waves, and the slope of the beach. McLachlan and co-authors (1981) therefore express energy flows per metre of shoreline rather than per square metre of surface area. The system is based on imports of carrion (estimated by regular surveys of strips of beach) and particulate detritus, and by phytoplankton. The proportions of phytoplankton production inside the system and imported to it have not yet been estimated, but the total amount of particulate matter has been calculated from the requirements of the filter feeders and the interstitial fauna. The carrion consists mainly of medusae and the Portuguese-man-of-war, *Physalia*, which are consumed by the scavenging whelk *Bullia* (Brown, 1971). The only other major invertebrate carnivore in the system is the crab *Ovalipes* which preys principally on molluscs. The remaining macrofaunal biomass consists chiefly of the bivalves *Donax serra* and *D. sordidus*, and the mysid *Gastrosaccus psammodytes*, all filter feeders (Brown, 1964; McLachlan, 1977). The energy requirements of the macrofauna were calculated from P/B ratios (McLachlan, 1979a; McLachlan and Hanekom, 1979) and from respiration rates Dye, 1979; Dye and McGwynne, 1980), giving an estimate of 109 421 kJ $m^{-1}$ $yr^{-1}$ required by filter feeders and 6500 kJ $m^{-1}$ $yr^{-1}$ by scavengers.

The sand has a rich interstitial biota of bacteria, protozoans, and meiofauna whose food relations are poorly understood in detail, but the biomass and respiration rates have been quantified (McLachlan, 1977; Dye, 1979). The interstitial food web is based on dissolved and particulate organic matter (McIntyre and co-authors, 1970) which is filtered through the sand column by wave action and by tides (Volume I: Riedl, 1971a; Riedl, 1971b, Riedl and Machan, 1972). The volume filtered has been quantified by McLachlan (1979b, 1980). Dissolved and particulate matter are utilized by bacteria

Fig. 3-14: Diagrammatic representation of the changing pattern of energy flow under upwelling and downwelling conditions in a southern Benguela kelp bed. (A) Upwelling conditions: negligible phytoplankton present as clear water wells up from below the photic zone and transports detritus out of the system. (B) Downwelling conditions: phytoplankton blooms from offshore are carried through kelp beds providing a reservoir of nitrogen-rich food. (C) Input of seasonally averaged upwelling/downwelling index, showing major summer upwelling season (S) and winter downwelling season (W) for a simulation of 5 yr. (D) Simulation-model output showing energy balance in a filter feeder community, which suffers net loss under upwelling and gain under downwelling conditions; horizontal line shows level of energy gain required to maintain observed filter feeder biomass. (After Wulff and Field; modified; reproduced by permission of Inter-Research, Halstenbek)

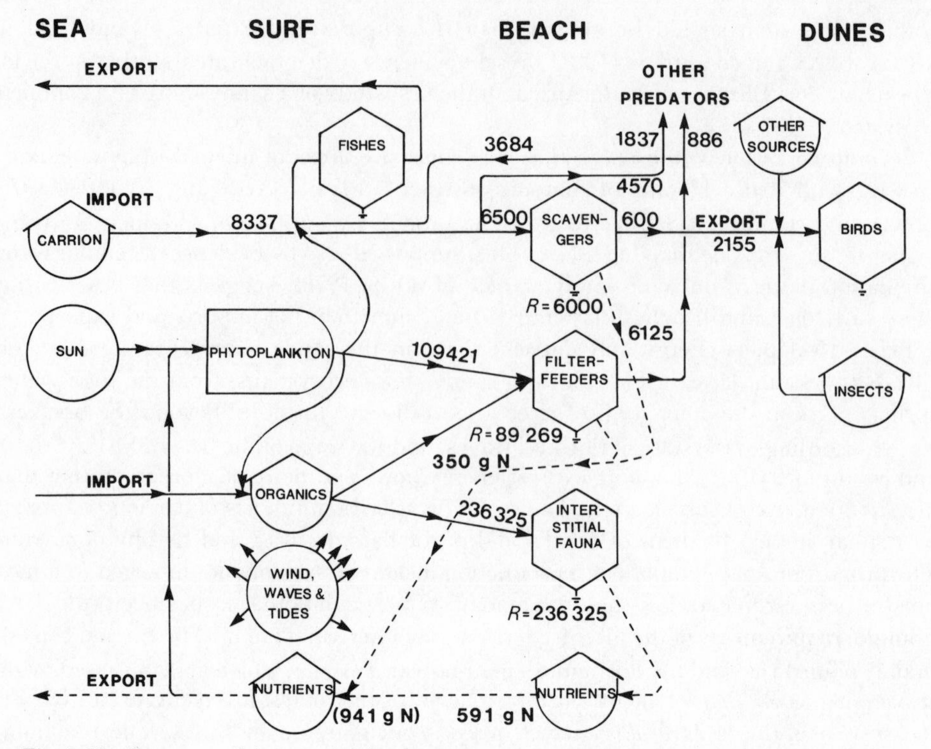


Fig. 3-15: Energy and inorganic nitrogen flows on a sandy surf-beach ecosystem at Port Elizabeth (South Africa) in relation to the sea on one side and sand dunes on the other. Energy flows (solid lines) given as kJ (m shore)$^{-1}$ yr$^{-1}$. Nitrogen flows given as g N (m shore)$^{-1}$ yr$^{-1}$. Cellular water circulation patterns are believed to retain phytoplankton in the surf zone, allowing much recycling of nutrients within the system. Symbols as in Fig. 3-7. (After McLachlan and co-authors, 1981; modified; reproduced by permission of Academic Press, London)

which in turn are preyed upon by bacteri-detritivores (principally protozoans) and carnivorous meiofauna, forming a 'closed' food chain within the interstitial community (McIntyre and co-authors, 1970), and exporting only the minerals excreted. Energy requirements of the bacteria and meiofauna were calculated from biomass and respiration measurements, using literature values for P/R and P/B ratios, respectively. The interstitial energy requirements were calculated to be 236 325 kJ m$^{-1}$ yr$^{-1}$, which combined with the requirement of 109 421 kJ m$^{-1}$ yr$^{-1}$ by filter feeders gives an estimated 345 745 kJ m$^{-1}$ yr$^{-1}$ of phytoplankton plus dissolved and particulate organic matter. When the imported carrion is added to this, the total energy input works out at 354 083 kJ m$^{-1}$ yr$^{-1}$ of which about 67% goes to the interstitial fauna, some 31% is consumed by filter feeders, 1·8% goes to scavengers, and 0·2% is taken by birds and fish. These estimates ignore the particulate matter filtered by fish such as mullet which migrate in and out of the surf zone.

Nitrogen mineralization has been estimated for the macrofauna from ammonia excretion rates published for bivalves (Bayne and co-authors, 1976; Lewin and co-authors, 1979). Nitrogen mineralization rates of the interstitial community were estimated by assuming that dissolved and suspended organic matter have an energy content of 16 kJ $g^{-1}$ and contain 4% by mass of nitrogen to give a yield of 591 g N $m^{-1}$ remineralized per year (McLachlan and co-authors, 1981). Nutrient runoff from the land is believed to be insignificant in the system studied. The total amount of nitrogen remineralized (941 g N $m^{-1}$) is either used within the system or is exported to the ocean. The relative proportions are unknown, although it is argued that the dense filter-feeding populations observed are only likely to occur when circulation patterns trap nutrients recycled by the beach biota within the system long enough to allow phytoplankton blooms to develop.

Observations of diatom blooms in the surf zone in various other parts of the world (Rapson, 1954; Lewin and Norris, 1970; Lewin and co-authors, 1975) suggest that this may not be an isolated phenomenon and that intertidal sandy beaches and the adjacent surf zones may function as ecosystems with some coherence. The extent to which they act as 'digesting and incubating systems . . . exporting minerals to the sea' (Pearse and co-authors, 1942) is likely to vary from beach to beach according to local water movements. On long open beaches of gentle slope there are wide surf zones, and cellular circulation patterns then tend to develop (McLachlan and co-authors, 1981), so that the physical forces determine whether there is an ecosystem with some integrity and its own populations of primary producers and nutrient cycling. Certainly sandy beaches are always open systems, and direct measurements are required to quantify the flows of water through and within such systems.

### (c) Rocky shores

Although there have been many studies of feeding, competition, predation, and community interactions on rocky shores in different parts of the world (e.g. Connell, 1970; Dayton 1971; Paine, 1974, 1977; Branch, 1975, 1981; Underwood, 1976), there have been no quantitative studies of energy or material flow through a complete rocky shore community. The reason for this is clear, for rocky shores are very complex communities, and in addition, much material is transported to and from rocky shores by waves and currents, with rates of transport that are very difficult to measure.

In an attempt to fill this gap, this section will utilize data at present being gathered on a rocky shore at Dalebrook, False Bay, near Cape Town (South Africa) by G. M. Branch and his students who are assembling information on energy flow through the system. McQuaid (1980) studied Dalebrook and five other rocky shores exposed to wave action and compared these with six rocky shores sheltered from wave action. Table 3-3 summarizes his data. The mean biomass of animals and seaweeds on exposed shores was over three times higher than that on sheltered shores and this applied to all trophic categories. Filter feeders dominate the biomass on exposed shores, exceeding even the primary producers, while on sheltered shores the more usual balance is of primary producers followed by herbivores and detritus feeders. Thus, it appears there is both a better supply of food to support more filter feeders, and a bigger biomass of algae on exposed shores. The sources of primary production on rocky shores are difficult to quantify although some seaweed production rates have been measured (e.g. Kanwisher, 1966; Khailov and Burlakova, 1969; Litter and Murray, 1974; Towle and Pearse, 1973; Brinkhuis and Jones, 1974; Brinkhuis, 1977a,b). Using $^{14}C$ uptake, Fielding and Branch (unpubl.) have estimated the net production of the four

Table 3-3

Biomass (g dry mass $m^{-2}$) of different trophic categories, comparing transects across the entire intertidal region at 6 exposed and 6 sheltered rocky shores near Cape Town, South Africa. Percentages are given in parentheses (Based on McQUAID, 1980)

| | Exposed | | | Sheltered | | |
|---|---|---|---|---|---|---|
| Category | Min. | Max. | Mean. | Min. | Max. | Mean. |
| Primary producers | 197 | 710 | 366 | 27 | 488 | 233 |
| | (21) | (81) | (37) | (31) | (89) | (79) |
| Herbivores/detritus feeders | 29 | 127 | 54 | 23 | 66 | 41 |
| | (1) | (24) | (5) | (3) | (52) | (13) |
| Filter feeders | 44 | 1361 | 565 | 0·3 | 19 | 10 |
| | (8) | (73) | (56) | (0·1) | (8·5) | (3·6) |
| Carnivores/scavengers/omnivores | 12 | 49 | 26 | 0·4 | 15 | 13 |
| | (1) | (5) | (2) | (0·6) | (18) | (4·4) |
| TOTAL | | | 1010 | | | 292 |
| | | | (100) | | | (100) |

locally commonest intertidal algae (*Ulva, Gelidium, Gigartina, Bifurcaria*) at the surface and at 1-m and 2-m depths. Using macrophyte biomass data, these rates give estimates of some 97 g C m shoreline$^{-1}$ d$^{-1}$ in summer and 23 g C m$^{-1}$ d$^{-1}$ in winter. If there is a 20-m wide intertidal zone of which the lowest 10-m band has macrophytes, this gives an estimated 1100 g C m$^{-2}$ yr$^{-1}$ of macrophyte production.

Using short-term clearance and exclusion experiments, BRANCH and BRITZ (unpubl.) have found that the surface film of sporelings and microalgae, normally grazed flat by limpets and other gastropods, grow to heights of 7 to 8 cm within 2 wk. They estimate a productivity at least as great as that of the seaweeds whose biomass is obvious, and one can thus estimate to be some 1100 g C m$^{-2}$ yr$^{-1}$. Presumably, when nutrient concentrations are low, these high rates of production by benthic algae are made possible by the rapid rate of ammonia and phosphate release by the animal community (SOLOVEVA and co-authors, 1977; KAUTSKY and WALLENTINUS, 1980; RAINE and PATCHING, 1980). On shores frequented by large colonies of birds, particularly islands, the contribution of bird guano to nutrient cycling can be important, resulting in enhanced seaweed growth on the shore (BRANCH and HOCKEY, unpubl.).

Finally, phytoplankton contributes to the diet of filter feeders. CLIFF (1982b) has shown that chlorophyll *a* concentrations decrease shorewards along a 6-km transect at Dalebrook during most seasons, suggesting that shallow-water filter feeders may deplete phytoplankton biomass. Chlorophyll *a* concentrations near the rocky shore ranged from 1 to 19 $\mu$g l$^{-1}$, but averaged a little over 2 $\mu$g l$^{-1}$. If we assume that some 20 m$^3$ of unfiltered water passes over each square metre of intertidal shore per day and that the phytoplankton carbon : chlorophyll ratio is 100 (average of local summer and winter values, ANDREWS and HUTCHINGS, 1980), then some 1500 g C m$^{-2}$ yr$^{-1}$ is available to filter feeders from phytoplankton. This is not an unreasonable estimate and is comparable with phytoplankton primary production in the nearby southern Benguela region (BROWN, 1981); however, it obviously depends very heavily on water exchange rates which have not been measured.

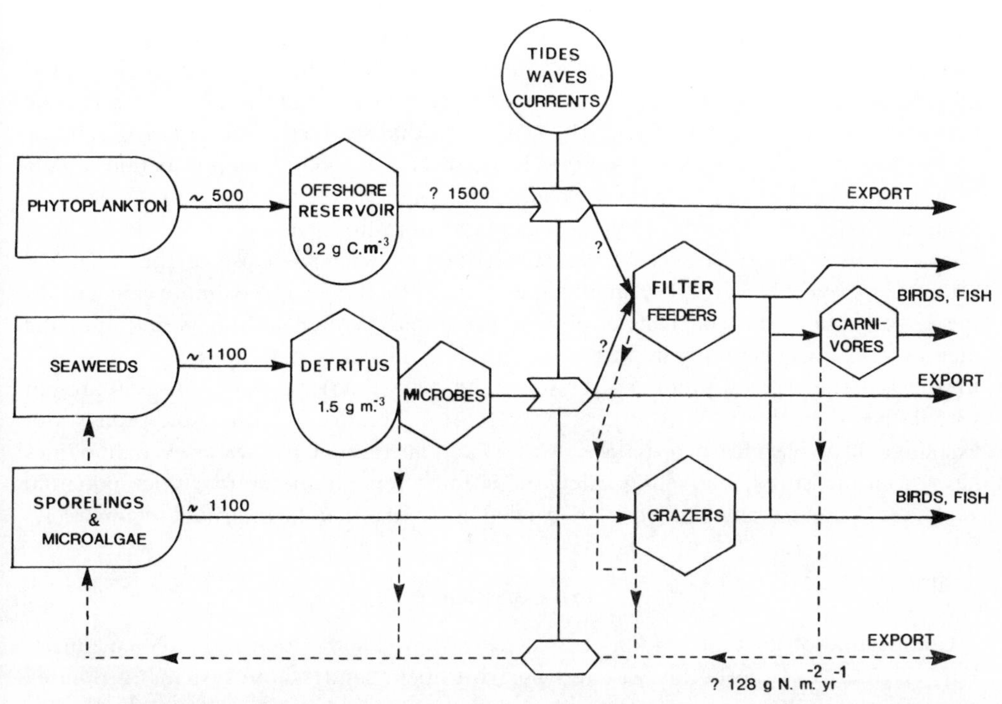


Fig. 3-16: Hypothetical diagram of possible carbon and nitrogen flows on an exposed rocky shore, based on observations near Cape Town, South Africa. Flows are expressed as g C $m^{-2}$ $yr^{-1}$ unless otherwise indicated, and standing stocks as g $m^{-3}$ of water. The influence of tides, waves, and currents in transporting phytoplankton, detritus, and excreted minerals (broken lines) indicated by means of directional work-gates. Flows are approximate and the relative proportions of plankton and detritus in filter feeder diets unknown. Symbols as in Fig. 3-7. (Original; partly based on BRANCH and BRANCH, 1981)

Fig. 3-16 shows a hypothetical carbon-flow diagram for an exposed rocky shore such as Dalebrook. Phytoplankton primary production is offshore and imported into the intertidal area by tides, waves, and longshore currents generated by waves. Thus, the rate of phytoplankton production is not as important to the intertidal shore as the rate of water exchange, which is indicated by 'work gates' in the flow diagram. Even a low concentration of phytoplankton or detritus can provide a large amount of food if the reservoir is large and the rate of transport fast. CLIFF (1982a) showed that detritus forms over 80% of the suspended particulate matter on this shore, but most of it was unidentifiable and therefore not likely to have been recently eroded from seaweeds on the same shore. If filter feeders preferentially assimilate phytoplankton, and egested faeces are re-suspended as detritus, it is easy to see that the activities of filter feeders would rapidly increase the proportion of detritus in suspension. At present we do not know the relative amounts of energy assimilated from detritus and phytoplankton by filter feeders on rocky shores. Grazers, however, eat mainly the thin covering of sporelings and diatoms on rock surfaces at mid-tide levels and account for nearly all this production by 'gardening' (BRANCH, 1981).

The rate of nutrient cycling can be very roughly guessed at from data on ammonium release by mussels (BAYNE and WIDDOWS, 1978; KAUTSKY and WALLENTINUS, 1980).

Assuming that the filter feeders which dominate the biomass on exposed rocky shores (Table 3-3) excrete $NH_4$-N at the same rate as mussels (5–40 $\mu$g $NH_4$-N $g^{-1}$ $h^{-1}$), and that this applies also to the other animals, the mean biomass of 645 g $m^{-2}$ may excrete between 28.5 and 228 g N $m^{-2}$ $yr^{-1}$. If one assumes that seaweeds require nitrogen in the atomic ratio of 20 C : 1N (KAUTSKY and WALLENTINUS, 1980), then production rates of between 485 and 3875 g C $m^{-2}$ $yr^{-1}$ could be supported on the shore by the animal ammonia released. The mean value calculated for nitrogen release would support 2175 g C $m^{-2}$ $yr^{-1}$ of seaweed primary production, which is close to the estimated 2200 g C $m^{-2}$ $yr^{-1}$ for large and small algae (Fig. 3-16). These approximate calculations therefore support the conclusions of KAUTSKY and WALLENTINUS (1980) that the macro-fauna are potentially important in recycling nutrients, and that they may supply the needs of the primary producers. However, on rocky intertidal shores this is at present speculative, since there have been few measurements of nutrient release, and rapid water exchange takes place caused by tides, waves, and currents. Until water exchange rates have been measured, budgeting calculations must remain speculative since potential food, faeces, and excreted minerals are carried into and out of the intertidal region freely.

### (5) Conclusion

Comparison of the nine coastal ecosystems reviewed shows that three (Narragansett Bay, Askö area of the Baltic Sea, and Port Elizabeth sandy shore systems) are based mainly on phytoplankton primary production. Four (the Georgia salt marsh, Florida mangrove forest, Nova Scotia, and Benguela kelp beds) are based largely on macrophytes and detritus while the Lynher mudflat is based principally on attached microalgae. The rocky shore example is not quantitative, but all three sources appear to contribute roughly equally.

When the food webs are compared, it is clear that filter feeders and/or deposit feeders dominate the consumers in all the systems and that faeces make an important contribution to detritus, forming a feedback loop, even in wholly phytoplankton-based systems such as Narragansett Bay. In macrophyte-based systems, faeces become yet more important because the structural components of macrophytes render the plant material less digestible (KRISTENSEN, 1972) and assimilation efficiencies are thus likely to be generally lower than on phytoplankton. However, it now appears that even in macrophyte-dominated systems like salt marshes and kelp beds, phytoplankton and attached diatoms are emerging as important foods of particle feeders (HAINES, 1975; CARTER, 1982; WULFF and FIELD, in press). These findings suggest that some of the distinctions between systems based on macrophytes and those based on phytoplankton may not be clear-cut, since macrophyte-based systems may have an important diatom component and the phytoplankton-based systems have considerable detrital components too.

Abiotic environmental factors such as waves, tides, and currents (Volume I, Part 2) are likely to play a much more important role in coastal systems than in the open ocean, because they are normally more marked and they tend to reach the bottom, affecting the sediments. In the first place, the degree of water movement (Volume I: RIEDL, 1971a,b), obviously determines the type of substratum—rock, coarse sand, fine sand, or mud. This in turn determines the physical structure and composition of the community. Thus, macrophytes can only become established if the substratum is stable: either rock for seaweeds or mud for mangroves and salt marshes. Water movements can also influence the flows of energy and matter through a community; for example, cellular water circulation patterns in a sandy shore surf zone can cause nutrients to be retained and rapidly recycled in a high-energy environment, allowing the development of dense phytoplankton blooms to support the animal communities (McLachlan and co-authors, 1981). Not only are the average and maximum conditions of water movement important, but changes in water movement can cause temporary changes in the flow of energy through a food web. Thus, Wulff and Field (in press) have shown by means of a simulation model that under calm conditions a kelp-bed community is essentially a closed system with a strong 'faeces loop' retaining detritus as the main food source and supporting the observed consumer biomass. Under upwelling conditions, water transports detritus away from the consumers and biomass cannot be maintained, while under downwelling conditions the food web switches to one based on phytoplankton, with a favourable C : N ratio (Russell-Hunter, 1970), and consumers also have an ample energy supply. In all the other coastal systems reviewed water movements have also been shown to be important in transporting food or nutrients into or within the system.

It is not only abiotic factors that control the structure of coastal communities and the patterns of energy flow through them. The history of kelp beds documented in Nova Scotia (see Fig. 3-14) suggests strongly that biotic interactions, and in particular predator control, can determine both the magnitude and the pattern of energy flow. It appears that the system is only stable within a certain range of energy flows and that man's activities may tip the balance of biological interactions out of the domain of stability, allowing it to change to less stable states with altered energy pathways and different flow rates. These biological interactions may be very complex and subtle, such as changes in the behaviour of invertebrates (Bernstein and co-authors, 1981) and have been shown to play an important role in rocky shore communities by Paine and co-workers (e.g. Paine, 1974).

There is less information on nutrient cycling than on energy (or carbon) flows in coastal ecosystems. In the few examples in which nutrient cycling rates have been estimated in whole coastal ecosystems there is no clear evidence as to whether microorganisms or the macro-fauna is mainly responsible for re-mineralization. In Narragansett Bay the benthos (including microbes) appears to be slightly more important than zooplankton (Table 3-2) but on the sandy beach the interstitial fauna re-mineralizes over 50% more than the macro-fauna (Fig. 3-15). The relative proportions are presumably directly related to the total metabolism of the heterotrophic populations. However, even omitting upwelling areas, coastal areas in general have higher primary production rates, and therefore nutrient cycling rates, than the open sea (Ryther, 1969). This is almost certainly brought about by the fact that physical forces (waves, tides, and currents) bring about mixing between the benthic community and the water column. Re-mineralization of nutrients by benthos has been demonstrated in several different types of coastal community (e.g. Nixon and co-authors, 1976; Kautsky and Wallentinus, 1980; Raine and Patching, 1980), and there is no reason to believe that this phenomenon is not important in all coastal areas.

In the water column, because of the roughly equal biomass distribution across the size spectrum from bacteria to fish (Sheldon and co-authors, 1972), and the faster growth and metabolism of smaller organisms (Fenchel, 1974) it follows that most nutrient recycling is via the 'microbial loop'. This loop involves the smallest organisms in the size range 0.2 to 100 μm (WILLIAMS, 1981), with microflagellates and other microzooplankton the major mineralizers. In a North Sea coastal area it is estimated that over 35% of the phytoplankton net production is utilized by bacteria (JOIRIS and co-authors, 1982). It is not known to what extent bacteria are net mineralizers, since they both release inorganic materials and take them up in competition with phytoplankton (AZAM and co-authors, 1983a,b).

In soft sediments, there is usually a large microbial community with bacteria and fungi acting as decomposers in both aerobic and anaerobic conditions; these in turn are kept in check by ciliates and other micro- and meiofauna. The many interwoven biological and chemical interactions which occur in the sediments are beyond the scope of this review; they have been reviewed by FENCHEL and BLACKBURN (1979). Thus, one would normally expect the smaller organisms to contribute most to re-mineralization in sediments, as in the water column (see Fig. 3-15). However, in exceptionally dense populations such as clam beds it is possible that the macro-fauna may so dominate the biomass that this compensates for the greater specific metabolism of microbes. On hard substrata there is less surface area on which a microbial community can become established and re-mineralization is likely to be mainly by the macro-fauna. Thus, mussels such as *Mytilus* (KAUTSKY and WALLENTINUS, 1980) and *Geukensia* (JORDAN and VALIELA, 1982) have been shown to release large quantities of mineral nitrogen used for photosynthesis (SOLOVEVA and co-authors, 1977; see also Figs. 3-10, 3-16), and for uptake by heterotrophic micro-organisms (NIXON and co-authors, 1976).

An important question in coastal ecosystems is whether the flows of energy and minerals are tightly or loosely coupled. It appears that in coastal systems, as in the open sea, nitrogen is the main limiting nutrient, and carbon flow is almost synonymous with energy flow, so the problem revolves around whether carbon and nitrogen flows are closely coupled. NEWELL and FIELD (1982) have compared carbon and nitrogen budgets in a kelp-bed community and show that nitrogen appears to be transferred from one trophic compartment to another much more efficiently than carbon at all levels in the food web. This may well be a manifestation of a shortage of nitrogen having forced the widespread evolution of nitrogen-conserving mechanisms. FENCHEL and BLACKBURN (1979) have demonstrated that if food and feeder have the same C : N ratio and remain in steady state, then carbon and nitrogen must also be given off as $CO_2$, faeces, and urine in the same C : N proportions, i.e. the processes are tightly coupled. This may occur generally in the open sea where the food web is phytoplankton-based and phytoplankton, bacteria, zooplankton, and fish all have C : N ratios in the region of 4 to 7 C : 1N. However, in coastal systems macrophytes may contribute, and they have widely ranging C : N ratios reaching 35 : 1 or even 75 : 1 (RUSSELL-HUNTER, 1970). To convert such material to food of suitable quality requires either extravagant use of carbon while conserving nitrogen, or inorganic nitrogen must be taken up from the water to make up the deficit, as bacteria can do very effectively (AZAM and co-authors, 1983a,b). The extent to which macrophyte detritus of high C : N ratio is processed, is thus probably an important factor in uncoupling carbon and nitrogen flows. It appears likely, therefore, that carbon and nitrogen flows are more tightly coupled in the open sea than in coastal systems, and those coastal systems with the highest detrital component are likely to have least coupled carbon and nitrogen flows.

*Acknowledgements.* I thank my colleagues C. L. Griffiths, D. W. Klumpp, and G. M. Branch for criticising the manuscript, and D. Gianakourass, S. Tolosana, and E. Klumpp for helping prepare the figures and manuscript.

## Literature Cited (Chapter 3)

ALDERDICE, D. F. (1972). Factor combinations: responses of marine poikilotherms to environmental factors acting in concert. In O. Kinne (Ed.), *Marine Ecology*, Vol. I, Environmental Factors, Part 3. Wiley, London, pp. 1659–1722.

ANDREWS, W. R. H. and HUTCHINGS, L. (1980). Upwelling in the southern Benguela current. *Prog. Oceanogr.*, **9**, 1–81.

ARTHUR, J. F. and BALL, M. D. (1979). Factors influencing the entrapment of suspended material in the San Francisco Bay–Delta Estuary. In T. J. Conomos (Ed.), *San Francisco Bay: The Urbanized Estuary.* Pacific Division, American Association for the Advancement of Science, San Francisco. pp. 143–174.

AYALA CASAÑARES, A. and PHLEGER, F. B. (Eds) (1969). *International Symposium on Coastal Lagoons (Origin, Dynamics and Productivity). (Lagunas Costeras, un Symposio; Memoria de Symposio International Sobre Lagunas Costeras).* Universidad Nacional Autonoma de Mexico.

AZAM, F., FENCHEL, T., FIELD, J. G., GRAY, J. S., MEYER-REIL, L. A. and THINGSTAD, F. (1983a). The ecological role of water-column microbes in the sea. *Mar. Ecol. Prog. Ser.*, **10**, 257–263.

AZAM, F., FENCHEL, T., FIELD, J. G., GRAY, J. S., MEYER-REIL, L. A. and THINGSTAD, F. (1983b). The role of free bacteria and bactivory. In M. J. Fasham (Ed.), *Flows of Energy and Materials in Marine Ecosystems.* Plenum Press, New York.

BARBER, R. T. and SMITH, R. L. (1981). Coastal upwelling ecosystems. In A. R. Longhurst (Ed.), *Analysis of Marine Ecosystems.* Academic Press, New York. pp. 31–67.

BAYLY, I. A. E. (1967). General biological classification of aquatic environments with special reference to those of Australia. In A. H. Weatherley (Ed.), *Australian Inland Waters and their Fauna.* National University Press, Canberra. pp. 1–78.

BAYNE, B. L., WIDDOWS, J. and THOMPSON, R. J. (1976). Physiological interactions. In B. L. Bayne (Ed.), *Marine Mussels. Their Ecology and Physiology.* Cambridge University Press, London. pp. 261–299.

BAYNE, B. L. and WIDDOWS, J. (1978). The physiological ecology of two populations of *Mytilus edulis* (L.) *Oecologia*, **37**, 137–162.

BERNSTEIN, B. B., WILLIAMS, B. E. and MANN, K. H. (1981). The role of behavioral responses to predators in modifying urchins' (*Strongylocentrotus droebachiensis*) destructive grazing and seasonal foraging patterns. *Mar. Biol.*, **63**, 39–49.

BOJE, R. and TOMCZAK, M. (Eds) (1978). *Upwelling Ecosystems.* Springer-Verlag, New York.

BRANCH, G. M. (1975). Mechanisms reducing intraspecific competition in *Patella* (spp): migration, differentiation and territorial behaviour. *J. Anim. Ecol.*, **44**, 575–600.

BRANCH, G. M. (1981). The biology of limpets: physical factors, energy flow and ecological interactions. *Oceanogr. mar. Biol. A. Rev.*, **19**, 235–380.

BRANCH, G. and BRANCH, M. (1981). *The Living Shores of Southern Africa.* C. Struik, Cape Town.

BREEN, P. A. and MANN, K. H. (1976). Changing lobster abundance and the destruction of kelp beds by sea urchins. *Mar. Biol.*, **34**, 137–142.

BRINKHUIS, B. H. (1977a). Seasonal variations in salt-marsh macroalgae photosynthesis. I. *Ascophyllum nodosum* ecad *scorpioides. Mar. Biol.*, **44**, 165–175.

BRINKHUIS, B. H. (1977b). Seasonal variations in salt-marsh macroalgae photosynthesis. II. *Fucus vesiculosus* and *Ulva lactuca. Mar. Biol.* **44**, 177–186.

BRINKHUIS, B. H. and JONES, R. F. (1974). Photosynthesis in whole plants of *Chondrus crispus. Mar. Biol.*, **27**, 137–141.

BROWN, A. C. (1964). Food relations on the intertidal sandy beaches of the Cape Peninsula. *S. Afr. J. Sci.*, **60**, 35–41.

BROWN, A. C. (1971). The ecology of the sandy beaches of the Cape Peninsula. Part 1. Introduction. *Trans. R. Soc. S. Afr.*, **39**, 247–277.

BROWN, P. C. (1981). Pelagic phytoplankton, primary production, and nutrient supply in the southern Benguela region. *Trans. R. Soc. S. Afr.*, **44**, 347–356.

BUECHLER, D. G. and DILLON, R. D. (1974). Phosphorus regeneration in fresh-water paramecia. *J. Protozool.*, **21**, 331–343.

CAMERON, W. M. and PRITCHARD, D. W. (1963). Estuaries. In M. N. Hill (Ed.), *The Sea*, Vol. II. Wiley, New York. pp. 306–324.

CAMMEN, L. M. (1980). The significance of microbial carbon in the nutrition of the deposit feeding polychaete *Nereis succinea*. *Mar. Biol.*, **61**, 9–20.

CARPELAN, L. H. (1978). Revision of Kolbe's System der Halobien based on diatoms of California lagoons. *Oikos*, **31**, 112–122.

CARTER, R. A. (1982). Phytoplankton biomass and production in a southern Benguela kelp bed system. *Mar. Ecol. Prog. Ser.*, **8**, 9–14.

CASPERS, H. (1950). Die Lebensgemeinschaft der Helgoländer Austernbank. *Helgoländer wiss. Meeresunters.*, **3**, 120–169.

CASPERS, H. (1959a). Die Einteilung der Brackwasser-Regionen in einem Aestuar. *Archo Oceanogr. Limnol.*, **11**, (Suppl), 155–169.

CASPERS, H. (1959b). Vorschläge einer Brackwassernomenklatur. *Int. Revue ges. Hydrobiol.*, **44**, 313–316.

CASPERS, H. (1967). Estuaries: analysis of definitions and biological considerations. In G. H. Lauff (Ed.), *Estuaries*. A.A.A.S., Washington, D.C. pp. 6–8 (*Publs Am. Ass. Advmt Sci.*, **83**).

CHAPMAN, A. R. O. (1981). Stability of urchin dominated barren grounds following destructive grazing of kelp in St. Margaret's Bay, eastern Canada. *Mar. Biol.*, **62**, 307–311.

CHAPMAN, A. R. O. and CRAIGIE, J. S. (1977). Seasonal growth in *Laminaria longicruris* relations with dissolved inorganic nutrients and internal reserves of nitrogen. *Mar. Biol.*, **40**, 197–205.

CHRISTIAN, R. R., HANSON, R. B., HALL, J. R. and WIEBE, W. J. (1981). Aerobic microbes and meiofauna. In L. R. Pomeroy and R. G. Wiegert (Eds), *The Ecology of a Salt Marsh*. Springer-Verlag, New York. pp. 113–160.

CLIFF, G. (1982a). Seasonal variation in the contribution of phytoplankton, bacteria, detritus and inorganic nutrients to a rocky shore ecosystem. *Trans. R. Soc. S. Afr.*, **44**, 523–538.

CLIFF, G. (1982b). Dissolved and particulate matter in the surface waters of False Bay and its influence on a rocky shore ecosystem. *Trans. R. Soc. S. Afr.*, **44**, 539–549.

COLLIER, A. and HEDGPETH, J. W. (1950). An introduction to the hydrography of tidal waters of Texas. *Publs Inst. mar. Sci. Univ. Tex.*, **12**, 123–194.

CONNELL, J. (1970). A predator-prey system in the marine intertidal region. 1. *Balanus glandula* and several predatory species of *Thais*. *Ecol. Monogr.*, **40**, 49–78.

CUSHING, D. H. (1971). Upwelling and the production of fish. *Adv. mar. Biol.*, **9**, 255–335.

DAHL, F. (1921). *Grundlagen einer Ökologischen Tierographie*, Vols. 1 and 2. Fischer-Verlag, Jena.

DAY, J. H. (Ed.) (1981). *Estuarine Ecology with Particular Reference to Southern Africa*. A. A. Balkema, Cape Town, Rotterdam.

DAY, J. H. and GRINDLEY, J. R. (1981). The estuarine ecosystem and environmental restraints. In J. H. Day (Ed.), *Estuarine Ecology with Reference to Southern Africa*. A. A. Balkema, Cape Town, Rotterdam. pp. 345–372.

DAYTON, P. K. (1971). Competition, disturbance and community organisation: the provision and subsequent utilization of space in a rocky intertidal community. *Ecol. Monogr.*, **41**, 351–389.

DIECKMANN, G. S. (1980). Aspects of the ecology of *Laminaria pallida* (Grev.) J. Ag. off the Cape Peninsula (South Africa). 1. Seasonal growth. *Botanica Mar.*, **23**, 579–585.

DUXBURY, A. C. (1979). Upwelling and estuary flushing. *Limnol. Oceanogr.*, **24**, 627–633.

DYE, A. H. (1979). Aspects of the respiratory physiology of *Gastrosaccus psammodytes* Tattersall (Crustacea, Mysidacea). *Comp. Biochem. Physiol. A*, **65**, 187–191.

DYE, A. H. and MCGWYNNE, L. (1980). The effect of temperature and season on the respiratory physiology of three psammolittoral gastropods. *Comp. Biochem. Physiol. A*, **66**, 107–111.

DYER, K. R. and RAMAMOORTHY, K. (1969). Salinity and water circulation in the Vellar estuary. *Limnol. oceanogr.*, **14**, 4–15.

EDWARDS, J. M. and FREY, R. W. (1977). Substrate characteristics within a holocene salt marsh, Sapelo Island, Georgia. *Senckenberg. maritima*, **9**, 215–259.

EISMA, D. and VAN BENNEKOM, A. J. (1978). The Zaire River and estuary and the Zaire overflow in the Atlantic Ocean. *Neth. J. Sea Res.*, **12**, 255–272.

FAIRBRIDGE, R. W. (1980). The estuary: its definition and geodynamic cycle. In E. Olavsson and I. Cato (Eds), *Chemistry and Biogeochemistry of Estuaries*. Wiley, Chichester. pp. 1–35.

FELL, J. and MASTERS, I. M. (1973). Fungi associated with the degradation of mangrove *Rhizophora mangle* L.) leaves in South Florida. In L. H. Stevenson and R. R. Colwell (Eds), *Estuarine Microbial Ecology*. University of South Carolina Press, Columbia. pp. 455–465.

FENCHEL, T. (1970). Studies on the decomposition of organic detritus derived from the turtle grass *Thalassia testudinum*. *Limnol. Oceanogr.*, **15**, 14–20.

FENCHEL, T. (1974). Instrinsic rates of natural increase: the relationship with body size. *Oecologia*, **14**, 317–326.

FENCHEL, T. (1977). Aspects of the decomposition of seagrasses. In C. P. McRoy and C. Hellferich (Eds), *Seagrass Ecosystems: A Scientific Perspective*. Marcel Dekker, New York. pp. 123–145.

FENCHEL, T. and BLACKBURN, T. H. (1979). *Bacteria and Mineral Cycling*. Academic Press, London.

FENCHEL, T. and HARRISON, P. (1976). The significance of bacterial grazing and mineral cycling for the decomposition of particulate detritus. In J. M. Anderson and A. Macfadyen (Eds), *The Role of Terrestrial and Aquatic Organisms in Decomposition Processes*. Blackwell Scientific Publications, Oxford. pp. 285–299.

FENCHEL, T. and JORGENSEN, B. B. (1977). Detritus food chains of aquatic ecosystems: the role of bacteria. In M. Alexander (Ed.), *Advances in Microbial Ecology*, Vol. 1. Plenum, New York. pp. 1–58.

FIELD, J. G., JARMAN, N. G., DIECKMANN, G. S., GRIFFITHS, C. L., VELIMIROV, B. and ZOUTENDYK, P. (1977). Sun, waves, seaweed and lobsters: the dynamics of a west coast kelp-bed. *S. Afr. J. Sci.*, **73**, 7–10.

FIELD, J. G., GRIFFITHS, C. L., GRIFFITHS, R. J., JARMAN, N., ZOUTENDYK, P. and BOWES, A. (1980a). Variation in structure and biomass of kelp communities along the south west Cape Coast. *Trans. R. Soc. S. Afr.*, **44**, 14–203.

FIELD, J. G., GRIFFITHS, C. L., LINLEY, E. A. S., CARTER, R. A. and ZOUTENDYK, P. (1980b). Upwelling in a nearshore marine ecosystem and its biological implications. *Estuar. coast. mar. Sci.*, **11**, 133–150.

FIELD, J. G., GRIFFITHS, C. L., LINLEY, E. A. S., ZOUTENDYK, P. and CARTER, R. A. (1981). Wind-induced water movements in a Benguela kelp bed. In F. A. Richards (Ed.), *Coastal Upwelling*. American Geophysical Union, Washington, D. C. pp. 507–513.

FISCHER, H. B. (1976). Mixing and dispersion in estuaries. *A. Rev. Fluid Mech.*, **8**, 107–133.

FONG, W. C. and MANN, K. H. (1980). Role of gut flora in the transfer of amino acids through a marine food chain. *Can. J. Fish. aquat. Sci.*, **37**, 88–96.

GAGNE, J. A. and MANN, K. H. (1981). Comparison of growth strategy in *Laminaria* populations living under different seasonal patterns of nutrient availability. In T. Levring (Ed.), *Proceedings of the International Seaweed Symposium, 10*. De Gruyter, Berlin. pp. 297–302.

GAGNE, J. A., MANN, K. H. and CHAPMAN, A. R. O. (1982). Seasonal patterns of growth and storage in *Laminaria longicruris* in relation to differing patterns of availability of nitrogen in the water. *Mar. Biol.*, **69**, 91–101.

GESSNER, F. and SCHRAMM, W. (1971). Salinity: plants. In O. Kinne (Ed.), *Marine Ecology*, Vol. I, Environmental Factors, Part 2. Wiley, London. pp. 705–820.

GIBBS, R. J. (1970). Circulation in the Amazon River estuary and adjacent Atlantic Ocean. *J. mar. Res.*, **28**, 113–123.

GIBBS, R. J. (1972). Amazon River estuarine system. *Mem. geol. Soc. Am.*, **133**, 85–88.

GREENWOOD, P. (1980). Growth, respiration and tentative energy budgets for two populations of the sea urchin *Parechinus angulosus* (Leske). *Estuar. coast. mar. Sci.*, **10**, 347–367.

GRIFFITHS, C. L. and KING, J. A. (1979a). Some relationships between size, food availability and energy balance in the ribbed mussel *Aulacomya ater*. *Mar. Biol.*, **53**, 217–222.

GRIFFITHS, C. L. and KING, J. A. (1979b) Energy expended on growth and gonad output in the ribbed mussel *Aulacomya ater*. *Mar. Biol.*, **53**, 217–222.

GRIFFITHS, C. L. and SEIDERER, L. J. (1980). Rock-lobsters and mussels: limitations and preferences in a predator-prey interaction. *J. exp. mar. Biol. Ecol.*, **44**, 95–109.

GUERINOT, M. L., FONG, W. and PATRIQUIN, D. G. (1977). Nitrogen fixation (acetylene reduction) associated with sea urchins (*Strongylocentrotus droebachiensis*) feeding on seaweeds and eel grass. *J. Fish. Res. Bd Can.*, **34**, 416–420.

HAGMEIER, A. and KÄNDLER, R. (1927). Neue Untersuchungen im nordfriesischen Wattenmeer und auf den fiskalischen Austernbänken. *Wiss. Meeresunters., N.F. Abt. Helgoland*, **16** (Heft 2), 1–90.

HAINES, E. B. (1975). Nutrient inputs to the coastal zone: the Georgia and South Carolina shelf. In L. E. Cronin (Ed.), *Estuarine Research*, Vol. 1. Academic Press, New York. pp. 303–324.

HAINES, E. B. and MONTAGUE, C. L. (1979). Food sources of estuarine invertebrates analysed using $^{13}C/^{12}C$ ratios. *Ecology*, **60**, 48–56.

HANSEN, D. V. (1965). Currents and mixing in the Columbia River estuary. In *Ocean Science and Engineering Transactions*. Joint Conference of the Marine Technological Society and the American Society of Oceanography and Limnology. pp. 943–955.

HANSEN, D. V. and RATTRAY, M. (1966). New dimensions in estuary classification. *Limnol. oceanogr.*, **11**, 319–326.

HARDING, L. W., Jr., COX, J. L. and PEQUEGNAT, J. E. (1978). Spring-summer phytoplankton production in Humboldt Bay, California. *Calif. Fish Game*, **64**, 53–59.

HARRIS, T. F. W. (1978). Review of coastal currents in southern African waters. *S. Afr. Natn. scient. Prog. Rep.*, **30**, 1–103 (C.S.I.R., Pretoria, S. Africa).

HARTOG, C. DEN (1960). Comments on the Venice-system for the classification of brackish waters. *Int. Revue ges. Hydrobiol.*, **45**, 481–485.

HARTOG, C. DEN (1964). Typologie des Brackwassers. *Helgoländer wiss. Meeresunters.*, **10**, 377–390.

HATCHER, B. G., CHAPMAN, A. R. O. and MANN, K. H. (1977). An annual carbon budget for the kelp *Laminaria longicruris*. *Mar. Biol.*, **44**, 85–96.

HAYES, M. O. (1975). Morphology of sand accumulation in estuaries: an introduction to the symposium. In L. E. Cronin (Ed.), *Estuarine Research*, Vol. II. Academic Press, New York. pp. 3–22.

HEALD, E. J. (1969). *The Production of Organic Detritus in a South Florida Estuary*. Ph.D. Thesis, University of Miami, Miami.

HEDGPETH, J. W. (1951). The classification of estuaries and brackish waters and the hydrographic climate. *Rep. Comm. Treatise mar. Ecol. Palaeoecol., Wash.*, **11**, 49–56.

HEDGPETH, J. W. (1953). An introduction to the zoogeography of the Northwest Gulf of Mexico with reference to the invertebrate fauna. *Publs Inst. mar. Sci. Univ. Tex.*, **3**, 107–224.

HEDGPETH, J. W. (1959). Some preliminary considerations of the biology of inland mineral waters. *Archo Oceanogr. Limnol.*, **11** (Suppl.), 111–141.

HEDGPETH, J. W. (1967). Ecological aspects of the Laguna Madre, a hypersaline estuary. In G. H. Lauff (Ed.), *Estuaries*. A.A.A.S., Washington, D. C. pp. 408–419 (*Publs Am. Ass. Advmt Sci.*, **83**).

HEDGPETH, J. W. and OBREBSKI, S. (1981). Willapa Bay: a historical perspective and a rationale for research. *U.S. Fish and Wildl. Serv.* (Office of Biological Services), **FWS/OBS-81/03**, 1–52.

HENSEN, V. (1887). Ueber die Bestimmung des Plankton's oder des im Meere treiben Materials an Pflanzen und Thieren. *Ber. Kommn wiss. Unters. dt. Meere, Kiel*, **5**, 1–102.

JANSSON, B. O. (1978). The Baltic—A systems analysis of a semi-enclosed area. In H. Charnock and G. Deacon (Eds.) *Advances in Oceanography*. Plenum, New York. pp. 131–183.

JANSSON, B. O., WILMOT, W. and WULLF, F. V. (1982). Coupling the sub-systems—the Baltic Sea as a case study. In M. J. Fasham (Ed.), *Flows of Energy and Materials in Marine Ecosystems*, Plenum, New York.

JANSSON, B. O. and WULFF, F. V. (1977). Ecosystem analysis of a shallow Sound in the northern Baltic—A joint study by the Asko Group. *Contrib. from Asko Lab.*, **18**, 1–60.

JOHANNES, R. E. (1964). Phosphorus excretion and body size in marine animals: microzooplankton and nutrient regeneration. *Science, N.Y.*, **146**, 923–924.

JOIRIS, C., BILLEN, G., LANCELOT, C., DARO, M. H., MOMMAERTS, J. P., BERTELS, A., BOSSICART, M., NIJS, J. and HECQ, J. H. (1982). A budget of carbon cycling in the Belgian coastal zone: relative roles of zooplankton, bacterioplankton and benthos in the utilization of primary production. *Neth. J. Sea Res*, **16**, 260–275.

JORDAN, T. E. and VALIELA, I. (1982). A nitrogen budget of the ribbed mussel, *Geukensia demissus*, and its significance in nitrogen flow in a New England salt marsh. *Limnol. Oceanogr.*, **27**, 75–90.

JØRGENSEN, C. B. (1966). *Biology of Suspension Feeding*. Pergamon Press, Oxford.

KANWISHER, J. W. (1966). Photosynthesis and respiration in some seaweeds. In H. Barnes (Ed.), *Some Contemporary Studies in Marine Science*. Allen and Unwin, London. pp. 407–420.

KAUTSKY, N. (1981). *On the Role of the Blue Mussel, Mytilus edulis L. in the Baltic Ecosystem*. Zoology Department and Asko Laboratory, University of Stockholm, Sweden.

KAUTSKY, N. and WALLENTINUS, I. (1980). Nutrient release from a Baltic *Mytilus*—red algal community and its role in benthic and pelagic productivity. *Ophelia*, **1** (Suppl.), 17–30.

KETCHUM, B. H. (1951). The flushing of tidal estuaries. *Sewage ind. Wastes*, **23**, 198–209.

KETCHUM, B. H. (Ed.) (1983). *Estuaries and Enclosed Seas*. Elsevier, Amsterdam.

KHAILOV, K. M. and BURLAKOVA, Z. P. (1969). Release of dissolved organic matter by marine seaweeds and distribution of their total organic production to inshore communities. *Limnol. Oceanogr.*, **14**, 521–527.

KINNE, O. (1964). The effects of temperature and salinity on marine and brackish water animals. II. Salinity and temperature—salinity combinations. *Oceanogr. mar. Biol. A. Rev.*, **2**, 281–339.

KINNE, O. (1971). Salinity: animals. Invertebrates. In O. Kinne (Ed.), *Marine Ecology*, Vol. I, Environmental Factors, Part 2. Wiley, London. pp. 821–995.

KINNE, O. (1978). Introduction to Volume IV. In O. Kinne (Ed.), *Marine Ecology*, Vol. IV, Dynamics. Wiley, Chichester. pp. 1–11.

KOLBE, R. V. (1927). Zur ökologie, morphologie und systematik der Brackwasser—Diatommen. *Pflanzenforschung*, **1927**, 1–145.

KOOP, K. and FIELD, J. G. (1980). The influence of food availability on population dynamics of a supralittoral isopod, *Ligia dilatata* (Brandt.). *J. exp. mar. Biol. Ecol.*, **48**, 61–72.

KOOP, K. and FIELD, J. G. (1981). Energy transformation by the supralittoral isopod *Ligia dilatata* (Brandt.) *J. exp. mar. Biol. Ecol.*, **53**, 221–233.

KREMER, J. N. and NIXON, S. W. (1978). *A Coastal Marine Ecosystem: Simulation and Analysis*. Springer-Verlag, New York.

KRISTENSEN, H. J. (1972). Carbohydrases of some marine invertebrates with notes on their food and on the natural occurrence of the carbohydrates studied. *Mar. Biol.*, **14**, 130–142.

KUENZLER, E. J. (1961). Phosphorus budget of a mussel population. *Limnol. Oceanogr.*, **6**, 400–415.

LANKFORD, R. R. (1976). Coastal lagoons of Mexico: their origin and classification. In M. Wiley (Ed.), *Estuarine Processes*, Vol. II, Circulation, Sediments, and Transfer of Material in the Estuary. Academic Press, New York. pp. 182–216.

LAUFF, G. H. (Ed.) (1967). *Estuaries*. A.A.A.S., Washington, D. C. (*Publs Am. Ass. Advmt Sci.*, **83**).

LEOPOLD, A. (1933). *Game Management*. Charles Scribner and Sons, New York.

LEWIN, J., ECKMAN, J. E. and WARE, G. N. (1979). Blooms of surf-zone diatoms along the coast of the Olympic Peninsula, Washington. XI. Regeneration of ammonium in the surf environment by the Pacific razor clam *Siliqua patula*. *Mar. Biol.*, **52**, 1–9.

LEWIN, J., HRUBY, T. and MACKAS, D. (1975). Blooms of surf zone diatoms along the coast of the Olympic Peninsula, Washington. V. Environmental conditions associated with the blooms. *Estuar. coast. mar. Sci.*, **3**, 229–241.

LEWIN, J. and NORRIS, R. E. (1970). Surf zone diatoms of the coasts of Washington and New Zealand (*Chaetoceros armatum* T. West and *Asterionella* spp.). *Phycologia*, **9**, 143–149.

LEWIS, J. R. (1980). Options and problems in environmental management and evaluation. *Helgoländer wiss. Meeresunters.*, **33**, 452–466.

LINLEY, E. A. S., NEWELL, R. C. and BOSMA, S. A. (1981). Heterotrophic utilisation of mucilage released during fragmentation of kelp (*Ecklonia maxima* and *Laminaria pallida*). I. Development of microbial communities associated with the degradation of kelp mucilage. *Mar. Ecol. Prog. Ser.*, **4**, 31–41.

LITTER, M. M. and MURRAY, S. N. (1974). The primary productivity of marine macrophytes from a rocky intertidal community. *Mar. Biol.*, **27**, 131–135.

LORENZ, J. R. (1863). Brackwasserstudien an der Elbemündung. *Sber. Akad. Wiss. Wien (Mathematisch-naturwissenschaftliche Klasse)*, **48**, 602–613.

LUCAS, M. I., NEWELL, R. C. and VELIMIROV, B. (1981). Heterotrophic utilization of mucilage released during fragmentation of kelp (*Ecklonia maxima* and *Laminaria pallida*). II. Differential utilization of dissolved organic components from kelp mucilage. *Mar. Ecol. Prog. Ser.*, **4**, 43–55.

LUGO, A. E., SELL, M. and SNEDAKER, S. C. (1976). Mangrove ecosystem analysis. In B. C. Pattern (Ed.), *Systems Analysis and Simulation in Ecology*, Vol. 4. Academic Press, New York. pp. 114–146.

LUGO, A. E. and SNEDAKER, S. C. (1974). The ecology of mangroves. *A. Rev. Ecol. Syst.*, **5**, 39–64.

MCINTYRE, A. D., MUNRO, A. L. S. and STEELE, J. H. (1970). Energy flow in a sand ecosystem. In J. H. Steele (Ed.), *Marine Food Chains*. Oliver and Boyd, Edinburgh.

MCLACHLAN, A. (1977). Composition, distribution, abundance and biomass of the macrofauna and meiofauna of four sandy beaches. *Zool. Afr.*, **12**, 279–306.

MCLACHLAN, A. (1979a). Volumes of seawater filtered by East Cape sandy beaches. *S. Afr. J. Sci.*, **75**, 75–79.

MCLACHLAN, A. (1979b). Growth and production of *Donax sordidus* Hanley (Mollusca: Lamellibranchia) on an open sandy beach in Algoa Bay. *S. Afr. J. Zool.*, **14**, 61–66.

MCLACHLAN, A. (1980). Exposed sandy beaches as semi-closed ecosystems. *Mar. environ. Res.*, **4**, 59–63.

MCLACHLAN, A., ERASMUS, T., DYE, A. H., WOOLDRIDGE, T. VAN DER HORST, G., ROSSOUW, G., LASIAK, T. A. and MCGWYNNE, L. (1981). Sand beach energetics: an ecosystem approach towards a high energy interface. *Estuar. coast. Shelf Sci.*, **13**, 11–25.

MCLACHLAN, A. and HANEKOM, N. (1979). Aspects of the biology, ecology and seasonal fluctuations in biochemical composition of *Donax serra* in the East Cape. *S. Afr. J. Zool.*, **14**, 183–193.

MACNAE, W. (1968). A general account of the fauna and flora of mangrove swamps and forest in the Indo-West-Pacific Region. *Adv. mar. Biol.*, **6**, 73–270.

MCQUAID, C. D. (1980). *Spatial and Temporal Variations in Rocky Intertidal Communities*. Ph.D. Thesis, University of Cape Town, South Africa.

MANN, K. H. (1972). Ecological energetics of the seaweed zone in a marine bay on the Atlantic coast of Canada. II. Productivity of the seaweeds. *Mar. Biol.*, **14**, 199–209.

MANN, K. H. (1977). Destruction of the kelp beds by sea-urchins: cyclic phenomenon or irreversible degradation? *Helgolander wiss. Meeresunters.*, **30**, 455–467.

MANN, K. H. (1980). The total aquatic system. In R. S. K. Barnes and K. H. Mann (Eds), *Fundamentals of Aquatic Ecosystems*. Blackwell Scientific Publications, Oxford. pp. 185–200.

MANN, K. H. (1982a). *Ecology of Coastal Waters: A Systems Approach*. Blackwell Scientific Publications, Oxford; University of California Press, Berkeley and Los Angeles.

MANN, K. H. (1982b). Kelp, sea urchins and predators: a review of strong interactions in rocky subtidal systems of eastern Canada, 1970–1980. *Neth. J. Sea Res.*, **16**, 414–423.

MANN, K. H., JARMAN, N. and DIECKMANN, G. S. (1979). Development of a method for measuring the productivity of the kelp *Ecklonia maxima* (Osbeck) Papenf. *Trans. R. Soc. S. Afr.*, **44**, 27–41.

MARGALEF, R. (1967). Some concepts relative to the organization of plankton. *Oceanogr. mar. Biol. A. Rev.*, **5**, 257–289.

MICHAELIS, H. (1981). Intertidal benthic animal communities of the estuaries of the Rivers Ems and Weser. In N. Dankers, H. Kuhl and W. J. Wolff (Eds), *Invertebrates of the Wadden Sea*. Final Report of the Section 'Marine Zoology' of the Wadden Sea Working Group. pp. 158–188.

canopies in South Florida. *Ecology*, **53**, 22–45.

MILLER, R. J., MANN, K. H. and SCARRATT, D. J. (1971). Production potential of a seaweed-lobster community in eastern Canada. *J. Fish. Res. Bd Can.*, **28**, 1733–1738.

MOEBIUS, K. (1877). *Die Auster und die Austernwirtschaft*. Hempel and Parey, Berlin.

MONTAGUE, C. L., BUNKER, S. L., HAINES, E. B., PACE, M. L. and WETZEL, R. L. (1981). Aquatic macroconsumers. In L. R. Pomeroy and R. G. Wiegert (Eds), *The Ecology of a Salt Marsh*. Springer-Verlag, New York. pp. 69–86.

MUIR, D. G. (1978). *The Biology of Talorchestia capensis* (Amphipoda, Talitridae) *Including a Population Energy Budget*. M.Sc. thesis, Zoology Department, University of Cape Town, South Africa.

MUNRO, A. L. S., WELLS, J. B. J. and MCINTYRE, A. D. (1978). Energy flow in the flora and meiofauna of sandy beaches. *Proc. R. Soc. Edinb.* (*Series B*), **76**, 297–315.

NEWELL, R. C. (1965). The role of detritus in the nutrition of two marine deposit feeders, the prosobranch *Hydrobia ulvae* and the bivalve *Macoma balthica*. *Proc. zool. Soc.*, **144**, 25–45.

NEWELL, R. C. and FIELD, J. G. (1982). The contribution of bacteria and detritus to carbon and nitrogen flow in a benthic community. *Mar. Biol. Lett.*, **4**, 23–26.

NEWELL, R. C., FIELD, J. G. and GRIFFITHS, C. L. (1982). Energy balance and the significance of micro-organisms in a kelp bed community. *Mar. Ecol. Prog. Ser.*, **8**, 103–113.

NEWMAN, G. G. and POLLOCK, D. E. (1974). Growth of the rock lobster *Jasus lalandii* and its relationship to benthos. *Mar. Biol.*, **24**, 339–346.

NIXON, S. W., OVIATT, C. A., GARBER, J. and LEE, V. (1976). Diel metabolism and nutrient dynamics in a salt marsh embayment. *Ecology*, **57**, 740–775.

ODUM, E. P. and DE LA CRUZ, A. A. (1967). Particulate organic detritus in a Georgia salt marsh—estuarine ecosystem. In G. H. Lauff (Ed.), *Estuaries*. A.A.A.S., Washington, D.C. pp. 383–388 (Publs Am. Ass. Advmt. Sci., **83**).

ODUM, E. P. and SMALLEY, A. E. (1959). Comparison of population energy flow of a herbivorous and a deposit-feeding invertebrate in a salt-marsh ecosystem. *Proc. Natn. Acad. Sci. U.S.A.*, **45**, 617–622.

ODUM, H. T. (1971). *Environment, Power, and Society*. Wiley Interscience, New York.

ODUM, H. T. and COPELAND, B. J. (1972). Functional classification of coastal ecological systems of the United States. Mem. geol. Soc. Am., **133**, 9–28.

ODUM, H. T., COPELAND, B. J. and MCMAHON, E. A. (Eds) (1969). *Coastal Ecological Systems of the United States.* The Conservation Foundation, Washington, D.C. (4 Volumes).

ODUM, W. E. (1970). Insidious alteration of the estuarine environment. *Trans. Am. Fish. Soc.*, **99**, 836–847.

ODUM, W. E. and HEALD, E. J. (1975). The detritus-based food web of an estuarine mangrove community. In L. E. Cronin (Ed.), *Estuarine Research*, Vol. 1. Academic Press, New York. pp. 265–286.

OFFICER, C. B. (1976). *Physical Oceanography of Estuaries (and Associated Coastal Waters).* Wiley, New York.

PAINE, R. T. (1974). Intertidal community structure: experimental studies on the relationship between a dominant competitor and its principal predator. *Oecologia*, **15**, 93–120.

PAINE, R. T. (1977). Controlled manipulations in the marine intertidal zone, and their contribution to ecological theory. In C. E. Goulden (Ed.) *The Changing Scenes in Natural Sciences, 1776–1976*. Philadelphia. pp. 245–270 (*Acad. Natn. Sci. Spec. Publ.* **12**).

PEARSE, A. S., HUMM, H. J. and WHARTON, W. G. (1942). Ecology of sandy beaches at Beaufort, N. C. *Ecol. Monogr.*, **12**, 135–190.

PERES, J. M. (1982). General features of organismic assemblages in pelagial and benthal. In O. Kinne (Ed.), *Marine Ecology*, Vol. V, Ocean Management, Part 1. Wiley, Chichester. pp. 47–66.

PERKINS, R. W., NIELSEN, J. M., ROESCH, W. C. and MCCALL, R. C. (1960). Zinc-65 and Chromium-51 in foods and people. *Science, N.Y.*, **132**, 1895–1897.

PETERSEN, C. G. J. (1918). The sea bottom and its production of fish food. A survey of the work done in connection with the evaluation of the Danish Waters from 1883–1917. *Rep. Dan. Biol. Stn*, **25**. 1–62.

PETERSON, D. H., CONOMOS, T. J., BROENKAW, W. W. and DOHERTY, P. C. (1975). Location of the non-tidal current null zone in northern San Francisco Bay. *Estuar. coast. mar. Sci.*, **3**, 1–11.

PILSON, M. E. Q. and NIXON, S. W. (1980). Marine microcosms in ecological research. In J. P. Geisy (Ed.), *Microcosms in Ecological Research.* Technical Information Centre, U.S. Dept. of Energy, Washington, D.C. pp. 724–740.

PLATT, T., MANN, K. H. and ULANOWICZ, R. E. (Eds) (1981). *Mathematical Models in Biological Oceanography.* UNESCO Press, Paris (*Monographs on oceanographic methodology,* **7**),

POLLOCK, D. E. (1979). Predator-prey relationships between the rock lobster *Jasus lalandii* and the mussel *Aulacomya ater* at Robben Island on the Cape West Coast of Africa. *Mar. Biol.*, **52**, 347–356.

POMEROY, L. R. (1960). Residence time of dissolved phosphate in natural waters. *Science, N. Y.* **131**, 1731–1732.

POMEROY, L. R., BANCROFT, K., BREED, J., CHRISTIAN, R. R., FRANKENBERG, D., HALL, J. R., MAURER, L. G., WIEBE, W. J., WIEGERT, R. G. and WETZEL, R. L. (1977). Flux of organic matter through a salt marsh. In M. Wiley (Ed.), *Estuarine Processes*, Vol. 2. Academic Press, New York. pp. 270–279.

POMEROY, L. R., DARLEY, W. M., DUNN, E. L., GALLAGHER, J. L., HAINES, E. B. and WHITNEY, D. M. (1981). Primary Production. In L. R. Pomeroy and R. G. Wiegert (Eds), *The Ecology of a Salt Marsh.* Springer-Verlag, New York. pp. 39–68.

POMEROY, L. R. and WIEGERT, R. G. (1981). *The Ecology of a Salt Marsh.* Springer-Verlag, New York.

PRATT, R. M. (1966). The Gulf Stream as a graded river. *Limnol. oceanogr.*, **11**, 60–67.

PRITCHARD, D. W. (1952). Salinity distribution and circulation in the Chesapeake Bay estuarine system. *J. mar. Res.*, **11**, 106–123.

PRITCHARD, D. W. (1955). Estuarine circulation patterns. *Proc. Am. Soc. civ. Engrs*, **81**, 717/1–717/11.

PRITCHARD, D. W. (1967). Observations of circulation in Coastal Plain Estuaries. In G. H. Lauff (Ed.), *Estuaries.* A.A.A.S., Washington, D. C. pp. 37–44 (*Publs Am. Ass. Advmt Sci.*, **83**).

RAINE, R. C. T. and PATCHING, J. W. (1980). Aspects of carbon and nitrogen cycling in a shallow marine environment. *J. exp. mar. Biol. Ecol.*, **47**, 127–139.

RAPSON, A. M. (1954). Feeding and control of toheroa (*Amphidesma ventricosum* Grey) (Eulammellibranchiata) populations in New Zealand. *Aust. J. mar. Freshwat. Res.*, **5**, 486–512.

RASMUSSEN, E. (1973). Systematics and ecology of the Isefjord marine fauna (Denmark). *Ophelia*, **11**, 1–507.

RAVANKO, O. (1969). Benthic algae as food for some invertebrates in the inner part of Baltic. *Limnologica*, **7**, 203–205.

REDEKE, H. C. (1922). Zur Biologie der niederländischen Brackwassertypen. *Bijdr. Dierk.*, **22**, 329–335.

REMANE, A. (1933). Verteilung und organisation des benthonischen mikrofauna der Kieler Bucht. *Wiss. Meeresunters., Kiel*, **21**, 161–221.

REMANE, A. (1934). Die Brackwasserfauna. *Zool. Anz.*, **7**, (Suppl.), 34–74.

REMANE, A. and SCHLIEPER, C. (1971). Biology of brackish water. In H.-J. Elster and W. Ohle (Eds), *Binnengewässer*, Vol. XXV, Ecology of Brackishwater, Part I. Wiley, New York. pp. 1–210.

RIEDL, R. (1971a). Water movement. General introduction. In O. Kinne (Ed.), *Marine Ecology*, Vol. I, Environmental Factors, Part 2. Wiley, London. pp. 1085–1088.

RIEDL, R. (1971b). How much sea water passes through sandy beaches? *Int. Revue ges. Hydrobiol.*, **56**, 923–946.

RIEDL, R. and MACHAN, R. (1972). Hydrodynamic patterns in lotic intertidal sands and their bioclimatalogical implications. *Mar. Biol.*, **13**, 179–209.

ROYER, T. C. (1981). Baroclinic transport in the Gulf of Alaska. 2. A fresh-water driven coastal current. *J. mar. Res.*, **39**, 251–266.

ROYER, T. C. (1982). Coastal fresh-water discharge in the Northeast Pacific. *J. geophys. Res.*, **87**, 2017–2021.

ROZENGURT, M. S. (1967). The organic substances in the water of lagoons, estuaries and harbors of the western part of the Black Sea. In Z. A. Vodyanitsky (Ed.), *Dynamika Vod i Voprosy Gidrokhimii Chernogo Morya*. Naukova Dumka, Kiev. pp. 167–176.

RUSSELL-HUNTER, W. D. (1970). *Aquatic Productivity.* Macmillan, London.

RYTHER, J. H. (1969). Photosynthesis and fish production in the sea. *Science, N.Y.*, **166**, 72–76.

SANDIFER, P. A. (1975). The role of pelagic larvae in recruitment to populations of adult decapod crustaceans in the York River estuary and adjacent lower Chesapeake Bay, Virginia. *Estuar. coast. mar. Sci.*, **3**, 269–279.

SEGERSTRÅLE, S. G. (1958). A quarter century of brackishwater research. *Verh. int. Ver. Limnol.*, xiii, 646–671.

SEGERSTRÅLE, S. G. (1959). Brackish water classification a historical survey. *Archo Oceanogr. Limnol.*, **11** (Suppl.), 7–33.

SEIDERER, L. J., HAHN, B. D. and LAWRENCE, L. (1982). Rock lobsters, mussels and Man: a mathematical model. *Ecol. Modelling*, **17**, 225–241.

SHAFIR, A. and FIELD, J. G. (1980a). Population dynamics of the isopod *Cirolana Imposita* (Barnard) in a kelp-bed. *Crustaceana*, **39**, 185–196.

SHAFIR, A. and FIELD, J. G. (1980b). Importance of a small carnivorous isopod in Energy Transfer. *Mar. Ecol. Prog. Ser.*, **3**, 203–215.

SHELDON, R. W., PRAKASH, A. and SUTCLIFFE, W. H. (1972). The size distribution of particles in the ocean. *Limnol. Oceanogr.*, **17**, 327–340.
SOLOVEVA, A. A., GALKINA, V. N. and GARKAVAYA, G. P. (1977). Experimental study of the effect of dissolved organic matter of mussel metabolites on the natural phytoplankton community of the White Sea. *Okeanologija,* **17**, 449–458.
STOMMEL, H. and FARMER, H. G. (1952). Abrupt change in width in two layer open channel flow. *J. mar. Res.*, **11**, 205–214.
STUART, V., FIELD, J. G. and NEWELL, R. C. (1982a). Evidence for the absorption of kelp detritus by the ribbed mussel *Aulacomya ater* (Molina), using a new $^{51}$Cr-labelled microsphere technique. *Mar. Ecol. Prog. Ser.*, **9**, 263–271.
STUART, V., LUCAS, M. I. and NEWELL, R. C. (1981). Heterotrophic utilization of particulate matter from the kelp *Laminaria pallida. Mar. Ecol. Prog. Ser.*, **4**, 337–348.
STUART, V., NEWELL, R. C. and LUCAS, M. I. (1982b). Conversion of kelp debris and faecal material from the mussel *Aulacomya ater* by marine micro-organisms. *Mar. Ecol. Prog. Ser.*, **7**, 47–57.
TANSLEY, A. G. (1935). The use and abuse of vegetational concepts and terms. *Ecology*, **16**, 284–307.
TEAL, J. M. (1962). Energy flow in the salt marsh ecosystem of Georgia. *Ecology*, **43**, 614–624.
TOWLE, D. W. and PEARSE, J. S. (1973). Production of the giant kelp *Macrocystis*, estimated by *in situ* incorporation of $^{14}$C in polythene bags. *Limnol. Oceanogr.*, **18**, 155–158.
TULLY, J. P. and BARBAER, F. G. (1960). An estuarine analogy in the subarctic Pacific Ocean. *J. Fish. Res. Bd Can.*, **17**, 91–112.
UNDERWOOD, A. J. (1976). Food competition between age classes in the intertidal neritacean *Nerita atramentosa* Reeve (Gastropoda: prosobranchia). *J. exp. mar. Biol. Ecol.*, **23**, 145–154.
VALIELA, I. and TEAL, J. M. (1979). The nitrogen budget of a salt marsh ecosystem. *Nature, Lond.*, **280**, 652–656.
VÄLIKANGAS, I. (1933). Über die Biologie der Ostsee als Brackwassergebiet. *Verh. int. Verein. theor. angew. Limnol.*, VI, 62–112.
VELIMIROV, B., FIELD, J. G., GRIFFITHS, C. L. and ZOUTENDYK, P. (1977). The economy of kelp bed communities in the Benguela upwelling system. Analysis of biomass and spatial distribution. *Helgolander wiss. Meeresunters.*, **30**, 495–518.
WARWICK, R. M., JOINT, I. R. and RADFORD, P. J. (1979). Secondary production of the benthos in an estuarine environment. In R. L. Jefferies and A. J. Davy (Eds), *Ecological Processes in Coastal Environments.* Blackwell, Oxford, pp. 429–450.
WARWICK, R. M. and PRICE, R. (1975). Macrofauna production on an estuarine mudflat. *J. mar. biol. Ass. U.K.*, **55**, 1–18.
WETZEL, R. L. (1976). Carbon resources of a benthic salt marsh invertebrate, *Nassarius obsoletus* Say (Mollusca: Nassariidae). In M. Wiley (Ed.), *Estuarine Processes,* Vol. 2. Academic Press, New York. pp. 293–308.
WHARTON, W. G. and MANN, K. H. (1981). Relationships between destruction grazing by sea-urchin *Strongylocentrotus droebachiensis* and the abundance of American lobster *Homarus americanus* on the Atlantic coast of Nova Scotia. *Can. J. Fish. aquat. Sci.*, **38**, 1339–1349.
WHITNEY, D. M., CHALMERS, A. G., HAINES, E. B., HANSON, R. B., POMEROY, L. R. and SHERR, B. (1981). The cycles of nitrogen and phosphorus. In L. R. Pomeroy and R. G. Wiegert (Eds), *The Ecology of a Salt Marsh*. Springer-Verlag, New York. pp. 163–182.
WIEBE, W. J., CHRISTIAN, R. R., HANSEN, J. A., KING, G., SHERR, B. and SKYRING, G. (1981). Anaerobic respiration and fermentation. In L. R. Pomeroy and R. G. Wiegert (Eds), *The Ecology of a Salt Marsh.* Springer-Verlag, New York. pp. 137–160.
WIEGERT, R. G., CHRISTIAN, R. R., GALLAGHER, J. L., HALL, J. R., JONES, R. D. H. and WETZEL, R. L. (1975). A preliminary ecosystem model of a coastal Georgia *Spartina* marsh. In L. E. Cronin (Ed.), *Estuarine Research*, Vol. 2. Academic Press, New York. pp. 583–601.
WIEGERT, R. G., CHRISTIAN, R. R. and WETZEL, R. L. (1981). A model view of the marsh. In L. R. Pomeroy and R. G. Wiegert (Eds), *The Ecology of a Salt Marsh*. Springer-Verlag, New York. pp. 183–218.
WIEGERT, R. G. and POMEROY, L. R. (1981). The salt-marsh ecosystem: a synthesis. In L. R.

Pomeroy and R. G. Wiegert (Eds), *The Ecology of a Salt Marsh*. Springer-Verlag, New York. pp. 219–230.

WILLIAMS, P. J. le B. (1981). Incorporation of microheterotrophic processes into the classical paradigm of the planktonic food web. *Kieler Meeresforsch.*, **5**, 1–28.

WOLFF, W. J. (1973). The estuary as a habitat. An analysis of data on the soft bottom macrofauna of the estuarine areas of the Rivers Rhine, Meuse and Scheldt. *Zool. Verh., Leiden*, **126**, 1–243.

WULFF, F. V. and FIELD, J. G. (1983). The importance of different trophic pathways in a nearshore benthic community under upwelling and downwelling conditions. *Mar. Ecol. Prog. Ser.*, **12** (3), 217–228.

ZENKEVITCH, L. A. (1959). The classification of brackish-water basins as exemplified by the seas of the USSR. *Estratto dall- Archivio di Oceanographia e Limnologia*, **10** (Supplemento).

ZENKEVITCH, L. A. (1963). *Biology of the Seas of the U.S.S.R.* Allen and Unwin Ltd, London.

ZERNOV, S. A. (1913). *Contribution to the knowledge of life of the Black Sea (Russ.)*. Chem. Acad. Imper. Sciences, St Petersburg.

Marine Ecology Vol. V, Part 2
Edited by Otto Kinne


# 4. WORLD RESOURCES OF MARINE PLANTS

## 4.1 BENTHIC PLANTS

G. MICHANEK

### (1) Introduction

The total world biomass of marine benthic plants has never been assessed. There does not even exist sufficient data for a rough guess. Only for species of economic importance are figures available regarding the quantities actually harvested and reported to national and international statistics. Official data on catches and landings of seaweed are published in the FAO yearbook of fishery statistics, and a first attempt to review world resources and harvesting of seaweed was undertaken by FAO (MICHANEK, 1975). More comprehensive and updated, this chapter reviews the world-wide resources of marine benthic plants with special emphasis on climatic regions and ecological aspects.

Differences in composition of marine floras following differences in climatic conditions were observed and investigated early (KJELLMAN, 1883; BØRGESEN and JONSSON, 1905). Later, much attention was given to the features of sociology and zonation (GISLÉN, 1930; STEPHENSON and STEPHENSON, 1972) and to the significance of floristic discontinuities (DRUEHL, 1970; HOEK, 1975). Not until recently, however, have marine plant distributions been related to climatic regions in coastal waters on a world-wide basis (MICHANEK, 1979a; Fig. 4-1).

It is obvious, but scarcely accounted for in research papers and textbooks, that biomass, productivity, grazing, and utilization by man is likewise showing different aspects in different climatic regions. In this chapter an attempt will be made to relate the world resources of benthic plants, as known from data on harvests, to the great climatic regions and to focus on special features where local conditions in a certain area bring about remarkably large or small quantities.

Rich natural growth of algae will by no means always form a harvestable resource. In areas like the Straits of Magellan, the Kerguelen Island, and the west coast of Spitsbergen extremely large quantities are found, but commercial utilization would encounter considerable problems: long distances to purchasers, sparse population, high cost of labour, faulty repair facilities, long periods of rough weather during which harvesting is impossible, and temperatures too low for sufficiently fast on-the-spot drying. Along the cold-temperate coasts of Europe, in the warm-temperate Mediterranean Sea, the tropical Caribbean, and in many other areas usually another problem prevails: the algal mass is composed of a large number of species—including epiphytes—entangled with each other, or distributed in a mosaic of patches too small for efficient harvesting. A good merchandise should consist of a single species with a minimum of mixtures

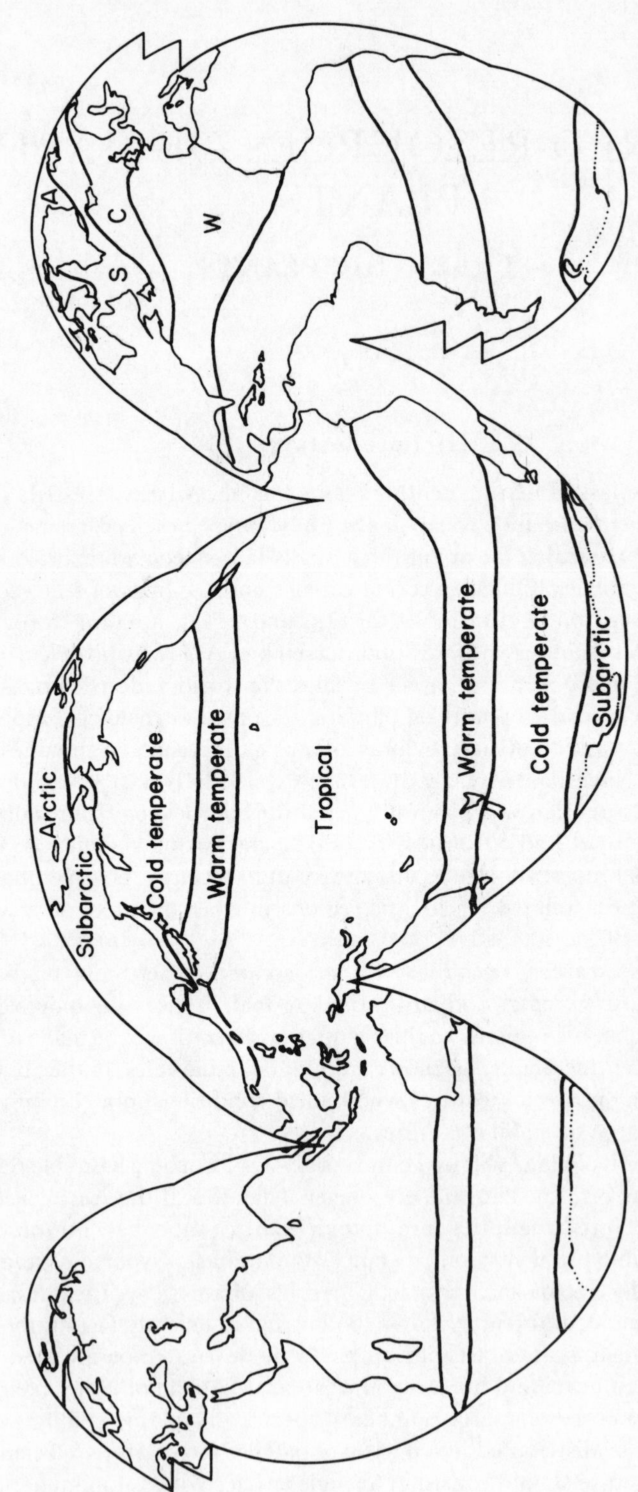


Fig. 4-1: Geographical distribution of marine algae. Delineations of climatic regions in coastal waters. A: arctic; S: subarctic; C: cold temperate; W: warm temperate. (After MICHANEK, 1979a; reproduced by permission of Walter de Gruyter & Co., Berlin)

and impurities. In other words, a commercial resource should, if at all possible, be monospecific. Otherwise, upgrading costs may be as high as the initial harvesting costs.

Of vital importance is the kind of product obtainable from the raw material. For human consumption by far the most important species is the Japanese kelp *Laminaria japonica*. Many times more expensive are the highest grades of nori produced from species of red *Porphyra* and green *Monostroma*.

An increasing part of the algal raw material harvested is used for technical products—all based on the polysaccharides of intercellular spaces or cell walls. These are not, as in higher plants, polymers of glucose but of other sugars or related acids, conveying particular properties to each kind.

In common terms, seaweed hydrocarbonates are referred to as phycocolloids. They form viscous gums, used as emulsifiers and stabilizers for a wide variety of purposes, depending on the particular consistency desired. In brown algae, the most important colloid is alginic acid. This is a heteropolysaccharide of D-mannuronic and L-guluronic acid residues arranged in a non-regular pattern along the chain. The proportion and arrangement of the acid components varies from species to species each having a typical range of composition and giving particular properties to its alginic acid and alginates. These are used in textile printing, for alginate fibres, for jellies and cakes, beer and fruit drinks, in pharmaceutical products, and papermaking. About 0.1% sodium alginate added to an ice-cream mix modifies the freezing procedure so that only small palatable ice crystals are formed.

A great number of phycocolloids are derived from red algae. The most famous are carrageenan and agar. Both are polymers of D-galactose and 3,6 anhydrogalactose—in carrageenan of D-type, in agar of L-type. Carrageenan is used in dairy products, for gelating materials in desserts, jams, and fish jellies. The advantage of carrageenan relative to other kinds of emulsifiers and stabilizers in foods is that it is possible to manufacture 'tailor-made' products, to control texture, mouth-feel, and stability in gels, liquids, pastes, and suspensions.

The term 'agar'—in its Malaysian form 'agar-agar'—is sometimes used for the phycocolloids from red algae, and also for the dried raw products which often form the commercial commodity. More often 'agar', applied in a more restricted sense, refers to bacterial agar, including the dried extracts from species of *Gelidium*, *Gelidiella*, *Pterocladia*, *Ahnfeltia*, and sometimes *Gracilaria*. Here we reserve the term 'agar' for phycocolloids from red algae which are insoluble in cold water but soluble in hot water. At 32° to 39 °C, a 1·5% solution settles into a solid gel which does not melt until at 85 °C. Related phycocolloids, like phyllophoran, which do not completely comply with the definition, are called agaroids, a term which often includes *Gracilaria* gum. In pharmacy, agar is used as a culture medium for bacteria and fungi; its extraordinary gelling capacity is of great importance in the production of cosmetics, dentistry objects, and a variety of foods. Even a 1% agar gives a stable gelation, allowing the final product to be cut into slices. For the biochemical purification of enzymes, polysaccharides, proteins, and other biological macromolecules, agar is used for 'gel filtration', a mild separation technique for macromolecular studies.

The remaining gums from red algae, 'carrageenans', include gums from species of *Chondrus*, *Gigartina*, and *Eucheuma*, usually also from *Furcellaria*, *Hypnea*, *Phyllophora*, and *Iridea*. Further breakdown is possible into substances such as 'furcellaran' and 'hypnean'. The botanical names of certain genera of red algae are also used as trade names for commercial products. In some cases the terms are unequivocal. *Furcellaria* always means *Furcellaria fastigiata*; *Chondrus* should mean *Chondrus crispus* which is also known under the trade name 'Irish moss'. However, a shipment of 'Irish Moss' from Canada may well contain a mixture of *C. crispus* and *Gigartina stellata* which has a very similar appearance and occurs in similar or identical localities. In each of the genera *Eucheuma* and *Hypnea* more than 20 species names are being used. Most species are difficult to determine taxonomically even for a specialist and some are known to be confusing. 'Zanzibar Weed' is the trade name for a mixture of *Eucheuma* species. One of these is referred to by commercial firms as *E. cottonii*, but is probably *E. striatum*. Another one, commercially sold as *E. serra*, was identified as *E. spinosum* (MSHIGENI, 1973). The agaroid *Gracilaria* is found in commercial quantities all around the world. The common type is called *G. verrucosa* (syn: *G. confervoides*). However, it is very much doubted whether the *Gracilaria verrucosa* found in Chile, Florida, the Mediterranean Sea, India, Thailand, Philippines, Taiwan, Japan, South Africa, and Australia is indeed the same species. The genus is in for taxonomic revision.

I have adopted here the names used in the pertinent literature—in a few cases corrected or modernized, in others, species names were omitted when in doubt or assumed not to have any informative value. In many cases only generic names are given in the original papers.

Products from algae, other than chemo-technicals, are algal meals used as food additives for cattle and poultry. The presence of trace elements and vitamins in algal meal facilitates improvement in health and gain in weight on these animals. Large quantities of seaweed are used as manure, some as fodder.

The conventional use of seaweed and seaweed products in crop production has been investigated in agriculture and horticulture. SENN and KINGMAN developed a method to test for plant growth hormones. They have reviewed more than 20 yr of research concerned with seaweeds for agriculture usage, as well as for greenhouse and other horticultural crops (SENN and KINGMAN, 1978). Among the numerous examples that can be mentioned, liquid spray of *Ascophyllum nodosum* resulted in significant increases in soluble solids of tomatoes, protein content of soyabean, and even in the quality of chrysanthemum. The observed physiological responses exceeded those explainable on the basis of known chemical seaweed composition.

The influence of environmental factors on marine algae is treated by SANTELICES (1977) and ROUND (1981). Chemical, physical, and biological factors affecting the production of biomass in mariculture were reviewed by JACKSON (1980). Pathology of marine algae was surveyed by OGATA (1975) and ANDREWS (1976). An important part of the pathological conditions is caused by environmental stress. Disease phenomena, those induced by organisms as well as those which are not, are often associated with nutrient deficiency, high temperatures, low salinities, and high plant densities (OGATA, 1975). The same background factors may be responsible, in many cases, for overgrowth of commercial species in cultures by undesirable epiphytes. It is of basic importance for the success of a marine culture to maintain optimal conditions, including optimal density. For many diseases there is still no cure, and if elimination of stress factors is not successful, removal of diseased fronds is necessary.

On a world-wide basis, 3·4 million tons of seaweed are harvested each year. According to NAYLOR (1976) the first-hand sale value of the world production of seaweeds was US $765 million and the total commercial value of the seaweed industry was approach-

Table 4-1

Annual seaweed harvest (world total) related to climatic zones and expressed in terms of thousands of tons wet weight (Original)

| Group of algae | Arctic and subarctic waters | Cold-temperate waters | Warm-temperate waters | Tropical waters | Percentage per algal group |
|---|---|---|---|---|---|
| Brown algae | — | 1585 | 900 | 5 | 74% |
| Red algae | — | 239 | 591 | 47 | 26% |
| Area percentage | — | 54% | 44% | 1·5% | |

*Note:* Not included are 2000 tons of green algae from warm-temperate waters (see South Korea) and 300 000 tons of 'maërl' which are corallinous red algae in cold-temperate waters—but which are not 'seaweed'.

ing some US $1000 million yr$^{-1}$. These figures are increasing as the volume grows and the amount of information becomes larger.

What is actually harvested (Table 4-1) may be compared to an estimate of potential resources (Table 4-2) founded on assessments of natural beds, but also including, to a minor degree, estimates of the outcome of a reasonably increased mariculture in areas where such a trend is discernible. These data are based on official records from countries contributing to international statistics (FAO, 1979) on information accounted for in seaweed literature (MICHANEK, 1975), and on personal communications from psychologists and officials.

According to present knowledge and estimates, some 50% of the resources of red algae—but only 16% of brown algae—are utilized. Five-sixths of the world's unexploited seaweed resources are made up of 12 million tons of brown algae in cold waters. This does not necessarily mean that seaweed utilization will develop along these lines. For fertilizer and cattle feed brown algae are as good as red. For gelling, emulsifying, and thickening purposes the alginates can often compete with colloids extracted from red algae, but not always, since they are chemically active, in contrast to agar. Actually, the present demand seems to favour tropical red algae from mariculture.

Table 4-2

Estimated annual potential of harvestable seaweed. Thousands of tons wet weight. World total (Original)

| Group of algae | Subarctic waters | Cold-temperate waters | Warm-temperate waters | Tropical waters | Percentage per algal group |
|---|---|---|---|---|---|
| Brown algae | 150 | 13 610 | 1136 | 1045 | 91% |
| Red algae | — | 522 | 942 | 106 | 9% |
| Area percentage | 0·9% | 80·7% | 11·9% | 6·6% | |

## (2) The Tropics

### (a) Ecology and Productivity

Tropical waters are characterized by relatively small changes during the year in temperature and sun inclination (Volume 1). Usually, monthly mean values of temperature do not vary more than 5 °C, and the mean temperature for the coldest month does not sink below 20 °C. In such waters, biotic ('coral') reefs develop on hard bottoms and mangrove swamps on soft bottoms (MICHANEK, 1979a). The tropical marine flora has been described by certain authors as being characterized by the absence of littoral marine algae. In many areas solar irradiation is so strong that algae of the types found in the intertidal belt on higher latitudes cannot tolerate the high temperatures in combination with low air moisture. On such shores exceptions may be found in some lithothamnia and other lime-incrusted algae adapted to intense irradiance and, to a certain degree, drought. In the same areas a rich flora may develop at a water-depth of 10 m.

Such conditions may be common in the tropics, but must not be generalized as being valid for the entire region. Where favourable conditions prevail, a rich littoral flora develops. Florideans are dominating both in number of species and in quantity. Usually they are dark purple to greyish brown. In comparison with other regions the number of short-lived species is low. The main part of the marine benthic plant biomass consists of perennial species which can tolerate sunlight all year round.

Even if the four seasons encountered in temperate waters are not met in the tropics, there is in large areas a distinct seasonality in algal biomass, fertility, and phycocolloid quality. This is due to the effects of the monsoon climate: seasonal protection from excessive sunshine, washing of the intertidal belt during low water, and dilution of surface water. There are also seasonable changes in turbidity as well as alternations between calm and rough weather.

The periodicity is displayed in the appearance of short-lived species, such as *Porphyra suborbiculata*, in the development of new branches on perennial species, e.g. in *Laurencia ceylanica*, or in seasonal fertility, e.g. in *Champia ceylanica* (SVEDELIUS, 1906).

Chemical composition and biomass also undergo changes (Table 4-3). For proper management of a seaweed resource, such changes must be known. The ecological dynamics of tropical marine waters do not favour the formation of large quantities of algae. The low proportion of harvested algae is striking—only 1·4% of the world total—most of which comes from mariculture. However, for technical purposes, many tropical red algae are in demand, because their phycocolloids have very pronounced gelling properties. Also economic considerations make tropical algae attractive to chem-

Table 4-3

*Hypnea musciformis*. Periodicity on Kathiawar peninsula NW India (Based on data by RAMA RAO, 1970)

| | Oct. | Nov. | Dec. | Jan. | Feb. | Mar. | Apr. |
|---|---|---|---|---|---|---|---|
| Reproduction month: | Oct. | | | | | | |
| Biomass is high only in | | | Dec., | Jan., | Feb. | | |
| Hypnean content is high only in | | | | (Jan.), | Feb., | Mar. | |
| Gel strength is high only in | | | | | | Mar., | (Apr.). |

ical industries. The buyer can have them at low prices because of low labour cost. The seller is eager to produce them because most countries in the area have unemployment problems and are much in demand of export articles. In the long run, however, production for human consumption should be given priority over export of raw technical materials.

The high value for public health of the protective substances, vitamins, proteins, and trace elements, which are all present in marine algae at remarkably high levels, corresponds to an increasing demand for such additives to the staple food of most populations in tropical areas. These demands are most pronounced among inland populations. This is a serious problem all over the world. Food deficiencies, especially in protective substances, increase with the distance from the sea, and are most pronounced in populations which concentrate on a staple food with few additives. The need for providing compensating additives to a diet deficient in some respect was known already during pre-Inca and Inca periods. Inhabitants of the Andean region obtained dried *Ulva lactuca* and *Porphyra columbina* from the coast through exchange, or sent for them by messengers. Dried stipes of *Phyllogigas* were used as 'goitre sticks'. These customs are still prevalent. In Peru, for example, dried *Porphyra* are brought in considerable quantities to the market, where endemic goitre has a high prevalence (ACLETO OSORIO, 1971; MICHANEK, 1979b), and *Gigartina chamissoi* is eaten fresh along the Peruvian coast.

### (b) Tropical Coasts of the Western Atlantic Ocean

The Caribbean Sea is extremely rich in number of species, but the quantities of algae are moderate, and seaweed utilization extremely poor. In terms of ecology, the main reasons for this are that the commercially utilizable algae do not occur in vast monospecific stands sufficient for large-scale harvesting. The continental coasts are dominated by sand beaches, the volcanic islands often have steep cliffs without fringing reefs; there are no extended flat bottoms at a suitable depth and the algal mat is often low and dense and made up of a mixture of various species.

Many members of the algal flora along tropical coasts of the western Atlantic Ocean are edible, and are appreciated in other parts of the world, for example, *Caulerpa racemosa*, *Hypnea musciformis*, and *Laurencia obtusa*. A few algae are consumed by man: species of *Ulva* are said to be eaten as a salad in Trinidad and Jamaica: in Barbados they are brewed into a 'bush tea' reputed to have medicinal properties. In Trinidad, jellies are prepared from *Gracilaria*, and in Antigua and Barbuda, *Eucheuma isiforme* is eaten. There is even some inter-island trade in *Codium decorticatum* dried in Barbados and exported to Grenada (RICHARDSON, 1958). Drift seaweed is used as crop fertilizers for coconut palms and cocoa plants, both in fields and in private gardens.

*Hypnea* species comprise the most important benthic plants for a possible carragenan production. In Venezuela, *H. cervicornis* and *H. musciformis* are especially abundant. As possible sources of phycocolloids, DIAZ-PIFERRER (1967) mentions 55 species, among which those of the genera *Eucheuma* and *Digenea* are known to give products of high quality. In Puerto Rico the most abundant agarophytes and carragenophytes are *Digenea simplex*, *Gracilaria verrucosa*, *Hypnea musciformis*, and *Bryothamnion triquetrum*. The latter is also observed in great quantities as a drift alga. During summer storms and winter high tides, hundreds of thousands of tons of drift algae are cast ashore (DIAZ-PIFERRER and CABALLER DE PEREZ, 1964).

Cuban waters are rich in *Sargassum* species, some of which may fringe the coast in pure stands. In addition to sessile forms, two species of pelagic *Sargassum* aggregate along the shores in considerable quantities all year round—in June and July in extraordinary quantities—while *Gracilaria domingensis* and *Hypnea musciformis* may be locally abundant in restricted parts of the coastline. Yet their biomass is nowhere sufficient for sustaining industrial productions of agar or carrageenan. Such discontinuity and low abundance of benthic marine plants seems to prevent commercial utilization in the entire Caribbean area. While the resources are sometimes considerable, they are never really sufficient for economic industrial exploitation. However, the present world-wide trend to neglect naturally growing seaweed, and to turn to mariculture, i.e. seaweed farming under controlled conditions (Volume III: BONOTTO, 1976), renders the prospects for commercial Caribbean seaweed production now more promising than ever before.

Industry considers marine benthic plant production in terms of profitability. In their search for raw materials they aim exclusively at products for phycocolloid production. However, there are alternatives: *Hypnea musciformis*, for example, also serves as vermifuge against species of the parasitic nematode genus *Ascaris*, which is estimated to infect one-third of mankind. In Cuba, as in other places, vermifuge production would require smaller quantities of raw material than local hypnean production.

The largest seaweed stocks have remained unutilized: the gigantic masses of free-floating Gulf weed, *Sargassum*. Within the Gulf of Mexico the *Sargassum* stock is estimated at 90 000 tons (PARR, 1939). The Gulf weed is probably concentrated in certain areas and hence easy to harvest. If half of it could be gathered, this would yield 1000 tons of algin $yr^{-1}$. In the Sargasso Sea the seaweed biomass ranges from 5 to 10 million tons. Through satellite technology it would be possible to direct a potential harvester to places of maximum *Sargassum* aggregations via the shortest route.

Turtle grass *Thalassia testudinum* is the most abundant among the marine plants of Cuba. A good half of the submerged coastal platform Northwest of Cuba is covered by turtle grass prairies. This phanerogam is found from about 0.5 m to 11 m depth with densest development at 3 to 5 m. The mean value for the biomass is 4 kg $m^{-2}$, which makes a total of almost 8 million tons fresh weight or half a million tons dry weight. Nitrogen components amount to 14% of the dry weight and growth rate corresponds to a duplication of the biomass in 200 d (BUESA, 1972).

During recent feeding experiments with sheep in Florida (USA), those which were given 10% turtle grass in their fodder grew faster and utilized their nourishment more effectively than the controls. Nevertheless, utilization of this rich resource in the Caribbean, as in so many other areas, still remains at the experimental stage.

Much richer than the Caribbean Sea in industrially utilizable raw plant material is Brazil. As unemployment is a severe problem in the northern states of Brazil, a seaweed company was organized some years ago with the obligation to provide work for 24 000 people. Following the official records from the company, in the early 1970s these employees collected 4000 tons of agarophytes, 50 000 tons of alginophytes, and 45 000 tons wet weight of unspecified seaweeds within 1 yr. This indicates the natural resource potential of the area (Fig. 4-2). Unfortunately, the 24 000 collectors were working without the guidance of ecologists who could have provided advice and controlled optimal harvesting and resource conservation. Ecological and technological problems, together with the economic burden of too many employees, led to the collapse of the enterprise. Is this an example of what we have to expect in other, comparable areas: seaweed harvest-

Fig. 4-2: Drying, sorting, and upgrading of seaweed in northern Brazil. (Reproduced by permission of Dr. de Sternberg)

ing directed not by supply and demand—and environment policy—but by employment policy considerations and by high production targets?

In the São Paulo region, OLIVEIRA, FILHO and PAULA (1979) estimated the standing crop of easily harvestable *Sargassum* spp., which equals a sustainable yield. They found a total amount of 388 tons fresh weight (corresponding to 7 tons alginate) and conclude that the southeast coast has enough raw material to support at least the internal consumption of algin in Brazil.

### (c) Tropical Coasts of the Eastern Atlantic Ocean

Few shorelines are as poor in benthic marine algae as the tropical parts of West Africa. This area provides an example for illustrating the ecological obstacles in the development of rich algal resources, natural or cultivated, in the tropical zone. The low biomass of benthic algae in the Gulf of Guinea is still more remarkable when compared to the adjacent deep sea, which is known to harbour a rich biomass of phytoplankton, zooplankton, and fish during peak upwelling periods. Phosphate-P, west of Congo, amounts to 9 to 22 mg $m^{-3}$. Evidently, the living conditions are strikingly different in the open-sea and in near-shore surface waters.

In addition to the inhospitable open coast without any archipelago, and the unsuitable substratum of laterite rocks, sand bottoms, and other unstable substrata, the main reason for the poor benthic algal flora must be due to the fact that attached algae are never, or are only occasionally, washed by the nutrient-rich waters of the Benguela Current. Heavy tropical rains dilute the surface-water layers, both directly and through large rivers. River water is very turbid, and does not allow light to penetrate deeply. In the uppermost water layer, where light may be sufficient, salinity fluctuations are strongest: in the intertidal belt these fluctuations oscillate between salinities close to zero during tropical rains and saturated salt brines during intense sunshine. In permanent or temporary lagoons behind coastal sand bars the salinities periodically fluctuate beyond the ranges endurable for the majority of these species. Thousands of kilometres of mangrove swamps line the lagoons and estuaries, and the intensity of grazing is tremendous. On rocky shores one may find virtually nothing but sea urchins.

In tropical West Africa no algae are eaten by man and there is only one area where algae are produced for industrial use: Senegal. Here the above-mentioned conditions do not prevail. The upwelling of phosphate-rich waters, characteristic of the warm-temperate part of the African coast from the Straits of Gibraltar to Cap Verde, are local and seasonal. During summer they are found in the northern part of that coastline and during winter in the southern part, and also south of Dakar off a small part of the tropical coast. This is where a submarine prairie is found, extending 120 km Southeast of Dakar, consisting primarily of *Hypnea* species and developing from January to April. The width of the prairie depends on the smoothness of the bottom slope and exceeds 15 km between Mbour and Point de Sangoma. The quantities available are indicated by the recent construction of a factory with an annual capacity of 12 000 tons wet weight. Twelve thousand workers will be employed along the whole Petit Côte. It is estimated that each man will collect 200 to 300 kg wet weight of *Hypnea* a day and earn about US $1·5 to 3 d^{-1}$. These surprising figures may have a simple explanation: seaweed collection is a way to provide money for unskilled workers in a society changing to a monetary basis. Deliberately, this chance is offered to as many hands as possible.

## (d) Tropical Coasts of the Indian Ocean

According to international statistics, the annual seaweed production of the entire Indian Ocean amounts to 427 tons of red algae. Even if the true figure were 10 or 50 times higher, as a contribution to world resources this is negligible. Nevertheless, the efforts in seaweed utilization are worth mentioning because they reflect significant trends, hopes, and difficulties.

Madagascar has a long tradition in the production of *Gelidium*, much of which was exported to Japan. However, from 1973 to 1976 the export declined from 3800 tons to zero. Tanzania is building up a production of industrially demanded carrageenan species, mainly of the genus *Eucheuma*. Actually there has been an export from Zanzibar since around 1935, but the most rapid expansion in this trade occurred since 1964. While the harvest was estimated at 3000 tons in 1968, official records are not available. France imports the largest quantity, followed by Denmark and Britain. *Eucheuma striatum* is foremost in quantities among the commercial species in Tanzania, followed in importance by *E. spinosum* and *E. platycladum*. There are good natural resources, but over-exploitation is threatening and modern mariculture methods have been introduced (MSHIGENI, 1973, 1976). In Djibuti a pilot attempt to cultivate *E. spinosum* on strings was successful and showed that farming is feasible in areas devoid of natural populations of *Eucheuma* (BRAUD and PEREZ, 1979).

Democratic Yemen and Oman have now turned their attention to a usually neglected marine plant resource, their sea grasses. In Khor Umaira Bay, considered to be a good pasturage for marine turtles, five kinds of sea grasses were found, belonging to the genera *Cymodocea*, *Halodule*, *Syringodium*, and *Thalassodendron*. The densest growths occurred at depths from 2 to 3·5 m. In the sample quadrats, where several types of sea grasses were growing together, wet weights of up to 4200 g $m^{-2}$ of grasses cropped at the ground surface, and up to 420 g $m^{-2}$ of 80 °C oven dried materials were recorded (FAO, 1973). During the turtle-grass survey in Oman previously unrecorded stocks of *Hypnea* and *Sargassum* species were observed with standing crops of 2500 tons and 28 000 tons, respectively.

India has for long made efforts to produce nationally as much as possible of the agar needed for cholera vaccine and bacteria cultures, and the alginates required for textile colour prints. There are two seaweed production areas: the Kathiawar Peninsula in the northwest and the Mandapam–Cape Comorin coast along the Gulf of Mannar in the southeast. The numerous estimates of quantities often differ considerably (MICHANEK, 1975).

For the 680 million inhabitants of India, algae cannot contribute significantly to food supplies and health—unless, with respect to the latter, the available resources are reserved for the population of the southern valleys and foothills of the Himalayas, where the highly prevalent goitre is said to be more formidable perhaps than in any other part of the world (MICHANEK, 1979b).

For a rational and beneficial use of the world's seaweed resources similar preferences should be given to all countries with endemic goitre; this applies among others to all countries with highlands. Areas with excessive rains suffer from such a leakage of soil iodine that all plant products show a deficit in iodine content; consequently, both animals and man exhibit a high prevalence of goitre.

While endemic goitre is the easiest of all known diseases to prevent, the world figure of incidents is in the vicinity of 300 million. Algae concentrate iodine from sea water: seaweeds have an iodine content 100 to 40 000 times higher than that of the ambient seawater. Areas all over the world where seaweed resources should primarily be distributed to populations with a high prevalence of goitre, are identified in Fig. 4-3 which also shows major seaweed resources.

Indonesia's old traditions and recent boom in seaweed production focus on the Pacific parts of the archipelago. On the coastline towards the Indian Ocean, however, a small production has started in 1974 along the southwest and southern coasts of Sumatra and Java, yielding 300 tons $yr^{-1}$ during the late 1970s.

The traditional seaweed consumption in the Far East, strongly developed in China and Japan, has declined in tropical countries for more than a century, and is now almost forgotten in Indonesia. Burma retains more of the tradition. The red alga *Catenella nipae* is a characteristic component of the algal flora found on the breathing roots of mangrove trees, to which it is attached by characteristic haptera. Together with *Bostrychia radicans*, *Caloglossa adnata*, and *C. leprieurii* it forms a pneumatophore-covering association, in particular co-existing with *Avicennia alba* and *A. marina* in the outermost fringe of the mangrove. The same species also occur on muddy rocks in salt marshes, and the seaweeds flourish even where they are only occasionally wetted by ocean spray. In Burma

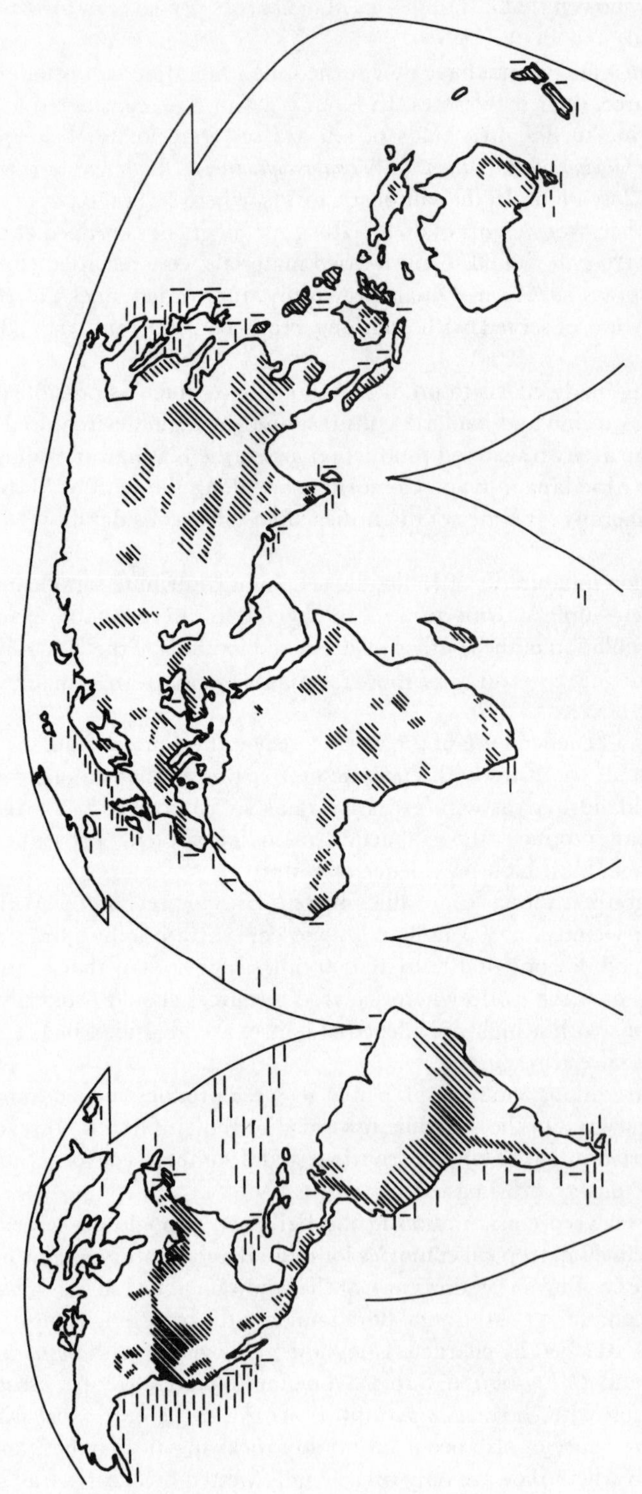

Fig. 4-3: Distribution of endemic goitre and of seaweed resoures. Hatched: areas where endemic goitre has been found. (After KELLEY and SNEDDEN, 1960); horizontal dashes: areas rich in resources of brown algae; vertical dashes: areas rich in red algae. (From MICHANEK, 1979b; Goode Homolosine Equal-area Projection; reproduced by permission of the University of Chicago)

they are eaten raw or after having been boiled for an hour, in both cases with sesame oil and spices. Parcels of dried *Catenella*, sold in the market of Rangoon, actually contained all members of the Bostrychietum association of mangrove algae (ZANEVELD, 1959).

Marine benthic algae are rich in iodine on account of their remarkable ability to concentrate this substance from sea water. The use of algae as a remedy for endemic goitre was established by the Chinese in 2700 BC. In the highlands of Thailand, Burma, and Vietnam, some regions show goitre incidences of well over 50%, often at very pronouned stages of development. Among the Kachins in the steep hill-sides along the north-east frontiers of Burma, vitamin A deficiency is regarded as the most important single goitrogenic factor (KELLEY and SNEDDEN, 1960). Of the algae eaten along the Burmese coast, all rich in iodine, species of *Gracilaria* and *Enteromorpha* are known also to be rich in vitamin A.

### (e) Tropical Coasts of the Pacific Ocean

Some Thai areas of endemic goitre also suffer from beriberi, or $B_1$ avitaminosis. Among the algae of Thailand, *Porphyra*, *Gracilaria*, and *Rhodymenia*—in addition to their iodine content—are also rich in vitamin $B_1$. As long as there are not sufficient amounts of algae for all, the resources available should be reserved for and channeled to such populations, where they can be expected to have a twofold beneficial impact on the consumers' health, in this case the areas known for both serious goitre and for vitamin A or B deficiencies (MICHANEK, 1979b).

Only a small part of China, the Hainan island in the South China Sea, is tropical. Here we find two *Eucheuma* farms, producing 300 to 400 tons $yr^{-1}$. If we disregard the Japanese production along the delineation line of the tropics, we will find almost three-quarters of the world's true tropical production in two countries only: Indonesia and the Philippines. Both are traditionally seaweed-consuming areas and exporters of edible algae. Recently, both increased their production and export considerably with an orientation towards industrially demanded species.

In the Philippines, *Eucheuma* harvesting has in recent years developed from a primitive gathering of wild crop material to a highly specialized mariculture. In the late 1960s research was started by the Marine Colloids company on simple biological applications at optimal salinity, temperature, water exchange, bottom conditions, plant depth, and plant distance, and on the two complicated problems of senescence and seasonality (DOTY and ALVAREZ, 1975). Overharvesting of natural resources led to a catastrophic decrease in the export of *Eucheuma* resulting in farm cultivation. Best results were obtained in salinities above 34‰ S, under good water-exchange conditions and with plants suspended 0.6 m above the bottom on nylon lines attached to stakes driven into the sand. If the lines reach the bottom they are climbed by predators which consume the algae. Excessive light might damage the thalli and induce premature 'aging'. For example, *Eucheuma* planted over light-reflecting sandy bottoms has frequently failed to develop properly. When the plants reach an average weight of 800 g, about 2 mo after planting, they are pruned back to 200 g again; hence, no replanting is needed (DOTY, 1973). The success was revolutionary, and as of mid-1974, there were over 1000 farms in the Sulu Sea and the Visayas Sea exporting over 600 tons $mo^{-1}$ of dry *Eucheuma*, or twice as much as the total world supply the year before.

The success led to attempts of a similar cultivation in Indonesia, Malaysia, China,

Fig. 4-4: *Eucheuma spinosum* is rinsed and dried on the platform of a farmhouse on poles. (Photo: courtesy of J. R. LIM, Genu Products, Philippines Inc.)

Djibouti, Tanzania, and other countries. In 1975 a competing enterprise, Genu Products, started research and initiated marine farms north of Bohol—all three cultivation areas mentioned are in the archipelago between Luzon and Mindanao, the two largest islands. On the Darajong Reef, 200 farm houses were constructed on poles in the sea and with platforms on which the product was dried (Fig. 4-4). In 1979, 500 ha were cultivated with 15 million plants in water; 2000 people were working daily on the farms (LIM and PORSE, 1981).

While biological and technical problems were solved faster than expected, economical problems were not. An exportation peak of 6000 tons dry weight was reached in 1974, i.e. 10 to 20 times more than the previous years. This exportation was only 60% of the annual farm production, and the price of *Eucheuma cottonii* dropped from US $0·40 to $0·05 kg$^{-1}$ at the source (RICHOHERMOSO and DEVEAU, 1979). Following the fall in price, production declined; later, both prices and production recovered slowly. In 1979 harvests of *E. spinosum* unexpectedly increased more than four times the 1978 production, i.e. to 3250 tons dry weight; again, prices declined drastically (LAITE and RICOHERMOSO, 1981).

The chief source of agar-agar, *Gracilaria*, has been harvested in Manila Bay in quantities of 35 tons d$^{-1}$. During dry seasons, when the natural growth of green algae in milkfish ponds is stunted due to high salinity, tons of *Gracilaria* are given as food supplement to pond-cultivated fish. Some 20 species of mostly red algae are used as food. In addition to *Eucheuma* and *Gracilaria*, *Porphyra*, *Hypnea*, and *Laurencia* are the most popular species besides brown *Sargassum* and green *Caulerpa*.

In Indonesia, as in the Philippines, areas away from densely populated regions are centres for seaweed production. This fact adds to the value of the seaweed trade for the development of the countries involved. The Sulu archipelago in the southwest of the Philippines has the most highly developed *Eucheuma* farms, with Zamboanga as the leading export harbour. In Indonesia, seaweed production is likewise spread over the entire island group, but among the provinces the Moluck Islands in the northeast provide 85% of the total production, which amounted to 8426 tons dry weight in 1975, out of which only one-fifth was exported.

Ujungpandang, the port of Macassar, processes three-quarters of the export, with Ambon, southwest of Ceram, being second. Importing countries are—by order of magnitude—Denmark, Hong Kong, Singapore, and Spain (SOEGIARTO, 1979).

In terms of climatic zones, the largest area to be dealt with is the tropical Pacific Ocean. However, to the east of Philippines and the Moluck Islands, there is no production of seaweeds.

As already mentioned, there is a hydrobiological and a biological definition of tropical marine waters: they are warm seas with minor temperature fluctuations where mean surface water temperatures during the coldest month do not sink below 20 °C and they are waters where reefs of corals and coralline algae develop on hard bottoms. In practice the two definitions give delineations which usally coincide in detail. In Japan, however, at the southern end of Kyushu, biotic reefs are found in waters with winter temperatures of only 17 °C (while summer temperatures exceed 27 °C). Along the relatively short coastline which forms this interesting exception it is essential to study the general character of the algal flora. This provides strong evidence for the conclusion that the southern-most part of the Japanese coast is essentially different from that of the rest of the main islands. For instance, the total production of *Digenea* is confined to the three southern-most protectorates on Kyushu, mainly to Kagoshima.

In times past, algae were an important part of food in the Pacific Ocean. Poor people along the coasts of Hawaii lived for long periods exclusively on fish and algae when they did not have access to taro their staple source of carbohydrates. For women, food was most onesided: up to the death of Kamehamea the Great in 1819 there was a death penalty for women who ate bananas, coconuts, turtles, pork, or certain fishes. During hard times there would not have been much more than seaweeds for them.

At the turn of the century there were still more than 100 words in use in Hawaii for various kinds of edible algae. But only the elder aborigines knew them; young people either were unfamiliar with the words, or confused the species. In 1907, more than 70 species of algae were still eaten in Hawaii. Today, only small quantities are consumed by people who like to keep traditions alive. Abandonment of seaweed-eating habits and oblivion of preparation methods and recipes is a trend observed in many old cultures.

### (f) Tropical Marine Vegetation

DAWSON (1962), a phycologist of many waters, has compared diversity and productivity on both sides of the Isthmus of Panama. His observations give an indication of some of the problems of the tropical marine flora (p. 377):

> 'The most generally reduced algal flora ever encountered by the writer occurs in the vicinity of Punta Catedral on the central Pacific Costa Rican coast. There, a mar-

vellously varied rocky shore occurs, replete with headlands, offshore rocks, islands, bays and coves of striking beauty. One can compare the physiography favourably with that of the Monterey Peninsula in California where one of the richest algal floras of the world lives. However, this most promising offshore area, as viewed from a distance, proved to be an algal desert in which only the most meager assortment of plants occurred at lowest tide levels, and then only in certain limited and protected small areas. Where one might have expected a hundred species, we found hardly more than a dozen.'

Similar species-poor floras were reported by DAWSON (1962) from the Perlas Islands in Panamá to Mazatlán, Sinaloa, Mexico. Throughout this region he found seaweeds to be absent or reduced, especially in the intertidal. Nowhere did calcareous coral–coralline reefs prevail, as in the opposite Atlantic Ocean or at the same latitude westward in the Pacific Ocean.

The Atlantic coast of Costa Rica, on the other hand, is dominated by sandy beaches that stretch monotonously along most of the shore, and only towards the southeast have some calcareous strata provided small areas for the development of coral–coralline reefs. Notwithstanding this apparently unhospitable character of the shorter Atlantic coastline, and the fact that the Pacific coastline has been more widely investigated, we find a larger number of species in Costa Rica's Atlantic flora: 196 species compared to 142 in the Pacific Ocean, and a much greater abundance of marine plant life in general.

DAWSON (1962) explains the striking paucity of the Pacific-coast seaweed flora as being influenced primarily by the tidal factor: Atlantic tides are of low amplitude, scarcely 1 m, and of mixed diurnal character with only one low tide daily. The Pacific tides are semidiurnal, with a marked low tide twice daily, and have an amplitude of about 3 m, which prohibits the development of fringing reefs with accompanying favourable algal habitats. The tidal dynamics always result in one low tide during the day during which the intertidal vegetation is exposed to intense insolation and desiccation, and also to fresh-water influence during the rainy season.

No attempts have been made to evaluate the relative impact of the possible reasons for the low productivity in tropical waters, e.g. desiccation, excess light, salinity fluctuations, low contents of nutrients in the water, and grazing.

From a short excursion on Curacao Island, TAYLOR (1942) mentions that large and broken rocks afforded good protection for the algae and, unfortunately, for an unusual multitude of *Diadema* sea-urchins; this fact made collecting very hazardous. In a large concrete bathing pool, apparently not in use, there was a heavy growth of *Caulerpa sertularioides* and of *Hypnea cervicornis*. This incidental observation is of interest since it suggests that profitable seaweed utilization in this area, as well as along most other tropical shores, might depend to a very large extent on the control of grazing.

## (3) Warm-Temperate Oceans

### (a) Ecology and Productivity

Warm-temperate waters exhibit seasonal temperature dynamics, but no production pause during winter. In this season, the relative productivity of phytoplankton—if it were dependent on radiation alone—would decrease by about 50 to 60% as compared to summer productivity within the same latitudes. On the contrary, for the intertidal flora, the winter decrease in insolation is advantageous; the danger of being killed within a few calm days due to intensive solar irradiation—often characteristic of the hot and dry summer and autumn—is less. Likewise, rough winter weather favours intertidal algae through higher surf and stronger splash intensities. Annual and ephemere species therefore exhibit a bloom in late winter.

Warm-temperate surface waters and intertidal beaches are strongly influenced by meteorological conditions in latitudes where dry and desiccating air masses sink down from higher atmospheric levels. Bringing little rain and much sunshine they cause desert belts on land and calm weather at sea which leads to weak water currents. Vertical water circulation is likewise faintly developed, and plankton production in the open sea is low. The transparent blue waters of the Mediterranean and the Sargasso Sea have been called the 'deserts of the sea'. Waters with desert characteristics extend in rather narrow belts on both sides of the wide tropical oceans. Also, benthic organisms feature a paucity of species on warm-temperate shores.

The quantities of seaweed harvested within the warm-temperate zones are very high. Brown algae, consisting largely of giant kelp, account for more than one-third of world production; red algae more than two-thirds. Since they are highest in price, the value of seaweed production from warm-temperate waters is higher than that from any other region. There is a confusion of terms. Some authors use 'subtropical' as a synonym to warm-temperate, others use 'subtropical' for the non-equatorial parts of the tropics.

### (b) Warm-Temperate Coasts of the Western Atlantic Ocean

There is no commercial production of seaweeds along the US coast, from Cape Cod to Rio Grande at the Mexican Border. Lack of rocky coasts is one reason for this—sand, pebbles, and other moveable substrata dominate. Labour costs is the other reason. Seaweed collection or cultivation demands a large amount of labour; hence, competition on equal terms with low-wage countries is not possible. During the war, species of *Gracilaria* and *Hypnea* were collected in North Carolina and on Florida's east coast (HUMM, 1951).

The US Gulf coast was described as barren taxonomically as well as quantitatively, until recently when imposing numbers and quantities of algae were reported. A previously unobserved flora was detected in the eastern Gulf, especially in offshore waters (EARLE, 1969). To the west of Mississippi along the Texas coast the number of species is likewise much higher than imagined before. At least in the bays behind the Barrier Islands species of *Gracilaria* occur in vast amounts (EDWARDS, 1969, 1970).

The fact that a seaweed flora rich in species and biomass could remain unreported within the US is indicative of the many gaps in our knowledge of world-wide seaweed resources. The warm temperate part of southern Brazil includes two very large coastal lagoons, possibly suitable for *Gracilaria* cultivation, while northern Argentina is poor in natural habitats for commercial algae.

### (c) Warm-Temperate Coasts of the Eastern Atlantic Ocean

Portugal and the Atlantic coasts of Spain, favoured by tidal changes along rocky coasts and with vast areas inside the 20 m isobate, produce some 30 000 tons of red algae. While *Gelidium sesquipedale* is the source of agar, *Chondrus crispus*, *Gigartina acicularis*, and *G. stellata* are sources of carrageen. In the Azores, another agarophyte, *Pterocladia capillacea*, with an annual harvest of about 1800 tons is utilized for the extraction of 325 tons $yr^{-1}$ of a high quality agar-agar. The standing crop seems to allow for an expansion to 400 tons agar-agar (FRALICK and ANDRADE, 1981).

In striking contrast, the entire Mediterranean Sea does not support any important and stable production of algae. In the Yugoslavian archipelago and elsewhere there are rich belts of at least 20 species of *Cystoseira*. The members of this genus, however, are not suitable for the production of alginates. There are also quite important monospecific masses of *Gracilaria verrucosa* in the coastal lagoons of Italy, especially in the southern part of the Po Delta. Sea grass and algae cast ashore are used as soil conditioners and manure in most countries, particularly in Greece. Some agar is extracted as a by-product of a food-canning industry in Egypt.

The most evident difference between the Atlantic beaches north and south of Gibraltar and those of the Mediterranean Sea east of the Straits is the absence of ecologically significant tidal fluctuations within almost the entire Mediterranean Sea. The area which appears to have the best prospects for seaweed utilization is the inner part of the Adriatic Sea which, as an exception from the general rule, also has some tidal amplitude.

There is further a most evident difference between nutrient loads in the Atlantic Ocean and Mediterranean Sea. The water deficit of the Mediterranean Sea is compensated through the Straits of Gibraltar, and the entering Atlantic surface water is very poor in nutrients. This has a well-known impact on phytoplankton, zooplankton, and fish production, and must also be expected to influence the productivity of benthic algae.

Besides the absence of an intertidal belt, the most conspicuous feature in the marine flora is the absence of large rockweed such as *Fucus vesiculosus*, *F. serratus*, *Ascophyllum nodosum*, and the kelp *Laminaria digitata*, *L. hyperborea*, and *Saccorhiza polyschides*, which fringe the temperate coasts of the Atlantic Ocean and are replaced in the Mediterranean Sea by a large number of *Cystoseira* species, which may reach a few dm in length, but may also stay at a few cm, much more often than the Atlantic rockweed.

Dense mats of cartilaginous algae comprising many species mixed with corallines like *Jania rubens*, *Amphiroa rigida*, or lime-incrusted green algae like *Halimeda tuna* and *Dasycladus clavaeformis* are also characteristic of exposed localities. Dead parts of *Posidonia oceanica* may enter as warp and weft in the algal cover on fringing reefs.

Less exposed localities are often dominated by *Padina pavonica* and various other Dictyotales. None of these biotopes is suitable for harvesting algae. A high degree of epiphytism and a rich mixture of species makes monospecific stands rare. It is significant that the famous 'Corsican moss', an efficient anthelmintic, marketed as 'Muscus corsicanus', is made up of a mixture of red algae of which only *Alsidium helminthocorton* is for certain a vermifuge.

The algal belt in the Mediterranean Sea is wide since the algae reach remarkable depths. In Malta, attached algae penetrate to depths of 100 m and more. In contrast to an over-simplification found in some textbooks, green algae dominate the submarine cliffs from about 15 m depth to the deepest point sampled, 75 m (DREW, 1969).

The few kelp species entering the Mediterranean Sea are found not in the upper sublittoral as in the Atlantic Ocean, but at depths where, in many coastal sea areas, there are no benthic algae at all. *Laminaria rodriguezii*, an endemic species, is reported off Stromboli, Montecristo, and other Thyrrenean isles far off the coast in depths from 50 to

90 m and *L. ochroleucha* in the Straits of Messina from 45 to 85 m in deep rheophilous biocoenoses (GIACCONE, 1969).

On the west coast of Corsica, *Laminaria rodriguezii* was found in very dense populations between 75 and 90 m when inspected by a diving saucer. While light penetration is responsible for the lower bathymetric limit it is harder to interpret the upper one which is able to ascend to 30 m in Tunisia.

As a resource of raw material, a relatively narrow kelp belt at such depth is uninteresting but rock lobsters *Palinurus vulgaris*, scattered in other biotopes, are numerous in biotopes populated by *Laminaria rodriguezii*, a fact well known to fishermen, which makes the discovery of *L. rodriguezii* populations in the Mediterranean Sea of real economic interest (FREDJ, 1972).

Below surface waters, conditions may sometimes favour the development of large seaweed quantities dominated by a single species. Such a situation is usually preferred by commercial utilizers. In the Gulf of Taranto, between Calabria and Apuglia, PARENZAN (1970) estimated for *Cladophora prolifera* a biomass of over 1 million tons at depths of 29 to 34 m. The northwestern Black Sea harbours 5 to 6 million tons of *Phyllophora nervosa* at depths of 30 to 60 m; here, in a huge mass, this alga covers a mud-shell gravel floor of about 15 000 km$^2$ in a region called Zernov's Phyllophora Sea. In 1978 not less than 8900 tons were collected for the production of the agaroid phyllophoran.

### (d) Warm-Temperate Coasts of the Western Pacific Ocean

The People's Republic of China is the world's largest producer of seaweed with a wet-weight harvest in 1978 of more than 1·5 million tons of kelp and 50 000 tons of laver, corresponding to about 275 000 tons of dry 'haidai' and 7200 tons of dry 'zicai'. These amounts result from a total of 328 tons in 1952, 146 000 tons in 1959, and 550 000 tons in 1970; they show what can be achieved when bold targets are firmly implemented.

There are really no specific ecological prerequisites for such a production, and the natural resources are rather reflected by 300 tons in 1952. Before 1949, as much as 40 000 to 50 000 tons of dried seaweed were imported annually, corresponding to 240 000 to 300 000 tons wet weight (CHENG, 1968). All recent figures refer to the whole of China. The early production, however, was restricted to the relatively short coast-line along the Yellow Sea, which belongs to the cold-temperate zone.

*Laminaria japonica* does not grow spontaneously in warm-temperate waters. If transplanted, it will die during the too warm summer. Nevertheless, millions of young plants are produced in nurseries in Lüda, in the northern-most coastal province of Liaoning; each autumn they are transported to aquaculture co-operatives along almost the entire coast of China. The transport requires control of temperature, oxygen, moisture, and micro-organisms. On the spot, the young sporophytes are all transplanted by hand to ropes or rafts. They are then allowed to develop during winter as long as the temperatures are low enough to favour growth, and are finally harvested before they deteriorate on account of the rising spring temperatures.

In the Yellow Sea, where the temperatures are low enough for year-round cultivation of *Laminaria japonica*, there are other drawbacks. The water is so turbid that around the Huang Ho and other estuaries transparency may be limited to 10 cm, and special arrangements must be made to keep the kelp lamina in the photic layer.

Another problem is that the water is too deficient in nitrogen for successful aquaculture without fertilization. In the Yellow Sea the content of nitrate-nitrogen rarely exceeds 5 mg $m^{-3}$, while in the warm-temperate provinces of Zhejiang and Fujian the corresponding values are 86 to 227 mg $m^{-3}$. A long-term fertilization method employing moderate quantities was developed. Porous clay cylinders containing nitrates are suspended among the plants and during the growth period nutrients seep out slowly through the walls of the jars.

The extraordinary development of aquaculture in China shows that the potential of a certain coastline for producing seaweed cannot be estimated from natural growths. The ultimate productive 'potential' can be drastically changed following sound research and management. Limitations due to unfavourable substrate, light, or nutrient conditions have been successfully compensated as in some other countries.

China, however, is the only country where, on a large scale, natural limits set by climatic regions have been trespassed. As a rough estimate one-quarter or possibly one-third of the total production of *Laminaria japonica* is obtained from seedlings transplanted each autumn to the central and southern provinces. If so, the harvest from the transplanted part of the Chinese *Laminaria* production is in the order of 380 000 tons or even 500 000 tons of a cold-temperate species grown in a warm-temperate region.

The People's Republic of China is also unique in directing a considerable part of the seaweed production towards inland areas where the positive effect on the consumer's health is much more pronounced. In general, coastal human populations are healthier than those of the inland areas (VELASQUEZ, 1953). Their food is more diversified, and the marine components are rich in trace elements. Following the rule that high mountain areas comprise a population with high goitre prevalence, the province in China with maximum goitre manifestation is Yunnan bordering Burma. Also around Beijing Peking), entire village populations were earlier reported to be almost 100% goitrous KELLEY and SNEDDEN, (1960).

The total area of zicai (*Porphyra*) cultivation in China was 3400 ha with a population of 7200 tons dry weight. The southern species, *P. haitanensis*, accounted for 90%, the northern species, *P. yezoensis*, for 10% of the total production (TSENG, 1981).

In Taiwan, seaweed cultivation has developed along another line. *Gracilaria* is grown extensively in brackish water ponds in the southern parts of the warm-temperate west coast. Along the tropical east coast there is no such pond cultivation. *G. verrucosa* is the species most commonly cultivated followed by *G. gigas* and *G. lichenoides*. Over 300 ha of ponds are cultivated, most of them smaller than 1 ha. The salinity of the pond water is usually kept at 10 to 20‰ S, but may sometimes be as low as 4‰ S or as high as 40‰ S. The yield is about 12 000 tons dry weight $yr^{-1}$, or roughly 100 000 tons wet weight. The product is used mainly for the domestic agar industry, export, and for the feeding of abalone, *Haliotis*, and other mariculture animals. The farming method mostly practiced is that of a polyculture system. Fronds of *Gracilaria* are torn into pieces and freely cultivated in ponds with milkfish, *Chanos chanos*, grass shrimps, *Penaeus monodon*, or crabs, *Scylla serrata*. These animals free the seaweed from its epiphytes and the polyculture increases the value of both plants and animals (CHIANG, 1981).

Japan is the leading seaweed-producing country with regard to the value of the products. It has the oldest unbroken traditions for a diversified large-scale production. Japanese seaweed fishery and aquaculture include a greater number of species and products than those of any other country. Number one in value among these is nori, the thin black-violet sheets obtained from *Porphyra yezoensis* and other *Porphyra* species through a complicated labour-intensive method. In 1979 some 8.8 billion sheets were prepared, mostly by hand. The raw material used accounted for less than one-third in weight of the Japanese seaweed harvest. The value of the final product exceeded three-quarters of the total for all algae. Close to 300 000 families were engaged in the production and each family produced 300 000 sheets. The old traditions—rich diversity and high production—of Japanese seaweed cultivation and preparation was comprehensively treated by several reviewers: OKAZAKI (1971), BARDACH and co-authors (1972), MICHANEK (1975), MIURA (1975), SAITO (1975), KORRINGA (1976), and CHAPMAN and CHAPMAN (1980). Along the warm-temperate coasts of Japan, *Porphyra* and agarophytes dominate in production. Of all kelp species in cold-temperate waters, only *Undaria pinnatifida* is found under natural conditions and is cultivated in warm-temperate areas, although this species has two-thirds of its production in the cold-temperate parts of Japan. In cold-temperate waters it attains a length of about 2 m or more, while in warm-temperate waters, its maximum length is 1 m. Two other species, *U. undarioides* and *U. peterseniana* are confined to the warm-temperate coasts of Japan; the former yielding a product of inferior quality.

In spite of the outstanding production of red algae in Japan (362 000 tons in 1978 of which some 265 000 tons are from warm-temperate waters) the yield does not meet the demand, and hence additional quantities are imported. Nori and other edible products are mainly taken from Korea, a country whose production is directed towards meeting Japanese standards. Algae for technical purposes are bought from Chile, Argentina, South Africa, and other countries.

South Korea is a leading nation in terms of utilizing green algae and is in fact the only nation with a production of over 2100 tons. Argentina produces 5 tons and Fiji 7 tons; this concludes the world list accounted for through FAO (1979). The leaf-like *Monostroma*, reminiscent of a thin sea lettuce, may be prepared into sheets like those of the red alga *Porphyra*. Such 'aonori' may command a higher price than any other algal product for consumption.

Korea is the world's third largest producer of both brown and red algae with landings in the order of 220 000 tons and some 650 000 tons, respectively. The brown algae are mainly *Undaria* kelp and *Sargassum* rockweed, previously derived from 'marine fisheries', i.e. collected from natural stands and nowadays cultivated to an increasing extent on ropes. Cultivated *Porphyra tenera*, *P. yezoensis*, *P. seriata*, and *P. kuniedai* are the dominating red algae species. The problem with attempting a precise account is that other species are collectively reported as 'miscellaneous aquatic plants'. This includes most of the species collected from their natural stands, e.g. *Hizikia fusiforme*, a member of the Sargassaceae with bulbous blades, which is one of the Far East algae most often sold in Western countries for 'Oriental cooking'. Another brown alga under the 'miscellaneous' entry is *Kjellmaniella crassifolia*, closely related to *Laminaria*. A number of red algae also collected include *Pachymeniopsis elliptica*, *Chondrus ocellatus*, and wild *Porphyra* spp. *Gelidium amansii*, both from aquaculture and from natural habitats, is included under this heading.

In per capita production, Korea thus surpasses its quantitatively leading neighbour countries. Korea is a prominent fishing nation, particularly in aquaculture. In contrast to China and Japan, its fishery is very much directed towards export. The main market for Korean-produced laver is Japan, and production depends on Japanese demand. In consequence, governmental efforts to develop the seaweed business are concentrating on improving the quality of the products to meet Japanese standards. In particular, there are great price differences for nori products between various quality grades.

The fishermen are often short of cash and have had problems in following the present trend which demands a fast switch from collecting natural stands to mariculture on nylon ropes. The south coast is the chief centre of production. The eastern coastline is steep, has rough weather, and there is no cultivation. The west coast has a turbid brackish surface water, poor in certain nutrients. Cultivation is feasible only with nutrient additions.

### (e) Warm-Temperate Coasts of the Eastern Pacific Ocean

Kelp does not always develop best in cold waters. In fact, the most productive forms, and at the same time those which are most easily harvestable, are the giant kelps distributed along both warm- and cold-temperate coastlines. Even these, however, show different patterns in the two climatic regions, as can be seen from a summary of the classical survey of US and Mexican kelp quantities (Table 4-4).

These figures are questioned. Their value lies in the clear demonstration of the importance of a distinct temperature change for the distribution of seaweed species and the development of resources. Bull kelp *Nereocystis lutkeana* disappears at Point Conception, California. Here occurs a sharp change in water masses. This is also the northern point for the abundant growth of the commercial red alga *Gelidium cartilagineum*. The exact point where the northern *Macrocystis integrifolia* is replaced by the southern *Macrocystis pyrifera* is at Monterey Peninsula, somewhat further to the north.

The harvestable quantities indicated in Table 4-4 should not be confused with biomass. They are based on the result of two annual harvests at 1 m below the sea surface. The difference between the figures of the survey and of quantities actually harvested are very great, and it has been concluded that the early estimates should be reduced to 5% of the figures given. The highest kelp harvests in California were taken in 1917 and 1918 with 395 000 tons each year. After a cessation in the 1920s, 10 000 tons were harvested in 1931, 100 000 tons in 1950, and 150 000 tons in 1971. In 1974, the harvest reached 170 000 tons (MCALLISTER, 1975; K. WILSON, pers. comm.)

Official records show that Mexico harvested 30 000 tons and the US 160 000 tons of

Table 4-4

Annually harvestable quantities of kelp following the optimistic 1911 to 1913 survey of US and Mexican Pacific coasts. All values expressed in tons wet weight (Original)

| | Bull kelp (*Nereocystis lutkeana*) | Giant kelp (*Macrocystis integrifolia*) |
|---|---|---|
| Cold-temperate waters (Washington, Oregon and two-thirds of California) | 3 350 000 | 749 000 |
| Warm-temperate waters S. California, N. and Central Baja California | — | *M. pyrifera* 35 175 000 |

brown algae in the eastern central Pacific in 1978. A part of this area lies within the cold-temperate region but no background data for a detailed assessment are available.

In Chile, there is a boom in seaweed production. The annual seaweed exports have increased almost ten times between 1969 and 1979. The number of fishermen involved in seaweed exploitation was almost 5250 in 1975. The export products were worth over US $17 million in 1979 (JOYCE and SANTELICES, 1978, SANTELICES and LOPEHANDIA, 1981).

Traditionally, *Gracilaria* is the most important genus and the vicinity of Concepción is the harvesting centre. As the southern end for distribution of *Macrocystis integrifolia* is found in this area, it has been regarded as the limit between the warm-temperate and the cold-temperate marine flora, although marine zoologists place the boundary somewhat further to the south, stating that a transition region lies north of Chiloé (KNOX, 1960).

Since the limit is still uncertain, it is less important to make a breakdown of the *Gracilaria* resource in terms of climatic regions. Available data on the estimated potential resource indicate that a fraction in the order of 41 000 to 44 000 tons dry weight is found in the warm-temperate area, while 38 000 to 39 500 tons dry weight or close to half the potential resource should be referred to the cold-temperate region. The actual export, however, from these two areas was 8360 tons and 1270 tons dry weight, respectively, or only 13% from the cold-temperate south; this may indicate that this area is not so well developed as the climatically more favoured central Chile.

Including the less important genera *Iridea*, *Gelidium*, and *Gigartina*, converted into wet weight and updated to 1979 we find a total harvest of some 58 000 tons red algae in Chilean warm-temperate waters and some 8000 tons red algae in cold-temperate areas. Recently, the harvest of the brown alga *Lessonia* has increased strongly to 6000 tons dry weight. The conversion factor is not given; most common for brown algae is 6:1; this would indicate a harvest of some 35 000 tons.

## (4) Cold-Temperate Oceans

### (a) Ecology and Productivity

Cold-temperate waters are characterized by a strong seasonality in temperature and a period of low radiation in winter, during which photosynthesis decreases to about half of the summer values in the southern parts and to zero in the northern parts. Many species of algae are fertile at the end of this production pause.

The vegetation is dominated by large quantities of brown algae. A rich belt of rock-weed covers the littoral and upper sublittoral regions. It is followed by a likewise well-developed kelp belt. In terms of biomass, most of the vegetation is found above 15 m. In a popular way, the cold-temperate coasts could be described as those, where during low water the solar radiation is no longer intensive enough to kill permanently or occasionally the littoral vegetation, as it may in the warm-temperate region, and where very low temperatures are not frequent enough to limit essentially the same littoral vegetation, as it may in subarctic regions. Annual and ephemere species reach their highest development during the later part of summer.

Available data on potential harvestable biomass indicate that the brown algae of cold temperate waters account for more than three-quarters of the world potential of harvestable marine algae. Certainly, the optimal conditions in the intertidal belt alone do not account for this outstanding resource.

The seaweeds harvested in cold-temperate waters amount to more than 50% of the total world harvest. Not less than 88% of the harvests from cold waters are made up of brown algae. Biomass production taken as an excess of assimilation over respiration is favoured by low temperatures. In the warm waters of tropical and temperate oceans respiration is considerably higher. Night and day it consumes a large part of the photosynthetic products. The very high biomass in cold-temperate waters must be credited to the light energy available—still sufficient to build large quantities of organic matter—but even more perhaps to the fact that temperatures are low enough to leave the primary producers an optimal remainder of the photosynthetic products due to minimal respiratory consumption.

It is also possible that the role of secondary producers is significantly different in the warmer regions of the oceans than it is in the colder regions. There are reasons to assume that grazing is a major cause for the low biomasses found in tropical coastal waters, which are known to have a high primary productivity and which sustain a rich stock of consumers.

### (b) Cold-Temperate Coasts of the Western Atlantic Ocean

Between the Labrador Current from the north and the Gulf Stream from the south, the cold-temperate province in the northwest Atlantic Ocean is confined to a restricted area between Newfoundland and Cape Cod. Most of it is a rocky indented coast, offering many intermediate stages between shores exposed to oceanic waves and really sheltered ones. Vast areas of flat rocks or glacial drumlins have moderate depths and are thus favourable for algal growth. South of Boston there are sandy beaches and consequently very few algae, never reaching commercial quantities.

The natural potential resources for kelp and rockweed are the rocky coasts of the Canadian Atlantic and Maine as exemplified by *Laminaria* species and *Ascophyllum nodosum*. Commercially, however, the area is dominated by a red alga *Chondrus crispus*, also known as 'Irish Moss'. The biomass of the mentioned brown algae might exceed that of the Irish Moss by more than ten times, but in harvest figures the relations are reversed. In 1969 a harvest of 44 000 tons of Irish Moss accounted for 82% of the quantities and 95% of the value of the total landings (FFRENCH, 1971), and in 1978 Canada harvested 32 000 tons of red algae and no brown algae at all. These proportions may reflect the great industrial demand for Irish Moss, the most important raw material for carrageenan. However, lack of data for brown algae in world statistics is probably due to faulty reports. In southwestern Nova Scotia, there is a commercial marine plant industry based on species of *Laminaria*, *Chondrus*, and *Ascophyllum*; it has been operating for approximately 45 yr, 40 yr, and 20 yr, respectively. The harvesting and extraction of algin from *Ascophyllum* is an important local industry (PRINGLE and SEMPLE, 1980). A kelp export of 1000 to 3000 tons lasted only from 1942 to 1949. In 1978, Marine Colloids purchased several 100 tons of kelp for a dried food product (SHARP, 1980). Carrageenan extraction in the United States is a $30 million industry, which is close to 90% based on imported raw materials.

The horizontal distribution of *Chondrus* coincides largely with the cold-temperate area with commercial quantities in Nova Scotia, Prince Edward Island, New Brunswick, and Maine, and with northern-most findings in Newfoundland. It is not distributed in the subarctic and, contrary to erroneous statements, not recorded from Labrador (WILCE, 1959; TAYLOR, 1962).

In the vertical range, maximum densities have been found from the low-water level to 7 m below. Along the exposure gradient, *Chondrus* is absent in maximal exposed sites, as well as in the most protected ones. It grows abundantly on semi-exposed open coasts and in estuaries with strong tidal currents. Carefully and moderately harvested plots allow regrowth of biomass after 5 to 6 mo (MATHIESON and PRINCE, 1973).

The maximum quantity landed was 53 000 tons in 1974 and the mean annual harvest during 1967–1976 was 34 000 tons $yr^{-1}$. In Nova Scotia 700 harvesters crop 65 distinct beds with handrakes. This method has been used since the 1940s, but with increasing demand in the mid-1960s the basket dragrake was introduced. It is feared that the increased effort and modified harvesting techniques have reduced the productivity of the beds—possibly permanently. A reason could be that dragraking removes the holdfasts; in a preliminary assessment of this technique 2000 $mm^2$ of holdfast was observed in each kg of *Chondrus crispus* harvested. Since it takes 2 yr to attain a holdfast of 4 $mm^2$, this annual removal of holdfasts has serious consequences on productivity (PRINGLE, 1979).

The fluctuations of landings during 30 yr including a period of low catches, another of high catches, and finally one of decreased landings, were related to fishing effort and to catch-per-unit-effort. Maximum sustainable yield was estimated at 12 000 to 14 000 tons for the harvesting district investigated (PRINGLE, 1981).

A total estimate of biomass in the Canadian Atlantic Ocean cannot be extrapolated from local assessments. For an evaluation of kelp resources, however, it is often of greater interest to know examples of quantities within areas of the size of a possible harvesting effort. The small Seal Islands southwest of Nova Scotia, distinguished by waters of high clarity and strong currents, harbour an abundance of kelp with *Laminaria longicruris* growing atop boulders, and *L. digitata* and *Alaria esculenta* often in depressions. The kelp growth covers about 5.7 $km^2$. In the late 1940s, MACFARLANE (1952) estimated an average yield of 25·1 kg $m^{-2}$, while—after a re-visit in 1977—MCPEAK (1980) arrived at a maximum production of 87 700 tons of kelp, corresponding to 15.3 kg $m^{-2}$. Sampling methods could be a possible reason for this difference, the earlier investigation being carried out with a Peterson grab and the latter with scuba divers. A change in density during the three decades is also possible. Finally, there is evidence for a strong annual variation in biomass of the kelp beds studied during consecutive years.

A problem which has attained increasing attention is that of the destruction of kelp beds by sea urchins. Beginning in 1968, sea urchins became locally abundant and over-grazed the kelp beds, converting large areas to urchin-dominated barren grounds. Almost all kelp beds in St Margaret's Bay (140 $km^2$) were destroyed within 6 to 7 yr. There is evidence in favour of the hypothesis that urchin-dominated barren grounds are a new, stable configuration of the ecosystem, and that a long-term decrease in primary and secondary productivity of these coastal waters can be expected (MANN, 1977).

Contrary to what was found in the North Atlantic Ocean, the littoral belt of the cold-temperate coast of Argentina is quantitatively poor (KÜHNEMANN, 1970). Species of *Porphyra*, *Ulva*, and *Enteromorpha* are encountered, but seldom in valuable quantities. On the other hand, in the sublittoral, gigantic seaweed prairies occur; these comprise, for example, the red algae *Gracilaria verrucosa*, *Iridea* sp., and *Gigartina skottsbergii*, and luxuriant kelp forests with *Macrocystis pyrifera*, *Lessonia nigrescens*, *L. flavicans*, *L. fuscescens*, *Desmarestia menziesii*, *Durvillea antarctica*, and *D. caepaestipes*.

According to an assessment of quantities available for industrial utilization, there are sufficient resources for an annual production of 15 000 to 20 000 tons of algal meal, 1000 to 1500 tons of alginates from *Macrocystis*, 1000 tons of agar from *Gracilaria*, and 100 tons of carrageenan from *Gigartina*. These end-products would correspond to harvests of roughly 200 000 tons of *Macrocystis*, 100 000 tons of *Gracilaria*, and 1500 tons of *Gigartina*. The present agar production of about 300 tons $yr^{-1}$ (from maybe 25 000 tons of fresh *Gracilaria*) is confined to a small production area in the territory of Chubut (OLIVEIRA, 1981).

### (c) Cold-Temperate Coasts of the Eastern Atlantic Ocean

Present official records from Norway, Iceland, France, and Scotland indicate that the brown algae dominate with a total of about 130 000 tons.

Iceland has old traditions of using the red alga *Palmaria palmata* for human consumption and of letting sheep and cattle graze on *Ascophyllum nodosum*. An industrial production has started just recently and was in its fourth year in 1978 using 12 000 tons.

Along almost the entire Norwegian coast there is kelp down to 20 to 30 m. In fact, kelp vegetation covers an area of *ca*. 10 000 $km^2$ or about the same as the total agricultural area of Norway. The biomass of kelp in Norway is estimated at 'several tens of million t' with a regrowth time of 4 to 5 yr (HALMØ and co-authors, 1981). The rockweed *Ascophyllum nodosum* grows abundantly mainly in the intertidal and is estimated at 1·8 million tons wet weight, *Fucus serratus* at 0·9 million tons, and *F. vesiculosus* at 0·3 million tons. *A. nodosum* requires 4 to 6 yr for full regrowth after harvesting; annual harvesting is suggested at 17 to 25% (BAARDSETH, 1970). Following official statistics the total Norwegian production of brown algae is now stabilizing at some 65 000 tons. Data from industrial sources indicate a greater harvest: three seaweed-meal factories produce a total of 10 000 tons with *A. nodosum* as the only raw material, and one factory for alginates produces more than 3000 tons $yr^{-1}$. If we assume a conversion factor of about 3·2 : 1 for wet rockweed to seaweed-meal and 15 : 1 for kelp to alginate we arrive at a total harvest of brown algae in the order of 80 000 tons.

Norway is now the world's second largest producer of alginate and one of the largest of seaweed-meal. *Ascophyllum nodosum* is the most harvested species. *Laminaria hyperborea* and *L. digitata* are likewise used for algin production. The factories are situated in southwest Norway; until 1974 the rich resources north of Trondheim were not utilized.

Far-reaching degradation of kelp beds in northern Norway have been reported; it is interesting to compare these with those observed in California, and in Atlantic Canada. In California the depletion of *Macrocystis* beds by *Strongylocentrotus purpuratus* and *S. franciscanus* was first explained as a consequence of eradication of the sea otter, which was its primary predator. Historical data, however, indicated that those portions of the beds close to sources of waste disposal were affected first (TEGNER, 1980). In Canada the corresponding population boom of *S. droebachiensis* was first ascribed to over-fishing of lobsters, their main predators. Also in this context it is now argued that the decrease in lobster numbers cannot be the only factor controlling sea-urchin populations; among other potential causes pollution has been mentioned. Against this background it is interesting to note that the recent boom of the same sea urchin species in Norway is observed along the coasts of northern Norway only—north of the natural distribution of the lobster and north of the area of a noteworthy pollution. It is hard to think of a predator which has changed sufficiently in numbers to explain the uncontrolled growth of *S. droebachiensis*. A possible clue could be that herring, or some other heavily depleted fish species, had been feeding previously at some stage on the planktonic larvae of the sea-urchins (HAGEN, 1981).

Potential resources are considerably higher than those actually harvested. A survey of 8500 km of coastline in Scotland arrived at 10 million tons of Laminariaceae, predominantly *Laminaria hyperborea* from which 1 million tons wet weight could be harvested each year. In 1978, 32 700 tons were harvested.

In France the *Laminaria* landings show strong fluctuations: 56 000 tons in 1950, 6000 tons in 1965, 16 000 tons in 1975, and 33 000 tons in 1978. Certainly the most recent figures indicate an upward trend for the alginate industry. The production areas are Normandy and Brittany (Fig. 4-5).

Harvests of red algae are reported from France only (2000 tons); France also reports 2100 tons of 'miscellaneous aquatic plants'. These figures could be based on or include 'Irish moss' from Normandy and Brittany but scarcely 'maërl', the calcareous red algae *Phymatolithon* (syn. *Lithothamnion*) *calcareum* and *Lithothamnion coralliodes*; these are col-

Fig. 4-5: Bretagne. *Laminaria* harvesting. (Photo: courtesy of J.-P. TROADEC)

lected off Glenan Islands in the south of Brittany and mainly in the Gulf of St Malo. The annual harvest of maërl is in excess of 300 000 tons (BLUNDEN and co-authors, 1975); apparently it is not reported at all in the French statistics, or not as algae. Maërl is composed of calcium carbonate and magnesium carbonate and used in agriculture and horticulture as a soil conditioner—smaller quantities also serve as animal food supplement and as filter material for treating acid drinking water.

Denmark and Ireland have some seaweed production, although not accounted for in international statistics. Denmark had for 20 yr a most particular locality for *Furcellaria fastigiata*, the excellent 'Danagar' raw material. North of the Peninsula Djursland, circulating currents gathered detached tufts in large amounts at moderate depths of mainly 3 to 4 m. In the early 1960s trawlers collected 20 000 to 30 000 tons annually, finally resulting in over-fishing and total stock depletion. After this ideal fishing ground had to be abandoned some of the trawlers continued to collect in various smaller areas yielding an annual total harvest in the order of 10 000 tons. *F. fastigiata* also penetrates into the Baltic Sea and is found attached to stones along the open coast between Klaipeda and Ventspils in quantities estimated at 80 000 to 90 000 tons, in the Gulf of Riga at some 10 000 tons. In 1960, loose-lying resources on soft bottoms at the islands of Hiumaa and Saarema were estimated at 150 000 tons. Data on landings are not available, but 25 000 tons would be a conservative estimate.

A century ago, when kelp was burned for iodine production, Ireland produced 6000 tons annually. Later 1000 to 1700 tons $yr^{-1}$ of air-dried stipes of *Laminaria hyperborea* were collected for alginate production. A governmental factory was unsuccessful, and now most of the harvest is exported to Scotland for the extraction of alginate. *L. hyperborea* production has dwindled from 2000 tons dry weight in 1973 to 325 tons in 1979. At the same time *Ascophyllum nodosum* has increased by 60% to 62 000 tons wet weight. *Chondrus crispus* the 'Irish moss' amounted to 165 tons dry weight in 1979, one-quarter of the 1942 peak value. A product called 'blackweed', consists of the total driftweed, mainly Phaeophyta: species of *Fucus*, *Ascophyllum*, and *Laminaria*. It is dried and milled and used for animal feed or for the manufacture of liquid seaweed extracts. The quantities sold in 1979 were 8250 tons dry weight (GUIRY and BLUNDEN, 1981).

In Iceland, Norway, and Scotland, *Chondrus crispus* and *Palmaria palmata* were used for human consumption—a custom which is now more or less abandoned. In Ireland, however, small quantities of these species are still consumed. The strongest survival of such traditions is met with in Wales, where laver porridge is a traditional food.

> 'There is still a considerable trade of *Porphyra* in South Wales, where consumption is in the order of 200 tonnes (wet weight) per annum. The raw material is almost all imported from Cornwall, the Solway Firth, North Wales and Dunbar, with small quantities imported from Ireland when supplies from more accessible collecting grounds fail. The *Porphyra* is washed, boiled, minced and prepared for eating by warming in fat, sometimes being made into cakes with oatmeal' (DIXON and IRVINE, 1977; p. 36).

In the southern hemisphere cold-temperate conditions are characteristic of Namibia and the west coast of South Africa. Localized inshore upwelling of cold water during spring and summer cause an extraordinary condition: within such upwelling regime, summer water temperatures are generally significantly lower than those in winter. *Laminaria pallida*, a major component of the subtidal kelp beds, is morphologically similar to *L. digitata* of the northern hemisphere (DIECKMANN, 1980). In the upper subtidal, *Ecklonia maxima* and *Macrocystis angustifolia* add substantially to the biomass of the kelp beds, but neither compares for size with the giant kelps of the Americas.

Following South African regulations, only drift kelp was collected. Recently for the first time, direct exploitation of the standing stock of large kelps has been allowed, which has also given opportunity for an assessment of the exploitability of the kelp beds. SIMONS and JARMAN (1981) investigated 700 ha of kelp field at Kommetje, west of Cape Town by means of aerial photographs. Under-water studies were carried out by SCUBA divers who collected 1-m wide transects through the kelp. It appeared that *Ecklonia maxima* covered 250 ha, *Laminaria pallida* 450 ha. The standing crop (excluding holdfasts) of *E. maxima* was 33 000 tons wet weight, that of all kelps 66 000 tons. Fronds of *E. maxima* increase their biomass by 1% $d^{-1}$, or produce about seven times their own weight in a year, total standing crop turning over about twice a year. When harvested from stranded material only, about 400 tons (air dry) was taken annually from the beach at Kommetjie. About 90% of this (i.e. 360 tons or, 2400 tons wet weight) would have been *E. maxima*, which is preferred to *L. pallida*. Maximum commercial harvest in this field is 3000 tons $yr^{-1}$ wet *Ecklonia* stipes at which rate the *E. maxima* field at Kommetjie would be worked once in 5·5 yr. Mean harvest per effort was *ca.* 1 ton $(\text{diver-hour})^{-1}$.

An extrapolation of the biomass found in the investigated area over the entire rocky west coast would give a standing crop of 5 million tons of kelp in the cold-temperate waters of the Republic of South Africa. Assuming that only 10% of this potential arrives on the shore, and that only 10% of the coast is commercially accessible, some 50 000 tons wet weight of harvestable kelp materializes (R. H. SIMONS, pers. comm., MICHANEK, 1981).

### (d) Cold-Temperate Coasts of the Western Pacific Ocean

*Laminaria* beds dominate the whole area; they consist mainly of *Laminaria japonica*. Along the Siberian coast they are estimated at 2 million tons, out of which 40% grows in Kamchatka. Probably this peninsula is situated too far away for a seaweed industry. Kamchatka has become well-known for being the only area where an alcoholic beverage is produced from algae. Raw material for this is dulse, *Palmaria*. The Kamschadals also use *Halosaccion*, *Fucus*, *Alaria*, *Chordaria*, and *Porphyra* as food.

*Ahnfeltia plicata* beds are roughly estimated at 100 000 tons near Vladivostok, 25 000 tons on Sakhalin, and 50 000 tons in the Kuriles. Harvests are processed locally or exported to Japan. Sakhalin and the Kuriles used to have a very important production of kelp. It is not known if this production is maintained under Russian hegemony.

In Japan, one-quarter of the red algae and a good nine-tenths of the brown algae are produced within the cold-temperate region: the north island Hokkaido and the three northern-most prefectures on the east coast of the main island. The Japanese total is in the order of 300 000 tons brown algae. The standing crop of alginophytes along the coasts of Hokkaido is estimated at about 1·7 million tons. There are at least six *Laminaria* species of commercial interest and also three very closely related genera: *Kjellmaniella*, *Arthrothamnus*, and *Cymathaere*.

Kelp production has been hampered by a phenomenon called 'reef burn', a growth of coralline algae which has spread dangerously in recent times. Vast areas, all around Japan, are now covered by such algae—crustose as well as articulate—on which *Laminaria* does not attach. Possibly, sea urchin grazing is the initial cause.

During drying each lamina is scrutinized for mussels, bryozoans, algae, and other epiphytes. High grades are sold for food as various kinds of kombu. Lower grades and a few species unsuitable for food are used as a raw material for alginates. Since supply is inadequate, alginate raw material is also imported, mainly *Macrocystis*.

Very little is known about seaweed production in the Korean Democratic Peoples Republic. However, there are reports of 100 tons *Gelidium* and 600 tons *Ahnfeltia* (YAMADA, 1976). Reasonably, there could also be *Porphyra* and *Laminaria*. The fact that algal biomass is essentially higher in cold-temperate waters, as compared to warmer or colder oceans, is illustrated well by numerical data recorded from the French Southern Territories. On the east coast of the Kerguelen Main Island an assessment of the quantities present has been made in the Baie du Morbihan (GRUA, 1964). In this bay, 45 $km^2$ are covered by *Macrocystis pyrifera* in stands estimated at 100 kg $m^{-2}$ in clear waters and 600 kg $m^{-2}$ in the least transparent waters. To this must be added a rich undergrowth estimated at up to 15 kg $m^{-2}$ in clear, and 5 kg $m^{-2}$ in medium transparent water. GRUA calculates the total biomass of *M. pyrifera* in the bay at 6·3 million tons, which would give a mean density of 140 kg $m^{-2}$.

DELEPINE (1976) investigated the biomass available for harvesting (expressed in terms of down to 0·5 m below surface). He found individual values from 3·4 to 22·5 kg $m^{-2}$ and mean values of different beds from 7 to 12 kg $m^{-2}$. After 6 mo there was a full regrowth. Kelp beds around Kerguelen Island covering over 200 $km^2$ would give a biomass in harvestable depths of 1 to 2 million tons. If a crop of about one-quarter of the standing stock could be sustained and two harvests a year be taken, the potential would be 0·5 to 1 million tons wet weight. Dry weight would be 10% of wet weight and alginates 20% of dry weight (DELEPINE, 1976; EVERSON, 1977). The ecological basis for this amazingly large standing crop is found in vast areas of favourable bottoms at moderate depths, optimal temperatures around 4 ° to 6 °C, and rich nutrients: 20 $\mu$g-at of N $l^{-1}$ and about 1·75 of P $l^{-1}$.

The two assessments quoted may differ substantially in methods and results; however, the essential point remains that even the largest biomasses recorded are not necessarily commercial resources. As Kerguelen Island primarily serves as a naval base, manpower for any work would have to be brought from very distant places. The possibilities for drying are doubtful and algin extraction would require much fresh water and energy. To bring untreated raw material to Japan or Europe would be expensive.

Tasmania is in a better position to develop seaweed resources since it is inhabited, has technical facilities, and is situated not far from a big market. On the east coast *Macrocystis pyrifera* covers an area of 120 $km^2$, growing at depths of 3·5 to 30 m, and is estimated to yield 355 000 tons fresh weight if three crops a year are taken; this corresponds to 3000 tons $km^{-2}$ or 3 kg $m^{-2}$ (in California, USA, 3 kg $m^{-2}$ may be removed on one harvesting occasion). Other *Macrocystis* beds are found on the Tasmanian west coast and in Bass Strait. A company working in southeast Tasmania had a specially designed vessel with blades similar to those on a hay mower; these cut the giant kelp at a depth of 1·2 m below the surface. The vessel had a crew of four men, a capacity of 20 tons $h^{-1}$, and a carrying capacity of 45 tons $trip^{-1}$. The alginate industry, however, has now ceased operating; it was replaced by a collecting company exporting dry *Macrocystis*. Nevertheless, Tasmania must be regarded one of the world's best-situated large, under-exploited resources.

Still higher figures are mentioned for South Australia, where the coast of Victoria lies in the cold-temperate area and is fringed by *Macrocystis*, estimated at 1 400 000 tons.

This quantity, however, cannot be regarded as a resource comparable to those of Kerguelen Island and Tasmania. The species growing on the coast of the Australian continent is *M. angustifolia*; it thrives close to the shore down to 3 m below the low-water level and cannot be harvested from a ship. The plants normally reach a length of 6 m as compared to 60 m in the Tasmanian *M. pyrifera* which grows at depths of 3·5 to 30 m.

In New Zealand, *Macrocystis pyrifera* grows in the Foveaux Strait and Cook Strait in beds which have been estimated at 8500 tons; they could yield 2800 tons annually. Another alginophyte of possible interest, *Durvillea antarctica*, grows abundantly along the east coast of the South Island. In five transects through the *Durvillea* band its standing crop ranged from 36 to 190 tons $km^{-1}$ of shore. The problems are that only a fraction of the shore is a suitable rocky habitat, and that regrowth after harvesting needs several years. It seems unlikely that there is sufficient *D. antarctica* in New Zealand to support an alginate industry with the desirable quantity of 30 000 tons $yr^{-1}$ (HAY, 1979).

### (e) Cold-Temperate Coasts of the Eastern Pacific Ocean

The giant kelp of the warm-temperate waters of southern California and Mexico, *Macrocystis pyrifera*, is estimated to produce almost 50 times as much harvestable quantities as *M. integrifolia* along the twice longer cold-temperate Pacific coastline of the US. These closely related species are so similar that they were not kept apart during early assessments. However, they respond differently, e.g. with regard to temperature preferences.

*Macrocystis pyrifera* grows at 10 to 20 m depth in southern California and in Baja California southwards to the end of upwelling at Pta San Hipolito. Normally it reaches a length of 20 to 30 m. The longest plant recorded along the Pacific Coast of North America measured 45·7 m (FRYE and co-authors, 1915). A closed vegetation creates enormous beds which do not reach the shoreline and in the shelter of which small boats are protected from the rough sea.

*Macrocystis integrifolia*, the species of cold-temperate waters in the northern Pacific Ocean, grows from zero tide level down to 9 m in sheltered inlets or other areas protected from full wave action by rocks or reefs or inside the outer protection fringe of the bull kelp. The lower limit is not usually set by the biology of the alga but by the grazing of the sea urchin *Strongylocentrotus franciscanus*. In British Columbia, the *M. integrifolia* kelp beds are typically restricted to 0 to 4 m below zero tide, which usually results in narrow (4 to 20-m wide) beds parallel to the shore. At these depths harvesting is difficult. An enhancement programme suggests planting on artificial substrate over soft bottoms where no sea urchins are found and, at a depth 11 m below the grazer-produced lower limit of adjacent natural kelp-beds (DRUEHL, 1979).

The first survey over the entire US Pacific coast was conducted in 1911 to 1913 due to a great demand of kelp for potash, later also for acetone, iodine, and a bleaching agent. For an area slightly smaller than the cold-temperate region, the estimate arrived at some 20 million tons of harvestable giant kelp. Conservative recent estimates stay at 1·5 to 1 million tons of commercially available giant kelp; other estimates range from 1·5 to 30 million tons, including all alginophytes.

The floating kelp beds in Oregon, Washington, British Columbia, and south east Alaska are dominated by *Nereocystis lutkeana*, which is recorded for 78 to 94% of the harvestable biomass.

Among the alginophytes, *Hedophyllum sessile*, *Alaria marginata*, and *Laminaria saccharina* flourish in northern latitudes and may reach 50 to 100% of the quantities of the giant kelp (SCAGEL, 1961). However great the potential resource, currently it is not utilized. While the US report 200 tons of brown algae in the northeast Pacific, Canada has not reported at all. As mentioned, part of the statistical area—the eastern central Pacific—is cold-temperate and possibly some of its 160 000 tons of *Macrocystis* are taken from within this part.

In southern-most Chile, there is an estimated potential of 200 000 tons dry weight of *Macrocystis pyrifera* in Magellan Strait and close to 300 000 tons in the Beagle Canal area (SANTELICES and LOPEHANDIA, 1981). The wet weight equivalence would be in the order of a total of 5 million tons. Studies on regrowth after harvesting have just started. This is one of the world's largest, untapped resources—untapped mainly for climatic reasons: the area is cold, moist, and stormy. It attracted the attention of Captain Cook, who gives the trustworthy figure of 40 m length for a single plant. Other early circum-navigators of the world gave 500 m, a pirate tale which entered serious scientific literature.

Possibly the South American beds of algae are now under the influence of changing ecological conditions. DAYTON and co-authors, (1973, p. 34) give some new observations:

> '. . . contrary to our original expectations, the Chilean sea urchins apparently have few important natural predators. As might be expected, these sea urchin populations usually consumed almost all the non-coralline macroalgae.'

The general pattern they observed was of rather small beds of what appeared to be young plants.

> 'The few larger *Macrocystis* beds we saw were in the southern region of the Gulf of Corcovado where man is in the process of harvesting *Loxochinus albus*, the sea urchin that most efficiently overexploits the kelp beds.'
> '. . . In contrast to the *Macrocystis* of California, which tends to grow at depths of 10 to 30 meters, the *Macrocystis* plants in Chile are much more shallow, usually occurring from the intertidal to about 5 meters and only rarely as deep as 10 meters. The algal understory beneath the *Macrocystis* canopy also was relatively poorly developed both in species diversity and in structural contributions.'

## (5) Subarctic and Arctic Oceans

### (a) Ecology and Productivity

Subarctic waters offer only a very short vegetation period—that of the ice-free months. Even during these, pancake ice may grind off all vegetation in the littoral belt. Consequently, rockweed like *Fucus* and other perennial algae are missing near the waterline or confined to protected areas. Frequent occasional frosts may kill most of the intertidal flora, including that of rockpools. Hence, the littoral may be void of algae or may have only ephimeral covers of green algae. The highest biomass values are found in sublittoral kelp beds at depths of 10 to 25 m; here the main biomass consists of perennial algae. These are favoured by high nutrient levels and low temperatures, seldom over 5 °C.

An excellent early summary of the characteristics of a subarctic marine flora was given by KJELLMAN (1877, p. 68) for the west coast of Novaja Semlja and Vaigach. The following is a slightly revised translation:

(i) In most places a littoral algal vegetation is missing completely. Where there is one, locally, it is poor in specimens and consists of small forms only. (ii) The main part of the algal vegetation is found in the lower sublittoral, but those parts of this level, which have an algal vegetation, are insignificant compared to those where algae are missing. (iii) Within the sublittoral there are various communities among which that of laminarians is the most frequent, most extended and richest in species. (iv) A particular Fucaceae-community is missing. (v) Green algae are few. (vi) There is a general poverty of specimens. (vii) The algal flora is monotonous. (viii) It is lush (e.g. where the vegetation is not hampered by the particular drawbacks of the subarctic environment it is luxuriant).

The extreme conditions of the subarctic region, in particular the need to compensate for a very long period of insufficient light, makes the vegetation sensitive to small changes in environmental conditions. Some of the richest biotopes of the world oceans are found in subarctic waters favoured by good water exchange; vast areas are completely deserted where glacier silt or river-mouth deposits create unfavourable substrates.

### (b) Subarctic Coasts of the Arctic Ocean and Adjacent Seas

Ungava Bay and Labrador exhibit typical subarctic features. Where mud flats occur they are covered in August and early September by a slippery turf of *Vaucheria* species. Rocky shores are generally poor in algae, but moderately exposed littoral areas may be covered by a brown band of the annual species *Chordaria flagelliformis*, *Petalonia fascia*, and *Scytosiphon lomentaria*. Kelp of immense size are found in sublittoral beds. In quiet areas of moderately exposed coasts, *Laminaria longicruris* emerges to the low tide level, seldom attaining a length of 15 m and a lamina width of 1·3 m. In more turbulent areas vegetation is restricted to considerable depths, beginning at approximately 9 m. Here, *Laminaria nigripes* and *Alaria grandifolia* grow in great abundance mixed with the deep-water species *L. solidungula*. Along fully exposed coasts small beds of relatively low kelps of the same species and *Agarum cribosum* were found at 8 to 16 m depth. Contrary to KJELLMAN's observations on Novaja Semlja, protected littoral shallows are overgrown by large populations of *Fucus vesiculosus* (WILCE, 1959).

In Discovery Bay, 1000 km further to the north, divers found forests of *Laminaria longicruris* so dense that they could not enter. Individual specimens with gas-filled stipes of 5 m plus lamina of 5 m length were measured (JENNEBORG, pers. comm.).

Even if the quantities are impressive, the region is far outside the harvestable area. Like all subarctic occurrences, only a high price for some particular species could render harvesting commercially feasible. One such species, which could be harvested if it would show, for example, outstanding pharmaceutical properties, is *Turnerella pennyi*.

There may be considerable quantities of *Ascophyllum nodosum* and kelp around southern Greenland. *A. nodosum* occurs from 65° N on the eastern side to 70° N on the western side, but quantitative data are wanting.

The subarctic east coast of Iceland is characterized by large forests of *Alaria* in exposed localities, meadows of *Halosaccion*, *Chordaria*, and *Acrosiphonia* and on protected slopes and in rocky lagoons, *Saccorhiza* and *Coilodesme*. MUNDA (1972) reports some species as being harvestable; enormous amounts of *Laminaria hyperborea* are washed ashore along the north and east coast and could possibly be used as raw material for alginates or as fodder for animals; *Alaria esculenta* is most abundant along the northern and eastern coast and could easily be harvested in the upper sublittoral of exposed coast-lines. Harvesting could also be feasible for *Saccorhiza dermatodea* and *Chordaria flagelliformis*. A *Porphyra* species of linear shape is characteristic of exposed sites where it is found without accompanying species. It has a protein content of 37 g $(100\ g)^{-1}$ dry weight.

Surprisingly, the subarctic area where exploitation of kelp resources would most likely be feasible in practice is the northern-most: Spitsbergen. The weather is usually clear and calm, communications are not so bad as in most other arctic areas, and technical skill and certain repair facilities are available. Finally, and most significantly, the coal resources—possibly the richest unutilized resources of the world—provide cheap energy for drying. Whether algin extraction could be carried out in Spitsbergen will depend not so much on energy costs as on the weighing of the high local prices for labour against the high transport costs for dry weed compared to algin, the former being at least ten times bulkier.

The resources are unexpectedly rich for the Latitude 77 to 81° N. While the shores of Greenland plunge steeply down hundreds of meters, Spitsbergen is surrounded by a flat shallow shelf. The area above 15 m is usually 1 to 2 km broad, often 5 km, and partly covered with a kelp forest. Exceptions are only some local white spots of sand, and outside the glaciers vast areas covered with silt. The kelp vegetation is dominated by *Alaria grandifolia*, *Laminaria digitata*, and *L. saccharina*. Remarkable advantages for a possible utilization are the facts that all three species have stipes of more than 1 m length and that the kelp stands are exceptionally free from epiphytes.

The many deep fjords will add very little to the possibly exploitable resources because of glacier silt or otherwise unsuitable bottoms. At their thresholds, on the other hand, skippers complain about kelp masses making it a problem to enter at low water. The lamina choke the propellers and smaller vessels have to be pushed through kelp forests covering the bottom of the fjord entrances. Such vessels also find it difficult to raise anchors when these are loaded with kelp. Standing crop in a well-developed kelp vegetation, sampled by diving on numerous places along the west coast, averaged 10 to 15 kg $m^2$ (JENNEBORG, unpubl.).

With an estimated standing crop of 10 kg $m^{-2}$, a shelf width of 1 km between 2 to 15 m depth (conservative estimate compensating for uncovered or unsuitable areas), and a coastline of 1000 km, the total standing crop would be in the order of 10 million tons wet weight. Even a large-scale utilization would consume only a negligible fraction of the biomass (50 000 tons would correspond to 5‰, which would not influence regrowth and would scarcely raise difficulties regarding an exception from conservation regulations, according to which also the sea around Spitsbergen is declared protected.

In contrast to the subarctic west coast of Spitsbergen the flora of the arctic east coast has not yet been investigated, nor even visited by a marine botanist; it could be expected to be extremely poor in species and quantities.

Richest in biomass of all subarctic seas is the Barent's Sea. The total phytobenthos was assessed to some tens of million tons, the annual productivity likewise to some tens of million tons. In *Ascophyllum* beds biomass values of up to 20 kg $m^{-2}$ have been recorded. Estimates from the Murman coast arrive at a total of 500 000 tons of rock-weed in the littoral and 500 000 to 600 000 tons of *Laminaria saccharina* and *L. digitata* in the sublittoral. The White Sea supply of marine plants exceeds that of the Murman Peninsula; there are 800 000 tons of *Laminaria*, 250 000 tons of *Fucus*, and 400 000 tons of the sea grass *Zostera marina* (ZENKEVITCH, 1963).

### (c) Subarctic Coasts of the Antarctic Ocean

There is a striking contrast between the severe conditions for life and the poor plant growth in terrestrial and intertidal grounds and the comparatively abundant subtidal growth of algae in many parts of the coasts of the Antarctic Ocean. In ice-free places underwater forests of large brown algae (species of *Desmarestia*, *Phyllogigas*, and *Himantothallus*) and many red algae (species of *Callymenia*, *Iridea*, *Gigartina*, *Plocamium*) are found in the sublittoral to considerable depths (SKOTTSBERG, 1907, 1941; ZINOVA, 1964). The seaweed flora, however, is out of reach for any possible commercial utilization. In addition to this it is protected by the Antarctic Treaty.

## (6) The Arctic Ocean

The true Arctic Ocean is the area which is either covered by ice all year round or where, if the ice melts, further solar energy will be used rather for melting more ice than for raising the sea temperature. In practice this equals most of the Arctic Ocean, with the exception of parts under the influence of the Gulf Stream.

The southern-most part of the Arctic Ocean is the Canadian Archipelago west of Baffin Bay; it may even include Hudson Bay. In the southern hemisphere only the Antarctic continent is high arctic, and the so-called Antarctic Ocean is subarctic, including the flora fringing the Antarctic continent.

From the definition of arctic conditions it could be expected that there are no algae at all, other than ephemerals, although nutrient conditions are sufficient to support a rich plankton development during ice-free periods. However, kelp has also been observed; in a year-round study CHAPMAN and LINDLEY (1981) determined biomass and followed development at Igoolik Island in the Canadian Northwest Territories at a latitude of 69°20′ N, where ice cover persists for up to 10 mo $yr^{-1}$. The average fresh biomasses of kelps were: *Laminaria solidungula*, 867 g $m^{-2}$; *L. longicruris*, 97 g $m^{-2}$; *Agarum cribosum* and *Alaria grandifolia*, 26 g $m^{-2}$ each. The productivity of *L. solidungula* was, according to different measurements, 27 to 73 g C $m^{-2}$ $yr^{-1}$. The biomass figures are far from those recorded in subarctic waters but confirm the impression that low respiration rates are more important than high assimilation for the production of biomass. Of course, as a commercial resource all arctic biomass is out of the question.

## (7) Unrecorded Uses of Algae

'From hand to mouth'—any direct use of the fruits of the seas, which is neither measured nor reported, will regrettably be neglected in a statistical context. Industries buying weed and producing liquid fertilizers or seaweed meal for fodder are included in production statistics, farmers spreading cartloads of drift weed for soil conditioning or letting their sheep feed on landed plants are normally not accounted for. There are few exceptions, e.g. France and New Zealand, where somebody has guessed at the extent of such activities. Marine benthic plants also include the vastly distributed, richly producing sea grasses, which have been unfairly neglected, as their utilization is normally not commercial.

In milkfish farms, the fish feed on green algae, cultivated for them in the ponds; other fish farming is dependent on phytoplankton. No figures are available, not even rough estimates are entered into assessments of the seaweed quantities involved.

The usefulness of seaweed and sea grasses reaches far beyond direct utilization by man. In coastal waters, the benthic flora nourishes the benthic fauna which is the basis for fish life. No coral polyps could build a 'coral' reef without the cementing of encrusting coralline algae; hence the appropriate term should be 'biotic' reefs. Sea grasses and mangrove trees play an important role in land formation. They are assisted by less conspicuous but most important filamentous algae, trapping finer soil particles, and plant debris. In many waters blue-green algae fix indispensable nitrogen; in rice fields

Table 4-5

World harvest of marine algae on oceans and climatic regions. Thousand tons wet weight (Original)

| | | Ocean | | | | | |
|---|---|---|---|---|---|---|---|
| Region | Algae | W. Indian | E. Indian | W. Pacific | E. Pacific | W. Atlantic | E. Atlantic |
| Cold-temperate | Brown | | | 1325 | 53 | 3·3 | 201 |
| | Red | | | 141 | | 32·6 | 32·4 |
| | Total | | | 1466 | 53 | 36 | 233 |
| | % red | | | 10% | 0% | 90% | 14% |
| Warm-temperate | Brown | | | 705 | 160 | — | 0·2 |
| | Red | | | 475 | 6·5 | | 52 |
| | Total | | | 1380 | 167 | | 52 |
| | % red | | | 34% | 4% | | 100% |
| Tropical | Brown | 5 | 0 | 0 | — | 0·2 | — |
| | Red | 5·6 | 1·1 | 39 | | 1 | |
| | Total | 11 | 1 | 39 | | 1 | |
| | % red | 53% | 100% | 100% | | 83% | |
| Warm-temperate | Brown | — | — | — | 35 | — | — |
| | Red | | | | 58 | | |
| | Total | | | | 93 | | |
| | % red | | | | 62% | | |
| Cold-temperate | Brown | — | — | — | 0 | 0 | 3 |
| | Red | | | | 8 | 25 | 0 |
| | Total | | | | 8 | 25 | 3 |
| | % red | | | | 100% | 100% | 0% |

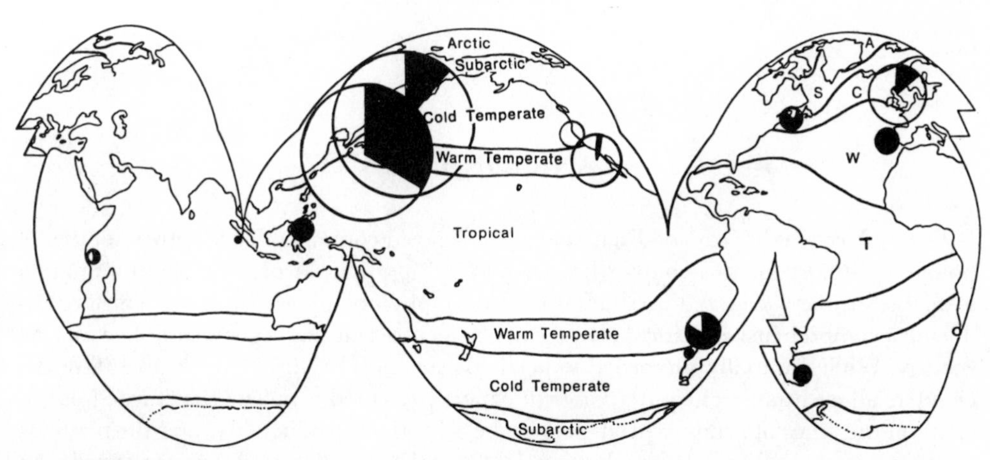


Fig. 4-6: Distribution of harvested quantities of marine algae in relation to climatic regions in coastal waters. Black sectors: red algae; white sectors: brown algae. A: arctic; S: subarctic; C: cold temperate; W: warm temperature; T: tropical. (Original)

they may be the only fertilizers. In China, present research considers the possibility of 'fertilizing' the nitrogen-poor waters of the Yellow Sea by introducing a suitable nitrogen–fixing blue-green alga in order to amend the conditions for *Laminaria* cultivation.

## (8) Summary

A breakdown of world figures for harvests of marine algae on oceans and climatic regions is given in Table 4-5 and graphically illustrated in Fig. 4-6.

## (9) Conclusions

Seaweed utilization is rapidly increasing, but only as raw material for industrial products and for human consumption—the latter within the three nations which are already the biggest consumers of seaweed food: the Japanese, the Chinese, and the Koreans. Two additional important lines are neglected: seaweed as feed and as medicine. Exploitation of the therapeutic significance of seaweeds could mean better health for billions of people. This obvious and urgent aim for a potential mutual effort of those with know-how and those with need has been obscured by overriding industrial needs for colloids and possibly also by the doubt in over-optimistic claims that seaweed could save mankind from starvation (which, of course, is not realistic). A rough estimate shows that if 4 billion people were to share the present world harvest and consume it all, this would give each of us three-quarters of a kg $yr^{-1}$ in fresh weight, or 100 g $yr^{-1}$ in dry weight. Only intensive mariculture, ocean farming, sewage reclamation, and harvesting in the coldest seas could add to these modest averages. Compared to cereals, seaweed will remain marginal in importance. However, this is no excuse for the present situation. A properly planned effort in production and distribution of marine benthic plants is likely to decrease the sufferings of individuals and raise the health standard and production capability of communities.

## 4.2 PLANKTONIC PLANTS

*Editor's Note:* Eight years of discussions and engaged correspondence with a number of potential authors for this chapter have convinced me that, at present, the information available on the world-wide resources of marine phytoplankters is insufficient for producing a comprehensive, critical chapter which focusses on this aspect of applied marine ecology. While I initially succeeded to persuade successively three experts to author the chapter, all eventually gave up. A fourth expert presented a review that turned out to focus on the types of primary producers in the seas, their productivity, and the methods used for measuring primary production (Volume IV), rather than—as requested—on the potential resources which phytoplankters constitute in the various parts of the World Ocean and on aspects of a potential management of such resources.

Of course, one cannot blame the experts for this. It was my fault to have insisted on a chapter for which there is not sufficient substance in the literature and on a topic obviously without much interest. I had adhered to my original intention to have each focal point reviewed in a concept developed over many years 'rather than to follow, as usual, the meandering path along which information happens to have accumulated' (Volume I: Foreword, p. viii). But then, in a comprehensive, integrated treatise it seems just as important to point out gaps in our knowledge than to summarize and critically evaluate the information at hand. Here, then, is such a gap; another one was the topic of Chapter 5.1 in Volume I.

The potential world-wide resources of marine planktonic plants are gigantic. However, marine phytoplankters cannot be harvested economically. In most cases, they are too sparsely, widely, and irregularly distributed. Hence, man can make use of them in economic terms only indirectly, i.e. after they have been collected, consumed, and transformed into animal bodies in the food web.

Although in principle harvestable, temporary plankton blooms are still difficult to predict and may contain detrimental or even poisonous substances (e.g. red tides: Volume V, Part 1, p. 356).

At present, the most promising way to exploit marine microalgae for man's use is to cultivate them (Volume III). However, even in aquaculture, large-scale use of marine phytoplankters for food and feed production still encounters a number of problems. Attempts to market them as human food have thus far been unsuccessful. Most phytoplankters appear to be characterized by tastes, textures, or odours that make it difficult to turn them into palatable foods.

What is needed is the development of new processing techniques, possibly involving (enzymatic) breakdown and subsequent re-synthesis of essential nutritive components prior to end-product marketing. Several marine planktonic plants have been shown to have high protein contents and amino-acid compositions that witness high nutritional values for animals and man. Also, the levels of carbohydrates, lipids, vitamins, and nutritionally important minerals are often high and, in principle, of potential economic interest. Here lies a wide open field of applied research, of interest to food producers and to searchers for raw materials in industry and medicine.

Resource utilization projects leading far into the future should further include attempts to isolate the physiological mechanisms of $CO_2$ and nitrogen fixation and of related biochemical systems involved in the production of organic material from their immediate cellular environment and to establish them in cell-free media (e.g. Volume III: Kinne and Rosenthal, 1977; Kinne, 1980, in press). Unicellular marine phytoplankters may hold the key for such modern bio-engineering projects.

Several hundred million tons of marine phytoplankters are presently used annually as food for commercially cultivated molluscs and fishes. In research cultivation (Volume III, Parts 1 and 2), the number of planktonic plant species utilized as food for a large variety of marine animals, especially at the larval stages, is rapidly increasing, and the use of phytoplankters as raw materials for a variety of pharmaceutical products and for other industrial uses appears at last to be attracting some attention now.

## Literature Cited (Chapter 4)

ACLETO OSORIO, C. (1971). Algas marinas del Perú de importancia económica. *Univ. Mayor de San Marcos, Perú, Museo Hist. Nat. 'Janvier Prado'. Ser. Divulg.*, (**5**), 1–85.

ANDREWS, J. H. (1976). The pathology of marine algae. *Biol. Rev.*, **51**, 211–253.

BAARDSETH, E. (1970). Synopsis of biological data on knobbed wrack, *Ascophyllum nodosum* (Linnaeus) Le Jolis. *FAO Fish. Synops.*, (**38**), Rev. 1.

BARDACH, J. E., RYTHER, J. H. and MCLARNEY, W. O. (1972). *Aquaculture. The Farming and Husbandry of Freshwater and Marine Organisms.* Wiley, New York.

BLUNDEN, G., BINNS, W. W. and PERKS, F. (1975). Commercial collection and utilization of maërl. *Econ. Bot.*, **29**, 140–145.

BONOTTO, S. (1976). Cultivation of plants: multicellular plants. In O. Kinne (Ed.), *Marine Ecology*, Vol. III, Cultivation, Part 1. Wiley, London. pp. 467–501.

BØRGESEN, F. and JONSSON, H. (1905). The distribution of the marine algae in the Arctic Sea and the northernmost part of the Atlantic. *Bot. Faeröes.*, **3**, 28.

BRAUD, J. P. and PEREZ, R. (1979). Farming on pilot scale of *Eucheuma spinosum* (Florideophyceae) in Djibouti waters. In A. Jensen and J. R. Stein (Eds), *Proceedings of the International Seaweed Symposium, 9*. Science Press, Princeton. pp. 533–539.

BUESA, R. J. (1972). La seiba, nuestra planta marina màs abundante. *Mar Pesca*, (**76**), 30–33.

CHAPMAN, A. R. O. and LINDLEY, J. E. (1981). Productivity of *Laminaria solidungula* J. Ag. on the Canadian High Arctic: a year-round study. In T. Levring (Ed.), *Proceedings of the International Seaweed Symposium, 10*. De Gruyter, Berlin. pp. 247–252.

CHAPMAN, V. J. and CHAPMAN, D. J. (1980). *Seaweeds and their Uses* (3rd ed.). Chapman & Hall, London.

CHENG, T.-H. (1968). Production of kelp: a major aspect of China's exploitation of the sea. *Econ. Bot.*, **23**, 215–236.

CHIANG Y.-M. (1981). Cultivation of *Gracilaria* (Rhodophycophyta, Gigartinales) in Taiwan. In T. Levring (Ed.), *Proceedings of the International Seaweed Symposium, 10*. De Gruyter, Berlin. pp. 569–574.

DAWSON, E. Y. (1962). Additions to the marine flora of Costa Rica and Nicaragua. *Pacif. Nat.*, **3**(13), 373–395.

DAYTON, P. K., ROSENTHAL, R. J. and MAHAN, L. C. (1973). Kelp communities in the Chilean Archipelago: R/V Hero cruise, 1972–1975. *Antarct. J. U.S.*, **8**, 34–35.

DELEPINE, R. (1976). Note préliminaire sur la répartition des algues marines aux Iles Kerguelen. *CNFRA*, **39**, 153–159.

DIAZ-PIFERRER, M. (1967). Los recursos marinos de Venezuela. Algas de importancia económica. *El Farol*, **29** (222), 18–221.

DIAZ-PIFERRER, M. and CABALLER DE PEREZ, C. (1964). *Taxonomía, Ecología y Valor Nutrimental de Algas Marinas de Puerto Rico: Algas Productoras de Agar*, Hato Rey, P. R., Administración de Fomento Económico de Puerto Rico y Universidad de Puerto Rico.

DIECKMANN, G. S. (1980). Aspects of the ecology of *Laminaria pallida* (Grev.) J. Ag. off the Cape Peninsula (South Africa). 1. Seasonal growth. *Botanica Mar.* **23**, 579–585.

DIXON, P. S. and IRVINE, L. M. (1977). *Seaweeds of the British Isles. I. Rhodophyta.* British Museum (Natural History), London.

DOTY, M. S. (1973). *Eucheuma* farming for carrageenan. *Sea Grant Advis. Rep.*, **UNIHI-SEAGRANT-AR-73-02**, 1–24.

DOTY, M. S. and ALVAREZ, V. B. (1975). Status, problems, advances and economics of *Eucheuma* farms. *J. mar. Technol. Soc.*, **9**(4), 30–35.

DREW, E. A. (1969). Photosynthesis and growth of attached marine algae down to 130 metres in the Mediterranean. In R. Margalef (Ed.), *Proceedings of the International Seaweed Symposium, 6.* Dirección General de la Pesca Maritima, Madrid. pp. 151–159.

DRUEHL, L. D. (1970). The pattern of Laminariales distribution in the northeast Pacific. *Phycologica*, **9**(3/4), 237–247.

DRUEHL, L. D. (1979). An enhancement scheme for *Macrocystis integrifolia* (Phaeophyceae). In A. Jensen and J. R. Stein (Eds), *Proceedings of the International Seaweed Symposium, 9.* Science Press, Princeton. pp. 79–84.

EARLE, S. A. (1969). Phaeophyta of the eastern Gulf of Mexico. *Phycologia*, **7**(2), 71–254.

EDWARDS, R. (1969). Field and cultural studies on the seasonal periodicity of growth and reproduction of selected Texas benthic marine algae. *Contr. mar. Sci.*, **14**, 59–114.

EDWARDS, R. (1970). Illustrated guide to the seaweeds and sea grasses in the vicinity of Port Aransas, Texas. *Contr. mar. Sci.*, **15** (Suppl.), 1–128.

EVERSON, J. (1977). The living resources of the Southern Ocean. *FAO Southern Ocean Fisheries Survey Programme*, **GLO/SO/77/1**, 30–32.

FAO (1973). Report to the People's Democratic Republic of Yemen on marine turtle management, based on the work of H. F. Hirth and S. L. H. Hollingworth, Marine Turtle Biologists. *Rep. FAO/UNDP (TA)*, (**3178**), 1–51.

FAO (1979). *Yearbook of Fishery Statistics. Catches and Landings, 46.* Food and Agriculture Organization of the United Nations, Rome.

FFRENCH, R. A. (1971). *The Irish Moss Industry. Canadian Atlantic. Current Appraisal.* Department of Fisheries and Forestry, Ottawa.

FRALICK, R. A. and ANDRADE, F. (1981). The growth, reproduction, harvesting and management of *Pterocladia pinnacea* (Rhodophyceae) in the Azores, Portugal. In T. Levring (Ed.), *Proceedings of the International Seaweed Symposium, 10.* De Gruyter, Berlin. pp. 289–295.

FREDJ, G. (1972). Compte rendu de plongée en S.P. 300 sur les fonds à *Laminaria rodriguezii* Bornet de la point de Ravellata (Corse). *Bull. Inst. océanogr. Monaco*, **71**, (1421), 1–42.

FRYE, T. C., RIGG, G. B. and CRANDALL, W. C. (1915). The size of kelps on the Pacific Coast of North America. *Bot. Gaz.*, **60**, 473–482.

GIACCONE, G. (1969). Note sistematiche ed osservationi fitosiociologiche sulle Laminariales del Mediterraneo occidentale. *Giorn. Bot. Ital.*, **103**(6), 457–474.

GISLÉN, T. (1930). Epibioses of the Gullmar Fjord. 2. Marine sociology. *Kristinebergs Zoologiska Station 1877–1927*, **4**, 1–380.

GRUA, P. (1964). Premières données sur les biomasses de l'herbier à *Macrocystis pyrifera* de Morbihan (Archipel Kerguelen). *Terre Vie* (issued also in IIOE Collected Reprints), **4**(209), 91–96.

GUIRY, M. D. and BLUNDEN, G. (1981). The commercial collection and utilization of seaweeds in Ireland. In T. Levring (Ed.), *Proceedings of the International Seaweed Symposium, 10.* De Gruyter, Berlin. pp. 675–680.

HAGEN, N. T. (1981). Kråkeboller og overbeiting av tareskogen i Nordatlanteren—en oversikt, hypotese og foreløpig rapport fra Vestfjorden. (Sea urchins and overgrazing of kelp forest in the North Atlantic: a review, hypothesis and preliminary report from Vestfjorden). *Nordland Distriktshøgskole Matematisk-naturvitenskaplig fagseksjon Rapport*, **1981**(1), 1–13.

HALMØ, G., INDERGAARD, M. and JENSEN, A. (1981). Utnyttelse av marin biomasse. *SINTEF Rep.*, **21**(A 80061), 1–109.

HAY, C. (1979). Growth, mortality, longevity and standing crop of *Durvillea antarctica* (Phaeophyceae) in New Zealand. In A. Jensen and J. R. Stein (Eds), *Proceedings of the International Seaweed Symposium, 9.* Science Press, Princeton. pp. 97–104.

HOEK, C., van den (1975). Phytogeographic regions in the northern Atlantic Ocean. *Phycologia*, **14**(4), 317–330.

HUMM, H. J. (1951). The red algae of economic importance: agar and related phycocolloids. In D. K. Tressler and J. Mcw. Lemon (Eds, *Marine Products of Commerce. Their Acquisition, Handling, Biological Aspects and the Science and Technology of their Preparation and Preservation*. Reinhold Publishing Corporation, New York. pp. 47–93.

JACKSON, G. A. (1980). Marine biomass production through seaweed aquaculture. In *Biochemical and Photosynthetic Aspects of Energy Production*. Academic Press, London. pp. 31–58.

JOYCE, L. and SANTELICES, B. (1978). Producción y explotación de algas en Chile, 1967–1975. *Biología Pesquera*, **10**, 3–26.

KELLEY, F. C. and SNEDDEN, W. W. (1960). Prevalance and geographical distribution of endemic goitre. In *Endemic Goitre*. World Health Organization Monograph Series, 44. pp. 27–233.

KINNE, O. (1980). Aquaculture. A critical assessment of its potential and future. *Interdisciplinary Science Reviews*, **5**, 24–32.

KINNE, O. (in press). Realism in aquaculture—the view of an ecologist. In *World Conference on Aquaculture, Venice 1981*.

KINNE, O. and ROSENTHAL, H. (1977). Commercial cultivation (aquaculture). In O. Kinne (Ed.), *Marine Ecology*, Vol. III, Cultivation, Part 3. Wiley, Chichester. pp. 1321–1398.

KJELLMAN, F. R. (1877). Ueber die Algenvegetation des Murmanschen Meeres an der Westküste von Nowaja Semlja und Wajgatsch. *Nova Acta R. Soc. Scient. Upsal. Series*, **3**(12), 1–86.

KJELLMAN, F. R. (1883). The algae of the Arctic Sea. *K. Sv. Vet. -Akad. Handl.*, **20**(5), 1–350.

KNOX, G. A. (1960). Littoral ecology and biogeography of the Southern Oceans. *Proc. R. Soc.*, **152**, 577–624.

KORRINGA, P. (1976). 'Nori' farming in Japan (cultivation of edible seaweeds of the genus *Porphyra*). In P. Korringa (Ed.), *Farming Marine Organisms Low in the Food Chain*, Developments in Aquaculture and Fisheries Science, 1. Elsevier, Amsterdam. pp. 17–48.

KÜHNEMANN, O. (1970). La importancia de las algas marinas en Argentina. *Contr. Tech. Cent. Invest. Biol. Mar. B. Aires*, **5**, 1–35.

LAITE, P. and RICOHERMOSO, M. A. (1981). Revolutionary impact of *Eucheuma* cultivation in the South China Sea on the carrageenan industry. In T. Levring (Ed.), *Proceedings of the International Seaweed Symposium, 10*. De Gruyter, Berlin. pp. 595–600.

LIM, J. R. and PORSE, H. (1981). Breakthrough in the commercial culture of *Eucheuma spinosum* in northern Bohol, Philippines. In T. Levring (Ed.), *Proceedings of the International Seaweed Symposium, 10*. De Gruyter, Berlin. pp. 601–606.

MACFARLANE, C. I. (1952). A survey of certain seaweeds of commercial importance in Southwest Nova Scotia. *Can. J. Bot.*, **30**, 78–97.

MANN, K. H. (1977). Destruction of kelp-beds by sea urchins: a cyclical phenomenon or irreversible degradation? *Helgoländer wiss. Meeresunters.*, **30**, 455–467.

MATHIESON, A. C. and PRINCE, J. S. (1973). Ecology of *Chondrus crispus* Stackhouse. In M. J. Harvey and J. McLachlan (Eds), *Chondrus crispus*, Proceedings of the Nova Scotian Institute of Science, **27**, (Suppl.), 53–79.

MCALLISTER, R. (1975). California marine fish landings for 1973. *Fish Bull. Calif.*, **163** (Appendix 3).

MCPEAK, R. H. (1980). A preliminary assessment of the *Laminaria* resource near Lower Wood Harbour, Nova Scotia, during July 1977. In J. D. Pringle, G. J. Sharp and J. F. Caddy (Eds), *Proceedings of the Workshop on the Relationship Between Sea Urchin Grazing and Commercial Plant/Animal Harvesting*. Fisheries and Oceans, Canada, Halifax, Nova Scotia. pp. 180–193.

MICHANEK, G. (1975). Seaweed resources of the ocean. *FAO Fish. Tech. Pap.*, **138**, 1–127.

MICHANEK, G. (1979a). Phytogeographic provinces and seaweed distribution. *Botanica Mar.*, **22**, 375–391.

MICHANEK, G. (1979b). Seaweed resources for pharmaceutical uses. In H. A. Hoppe, T. Levring and Y. Tanaka (Eds), *Marine Algae in Pharmaceutical Science*. De Gruyter, Berlin, New York. pp. 203–235.

MICHANEK, G. (1981). Methods and terminology in phycology. In T. Levring (Ed.), *Proceedings of the International Seaweed Symposium, 10*. De Gruyter, Berlin. pp. 751–754.

MIURA, A. (1975). *Porphyra* cultivation in Japan. In J. Tokida and H. Hirose (Eds), *Advance of Phycology in Japan*. Fischer-Verlag, Jena. pp. 273–304.

MSHIGENI, K. E. (1973). Exploitation of seaweeds in Tanzania: the genus *Eucheuma* and other algae. *Tanzania Notes and Rec.*, **72**, 19–36.

MSHIGENI, K. E. (1976). Seaweed farming: a possibility for Tanzania's coastal Ujamaa villages. *Tanzania Notes and Rec.*, **79/80**, 99–105.

MUNDA, I. (1972). On the chemical composition, distribution and ecology of some common benthic marine algae from Iceland. *Botanica Mar.*, **15**, 1–45.

NAYLOR, J. (1976). Production, trade and utilization of seaweeds and seaweed products. *FAO Fish. Tech. Pap.*, **159**, 1–73.

OGATA, E. (1975). Physiology of *Porphyra*. In J. Tokida, H. Hirose (Eds), *Advance of Phycology in Japan*. Fischer-Verlag, Jena. pp. 151–180.

OKAZAKI, A. (1971). *Seaweeds and their Uses in Japan*. Tokai University Press, Tokyo.

OLIVEIRA, E. C. de. (1981). Marine phycology and exploitation of seaweeds in South America. In T. Levring (Ed.), *Proceedings of the International Seaweed Symposium, 10*. De Gruyter, Berlin. pp. 96–112.

OLIVEIRA FILHO, E. C. de and PAULA, E. J. de (1979). Potentially for algin production in the Sao Paulo (Brazil) littoral region. In A. Jensen and J. R. Stein (Eds), *Proceedings of the International Seaweed Symposium, 9*. Science Press, Princeton. pp. 479–485.

PARENZAN, P. (1970). La *Cladophora prolifera* Kütz. dal golfo di Taranto e possibilità di una sua valorizzazione economica. In *Possibilità di Utilizzazione Industriale delle Alghe in Italia*, Vol. 5, Dagli Incontri Tecnici, C.N.R. Laboratorio di Tecnologia della Pesca ed Ente Fiera Internazionale della Pesca, Ancona, 5 Luglio 1970. Falconara Marittima, Tecnografica. pp. 14–21.

PARR, A. E. (1939). Quantitative observations on the pelagic *Sargassum* vegetation of the western North Atlantic. *Bull. Bingham oceanogr. Coll.*, **6**(7), 1–94.

PRINGLE, J. D. (1979). Aspects of the ecological impact of *Chondrus crispus* (Florideophyceae) harvesting in eastern Canada. In A. Jensen and J. R. Stein (Eds), *Proceedings of the International Seaweed Symposium, 9*. Science Press, Princeton. pp. 225–232.

PRINGLE, J. D. (1981). The relationship between annual landings of *Chondrus*, dragrakers, effort, and standing crop in the southern Gulf of St. Lawrence. In T. Levring (Ed.), *Proceedings of the International Seaweed Symposium, 10*. De Gruyter, Berlin. pp. 719–724.

PRINGLE, J. D. and SEMPLE, R. E. (1980). The benthic algal biomass, commercial harvesting, and *Chondrus* growth and colonization off southwestern Nova Scotia. In J. D. Pringle, G. J. Sharp and J. F. Caddy (Eds), *Proceedings of the Workshop on the Relationship Between Sea Urchin Grazing and Commercial Plant/Animal Harvesting*. Fisheries and Oceans, Halifax, Nova Scotia. pp. 144–169.

RAMA RAO, K. (1970). Studies on growth cycle and phycocolloid content in *Hypnea musciformis* (Wulf) Lamour. *Botanica Mar.*, **13**, 163–165.

RICHARDSON, W. R. (1958). Preliminary investigations on the utilization of marine algae in the Caribbean. *Abstr. Int. Seaweed Symp.*, **3**, 49–50.

RICOHERMOSO, M. A. and DEVEAU, L. E. (1979). Review of commercial propagation of *Eucheuma* (Florideophyceae) clones in the Philippines. *Proc. Int. Seaweed Symp.*, **9**, 525–531.

ROUND, R. E. (1981). *The Ecology of Algae*. Cambridge University Press, Cambridge.

SAITO, Y. (1975). *Undaria*. In J. Tokida and H. Hirose (Eds), *Advance of Phycology in Japan*. Fischer-Verlag, Jena. pp. 304–320.

SANTELICES, B. (1977). Ecología de algas marinas bentónicas—efecto de factores ambientales. *Documento de la Dirección General de Investigaciones*. Pontifica Universidad Católica de Chile, Santiago.

SANTELICES, B. (1981). Production ecology of Chilean Gelidiales. in T. Levring (Ed.), *Proceedings of the International Seaweed Symposium, 10*. De Gruyter, Berlin. pp. 351–356.

SANTELICES, B. and LOPEHANDIA, J. (1981). Chilean seaweed resources: a quantitative review of potential and present utilization. In T. Levring (Ed.), *Proceedings of the International Seaweed Symposium, 10*. De Gruyter, Berlin. pp. 725–730.

SCAGEL, R. F. (1961). Marine plant resources of British Columbia. *Bull. Fish. Res. Bd Can.*, **127**, 1–39.

SENN, T. L. and KINGMAN, A. R. (1978). *Seaweed Research in Crop Production, 1958–1978*. Clemson University, Clemson.

SHARP, G. J. (1980). History of kelp harvesting in south-western Nova Scotia. In J. D. Pringle, G. J. Sharp and J. F. Caddy (Eds), *Proceedings of the Workshop on the Relationship Between Sea Urchin Grazing and Commercial Plant/Animal Harvesting*. Fisheries and Oceans, Halifax, Nova Scotia. pp. 170–179.

SIMONS, R. H. and JARMAN, N. G. (1981). Subcommercial harvesting of a kelp on a South African shore. In T. Levring (Ed.), *Proceedings of the International Seaweed Symposium, 10*. De Gruyter, Berlin. pp. 731–736.

SKOTTSBERG, C. (1907). Zur Kenntnis der subantarktischen und antarktischen Meeresalgen. 1. Phaeophyceae. *Wiss. Erg. der Schwedischen Südpolar-Expedition, 1901–1903*, **IV**, 1–172.

SKOTTSBERG, C. (1941). Observations on the vegetation of the Antarctic Sea communities of marine algae in subantarctic and antarctic waters. *K. Sven. Vetenskapsakad. Handl.*, **19**(4), 1–92.

SOEGIARTO, A. (1979). Indonesian seaweed resources: their utilization and management. In A. Jensen and J. R. Stein (Eds), *Proceedings of the International Seaweed Symposium, 9*. Science Press, Princeton. pp. 463–471.

STEPHENSON, T. A. and STEPHENSON, A. (1972). *Life Between Tidemarks on Rocky Shores*. Freeman and Co., San Francisco.

SVEDELIUS, N. (1906). Über die Algenvegetationen eines ceylonischen Korallenriffes mit besonderer Rücksicht auf ihre Periodizität. in F. R. Kjellman (Ed.), *Botaniska Studier tillägnade. F. R. Kjellman*. Uppsala. pp. 184–221.

TAYLOR, W. R. (1942). Caribbean marine algae of the Allan Hancock Expedition, 1939. *Rep. Allan Hancock Atlant. Exped.*, **2**, 1–193.

TAYLOR, W. R. (1962). Marine algae of the northeastern coast of North America. *Univ. Mich. Stud. scient. Ser.*, **13**, 1–509.

TEGNER, M. J. (1980). Multispecies considerations of resource management in southern California kelp beds. In J. D. Pringle, G. J. Sharp and J. F. Caddy (Eds), *Proceedings of the Workshop on the Relationship Between Sea Urchin Grazing and Commercial Plant/Animal Harvesting*. Fisheries and Oceans, Halifax, Nova Scotia. pp. 125–143.

TSENG, C. K. (1981). Marine phycoculture in China. In T. Levring (Ed.), *Proceedings of the International Seaweed Symposium, 10*. De Gruyter, Berlin. pp. 123–143.

VELASQUEZ, G. T. (1953). Seaweed resources of the Philippines. *Proc. Int. Seaweed Symp.*, **1**, 100–101.

WILCE, R. T. (1959). The marine algae of the Labrador Peninsula and Northwest Newfoundland (ecology and distribution). *Bull. natn. Mus. Can.*, **158**, 1–101.

YAMADA, N. (1976). Current status and future prospects for harvesting and resource management of the agarophyte in Japan. *J. Fish. Res. Bd Can.*, **33**, 1024–1030.

ZANEVELD, J. S. (1959). The utilization of marine algae in tropical and East Asia. *Econ. Bot.*, **13**(2), 89–131.

ZENKEVICH, L. (1963). *Biology of the Seas of the USSR*. Allen and Unwin, London.

ZINOVA, A. D. (1964). The composition and character of algal flora at the Antarctic coast and in the vicinity of Kerguelen and Macquaire Islands. In *Information Bulletin of the Soviet Antarctic Expedition*, I. Elsevier, Amsterdam. pp. 123–125.

Marine Ecology Vol. V, Part 2
Edited by Otto Kinne


# 5. WORLD RESOURCES OF FISHERIES AND THEIR MANAGEMENT

J. A. GULLAND

## (1) An Overview of World Fisheries

### (a) Introduction

Fishing is one of man's oldest activities. Excavations of the ancient Middle East sites have found tablets giving records of the catches of individual types of fish which can easily be recognized as the same species supporting the artisanal fisheries of today. Even with the increasing use of the sea for transportation, for recreation, and, especially in the last few years, as a source of oil and other minerals, fishing is still one of man's main uses of the sea. Though the detailed characteristics of fishing have changed, the basic nature of fishing has not changed. The fisherman is still a hunter, depending on the wild stocks of fish to provide his livelihood. He has therefore to be a practical ecologist, understanding the populations he pursues—where they are most abundant, the pattern of their movements, and how they react to the weather and the changing seasons.

As fishing grew from a purely subsistence activity, providing food for the fisherman and his family, the fisherman also came to depend on markets for his fish. Fish are very perishable, and even in temperate climates last at best a few days unless some form of preservation—until recently drying or salting—is used.

The growth of fisheries throughout the world has thus been governed by the interactions of several different factors—a high natural production of fish, the ability to catch a desirable species in good numbers, methods of processing the fish after capture, and the presence of markets to absorb the catches. Especially in the two decades between 1950 and 1970 technical improvements and greater demand led to an enormous increase in world-wide catches, though this was preceded by more localized jumps in production, for example the growth of steam trawlers in the North Sea towards the end of the nineteenth century. The technical advances at sea, especially in freezing fish at sea, and in the supply and operation of long-distance fishing fleets, together with the improved network of world trade (a bluefin tuna can be on the Tokyo market for sashima within a day of leaving the waters on the Canadian east coast), have removed most of the technical limits to the spread of fishing throughout the world. Trawlers are fishing all the shallow areas from Spitzbergen to South Georgia, and tuna long-liners scattered over all the warmer oceans. Nevertheless, the biological constraints mean that the bulk of the world catch comes from a few relatively restricted areas which are particularly favourable for fishing. The most outstanding of these are the coastal upwelling areas of the subtropical belt, off the semi-desert coasts of northwest Africa, Namibia, Peru/Chile, and California (USA). The shallow shelf areas of temperate waters, which have been the cradle of most types of fishing, also remain one of the most productive regions.

### (b) Distribution of Catches

The nature of the fisheries in each region of the world is discussed in detail in the following sections. A good graphical presentation is also given in the FAO *Atlas of the World's Fisheries* (FAO, 1981a). Here only a broad overview will be given of where the fish are caught, who catches them, what species are caught, and with what gears. These can be summarized in a series of tables, taken mostly from the annual FAO *Yearbook of Fishery Statistics* (FAO, 1981c).

The first of these gives the catches from the major areas into which the oceans have been divided for statistical purposes (Table 5-1). The boundaries between these areas correspond to the divisions used in the later detailed discussion, but in addition three of the areas (eastern central Atlantic, eastern central Pacific, and the southeast Pacific Oceans) have been divided to treat the major upwelling areas separately from the less-productive parts of the same region. In addition to the marine catches, the catches from inland waters are also shown in Table 5-1 (and in the totals of national catches in Table 5-2 below). To the consumer the difference between marine and freshwater fish is not important—far less than the difference, say, between herring and lobster or sole. Consideration of the freshwater catch is important to give a complete picture of fish and fishing in human affairs, especially in some parts of the developing world (central Africa, parts of south and southeast Asia). No distinction is made in these tables between production from aquaculture and catches from wild stocks. Cultivation accounts for a significant fraction of the production from fresh waters in some parts of the world (notably China and India), though it must be stressed that there is a considerable 'grey area' associated with stocking and moderate supplemental feeding between intensive aquaculture on the one hand, and completely unmanaged fisheries on wild stocks on the other. So far as marine fisheries are concerned, however, cultivation does, with a few exceptions, contribute little in terms of weight to the world fish production. The exceptions are the cultivation of molluscs (oysters, mussels, etc.), mainly in western Europe and Japan.

The striking feature of Table 5-1 is that although the ocean areas in each of the regions are approximately equal, there are very large differences in the catches taken. Just two regions (out of 15, excluding the Antarctic)—the northwest Pacific, and northeast Atlantic—account between them for close to half the world catch. The catches in the northwest Pacific—from the south coast of China to the western Pacific—exceed those from the rest of the Pacific and the Indian Ocean combined; the catches in the Atlantic are dominated by those in the northeast—from Gibraltar to Greenland and Spitzbergen—in the same way.

This discrepancy arises essentially from two causes—the productivity of harvestable fish is not the same in all areas, and some areas are more heavily fished than others (though this difference is decreasing). Many factors determine how productive an area of ocean is of fish suitable for man to harvest. RYTHER (1969) has dramatically illustrated the differences that can occur between broad types of marine systems—the open ocean, the coastal zone (approximately the continental shelf), and upwelling areas. Criticisms can be made about the details of his results, especially concerning the differences between different zones in the number of trophic levels between primary production and harvestable fish, and in the transfer efficiency between each level. Also the division between the zones is not sharp; there are many areas where there is weak or seasonal upwelling which are not classed by him as upwelling zones. Nevertheless, the main conclusions are valid: the open ocean is no place for commercial fishermen except for a few pursuing tuna, which lie near the top of the ecological pyramid, and allow fishermen in a few hours' work to harvest the concentrated production from many square miles of ocean, and their main interests must be the continental shelves and the coastal upwelling areas.

Table 5-1

Total landings of fish (including molluscs and crustaceans but excluding whales and seaweeds) from the major water bodies of the world (million tons) (Based on FAO *Yearbook of Fishery Statistics*)

| Area | 1950 | 1955 | 1960 | 1965 | 1970 | 1975 | 1978 | 1981 |
|---|---|---|---|---|---|---|---|---|
| Atlantic — northwest‡ | 1·7 | 1·8 | 2·3 | 3·9 | 4·2 | 3·8 | 2·8 | 2·8 |
| — northeast | 6·0 | 7·6 | 7·5 | 9·5 | 10·7 | 12·0 | 11·7 | 11·7 |
| — western central‡ | 1·4 | 0·7 | 0·8 | 0·9 | 1·4 | 1·5 | 1·9 | 1·9 |
| — eastern central | 0·4 | 1·8 | 2·0 | ·2·5 | 2·8 | 3·5 | 3·3 | 3·2 |
| Mediterranean | 0·7 | 0·4 | 0·8 | 1·1 | 1·1 | 1·3 | 1·2 | 1·7 |
| Atlantic — southwest | 0·2 | 0·3 | 0·4 | 0·5 | 1·1 | 0·8 | 1·3 | 1·3 |
| — southeast | 0·4 | 0·9 | 1·1 | 2·1 | 2·5 | 2·5 | 3·3 | 2·3 |
| Total Atlantic Ocean | 10·8 | 13·5 | 14·3 | 20·5 | 23·8 | 25·5 | 25·3 | 24·8 |
| Indian Ocean — west | 0·7 | 0·8 | 1·1 | 1·2 | 1·7 | 2·1 | 2·3 | 2·0 |
| — east | 0·5 | 0·5 | 0·7 | 0·7 | 0·8 | 1·1 | 1·3 | 1·5 |
| Total Indian Ocean | 12·0 | 1·3 | 1·8 | 1·9 | 2·5 | 3·2 | 3·7 | 3·5 |
| Pacific — northwest } | 2·6 | 3·1 | 4·1 | 5·4 | 13·0 | 17·2 | 18·4 | 19·8 |
| — northeast } | | | | | 2·6 | 2·2 | 1·9 | 2·3 |
| — western central | 3·0† | 6·3† | 8·3† | 9·8† | 4·2 | 5·1 | 6·0 | 5·8 |
| — eastern central | 0·7 | 0·4 | 0·5 | 0·5 | 0·9 | 1·3 | 1·8 | 2·6 |
| — southwest | 0·1 | 0·1 | 0·1 | 0·2 | 0·2 | 0·3 | 0·3 | 0·4 |
| — southeast | 0·2 | 0·4 | 3·9 | 8·2 | 13·8 | 4·4 | 5·5 | 6·9 |
| Total Pacific Ocean | 6·6 | 10·3 | 16·9 | 24·1 | 34·7 | 30·5 | 34·0 | 37·8 |
| Southern Oceans | — | — | 0 | 0 | 0 | 0·0 | 0·4 | 0·6 |
| Total Marine | 18·6 | 25·1 | 33·6 | 46·5 | 61·0 | 59·3 | 63·4 | 66·7 |
| Inland — Africa | 0·4 | 0·5 | 0·6 | 0·7 | 1·2 | 1·4 | 1·5 | 1·4 |
| — North America | 0·1 | 0·1 | 0·1 | 0·1 | 0·1 | 0·1 | 0·1 | 0·2 |
| — South America | 0·1 | 0·1 | 0·2 | 0·2 | 0·1 | 0·2 | 0·3 | 0·3 |
| — Asia (excl. USSR) | 1·3 | 2·0* | 2·5* | 2·8* | 2·9* | 4·2 | 4·2 | 5·0 |
| — Europe (excl. USSR) | 0·1 | 0·1 | 0·2 | 0·2 | 0·2 | 0·3 | 0·3 | 0·4 |
| — Oceania | + | + | + | + | + | + | + | + |
| — USSR | 0·6 | 0·8 | 0·6 | 0·6 | 0·8 | 0·9 | 0·7 | 0·8 |
| Total Inland Waters | 2·5 | 3·6 | 4·2 | 4·6 | 5·3 | 7·2 | 7·1 | 8·1 |
| Total | 21·2 | 28·7 | 37·8 | 51·1 | 66·3 | 66·5 | 70·5 | 74·8 |

*Using reduced figures for Chinese production.
†Includes areas in southern Japan, and around China now included in northwest Pacific Ocean.
‡Area 6 of ICNAF/NAFO included in northwest Atlantic Ocean from 1965 onwards.

These are by no means evenly distributed. The northeast Atlantic Ocean contains over one-sixth of the shelf area of the world with the North Sea being one of the great traditional fishing grounds, while the arc of western Pacific Ocean from the Java Sea through the south and east China Seas to Japan account for about one-third. In contrast, the Indian Ocean, and most other parts of the Atlantic and the Pacific Oceans, have narrow shelves. In parts of the eastern sides of the latter oceans the lack of shelf is balanced by the presence of strong upwelling. The Peruvian upwelling and the associated anchoveta fishery, once the largest in the world, is probably the best known of these, but analogous systems exist off California, and northwest and southwest Africa. In each case the richness of the seas is in sharp contrast to the deserts, or near deserts on land, both being caused by the same patterns of winds.

### (c) National Catches

Another way of looking at the distribution of catches is according to the country catching the fish. Summary data on national fish catches are given in Table 5-2. This gives the catches by the major countries (those that have taken over a million tons of fish in one or more recent years), and also by the main geographical or political groupings. In addition to the catches in 1978, figures are also given for 1950 to provide some idea of the pattern of development in different areas.

Looking first at the figures for 1950 it can be seen how the greater part of the catches (nearly two-thirds of the total of 21 million tons) were taken by the developed countries of the Western world (including Japan in this group). Generally these catches were taken from the rich resources around the coasts of the countries concerned, which in the case of Japan and the United States with both large resources and large local markets provided catches of several million tons. Other developed countries with plenty of fish but smaller populations, notably Norway and Canada, had developed fisheries based mostly on exports, to a large extent in traditional forms such as dried or salted cod. The United Kingdom also had catches of around a million tons, but while much of this came from local waters, the British fishery (or more strictly the English fishery) also took a substantial part of its catch from more distant waters (e.g. around Iceland), a characteristic which became a feature of other countries in later years.

In contrast the catches by developing countries as a whole were still low in 1950. Fishing was carried out virtually entirely by artisanal fishermen, usually with canoes and other non-mechanized boats. Nevertheless, in countries with large populations and extensive areas of inland and shallow coastal waters the number of fishermen were enough to produce substantial catches; India and China both recorded catches of over three-quarters of a million tons (statistics of these fisheries are by their nature not very precise, and the true catches may have been more). These totals include both freshwater and marine fish, and in many developing countries, including both India and China, as well as many African countries, freshwater catches made up a major part of the total catch. Indeed, where the fishery population was high, and the fish highly priced or particularly vulnerable to fishing, declines in some inland fish stocks due to fishing were already becoming noticeable, e.g. in the fishery for the *Sarotherodon esculentus* (*Tilapia*) in Lake Victoria (GARROD, 1961).

Developments since 1950 have followed some very different patterns. In the three decades since 1950 world catches have nearly trebled. Most countries have shared in this

Table 5-2

Total landings, including landings from inland fisheries, in 1979 (1950 in brackets) by major countries and groups of countries (thousand tons) (Based on FAO *Yearbook of Fishery Statistics*, Vol. 48)

| Developed countries | | |
|---|---|---|
| USA | (2957) 3511 | |
| Canada | (1048) 1332 | |
| Japan | (3374) 9966 | |
| EEC | (2819) 4623 | |
| Denmark | | (251) 1738 |
| France | | (509) 732 |
| Germany | | (551) 365 |
| Italy | | (186) 427 |
| Netherlands | | (258) 324 |
| UK | | (989) 905 |
| Other western Europe | 6535 (3213) | |
| Iceland | | (373) 1645 |
| Norway | | (1468) 2652 |
| Portugal | | (307) 242 |
| Spain | | (598) 1205 |
| South Africa | 659 (255) | |
| USSR | 9114 (1627) | |
| Other eastern Europe | 1148 (130) | |
| Eastern Germany | | (30) 224 |
| Poland | | (80) 601 |
| Others | 263 (77) | |

| Developing countries | | |
|---|---|---|
| Africa | 3701 (805) | |
| Morocco | | (134) 280 |
| Ghana | | (20) 230 |
| Nigeria | | (40) 535 |
| Senegal | | (50) 308 |
| Tanzania | | (35) 344 |
| Latin America | 10 128 (665) | |
| Argentina | | (58) 566 |
| Brazil | | (153) 843 |
| Chile | | (88) 2633 |
| Cuba | | (10) 153 |
| Mexico | | (74) 875 |
| Peru | | (114) 3682 |
| Asia | 20 203 | |
| Bangladesh | | *(a)* 640 |
| China | | (912) 4054 |
| India | | (817) 2343 |
| Indonesia | | (450) 1732 |
| N. Korea | | (275) 1330 |
| S. Korea | | (219) 2162 |
| Malaysia | | (150) 698 |
| Pakistan | | (240) 300 |
| Philippines | | (226) 1476 |
| Thailand | | (158) 1716 |
| Oceania | 103 (18) | |

| | |
|---|---|
| Total—Developed countries | 37 151 (15 140) |
| —Developing countries | 34 135 (5760) |
| World total | 71 285 (20 900) |

*(a) included in Pakistan*

expansion, but there have been exceptions. These exceptions are mostly those countries who were already, by 1950, taking much of the potential harvest from the waters around their coasts. The obvious example is the United Kingdom whose catch has removed very close to a million tons for the whole period. Canada and the United States have experienced a somewhat similar stagnation. Although the catches off their coasts have grown since 1950, these increased catches have been taken mainly by foreign vessels because the fish concerned have been too low priced, or too expensive to catch, to attract North American fishermen.

In contrast there have been many countries whose catches have greatly increased since 1950. The most remarkable, and best known, has been the Peruvian anchoveta fishery. Before 1955 the rich resources of the Peruvian upwelling system had hardly been used, except for the production of guano. Around 1955 a fish-meal fishery started, partly through using equipment made idle by the collapse of the Californian sardine stock. Thereafter, helped by the large and growing world market for fish-meal to supply the rapidly developing broiler chicken industry, catches of anchoveta doubled each year until reaching nearly 8 million tons in 1965. Later, catches continued to grow, but more slowly, to reach a peak of over 12 million tons in 1970. Other countries whose catches have also grown rapidly to supply the fish-meal market include Chile (based on the southern part of the same upwelling system as Peru, with anchoveta also supplying nearly all the catch), Norway (on a succession of pelagic species, shifting from herring to mackerel to capelin as stocks decline), and Denmark (based on a variety of small species—herring, sprat, and Norway pout—in the North Sea).

Almost as striking as the growth of the Peruvian fishery, and at least as important in terms of value, though less well known, has been the growth of trawling by Thailand. Prior to 1960 the large quantities of demersal fish existing in the wide shelf areas of the South China Sea, including the Gulf of Thailand, had only been exploited to any significant extent in the shallowest areas near the coast, by a variety of relatively inefficient gears. In the early 1960s otter-board trawling was introduced by German technical assistance, and the subsequent growth increased the catch of demersal fish by a factor of ten, to reach over a million tons by 1970. Their catches have continued to grow, if more slowly, since then reaching over 2 million tons in 1978, putting Thailand in the top dozen fishing countries in the world, ahead of all European countries except Norway.

Though their growth has been less spectacular than those of Peru or Thailand most other developing countries have seen their catches grow rapidly since 1950. A part of this increase has been due to the growth of export-directed industries. Peru and Chile are exceptional among developing countries in being major producers and exporters of fish-meal. Many tropical countries, however, notably Mexico and India, produce large quantities of shrimp. Because of their high price, shrimp earn important amounts of foreign currency, and the development of the shrimp fisheries in several countries has been pushed forward by outside interests, often in the form of joint ventures. The larger part of the increase in the catches of developing countries has come from the steady growth and expansion of the existing local fisheries, without a great deal of outside assistance. Improvements in gear and vessels, particularly mechanization, and, often equally important, improvements in marketing and distribution, have allowed the catches of countries like the Phillippines and Indonesia, with great resources around their coasts, to grow to over a million tons.

While these developments of local fisheries have been striking, the feature of the world's fisheries in the 1960s and 1970s which distinguishes them from earlier periods—and almost certainly from future periods—has been the importance of long-range fishing. The invention of freezing at sea, and particularly of the large factory trawler pioneered by the British *Fairtry*, has enabled countries with few resources around their coasts to become major fishing countries, as well as allowing those with good, but still limited, local resources to expand their catches beyond those limits. The most noticeable examples have been the countries of eastern Europe; Russian catches increased nearly six-fold between 1950 and 1975, reaching over 10 million tons in 1976, and the expansion of Polish catches (to over 800 000 tons in 1975) has been almost as striking. Another important long-range fishing country has been Japan—particularly for tuna—but others, including some developing countries (e.g. Republic of Korea, Ghana, and Cuba), have their own long-range fleets.

The importance of these long-range fleets is shown in Table 5-3, which gives the summary information on the catches by non-local vessels (i.e. ships working off the coasts of countries other than that of their home port). Table 5-3(A) shows that this non-local fishery is concentrated in three main areas—the North Pacific Ocean, the North Atlantic Ocean and West African waters (principally in the upwelling areas off the desert coasts of northwest and southwest Africa)—where catch rates are high enough to attract ships from far away. The data are given for 1972, since from that time onwards countries were, in growing numbers, extending their jurisdiction over fisheries from the traditional 3 to 12 miles out, usually to 200 miles, in accordance with the new legal regime that was emerging from the long drawn-out United Nations Conference on the Law of the Sea. Despite the importance of local fishing in the northern temperate waters, the rich grounds, such as off Newfoundland or in the Bering Sea, were attractive enough to long-range fleets, for these vessels to take up to a quarter of the total catch from these waters. Off West Africa, local-based fishing—particularly in the poor, sparsely inhabited coasts bordering the richer fishing grounds—has been much less, and long-distance fishing has accounted for nearly two-thirds of the total. In most other parts of the world catch rates have not, except for tuna, been high enough to attract much non-local fishing.

Another way to look at the impact of non-local fishing is in terms of the countries who are doing the fishing, or off whose coasts they are fishing. This is shown in Table 5-3(B). Contrary to what might be expected, this is not simply a matter of fleets from the advanced developed countries fishing off the coasts of the poorer countries. As already noted, a number of developing countries (Korea, Cuba, Thailand) depend on fishing in

Table 5-3

(A) Catches by long-range vessels in 1972 according to region of capture (thousand tons) (Based on GULLAND, 1979a, Table 1)

| Area | Total | Non-local | |
|---|---|---|---|
| | | All species | Tuna |
| North Atlantic Ocean | 15 026 | 5959 | 11 |
| North Pacific Ocean | 17 305 | 5190 | — |
| West African waters | 6124 | 3701 | 209 |
| Other areas | 17 654 | 1509 | 621 |
| Total | 56 109 | 16 359 | 841 |

(B) Catches by long-range vessels according to country (thousand tons) (Based on GULLAND 1979a; Table 2)

| Nationality of fishing vessel | Location of fishing ground | | |
|---|---|---|---|
| | Off developed countries | Off developing countries | Total |
| Developed countries | 10 716 | 4572 | 15 288 |
| Developing countries | 430 | 641 | 1071 |
| Total | 11 146 | 5213 | 16 359 |

the waters off other countries for a significant part of their catch. Equally, much of the non-local fishing is done off developed countries, notably USA and Canada in the northern temperate waters.

### (d) Species Caught

The third way of looking at the world catches is in terms of the species caught. Summary data for selected gears from 1948 to 1978 are given in Table 5-4. The total number of species of fish, crustaceans, and molluscs that appear in commercial or subsistence fisheries in one part of the world or another is unknown, but is certainly large—probably thousands or tens of thousands. It would be unpracticable to list catches of them all, even if the data existed. In fact, in most tropical countries—where the largest number of species are caught—data on the species composition of the catches are sparse and sometimes are completely lacking—a matter which makes the scientific study of the fish stocks and their state of exploitation very difficult.

For the large industrial fisheries of the developed countries, and for a growing number of other fisheries, species data are available. Table 5-4 gives the breakdown of the total catch according to the broad groupings used by FAO. It also lists the catches for a selection of the major species (roughly those for which annual catches have exceeded half a million tons).

Looking at the individual species in Table 5-4 it can be seen that the two species that were the mainstay of the traditional fisheries of northern and western Europe—the herring and the cod—figure largely throughout the period, but their relative importance has fallen more or less steadily. In 1948, these two species accounted for over 4 million tons, or for over 20% of the total (if figures were available they would have shown an even greater share in earlier years and earlier centuries; for a general history of the cod fisheries see, for example, INNIS, 1954). Since 1948 the relative importance of these species has declined, and indeed since the early 1970s the actual catches have fallen, following the collapse of many of the herring stocks, especially on the eastern side of the Atlantic, so that in 1978 they accounted for less than 5% of the total world catch.

Of the other major species, most are closely analogous to either cod (i.e. demersal fish, mostly gadoids, of the temperate shelves) or herring (i.e. shoaling pelagic fish of either the temperate shelves, or the major upwelling areas). Among the closest relatives to the cod are the hakes (various species of *Merluccius*), which occur in nearly all the temperate waters of the world. The European species (*M. merluccius*) is less abundant (recent catches only about 150 000 tons) than several other species, notably the Cape hakes (actually two species, *M. capensis* and *M. paradoxus*) from the Benguela upwelling area off southwestern Africa, for which the catches are shown in the table, and which reached a peak of over 800 000 tons. The most important of the relatives of cod, in terms of weight caught, is now the Alaska pollock *Theragra chalcogramma*. This species occurs in all colder shelf waters of the North Pacific Ocean, from northern Japan to Alaska, and in particular on the large shelf area of the Bering Sea. Catches have reached over 5 million tons, making this—following the decline of the Peruvian anchoveta—the biggest single species fishery in the world.

Despite the decline of the anchoveta, however, the herrings and their relatives (sardines, anchovies, etc.) still comprise the biggest single group of species, and include several of the most important individual species. A feature of many of these species has

Table 5-4

World catches of main species and species groups (million tons) (Based on FAO *Yearbook of Fishery Statistics*)

| Group | | 1948 | 1958 | 1964 | 1970 | 1975 | 1978 | 1979 | Ranking of individual species in 1981 |
|---|---|---|---|---|---|---|---|---|---|
| 31 | Flounders | 0·05 | 0·79 | 0·99 | 1·32 | 1·16 | 1·26 | 1·15 | |
| 32 | Cods, etc. | 3·61 | 4·49 | 6·02 | 10·52 | 11·86 | 10·41 | 10·59 | |
| | — Atlantic cod | 2·09 | 2·56 | 2·68 | 3·14 | 2·43 | 2·17 | 2·03 | 5 |
| | — Alaska pollock | 0·19 | 0·34 | 0·92 | 3·06 | 5·02 | 3·90 | 3·95 | 1 |
| | — Blue whiting | 0·05 | 0·04 | 0·03 | 0·04 | 0·06 | 0·55 | 1·12 | 11 |
| | — Cape hakes | 0·04 | 0·08 | 0·14 | 0·76 | 0·66 | 0·52 | 0·45 | 38 |
| | — Haddock | 0·41 | 0·44 | 0·59 | 0·91 | 0·53 | 0·34 | 0·34 | 27 |
| | — Norway pout | — | — | 0·11 | 0·31 | 0·69 | 0·42 | 0·44 | 34 |
| 33 | Misc. demersal | 1·19 | 2·23 | 2·93 | 3·95 | 5·21 | 5·60 | 5·29 | |
| | — Sand eels | — | 0·10 | 0·12 | 0·19 | 0·52 | 0·81 | 0·64 | |
| | — Atlantic redfishes | 0·18 | 0·55 | 0·70 | 0·64 | 0·60 | 0·32 | 0·36 | |
| 34 | Jacks, mullets | 0·55 | 1·74 | 1·97 | 4·07 | 5·89 | 8·09 | 7·85 | |
| | — Capelin | 0·02 | 0·10 | 0·04 | 1·51 | 2·25 | 3·16 | 2·93 | 4 |
| | — *T. capensis* | 0·03 | 0·06 | 0·03 | 0·06 | 0·32 | 0·55 | 0·47 | 16 |
| | — *T. murphyii* | 0 | 0 | 0·01 | 0·12 | 0·30 | 1·01 | 1·29 | 7 |
| 35 | Herring, sardines | 4·72 | 7·36 | 18·72 | 21·57 | 13·74 | 14·35 | 15·64 | |
| | — Atlantic herring | 2·23 | 2·55 | 3·54 | 2·32 | 1·53 | 0·94 | 0·84 | 10 |
| | — Pacific herring | 0·58 | 0·60 | 0·73 | 0·58 | 0·48 | 0·17 | 0·19 | |
| | — Japan pilchard | — | 0·14 | 0 | 0·02 | 0·53 | 1·93 | 2·00 | 2 |
| | — Chil. pilchard | 0 | 0·02 | 0·05 | 0·07 | 0·23 | 1·86 | 3·35 | 3 |
| | — Europe pilchard | 0·30 | 0·45 | 0·51 | 0·48 | 1·07 | 0·80 | 0·73 | 9 |
| | — Anchoveta | 0 | 0·78 | 9·80 | 13·06 | 3·32 | 1·18 | 1·41 | 8 |
| | — Gulf menhaden | 0·46 | 0·70 | 0·71 | 0·55 | 0·54 | 0·82 | 0·81 | 19 |
| 36 | Tunas | 0·38 | 0·99 | 1·20 | 1·60 | 2·07 | 2·52 | 2·42 | |
| | — Skipjack | 0·08 | 0·24 | 0·23 | 0·39 | 0·54 | 0·79 | 0·70 | 13 |
| | — Yellowfin | 0·10 | 0·20 | 0·25 | 0·33 | 0·51 | 0·50 | 0·58 | 20 |
| 37 | Mackerels | 0·57 | 0·99 | 1·35 | 3·14 | 4·16 | 4·76 | 4·52 | |
| | — *S. japonicus* } — *S. scombrus* } | 0·32 | 0·54 | 0·92 | 1·79 | 1·87 | 2·56 | 2·57 | 6 |
| | | | | | 0·68 | 1·10 | 0·72 | 0·76 | 18 |
| 38 | Sharks, etc. | 0·33 | 0·34 | 0·40 | 0·51 | 0·60 | 0·61 | 0·57 | |
| 39 | Unspecified | 2·92 | 4·94 | 7·08 | 7·55 | 7·35 | 7·62 | 7·10 | |
| 45 | Shrimp | 0·33 | 0·44 | 0·56 | 0·98 | 1·33 | 1·65 | 1·53 | |
| 46 | Krill | — | — | — | — | 0·04 | 0·14 | 0·39 | 25 |
| | — Other crustaceans | 0·29 | 0·41 | 0·61 | 0·67 | 0·66 | 1·07 | 1·07 | |
| 53 | Oysters | 0·35 | 0·64 | 0·80 | 0·71 | 0·85 | 0·90 | 0·87 | |
| 54 | Mussels | 0·13 | 0·17 | 0·30 | 0·56 | 0·47 | 0·56 | 0·58 | |
| 57 | Squids | 0·33 | 0·47 | 0·47 | 0·93 | 1·19 | 1·34 | 1·56 | |
| 56 | Clams | 0·16 | 0·32 | 0·44 | 0·71 | 0·94 | 1·12 | 1·08 | |
| | — Other molluscs | 0·32 | 0·32 | 0·43 | 0·31 | 0·45 | 0·64 | 0·55 | |
| Freshwater species | | 2·49 | 5·31 | 6·00 | 6·41 | 7·39 | 7·50 | 7·92 | |
| Total* | | 19·3 | 32·3 | 51·4 | 66·2 | 66·5 | 70·5 | 71·3 | |

*Includes small quantities of miscellaneous animals not specified in the table.

been the high variability in the catches (e.g. MURPHY, 1977). This variability is best known from the collapse of several major fisheries, e.g. that of the Californian sardine in the 1940s (MURPHY, 1966), of the Peruvian anchoveta in 1972 (GULLAND, 1975), and of many North Atlantic herring stocks in the last decade. Changes are, however, not all downwards. NAGASAKI (1973) has pointed out that around Japan the fluctuations in different species of pelagic fish have tended to balance, so that the total catch has not changed much. Of the species concerned, the most striking changes have been in the catches of sardines *Sardinops melanosticta*. Catches of this species reached over 2 million tons in 1939, collapsed to almost nothing by the 1960s, but very recently have recovered to pass the 1 million mark in 1976 and over 3·5 million tons in 1981 (Table 5-4).

It is not clear how much of these changes in pelagic fish are due entirely to natural fluctuation, and how much to the additional stress caused by fishing. Certainly the fluctuations in the abundance of scales of sardine and anchovy observed in bottom deposits off California—and hence presumably of these species in the waters above—over a period of several centuries (SOUTAR, 1967) must have been due to purely natural causes. Equally, the collapses of several pelagic stocks after short periods of intense fishing can hardly be ascribed to coincidence. Elucidation of the factors causing declines in these stocks is necessary for the identification of the measures to achieve their proper management and, as far as possible, maintain catches from them at a high level. It is therefore a matter of both great scientific interest and practical importance.

A third group of species that deserves mention are the tunas. Though the weight of tunas caught is only about 4% of the total of all species, their high price means that their contribution to the total value is much greater. Apart from the two species picked out in Table 5-4—skipjack and yellowfin—which are both strictly tropical, the major commercial species of tuna include albacore and bluefin, which are also widely distributed in temperate waters. Indeed, the southern bluefin—believed to be a separate species—goes deep into subantarctic waters in the southern Indian Ocean. Between the different species tunas therefore cover nearly all the warmer parts of the ocean waters. The distribution of tuna fishery—and more particularly of the fishery with long-lines by Japan, Korea, and Taiwan—is equally wide. It therefore provides the exception to the general picture of commercial fishing being concentrated on—and indeed almost confined to—the shelf areas, and particularly the upwelling areas.

## (e) Gears Used

The fourth way of looking at the fish catches is in terms of the gears used in catching them. There are no comprehensive statistics that describe just how much fish is taken by each type, but it is clear that the world production is dominated by two main types of gear: trawl and purse-seine. In round figures, the various types of trawl probably account for about half the world catch; ca. one-third (including much of the 20 million tons that go into fish-meal) is taken by purse seine, the rest by other gears.

This dominance by just two basic types of gear is relatively new. The rising importance of these active gears dates only from the time when it was found possible to use machinery to replace manpower in handling fishing gear. Historically, fisheries have relied much more on less active gears—especially on hook and line. Even now these gears are the mainstay of the artisanal or small-scale fisheries of the developing countries, and provide employment for far more people than the technically more advanced methods.

It is impossible here to more than touch on some of the more interesting types of gear used. The number of different ways that someone, somewhere, has managed to devise to catch fish, taking advantage often of special local conditions, is bewildering. In one of the standard descriptions of the fishing gears of the world, VON BRANDT (1964) gives 15 basic classes of gear, each of which is further divided into a dozen or more different types. The main outline of this classification, based originally on the discussions at the FAO Fishing Gear Congress in 1957 (KRISTJONSSON, 1959) is as follows:

*Collecting by hand*

Although this is the simplest possible method of fishing, and normally limited to the poorest fishermen collecting shellfish off the beach at low tide, this class includes highly successful fishermen of southeast Australia (and elsewhere), who collect abalone while using SCUBA gear. However, their contribution to the world food supply is small.

*Harpoons and spears*

This method is best known (or notorious) as the method used by the whaling industry (see also Volume III: KINNE, 1977); otherwise it is virtually only used by individual fishermen, probably as much for subsistence as a commercial activity, but in various forms is widespread throughout the world.

*Stupefying methods*

The total production from these methods is small, but the uncontrolled use of poison or of explosives (especially on coral reefs) can cause more long-term damage to the resource than less indiscriminate methods which catch much more fish.

*Hooks and lines*

This is generally thought of as a typical method of fishing of the amateur sportsman. As such the contribution to the total catch is far from negligible even in the sea, e.g. in the coastal waters of the United States the sports catch exceeds the commercial catch of all but a handful of species. In addition catches by long-lines, trolling, squid jigs, etc. account for a good proportion of the commercial catches of some of the larger and more valuable species, such as tuna or halibut.

*Traps*

Though widely used, in forms that range from enormous madragues used for tuna in the Mediterranean Sea to small devices that are reminiscent of nothing so much as aquatic mouse-traps, the total catch from traps is small. On an industrial scale they are mostly used for lobsters and crabs, but they are also important as one of the main ways (with lines) of harvesting bottom-living fish on rough ground, including coral reefs.

*Aerial traps*

These depend on the fact that many fish attempt to escape enemies by leaping out of the water. Once out of the water, they lack control, and can be caught fairly easily. Several forms of this type of trap, e.g. in Thailand and on Lake Chad, even arrange for the fish to fall directly into the fisherman's boat. Aerial traps are not a method for bulk catches.

*Scoop nets and bag nets*

Kept open by a frame, these are the predecessors of the trawl. The larger types (stow nets) depend on the flow of the tide; the smaller ones are worked by hand. They are now not important even in the artisanal sector.

*Dragged gear*

This class contains dredges, and the many forms of trawl (one-boat or two-boat, mid-water or bottom, etc.). Despite its importance to modern industrial fishery it is not widely used in less advanced fisheries.

*Seine net*

Beach seines were probably the most powerful type of gear in the pre-industrial age. Some of them, for instance along the west coast of India, could reach enormous sizes, and it is said (PHILLIPS, 1966, quoted by VON BRANDT, 1972) that Maori seines in New Zealand were a mile long and required 500 people to haul them. Even today seines account for a large part of the catch in several developing countries, but are not important in industrial-scale fisheries.

*Surrounding nets*

This class includes the modern purse seine, which alone accounts for a big share of catches in industrial fisheries for all types of pelagic fish. The ancestors of the purse-seine—lampara nets and ring nets of various types—have a long history, and are still commonly used in many parts of the world. The use of lights to attract and hold fish while the net is set around them is very common.

*Drive-in nets*

Though fish can be driven into many types of gear, these relatively uncommon gears rely on driving the fish by splashing on the surface, by the fishermen swimming, etc.

*Lift nets*

Various devices which wait for the fish to collect over the net, and then lift them out of the water are widespread. Because they usually require a large framework of poles they are a conspicuous feature of harbours and river mouths, particularly in the developing countries, though the total catch taken is small.

*Falling gear*

These work on the opposite principle, coming at the fish from above. The picturesque cast net is a typical example of this class of gear. It provides widespread employment but not much fish in poorer countries.

*Gill-nets*

With the replacement of drift nets by purse-seines and pelagic trawl as the chief way of catching herring and other pelagic fish, catches by this class of gear have declined. While no longer important in the developed countries, except locally, gill-nets of various types are still widely used in the poorer areas.

*Tangle-nets*

These relatives of the more size-selective gill-nets are moderately widely used when a variety of sizes or species of fish are being pursued, but are nowhere very important.

This very brief catalogue witnesses that it is impossible to talk about a typical fishing gear. Fishermen have devised a wide variety of gears to take advantage of the behaviour of different species, varying physical conditions of estuaries and rivers, etc. What is less clear from such a summary, but emerges from looking at gears from different parts of the world, is the degree to which fishermen in different places have found very similar solutions to similar problems. Large beach seines are found in most places where open shelving beaches and the presence of big schools of fish close inshore give favourable physical and biological conditions for this sort of gear. The type of basket trap used in the coral reefs of the Indian Ocean is almost identical to the traps used in the Caribbean Sea.

To some extent the replacement of wind and human muscles by engines as the prime source of energy for fishing, and the subsequent development of the trawl and purse-seine, has tended to reduce the variety of gears. Certainly, anyone whose introduction to commercial fisheries was the thriving trawling ports of the Humber or the Elbe in the early 1970s might be forgiven for feeling that fishing had become syonymous with trawling. The successive oil crises, the need to be less indiscriminate in order to protect the more vulnerable stocks, and the reduced access to many grounds as a result of changes in the law of the sea, are all helping to reverse this trend towards the large super-trawler or super-seiner as the ultimate fishing machine. The present demand is for something that is more flexible and requires less fuel. Some of the more successful technological advances in recent years have concerned the less energy-consuming types of gear, e.g. long-lining or squid jigging, making them less labour intensive (e.g. through machines to bait the hooks automatically) while maintaining the other advantages.

### (f) Processing and Marketing

The final assessment of the world fish catch is in terms of how the fish is treated once it is landed. Table 5-5 shows the changes in the disposition of the world catches since 1938. Five main methods are distinguished: utilization of fresh fish, freezing, canning, and curing for direct human consumption, and reduction to meal and oil for feeding to broiler chickens and other farm animals. Fifty years ago more than half the catch was

Table 5-5

Disposition of world landings of fish, according to main uses. Average annual quantities (million tons) (Based on FAO *Yearbook of Fishery Statistics*)

| Disposition | 1938 | 1948–52 | 1953–57 | 1958–62 | 1963–67 | 1968–72 | 1973–77 | 1978 |
|---|---|---|---|---|---|---|---|---|
| Fresh } | 11·1 | 10·1 | 12·6 | 15·9 | 17·9 | 18·9 | 19·1 | 20·3 |
| Freezing } | | 1·2 | 1·9 | 3·5 | 6·0 | 9·5 | 12·5 | 13·2 |
| Curing | 5·7 | 5·7 | 7·1 | 7·6 | 8·2 | 8·1 | 7·9 | 8·1 |
| Canning | 1·5 | 1·9 | 2·7 | 3·6 | 4·7 | 6·2 | 9·3 | 9·7 |
| Meal and oil | 1·7 | 2·0 | 3·6 | 8·0 | 16·2 | 22·7 | 19·0 | 20·0 |
| Miscellaneous | 1·0 | 1·0 | 1·0 | 1·0 | 1·0 | 1·0 | 1·0 | 1·0 |
| Total | 21·0 | 21·9 | 28·9 | 39·6 | 54·1 | 66·3 | 69·1 | 72·4 |

consumed fresh (so little was frozen that the amounts were not distinguished in most statistics), and more than half of the rest was cured—smoked or salted. Since then these uses have declined in relative importance; the amount consumed fresh has grown fairly steadily, though less rapidly than the total catch, while the amount cured has not changed much. In contrast the amounts frozen and canned have increased steadily. The quantities used for fish-meal have increased even more, but less steadily; after growing extremely quickly in the late 1950s and 1960s they reached a peak of around 25 million tons in 1970 and have since declined. This rapid rise and partial decline coincides with the rise and collapse of several fisheries directed almost entirely at the production of fish-meal (e.g. Peruvian anchoveta; herring, fished in Norway by purse seine). As these fisheries collapsed there have been few other alternative stocks which offered the necessary high catch rates that could support new fish-meal fisheries, and maintain the overall world production. The growth of new large-scale fish-meal fisheries, where they might otherwise have occurred (e.g. on anchovy in the southwest Atlantic Ocean), has also been discouraged by increased production of soya-bean meal. This provides an alternative source of protein for supplementing animal feeding stuffs; fish-meal can only be marketed if it can be sold at a price competitive with soya-bean meal.

These changes have been reflected in changes in the pattern of world trade in fisheries. Until recently the only form in which fish could last long enough for international trade was dried or salted. The traditional international fish trade was the export of salt herring from the North Sea countries to eastern and southern Europe, and of salt and dried cod from the North Atlantic countries (Newfoundland, Canada, Iceland, Norway), to southern Europe, as well as to West Africa and the Caribbean. While some remnants of this trade—at least in salt cod—remain, most curing is now done for trade within the country. In the developed countries this is mainly done to make a more attractive product, but in many developing countries smoking or sun-drying still remains the most usual way of enabling fish to reach consumers away from the coast or productive inland waters.

The flow to developing countries from the big fishing countries of temperate waters is now more often in cans. Canned mackerel from Japan has become a standard source of cheap protein in many parts of southeast Asia and Oceania. With the collapse of many herring stocks, herring has become a much less important item of international trade, particularly as a low-cost product for poorer people. Similarly, most of the cod catch is now frozen. In the form of fish fingers, or similar products, it appears in the richer countries' supermarkets. The main flow of cod is now from Canada, Iceland, and Norway to the United States and western Europe.

In some cases the flow has actually been reversed, fish going from the poorer countries (generally short of high-grade protein) to the richer countries. The commonest case is that of shrimp. Penaeid shrimp of various species are widely distributed throughout the tropics, and are common whenever conditions are favourable. Only a simple freezing plant is required to produce a high-valued product which can be relatively easily marketed. Many tropical countries have important shrimp fisheries, and the exports of Mexico, India, Indonesia, and Thailand have each exceeded $100 million annually (including other crustaceans and molluscs). The direct contribution to local food supply by these fisheries may be small. Though in some areas, especially in southeast Asia, it is the contribution of shrimp to the catch that maintains the economic viability of trawl fisheries for shrimp and finfish; the fish are used for local consumption. However, in most cases the social value of the fisheries as provider of employment and generator of cash is considerable.

Tuna also are distributed from tropical waters to the consumer in temperate areas—notably Japan and the United States. Some are exported in cans, but most are frozen, in part for subsequent canning in the recipient country, and in part for direct consumption. In the case of Japan this may involve freezing at very low temperatures (−30 ° to −40 °C) for the high-valued sashimi market. For these species of larger tuna the direct involvement of developing countries relatively close to the fishing grounds may be small, since much tuna is caught by long-range fleets of purse-seiners or long-liners.

## (2) Regional Resources and Present Fisheries

### (a) Introduction

The fisheries resources are grouped here in broad regions characterized by generally homogeneous ecological conditions: northern temperate waters; upwelling areas off the eastern continental coasts; tropical areas; southern temperate areas, and other resources (Southern Ocean, open ocean). For the most part subdivisions of these broad regions correspond to the statistical regions used by FAO in its *Yearbook of Fishery Statistics* and other publications. They also correspond in most cases to the areas of responsibility of the various regional fishery bodies. The exceptions are some of the major upwelling areas (Peru, California, and northwest Africa) which have been treated separately from the rest of the region. For each region a description is given of the nature of the resources, a brief history of the major fisheries, the present pattern of fishing, and the possibilities for increased production or improved management. The material is largely drawn from the reports of working parties and other activities of the various regional fishery bodies, much of which is produced in a summary form in the annual reports prepared by FAO for its Committee on Fisheries (FAO, 1981).

### (b) Northern Temperate Waters

#### *Northeast Atlantic Ocean*

*Resources*

This is one of the richest regions of the world, ranking second in weight caught only after the northwest Pacific Ocean, and accounting for about one-fifth of the total world marine fish catch. It is also among the most complicated regions. Four major subregions can be distinguished: northern waters (from Greenland and Iceland to Norway and north Russia); Baltic Sea; waters around the British Isles; and southern waters from Ushant to Gibraltar.

All subregions, except the last-mentioned, have large areas of relatively shallow water—altogether nearly one-fifth of the total shelf area of the world. This, and the high primary productivity of much of the region, explains the richness of the local fish stocks, while their closeness to some of the world's biggest markets accounts for the early and sustained development of the fisheries on these stocks.

In the northern region, the fisheries have been dominated by two species: the cod and, a little way behind, the herring. Even now when some of the stocks have been depleted by over-fishing, and every effort has been made to make full use of the other species (among which haddock and redfish are the most important) cod still make more than half of the total demersal catch from the subregion. While herring is the only pelagic species that has been used to any significant extent for human consumption, the biggest pelagic catches are now those of capelin, used mainly for production of meal and oil.

Cod and herring are also important around the British Isles, but the more southerly species—especially whiting, plaice, and sole among the demersal species, and mackerel among pelagic species, as well as smaller species such as sand eels, Norway pout, and sprat which are chiefly used for fish-meal—are also important. In their review of the west of Scotland stocks, GORDON and DE SILVA (1980) list about 40 species for which some biological information is available, as well as twice as many for which there are published data on distribution. Thus, no one species dominates the catches in the subregion as a whole, and there are some ten species which contribute at least 5% to the total. More significantly, each of these ten is sufficiently common, at least locally, and sufficiently distinct in its distribution to be the main support to some group of fishermen.

The Baltic Sea is a peculiar region. It is connected with the open ocean by narrow and very shallow passages between the Danish islands and between Denmark and Sweden. Salinity decreases steadily from the Belts, where it is 20‰ S on the surface and 30‰ S at the bottom, to not more than 2‰ S at the ends of the Gulfs of Bothnia and Finland. The lack of strong circulation also results in the periodic build-up of low oxygen concentrations in the deeper parts of the Baltic Sea. This build-up is assisted by the development of a strong halocline, with the surface waters being almost fresh, and the inflow of high salinity water being restricted to the bottom layers. The amount of inflow, which seems to be governed by the strength and persistence of westerly winds, seems to vary appreciably (FONSELIUS, 1962) and this has an important impact on the fish stocks, and marine life in the Baltic Sea generally. Only a few marine species have coped sufficiently successfully with the unusual conditions in the Baltic Sea to become commercially important. Three species—cod, herring, and sprat—account for some 80 to 90% of recent catches, and much of the rest of the catch is of a variety of catadromous, anadromous, and freshwater species.

South of the British Isles there is a marked change in species composition, with the typical northern species (cod, herring, etc.) giving way to a more southerly fauna (sardines, hake, sea breams). A number of species, notably hake and mackerel, occur also in the waters around the British Isles, particularly in the more oceanic waters to the west, but to a large extent this part of the region (from Ushant south to Gibraltar) has at least as much in common with the northern part of the eastern central Atlantic Ocean, from Gibraltar southwards as with the waters around the British Isles. Unlike the rest of the region, the continental shelf is narrow, except in the northern part of the Bay of Biscay, and the total potential is low compared with the more northern waters, even though the upwelling system, which is responsible for the rich resources off northwest Africa, extends seasonally as far north as Portugal.

*The fisheries*

The history of the fisheries in the northeast Atlantic Ocean is to a large part, and especially up to the last couple of decades, the history of world fisheries. Excavations of waste dumps in ancient sites have shown the importance of fishing in northwestern Europe, especially around the North sea and Baltic seas, even in prehistoric times. The later importance of fishing, and especially of salted and dried cod and herring in the regional economy and in international trade, is borne out in the history of the region. For example, a major dispute between England and the Low Countries in the seventeenth century and one of the causes of the Anglo–Dutch wars of the middle of the century, was over fishing rights for herring in the North Sea. Again the changes in fortune in the cities of the Hanseatic League during the Middle Ages can be ascribed in part to changes in the distribution of herring in the northern North Sea, the Baltic Sea, and the southwestern coast of Norway. Because in several of the fisheries the fishermen could not work far from the coast, they were particularly vulnerable to changes in the behaviour or migration pattern of the fish, which could take the fish out of their reach. HÖGLUND (1978), for example, has described the changes in the herring fishery off the western coast of Sweden, which has been subject to large fluctuations. The exact periods seem subject to argument, though certain periods (1556 to 1589, 1747 to 1808, and 1877 to 1906) clearly featured a high local availability of herring.

For the purpose of understanding the fisheries in the 1980s and their problems, the important part of fisheries history starts with the application of industrial technology to fishery, and especially with the replacement of human muscle by machines. Here is a clear difference between the demersal fisheries (for cod, plaice, etc.) and the pelagic fisheries (particularly for herring). The present phase of demersal fishing started a little over a century ago, around 1875, with the start of the English steam trawler fishery in the southern North Sea. The history of this development has been well described by ALWARD (1911) and summarized by CUSHING (1966). Coupled with the availablity of ice that kept the fish fresh on board for many days, and a good network of railways that took the fish quickly from the Humber ports of Hull and Grimsby to the inland markets, the fishery was initially highly successful (ALWARD, 1932). This success was repeated stock after stock throughout the world. While the success lasted, the fleet grew in numbers and size of vessels. This growth was inevitably followed by a fall in abundance of the favoured species and a drop in catch rates and in profits. The more enterprising fishermen then looked further afield, first to the more northern parts of the North Sea, and then outside the North Sea. For some three-quarters of a century there was a steady expansion, mostly northwards to the Faroes, Iceland, northern Norway, and the Barents Sea, and westwards to Greenland and Labrador, but also southwards. In the early years of the century, English trawlermen—who took the lead in this expansion—were fishing for plaice off northern Russia (referred to by them erroneously as the White Sea; ATKINSON, 1908), and for hake off Morocco.

As the expansion progressed, the fisheries on the grounds first exploited tended to stagnate for a time. This was especially the case when the fleet had grown in numbers too much, but the individual vessels had not the size or range to move to other grounds. The best example was the British North Sea fleet of steam trawlers during the interwar period. The high catches which lasted for a couple of years (1919–1920) because the stocks had enjoyed 4 yr of very little fishing initiated the building of numerous new ships. These joined the large number of vessels which had been built, on trawler designs, for minesweeping during the war, to give a fleet unsuitable for anything except trawling in the North Sea, and so large that the individual vessel could earn little more than its basic running costs. As a result the fleet changed little, except for a few losses, between

1920 and 1939—a fact which had an important side-effect on the development of fish population dynamics. The classic work of BEVERTON and HOLT (1957), based mainly on the plaice fishery between the wars, gives the impression that a long period of roughly constant fishing effort is typical of fisheries. This aspect of the plaice fishery is very untypical and there are very few fisheries—other than the traditional artisanal fisheries—that have remained steady, even over a whole decade.

Ultimately even the British North Sea fleet adjusted to the local conditions—weather and sea, distance from port to the fishing grounds, and catch rates (and hold capacity required). This resulted in a fleet of trawlers appropriate to these conditions. The vessels now range from large stern trawlers (usually with freezing capacity) of as much as 1000 tons GRT or more working in the harsh waters of East Greenland or the Barents Sea, to small coastal trawlers, with a crew of 2 or 3, working on inshore waters of the Baltic Sea, or in the English Channel. In either case a fairly smooth line of descent can be traced from the original steam trawlers of the North Sea in the late nineteenth century. In this descent changes are noticeable beyond the slow evolution: these include the transition from steam to motor propulsion; the introduction to the Vigneron-Dahl gear in the 1920s; the change from side-trawling to predominantly stern trawling (first in the very small vessels, and the very largest factory trawlers, and now for nearly all sizes); and the reversion to beam trawling for soles and some other flatfish. Special mention must be made of the growth of fishery for small species (sand eels, Norway pout, etc.) for fish-meal and other industrial purposes (pet food, or food for trout and mink farms; POPP MADSEN, 1978). This growth in the 1950s and 1960s paralleled the growth of the great purse–seine fisheries for pelagic species, also in terms of catch reduction to meal and oil. This has taken place mainly in the North Sea, with Denmark being the leading country. With minimal handling on board, small boats and small (2 or 3 men) crews can deal with large quantities of fish. Though the main target species have little or no value for direct human consumption, there is inevitably some by-catch of more valuable species (especially haddock and whiting), which has given rise to considerable argument, at all levels up to Prime Minister, between the countries engaged in these fisheries, and those for direct human consumption.

The pelagic fisheries—for herring, mackerel, pilchard, etc.—were slower to emerge from the traditional pattern into the age of modern technology. As late as the 1950s the bulk of the British herring catch was taken in drift nets in a manner (except for the replacement of sails by steam for the means of propulsion) which had changed little for ages. Trawling for herring—at first mostly with bottom trawls rather than mid-water trawlers—was started before the war, but the modern age of the pelagic fisheries really started with the use of purse-seine and power block. In the North Atlantic this was first used on a large scale by the Norwegians. The Norwegian fishery, working mainly for fish-meal, grew rapidly in the 1950s and 1960s, directed first on herring (PARRISH and SAVILLE, 1967), then on mackerel (HAMRE, 1978) and, more recently, on capelin; it reduced several of these stocks to very low levels before adequate controls were introduced. Iceland and other countries have also adopted the purse-seine, and in northwest Europe there is now a fleet of highly efficient vessels, capable of catch rates well above the tolerance capacity of the resources, if they were to be maintained throughout the year. The pressures of these technologically efficient and highly mobile fleets are among the major problems of management in these regions.

The rise of trawl and purse-seine has not completely eclipsed other methods of fishing.

In the North Sea the Danish seine, in either its Scottish ('fly-dragging') or Danish (from anchor) form has remained an effective alternative to the trawl for haddock and whiting, or plaice and cod, respectively, requiring small crews and (especially from anchor) less fuel. Most of the miscellaneous fishing methods are with small boats for the less common types of sea food, e.g. traps for crabs or lobsters, but in Iceland and northern Norway the traditional fisheries for spawners with gill-nets and other gear are still a significant component of the cod fishery.

The result has been that the northeast Atlantic includes as great a variety of fishing methods and fishing vessels as any part of the world. Social factors have maintained some of the simpler methods in the more isolated areas, while elsewhere economic forces have caused the evolution of highly efficient and technically advanced methods. This only rarely has meant the use of very large and expensive vessels. Only in the extreme west and north of the region, off Greenland and in the Barents Sea, has the weather and the distance from port allowed the long-term success of the large factory-trawler—the popular image of modern fishing. Elsewhere it is the small to medium-sized vessel, often still family owned, that represents efficient modern harvesting.

*The state of the stocks*

The state of the stocks is regularly reviewed by a bewildering array of working groups of ICES,* dealing either with individual stocks or, more usually, groups of stocks in a particular area. Most of these now meet annually, and their reports, available as working papers at the annual meeting of ICES, are reviewed and summarized by ICES's Advisory Committee on Fishery Management (ACFM). The ACFM report is presented to the Baltic Sea Fishery Commission and to the Northeast Atlantic Fishery Commission, as well as to national governments, and is the basis for the management measures taken, especially for the level of the Total Allowable Catches for the forthcoming year. While this report is readily available within ICES, it is not so well circulated outside. Also the pressure on those scientists carrying out the assessments is such that they have little time to bring the work to the stage of formal publication. A notable exception is the report of the symposium on the North Sea fisheries held in 1975 (HEMPEL, 1978). Otherwise this review of the state of the stocks is largely taken from the unpublished reports of the ICES working parties, and especially the annual reports of ACFM.

*The northern region*

The situation in this region is most easily described, since the events in each major stock have largely followed the simple stock assessment models. Fig. 5-1 shows trends in the total catch of cod from the so-called Arcto-Norwegian stock, and the estimate of the total amount of fishing on it. This fishery takes place in three main areas: (i) at the beginning of the year and on the approaches to the spawning area around the Lofoten Islands; (ii) in the feeding grounds to the north—from Bear Island to Spitzbergen; (iii) to the east in the Barents Sea from the Russian coast northwards. Before the Second World War, catches rapidly expanded (as did the amount of fishing, particularly by the growing English distant-water fleet, though full statistics are not available), reaching a peak in 1937. Catches and effort dropped during the war, but recovered rapidly to the

*International Council for the Exploration of the Sea.

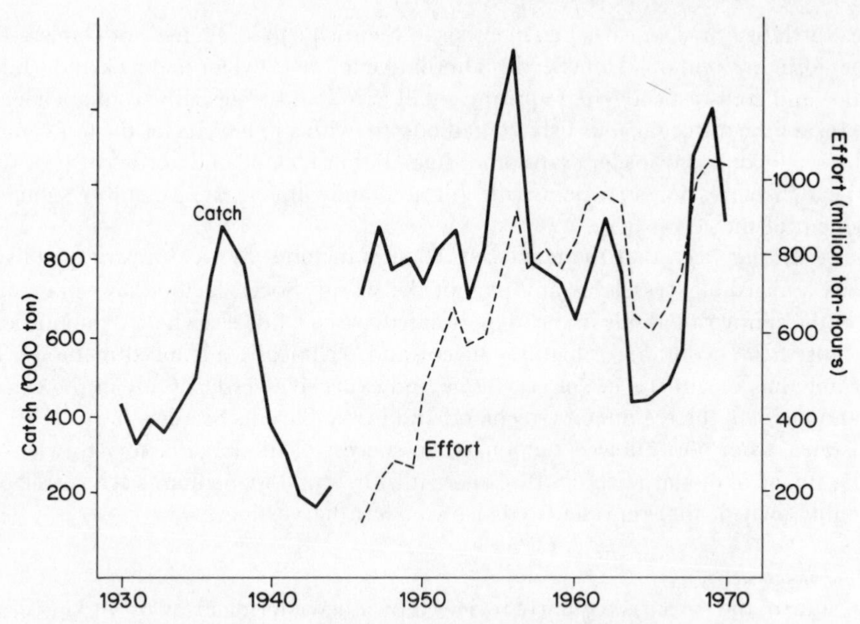


Fig. 5-1: Trends in total catch and amount of fishing on the Arcto-Norwegian cod. (Reproduced by permission of the International Council for the Exploration of the Sea)

pre-war level. Thereafter the amount of fishing grew steadily, tripling between 1950 and 1960, but the total catch remained at around the same level. The rise in the amount of fishing was accompanied by a great increase in total mortality rate. This is shown in Fig. 5-2, though it must be stressed that Virtual Population Analysis (GULLAND, 1965) has shown that the fishing mortality is by no means constant with age (as is assumed by the simple analysis) and in fact the fishing mortality is high in the younger ages (around 3 to 7 yr) exploited in the trawl fisheries, drops in the larger, but still immature fish, and then rises again in the mature ages exploited in the mixed-gear spawning fisheries. Given the long life of the Arctic cod and the great potential growth of more than an order of magnitude between the time the 2 to 3 year-old cod first become vulnerable to trawling in the feeding areas, and the time when the 8 to 10-year-old cod first spawns, the events in this stock can be broadly accounted for by yield-per-recruit calculations. The pattern of exploitation since 1950—low size at first capture, and high fishing mortality—does not allow the individual fish, on the average, to survive long enough to grow to a good size. Though the number caught is high, the weight landed is less than could be achieved by a more rational pattern of fishing. The distinction between catches and landings is important because a large but unquantified number of small cod are known to have been caught but discarded by the trawlers in the feeding area.

The same remarks apply to the other stocks of larger demersal stocks—haddock, plaice, saithe, redfish, as well as cod at Iceland, East Greenland, and around the Faroes—though the time at which they become fully exploited has varied. The plaice stocks were smaller, and plaice responded better to being kept on ice for long periods (or at least somewhat stale plaice was better accepted on the English markets), and the

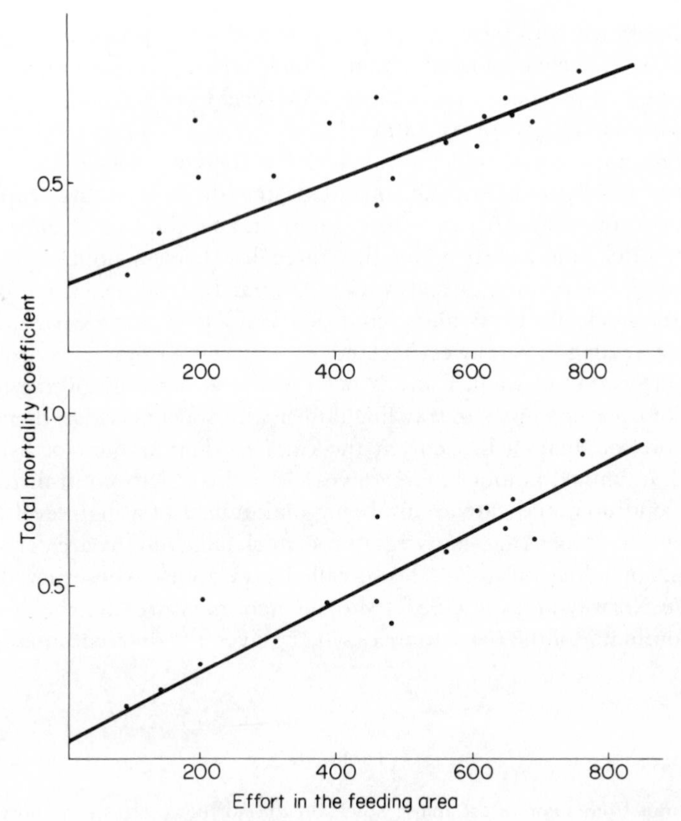


Fig. 5-2: The relation of the estimated mortality in Arcto-Norwegian cod to total fishing effort for 7-year-old fish (above) and 6-year-old fish (below). (Based on Working Party Reports of the International Council for the Exploration of the Sea)

plaice stocks in southwest Iceland, and off the Murman Coast (the 'White Sea' of the Humber fishermen) were probably fully exploited before the First World War; Iceland cod, and haddock (and probably also the Faroe stocks; cf. PARRISH and JONES, 1953) were fully exploited in the 1930s. Saithe only became fully exploited in the 1960s and 1970s, particularly as a result of purse-seining for the younger fish.

'Recruitment over-fishing' does not seem to have been such a serious problem for the northern demersal stocks. More accurately: it has not yet been so generally accepted as a serious problem. In both the Icelandic and Arcto-Norwegian stocks, there have been fluctuations in abundance, over and above the general trend due to changes in the total amount of fishing, which are clearly due to variations in year-class strength, often with several good year classes (e.g. those of 1948, 1949, and 1950 and of 1963 and 1964 in the Arcto-Norwegian stock) occurring together. These often correspond roughly with high adult abundance at the time of spawning. GARROD (1977) has examined the stock–recruit relation for this stock. Though, given the usual scatter of points, the results are no more convincing than for other sets of stock and recruitment; they are highly suggestive. Combined with yield-per-recruit calculations, expressed as egg production-per-recruit to be expected under different fishing regimes, these indicate that fishing mortality as high as has been observed in some recent years could not be maintained indefinitely without collapse of recruitment.

Management of the northern demersal stock has therefore concentrated on measures to increase the yield-per-recruit, i.e. on increasing the size at first capture through mesh-size regulations and other measures, and reducing the fishing mortality, particularly through catch quotas. In doing this there has been a noticeable difference in success regarding the two major cod stocks. Around Iceland, extension of jurisdiction has brought the stock effectively under control of a single country (apart from those fish which, while spawning in southwest Iceland, spend their earlier years around southern Greenland). Protection of small fish has been achieved not only by the use of a large mesh, but also by closing areas to trawling, at least for short periods, whenever undesirable proportions of small fish occur in the catches. The amount of fishing by both Icelandic and, in limited numbers, foreign vessels is also strictly controlled. The stock is in a healthy condition, and catches are being maintained at a high level (Table 5-6).

The distribution of the Arcto-Norwegian cod stock falls into the area of jurisdiction of two countries, including an area—the so-called grey zone—whose jurisdiction is disputed between Norway and the USSR. Management measures are therefore the lowest common denominator of the two countries which, given the different interests (USSR in

Table 5-6

Reported landings from some of the major stocks in the northern part of the northeast Atlantic Ocean (thousand tons) (Based on ICES Cooperative Research Reports)

| Principal state | 1962 | 1966 | 1970 | 1974 | 1977 | 1979 |
|---|---|---|---|---|---|---|
| *Region 1* | | | | | | |
| Herring: | | | | | | |
| I + II* | 601 | 1520 | 62 | 8 | 18 | 4 |
| Va + b† | 657 | 491 | 19 | 9 | 29 | 45 |
| Capelin: | | | | | | |
| I + II | 4 | 389 | 1314 | 1147 | 2941 | 1829 |
| Va + b | — | 125 | 192 | 462 | 761 | 868 |
| Cod: | | | | | | |
| I + II | 928 | 557 | 956 | 1145 | 943 | 485 |
| Va + b | 411 | 380 | 506 | 401 | 377 | 397 |
| Haddock: | | | | | | |
| I + II | 184 | 130 | 86 | 231 | 112 | 110 |
| Va + b | 147 | 79 | 66 | 57 | 65 | 68 |
| Saithe: | | | | | | |
| I + II | 121 | 203 | 265 | 264 | 182 | 164 |
| Va + b | 61 | 78 | 146 | 144 | 97 | 91 |
| Redfish: | | | | | | |
| I + II | 36 | 35 | 29 | 97 | 186 | 113 |
| Va + b | 75 | 107 | 80 | 77 | 69 | 77 |

*Regions I and II: Barents Sea and Norwegian Coasts.
†Regions Va and b: Iceland and Faeroese of ICES.

smaller cod, Norway mainly in spawning cod), tends to be very low indeed. While it has been agreed for a long time, at least at the level of science, that a mesh size of 150 mm would be desirable, the legal mesh in 1980 was still only 120 mm. The effective mesh size, as estimated from the sizes of fish actually landed, was even smaller. ACFM (ICES, 1980: Table 7) gave the following estimates:

| | Area I (Barents Sea) | Area II (Norway Coast and Bear Island) |
|---|---|---|
| USSR trawl | 90 mm | 111 mm |
| UK trawl | 113 mm | 113 mm |
| Norway trawl | 119 mm | 120 mm |

The amount of fishing has been controlled by catch quotas since 1976, but the quota set has often been in excess of that recommended by scientists. Also, the actual quantities caught are believed to have exceeded the quotas, particularly since there have probably been considerable quantities of small fish discarded at sea, which are not included in the landings or quotas, and for which the data are very poor. As a result the fishing mortality has remained at a very high level, the stock is in a poor condition, and the most recent catches have been very low.

In contrast, the principal pelagic stocks—Atlanto-Scandian and Icelandic herring—provide good examples of 'recruitment overfishing', at least in broad outline. The Atlanto-Scandian herring, which has a very long potential life span (more than 20 yr) has long been noted for great variation in year-class strength (HJORT, 1914; DRAGESUND, 1970). Outstanding year classes only occur at intervals, perhaps as much as 10 or 15 yr apart, but if they were not fished too hard, they were capable of supporting a viable fishery until the next good year class came along. The last outstanding year class was that of 1959, though those of 1960 and to a lesser extent 1961 were also good. Instead of being fished at a moderate level, corresponding to the historical pattern, these year classes were fished extremely hard, particularly for fish-meal by the new Norwegian purse-seine fleet and by the large USSR drift net fleets. Over $5 \times 10^9$ fish of these year classes were caught in 1966 and, despite falling abundance, $3 \times 10^9$ in 1967. Thereafter catches dropped catastrophically from a total, of all ages, of over 1·7 million tons in 1966 to less than 25 000 tons in 1969. None of the year classes since 1961, at least up to 1980, has provided more than a handful of fish. What was one of the world's leading fisheries has disappeared.

A short run of poor year classes would have been nothing unusual for this stock, but such a long period without anything even approaching the previous long-term average must indicate something more than chance. It seems that even if the spawning stock is reasonably abundant, unless some so far unidentified set of environmental factors are favourable, the year class will not be average or better. However, unless the spawning stock is adequate, even favourable environmental conditions will not produce a good year class. The prognosis for this stock is that recovery will be slow; good environmental conditions will be required for the current very low-spawning stock to produce even a moderate year class. The stock must be allowed to reach maturity and to spawn for a sufficient period of years until another year of favourable conditions occurs to produce a good year class and thus rebuild the stock.

Much the same has happened to the main stock of herring around Iceland. Catches around Iceland (which included some fish from the Atlanto-Scandian stock) reached over half a million tons in the late 1950s and early 1960s, but collapsed from 480 000 tons in 1966 to only 30 000 in 1968. They have remained below even this level ever since. The only other significant pelagic fish stock in the northern region is the capelin. The two main stocks (around Iceland and in the Barents Sea) both seem to be heavily fished. Catches rose to 3·7 million tons in 1977 and have since fallen slightly, but the stocks appear to be in a reasonably healthy state.

*The Baltic Sea*

These stocks have been reviewed by THUROW (1978). More recent information is provided in the section of the ACFM report dealing with Baltic stocks; it provides advice to the International Baltic Sea Fishery Commission. More than 80% of the current catches, as well as of the estimated potential, is made up of the three major species (cod, herring, and sprat). All three seem to be heavily fished, with catches being around the level of the maximum sustainable yield. Many of the minor species are taken incidentally in the trawl fishery for cod, and are probably also fully exploited. There may be some potential though for expanding the catches of minor demersal species. Very recently the sprat stock has practically collapsed, although catches have been restricted by the TACs agreed upon by the International Baltic Sea Fisheries Commission. The reason is not clear, but a likely explanation is increased predation by rich year classes of cod. This is a striking example of the weaknesses of single-species management (LINDQUIST, pers. comm.).

*Waters around the British Isles*

While substantial catches (over 1 million tons in recent years) are taken to the west and south of the British Isles (Areas VI and VII of the ICES statistical system), the bulk of catches in this part of the region are taken in the North Sea. Possibly more significantly, most of the major fishery laboratories of the countries of northwest Europe (Aberdeen, Bergen, Hamburg, Ijmuiden, Lowestoft) lie around the perimeter of the North Sea. The North Sea and its commercial fish stocks are therefore much better studied than the rest of the subregion, and indeed better studied than those in virtually any other area of comparable size in the world. North Sea fish stocks—particularly the plaice—have provided the raw material of fish population studies, from the time of BARANOV's (1918) analysis of the length composition of plaice catches onwards. In particular the English plaice fishery during the interwar period provided most of the raw material for BEVERTON and HOLT's (1957) major study of fish population dynamics. Much of this research, especially that of recent years, was reviewed at the ICES Symposium on 'North Sea fish stocks—recent changes and their causes', held in Aarhus in 1975. The published papers of this Symposium (HEMPEL, 1978) provide a very convenient entry into the massive literature on these stocks. The paper by HOLDEN (1978) presents compact descriptions of the trend in total catches of most species.

This literature, and the evolution of the catches from the various stocks, shows that in some respects the dynamics of these stocks are well understood. An early output of research was knowledge of the growth pattern of each species. This could readily be combined with some estimate of natural mortality (0·1 or 0·2 were often taken as convenient approximations, though not always with convincing evidence) to provide yield-per-recruit curves. The history of the changes in the yield-per-recruit from these stocks subsequent to the studies has generally confirmed the predictions—explicit, or implicit in the results—of these studies. For example, GULLAND (1968a) showed that the increase in North Sea plaice landings between the interwar period and the late 1950s and early 1960s could largely be explained by an increase in yield-per-recruit, brought about by a reduction in fishing mortality, and an increase in size at first capture. These in turn were due to a reduction in the size of the fleet (mostly through war-time losses) and a shift in the position of the main fishing grounds to the northern and deeper waters where small fish are less abundant. This increase matched closely the calculations of BEVERTON and HOLT (1957).

The various studies showed quite clearly that at least by the 1930s and probably much earlier the three main and best studied species of demersal fish (cod, plaice, and haddock) as well as the less abundant, but still medium-to-high priced species (sole, turbot, etc.) were heavily fished, and the only way to increase the yield from a year class, once it was recruited, was to fish less, and protect the smaller fish and thus give the individual fish a reasonable chance to grow to a tolerable size.

For the principal pelagic stocks (herring and mackerel) yield-per-recruit calculations gave more equivocal results. In the main herring fisheries, i.e. excluding the industrial fishery for small immature herring, recruitment does not occur until a relatively large size; combined with what is commonly supposed to be a fairly high natural mortality this gives a very flat-topped yield-per-recruit curve, with the maximum, if any, occurring at high fishing mortality rates. On the basis of this type of calculation it was, until the late 1960s, not uncommonly believed that it was impossible to 'over-fish' herring, and that though the stocks might be becoming heavily fished, there was little risk of total herring catches falling.

The recent history of the North Sea fisheries shows how poor a description of the events in these stocks is provided by yield-per-recruit calculations, and how incomplete is our understanding of the total dynamics of these stocks. Attempts to make prognoses about future trends in total catches or catches of particular species have often been proved wrong extremely promptly. To take examples by scientists who have made major contributions to our knowledge of the North Sea and its biological systems, BEVERTON (1962) noted that haddock catches, in addition to a high variability also showed, between 1905 and 1960, a definite downward trend. The response of the stock was to undergo a remarkable rise that brought catches from a low point of little more than 50 000 tons in 1952 to well over 600 000 tons in 1970—nearly three times the peak pre-1960 catches, taken in 1920 when the war-time accumulation was being fished down. STEELE (1965) notes that despite over-fishing and war-time respites the total demersal catch in the North Sea had kept remarkably constant—at around 400 000 tons—for some 50 yr, and that this total appeared to agree well, given reasonable transfer efficiencies, with the total primary production in the North Sea. The immediate response of the stocks was for the catches to increase to 1 million tons and more, due to increases in all the main demersal species, including haddock.

The point here is not to remind the reader of the ability of natural systems to make fools of even the most eminent scientists, but to contrast our good understanding of the behaviour of the yield-per-recruit with the lack of understanding of the behaviour of recruitment. In the North Sea since the 1960s there have been two major recruitment changes: the collapse of herring recruitment, and increases in recruitment of many demersal fish (especially gadoids such as cod, haddock and whiting), and probably also sprat.

The collapse in herring recruitment, which first occurred in the most southern (and probably most heavily fished) Downs stock, and later in the northern Banks stock, can be reasonably ascribed to the existence of a stock–recruitment relation for which the chances of average or better recruitment are poor at low stock sizes. For such stocks, which probably also include the Atlanto-Scandian herring stock, and several other pelagic stocks, a high fishing mortality can only be maintained for a short period.

The rise in recruitment of other species is less easy to explain. It is tempting to ascribe this to some sort of interspecific mechanism, and certainly the timing of the so-called gadoid outburst agrees well with the decline in the herring. However, the exact mechanism is not clear. Competition between species (in some ill-defined form) or the effect of predation of adult herring (and also mackerel) on larval cod or haddock have been suggested, but the details do not match well—neither herring nor mackerel are feeding in numbers in the areas and at the times when cod larvae are abundant.

Nevertheless it is reasonable to suppose that some kinds of interspecific actions play some role; and these interactions have received increasing attention. Most notable among these studies are those of the Danish group (e.g. ANDERSEN and URSIN, 1977) who have developed complicated models to take account of some of the interactions between species. These mainly concern the effect of feeding, i.e. how much herring is eaten by cod to maintain their growth, and the effect of this on herring abundance. These show, for example, that measures to increase the yield-per-recruit of cod by fishing less, and thus having more and larger cod in the sea, would mean much higher predation by cod on smaller species and hence fewer of these species to harvest. The result could mean, taking all species together, a smaller total catch from the sea. However, these models, though complex, still provide no quantitative prediction of the effect on recruitment of one species of changing abundance of other species.

A quantitative description has been attempted by GULLAND (1981). He pointed out that the price of fish from the North Sea is far from constant, and that the overall effect of the developments in North Sea fisheries in the last 20 yr has been to increase the proportion of the lower value species, particularly the sand eel and Norway pout which are used only for fish meal. The result has been that while the total weight taken from the North Sea increased greatly in the 1960s and 1970s, the total value (at constant prices) has not. At present each individual species of fish capable of being exploited with present technology, costs, and prices is fully expoited, i.e. with a fishing mortality approaching or beyond that giving the maximum yield-per-recruit. However, the nature of yield may be affected by adjusting the intensity of fishing on each species so as to shift the recruitment in a favourable direction. It is reasonably certain that reducing the fishing effort on herring, and so far as possible keeping the spawning stock at a high level, will result in better herring recruitment. It should be possible to restore the stocks to the levels of the 1950s and earlier, though the catches that can be maintained will be lower than those taken in these years—perhaps around 500 000 tons, compared with 700 000 tons in most years, and over 1 million tons in 1965 (BURD, 1978; the rather higher figures by HOLDEN, 1978, in the same volume include catches in the northeastern North Sea which came from the Atlanto-Scandian stock, and not the North Sea stocks). GULLAND (1981) therefore suggests that the value of the total catch might be increased, at the cost of some reduction in the total weight caught, by patterns of fishing that encouraged high recruitment of valuable species. This might involve deliberately 'over-fishing' sand eel and Norway pout, and keeping some of the pelagic stocks moderately low so as to facilitate higher recruitment of plaice, haddock, and cod.

*Southern region*

The dynamics of the stocks in this area have been less well studied than those further north, and there are fewer explicit assessments of their state of exploitation. Probably the best studied has been the hake; and an ICES working group has reported on this stock. The catch statistics of some countries, at least until recently, did not show where catches were taken, so that some catches from further north (from the entrance to the English Channel as far north as the west of Scotland) have been reported as coming from the stocks in the Bay of Biscay and the Portuguese coast. With these reservations it does appear that these hake stocks are very heavily fished, and that there has been a steady decrease in total catch over the last few decades. This decline is probably due to a high fishing mortality and a small size at first capture. The effective mesh size used in most of the trawl fisheries is very small, and very large numbers of extremely small hake are taken in the Bay of Biscay.

Hake have, on the average over the years, made up between a quarter and a half of the total demersal catch. No other species approach hake in importance, and their dynamics are less well studied. Presumably the same factors—very high fishing effort and small mesh size—that affect hake affect these other species too, and they are probably also heavily fished.

Catches of the main pelagic species (pilchard, sardine) have declined from around 200 to 250 000 tons in the early 1960s to around 150 000 tons in the 1970s. The reasons for this decline, which has been most marked in the catches by Portugal, rather than Spain, are unclear, but do not appear to be related to over-fishing. At the same time, the shelf area is narrow and the waters not particularly productive, so the potential catch would not be expected to be high. Probably therefore the current catches are a significant proportion of the potential, and the stocks of sardine (and horse mackerel, the other important pelagic species) are at least moderately heavily fished.

*Future prospects*

Through the northeast Atlantic the stocks are nearly all heavily fished. Catches continued to expand until 1976, when over 12·5 million tons (excluding molluscs and crustaceans) were taken. This was achieved by continued expansion onto previously little used species, usually smaller and of less value. Though this is still continuing—for example the large stock of blue whiting to the north and westward of the British Isles is only just becoming exploited on a large scale, catches in 1979 passing a million tons compared with only 100 000 tons in 1976—the possibilities for further expansion are very limited. Possibly, some of the pelagic species in the southern zone could support higher catches. The opportunities for improving the performance of the fisheries (high value of catches, lower costs, or better allocation of benefits between participants) must come from better management. This in turn requires adequate ecological knowledge to know what to do, as well as the (essentially political) capacity to act on this knowledge.

In the single-species fisheries of the northern region, the basic ecological knowledge is there—cod (and other demersal fish) should be fished at a lower fishing mortality, thus avoiding the catching of smaller fish. The adult stocks of herring should be maintained at a high enough level to ensure adequate recruitment. Around Iceland (where only one country is involved) it has been possible to introduce effective measures along these lines. The cod stock (and the fishery on it) is in a healthy state. The relatively small summer-spawning herring stock is recovering, though the larger spring-spawning stock—like the spring spawning Atlanto-Scandian stock—is still at a low level, and may take many years to rebuild. In contrast it has been less easy to introduce effective measures for the Arcto-Norwegian cod stock, which is at a low level.

Further south, both the scientific and political problems are difficult. The biggest scientific problem is how to deal with interactions between species. The closest approach to a quantitative description—the model of ANDERSEN and URSIN (1977)—is far from complete, but still involves so many parameters that their estimation within acceptable confidence limits is extremely difficult. It is also difficult to confine attention to a single species, to say what size of spawning is necessary to ensure good recruitment, apart from a general acknowledgement that the current herring spawning stock is too small. There are two political problems. The more obvious one is that for most stocks there is no clear single authority. Around the North Sea there are uncertainties on the relative authority of the European Economic Community (EEC) and the individual countries, while many stocks in the North Sea, as well as elsewhere, are shared between two or more countries, or between the EEC and Norway or other non-member countries. The less obvious but at least as significant problem is that there has been little discussion of the various long-range and short-term objectives of management, nor is there any obvious forum in which such discussions could take place.

ICES is purely a scientific (essentially biological) body, and has deliberately avoided getting entangled in economic or social aspects of fishery management although ICES has arranged for 'dialogue meetings' between biologists and managers to improve mutual understanding. The result has been that arguments have concentrated on short-term issues, and the only agreed technical advice from international groups has been on biological matters. The decisions taken have therefore generally followed closely the biological advice, apart from delays, and a tendency for catch quotas to slip upwards slightly. This has been a pity in several ways. First, the scientific difficulties have meant that the recommendations of the biologists are mostly based on single species yield-per-recruit models, and are for a number of stocks specifically aimed at giving the maximum yield-per-recruit. This is very likely to be above the level giving the greatest total yield, and virtually certain to be above the level giving the greatest social and economic benefits, however these are determined. Secondly, by receiving a single recommendation (usually for a Total Allowable Catch) from the biologists, the administrators, and politicians can avoid taking the difficult decisions on policies—e.g. on whether to aim at a maximum catch or maximum net economic returns, and the trade-off between short-term interests (high catches now) and long-term objectives (low catches now so as to re-build the stocks most quickly)—which need to be taken, and can be made only at the political level.

The immediate prospect is therefore uninspiring. Arguments, often at the highest political level, are likely to continue to be concentrated on short-term matters. Probably the biologists now have sufficient information, and everyone has acquired sufficient sense, to avoid further serious declines in stocks, and there may be a slow rebuilding of some stocks. Catches will remain around, or perhaps a little below, their present level. However, the size (and costs) of the fleets involved in taking these catches will remain much too high.

In the long run there is cause for optimism. The potentially very wide scope of fishery management, and the great variety of benefits that can be obtained, are being much more generally discussed (FAO, 1980). While the total catch from the northeast Atlantic cannot be increased much, other types of benefits might be increased greatly. Greater total value might be obtained by shifting the balance in the species caught towards those fetching a higher price, and for many species by catching fewer but larger fish; costs could be greatly reduced, and there might be a politically more acceptable distribution of benefits. Although up to 1983 the existence of the EEC and the centralizing of many fishery decisions in Brussels seem merely to have complicated the problems, it may well be that the EEC structure in the long run makes it easier to take the necessary decisions.

### *Northwest Atlantic Ocean*

This region has many similarities with the northeast Atlantic Ocean; many of the species are the same, and if different are mostly replaced with closely related species. The fisheries are similar, except that very large long-range vessels have played a much more important role in the western Atlantic Ocean, and the research (and until recently the management) has been co-ordinated and stimulated through an active international body. The mode of operation of the International Commission for the Northwest Atlantic Fisheries (ICNAF) differ in detail from those of ICES (in respect of research, and the provision of scientific advice to managers) and NEAFC (in respect of management), but the broad pattern was very similar. ICNAF documents, and especially the reports of the Standing Committee on Research and Statistics (STACRES), provide an excellent guide to the fisheries and the state of exploitation of the stocks which is invaluable to those familiar with, and with ready access to, this material. Like the reports of the ICES working groups and ACFM, STACRES material is not easily available to the general scientific community.

#### *The stocks*

The region considered here extends from the polar regions southward to 35° N and from the shores of North America eastward to the southern tip of Greenland. In the north, the shelf (off Labrador and West Greenland) is moderately narrow, but further south off Newfoundland, Nova Scotia, and New England it widens to give some of the larger areas of shelf in the world. The southeastern tip of the Grand Banks of Newfoundland, and the isolated Flemish Cap westward, are among the few areas of shelf that lie more than 200 nautical miles seaward of the accepted baselines. This central area also receives the benefit of mixing of warm water from the northwestern flank of the Gulf Stream, and cold waters of the Labrador Current. The area is therefore highly productive, and has for centuries supported large fisheries.

The main species in the northern half of the region—from Nova Scotia northwards—are the same as in the northeast Atlantic Ocean, and indeed there are some common stocks. Cod from southwest Greenland migrate round Cape Farewell to spawn at Iceland, and the exceptional individual cod has moved from the North Sea to the Grand Banks (GULLAND and WILLIAMSON, 1962). The Greenland cod stock, including the more northerly group that stays at Greenland, is at the northern edge of cod distribution and is highly sensitive to changes in climate (CUSHING and DICKSON, 1976). Prior to the 1920s only few cod were found there, but following a rise in temperature a major stock was established, which provided annual catches of up to 400 000 tons. In the 1960s the water temperature dropped; since the good 1957 year class, recruitment has been very poor, and annual catches have dropped to 30 to 40 000 tons. Apart from cod, a few other demersal fish (redfish and flatfishes) are caught, but by far the most valuable species is the deep water shrimp *Pandalus borealis*. The coastal waters of West Greenland are also notable as being one of the ocean feeding grounds of salmon from both sides of the Atlantic.

Further south, from Labrador to Nova Scotia, comes the kingdom of the cod. In the scattered communities along the coast of a harsh and generally barren land not only is fishing the principal way of life, but fishing means fishing for cod—so much so that in common parlance 'fish' without further qualification means cod. Only for the inferior species is it necessary to specify what is being talked about—capelin, halibut, etc.

In fact cod makes up only about half the harvestable potential of all commercially useful species, the proportion decreasing further south. A number of more or less discrete cod stocks exist (GARROD, 1977). The biggest is that in Labrador and off the east coast of Newfoundland (Divisions 2G-J and 3K and 3L of ICNAF). Recuitment (at age 2) to this stock is given by GARROD (his Table 18) as some $2 \times 10^9$ individuals annually—nearly twice that of the biggest stock in the eastern Atlantic Ocean. The next biggest, with a potential perhaps a quarter as big, is the Grand Bank stock (Divisions 3N and O). Other stocks are found in the Gulf of St Lawrence, off Nova Scotia, and on Georges Bank. None of these stocks makes such long migrations as the Arcto-Norwegian cod, but the limited migrations, more or less perpendicular to the coast, are economically and socially very important. The inshore movement for feeding in the summer makes the cod available to the various inshore gears of the coastal communities.

Apart from cod the most valuable demersal species in these more northern areas are redfish, grenadiers, and a variety of flatfishes, though none of these approaches cod in importance. Pelagic species are not important off Labrador, but off Newfoundland, especially on the Grand Banks, capelin are abundant. Herring occur in large numbers off southern Newfoundland as well as in the Gulf of St Lawrence.

Further south, off Nova Scotia and New England, some of these northern species, notably cod and herring, also occur; however, they no longer dominate the ecosystem. Among demersal fish, haddock, silver hake, and a variety of flatfishes are all important. Important pelagic species include mackerel, and in the south of the region, menhaden (though this species is better discussed in the section on the western central Atlantic Ocean). Squid, which seem to be increasing, should also, so far as their ecological position is concerned, be included in this group. Other molluscs of the more familiar types—scallops and clams—also figure highly in the statistics, though because they are based on whole live weight (including shells) the weights over-estimate their importance.

*The fisheries*

The history of the fisheries in the northwest Atlantic Ocean is at least as long as the history of European contact with this region. Fishing was undoubtedly important to the coastal Indians, but details are little known. The voyages of early European fishermen, too, are much less well recorded than the voyages of exploration of Columbus, Cabot, and Cartier—traditionally fishermen are secretive about where they get the best catches. Nevertheless, it is clear that even by the first decade of the sixteenth century the riches of the Grand Banks were becoming known, and fishermen were flocking to the area from

Portugal, France, and England (MORISON, 1971). The fishermen soon took to going ashore to salt their fish, and in due course the coastal settlements became more permanent. For the next 400 yr the fisheries in the region were almost entirely for cod, and more specifically for salt cod for Europe and, to a growing extent, for the Caribbean. It was also predominantly a hand-line fishery, either from small boats working directly out from the numerous small fishing ports scattered along the coasts of Newfoundland and Nova Scotia, or from dories (small one-man boats carried on larger sailing schooners of the type so well described by Kipling in 'Captain Courageous'). A pattern gradually emerged of three scales of fishery—local fishing by small boats; medium-range fishing from ports of New England, particularly Gloucester; and long-range fishing by European vessels, particularly from France, Spain, and Portugal.

These patterns of fishing, though definitely of the pre-industrial age, proved very long lasting. In the 1950s Portuguese schooners, with their decks loaded with dories (Great White Fleet) were still sailing, albeit in small numbers, to the Grand Banks and to West Greenland. Some of the small fishing communities of Newfoundland changed even more slowly. Nevertheless, by the second quarter of the twentieth century modern fishing methods, as represented by the otter trawl, were changing the pattern. In the European-based salt-cod fleet trawling steadily took over from the hard conditions of the dory vessels and, except in Portugal, had become the dominant method by the Second World War. However, salting fish is labour-intensive, requiring large (and increasingly expensive) crews even when the fish are caught by trawl; hence, the salt-cod fleets have steadily declined from their peak in the 1950s. The last French salt-cod trawler left service in 1980.

The pattern of inshore fishing has changed even more slowly, particularly along the coasts of Newfoundland. Here the same types of gear—cod traps, hand-lining, etc.—are still continuing, though the catches of cod are now mainly processed in the local freezing plants rather than being dried and salted. The modern, highly mechanized forms of fishing—particularly trawling and purse-seining—grew very slowly in the local fisheries. The chief developments came in the southern half of the region. In New England a North Sea type of trawling became established during the interwar period, principally on haddock on Georges Bank, though cod and yellow-tail flounder and other assorted flatfish also featured in the catch. By volume the biggest fishery was on menhaden in the southern limit of the area, though since this species is only used for fish-meal the value was small.

The modern age of fishing in the northwest Atlantic Ocean really began with the factory trawler. The prototype of this vessel was the British *Fairtry*. These trawlers were never very successful under British conditions of wages and market prices, but the type—a large stern trawler of 1000 gross tons upwards with ability to process and freeze some tens of tons of fish per day—was very successful adopted by the eastern European countries, and also Spain, where crew costs were less.

The freezer trawler, accompanied in the case of the Soviet fleets by a variety of support ships, including large mother-ships which processed the catches of the smaller vessels in the fleet, encouraged the growth of very large fleets which could move easily from species to species, or from ground to ground. While the countries that traditionally had long-range fleets in the northwest Atlantic were very conservative in their taste—the Portuguese in particular sticking so strictly to cod that any incidental catches of halibut (a fish which fetches several times the price of cod on most markets) by their dory vessels were used for bait—the newer fleets were quite happy to make abrupt changes from one species to another.

Some idea of the changes that took place in the last quarter century are given by the summary catches by country (Table 5–7) and by major species (Table 5–8). These figures are taken from the Statistical Bulletins of ICNAF, and its successor NAFO (the Northwest Atlantic Fisheries Organization). These Bulletins provide an invaluable source of data on the fisheries of the northwest Atlantic, including very detailed data on catches and fishing effort. Until recently published data referred to the area as established by the ICNAF Convention—roughly southward as far as Georges Bank. This was divided into five major areas (1: West Greenland; 2: Labrador; 3: Newfoundland; 4 Nova Scotia; 5: New England, including Georges Bank), which in turn were further subdivided so that statistics can be collected and published by areas which match what is known about the stock divisions of the major species. Recently the statistical coverage

Table 5-7

National catches in the northwest Atlantic Ocean (thousand tons) (Compiled from the sources indicated)

| | 1955 | 1960 | 1965 | 1970 | 1975 | 1978 | 1981 |
|---|---|---|---|---|---|---|---|
| Canada (M + Q)† | 364 | 529 | 581 | 754 | 592 | 697 | 1163 (Canada M + Q and N combined) |
| Canada (N)‡ | 294 | 294 | 306 | 471 | 256 | 450 | |
| USA (Areas 1–5) | 504 | 477 | 350 | 273 | 990 (USA Areas 1–5 and Area 6 combined) | 383 | 1236 (USA Areas 1–5 and Area 6 combined) |
| USA (Area 6) | — | — | 719 | 727 | | 775 | |
| Denmark | 60 | 94 | 121 | 64 | 71 | 90 | 115 |
| France | 143 | 151 | 140 | 73 | 40 | 39 | 31 |
| Spain | 161 | 177 | 234 | 276 | 122 | 36 | 29 |
| Portugal | 206 | 185 | 197 | 163 | 100 | 23 | 12 |
| Germany (FR) | 22 | 97 | 181 | 206 | 80 | 52 | 5 |
| Germany (DR) | — | — | 93 | 89 | 113 | 7 | 22 |
| Poland | — | — | 57 | 216 | 188 | 18 | 3 |
| USSR | — | 258 | 886 | 813 | 1167 | 211 | 114 |
| Iceland | 28 | 40 | 9 | — | 16 | — | — |
| Italy | 10 | 2 | — | — | 4 | 7 | 16 |
| Norway | 44 | 38 | 44 | 47 | 53 | 17 | — |
| Romania | — | — | 3 | 10 | 2 | 4 | — |
| UK | 9 | 25 | 56 | 7 | 2 | 1 | — |
| Japan | — | — | — | 38 | 25 | 37 | 24 |
| Coastal states* | (1222)§ | (1394)§ | 2077 | 2366 | 1909 | 2395 | 2514 |
| Traditional countries | 510 | 513 | 571 | 512 | 262 | 98 | 72 |
| Other countries‖ | 113 | 460 | 1329 | 1388 | 1650 | 357 | 185 |
| Total | 1845§ | 2367§ | 3977 | 4266 | 3821 | 2850 | 2771 |

—Data not available.

*Figures include Faroese catches as well as those of Greenland, but exclude French catches landed in St Pierre and Miquelon.

†Catches in Maritime provinces and Quebec.

‡Newfoundland.

§Excluding catches in Area 6.

¶Data for 1955–1978 from ICNAF Statistical Bulletins; 1979 data from FAO *Yearbook of Fishery Statistics*, Vol. 48.

‖Including data not listed separately.

Table 5-8

Landings of major species from the northwest Atlantic Ocean (thousand tons) (Based on FAO, 1981, and ICNAF Statistical Bulletins)

| Species | 1955 | 1960 | 1965 | 1970 | 1975 | 1981 |
|---|---|---|---|---|---|---|
| Cod | 902 | 1134 | 1462 | 1199 | 639 | 588 |
| Haddock | 198 | 159 | 249 | 48 | 29 | 83 |
| Silver hake | 46 | 47 | 383 | 222 | 232 | 61 |
| Redfish | 123 | 288 | 237 | 230 | 216 | 130 |
| Flatfishes | † | † | 209 | 294 | 247 | 215 |
| Capelin | § | § | 7 | 6 | 367 | 39 |
| Herring | 149 | 180 | 266 | 851 | 448 | 224 |
| Mackerel | 16 | 31 | 16 | 229 | 287 | 28 |
| Menhaden | † | † | 232 | 223 | 200 | 263 |
| Other fish‡ | † | † | 398 | 390 | 389 | 219 |
| Crustaceans | † | † | 79 | 100 | 120 | 195 |
| Squids | § | § | 10 | 16 | 70 | 49 |
| Other molluscs* | † | † | 479 | 519 | 519 | 678 |
| Total | † | † | 4027 | 4327 | 3764 | 2771 |

*Includes small quantities of other invertebrates.
†No data available for statistical Area 6. Figures given only for species confined to Areas 1–5.
‡Figures include other pelagic fishes.
§Catches very small, no separate data available.

has included a Southern Area 6, from Georges Bank to 35° N, i.e. to include the whole of the northwest Atlantic Ocean as used in FAO global statistics.

Table 5-7 witnesses the dramatic rise of Soviet catches; in several years around 1975 they exceeded those by either Canada or the United States. The rise in Polish catches, at least relative to the size of the country, were almost as dramatic. In contrast, only Spain—among the traditional countries—managed to increase its catches, while those of France steadily declined.

Table 5-7 also shows, in the data for 1978, some of the great changes that have taken place in the region as a result of the emerging new regime of the sea, and the establishment of 200-mile zones of fisheries jurisdiction. The reasons for this, and implications for the exploitation and management of the resources, are discussed in the following sections. Here it is sufficient to note that foreign fishermen are only allowed to operate within the new EEZs (Exclusive Economic Zones) when they can do so without damage to the stocks, and are under the strict control of the coastal states. The result has been that non-local catches have dropped enormously.

Table 5-7 and Fig. 5-3 show similar data for the major species. Several types of trends can be distinguished. First there are the species caught only by the coastal states, these show no obvious trend. The prime example is the menhaden–though over its entire distribution there have been big changes, as discussed in the section on the western central Atlantic Ocean (p. 962). The molluscs, other than squids, also as a whole show little trend. This is also true of some of the clams taken in the southern area by the United States, but there have been ups and downs in the scallop fishery on Georges Bank, in which both the United States and Canada take part. Then there are the traditional species, such as cod, haddock, redfish, and herring, in which catches built up

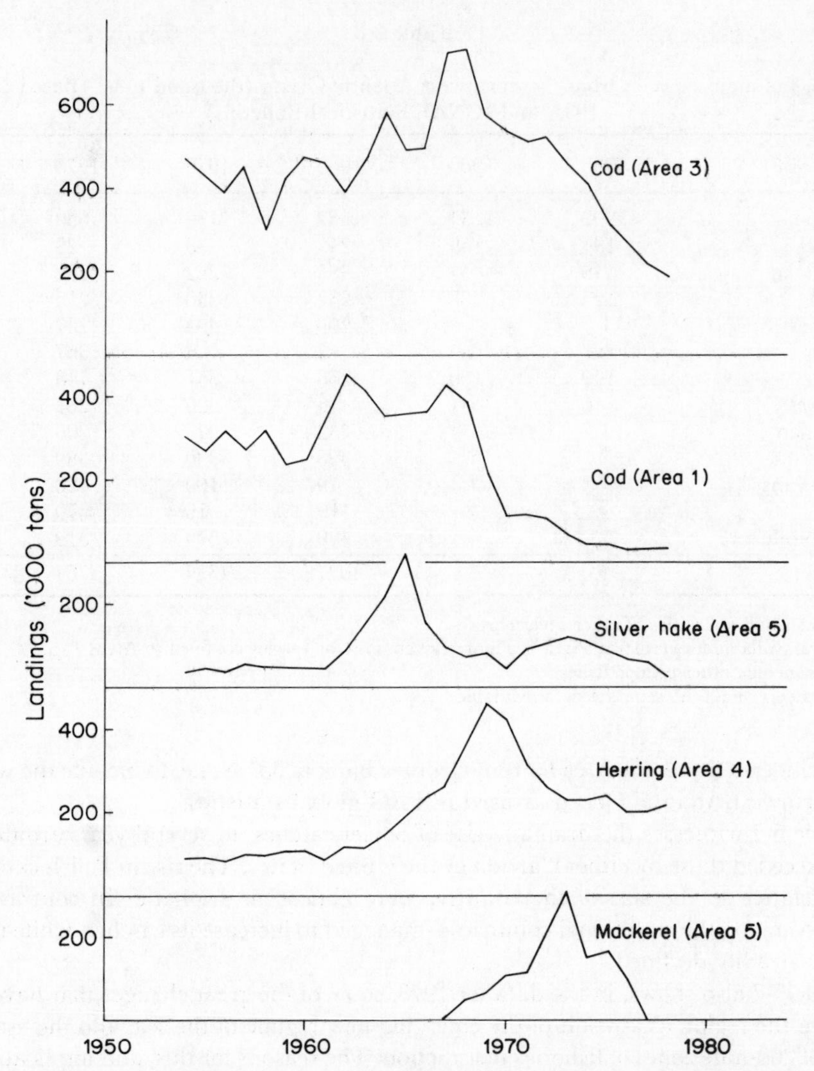


Fig. 5-3: Trends in catch from selected stocks in the northwest Atlantic. (a) cod in Subarea 3 (Newfoundland); (b) cod in Subarea 1 (West Greenland); (c) silver hake in Subarea 5 (New England); (d) herring in Subarea 4 (Nova Scotia); (e) mackerel in Subarea 5. (Based on data from the International Commission for the Northwest Atlantic Fisheries Statistical Bulletins)

from a moderate level, reached a peak, and then declined. This decline has been for two interconnected reasons. In part there was a real, and sometimes severe, decline in the real stock abundance, but there have also been switches away from the species concerned to other more abundant species. Finally, there are the species such as silver hake, mackerel, capelin, and squid which in the 1950s were not exploited, or were only exploited at a moderately low level, and on which new and large-scale fisheries developed. Some of these have not been long-lasting—notably capelin—but others, e.g. squid, have up to the present been able to sustain a high rate of exploitation.

*State of the stocks*

The activities of the long-range fleets, searching out any species or areas where catch rates are high, has meant that all abundant stocks in the region have become heavily fished, as have most of the other demersal species that are caught by the bottom trawl. Even when they have not been the subject of directed fishing, they have been caught incidentally, but in quantities thay may be high relative to their abundance, in fisheries directed at other species. The effects of heavy fishing, which has led to the collapse of some stocks and the partial or almost complete switch of fishing away from others, has led to very great differences between the catches in the peak year of fishing on a stock, and the recent catches.

Some examples of this, mostly for stocks from which the peak catches have exceeded 100 000 catches, are given in Table 5-9. For convenience, and to enable the widest range of data to be used, the data are presented for catches of a species in a particular statistical area. This often corresponds to a stock, but not always exactly. The greatest discrepancy occurs in cod. The identity of the stocks of this species, and their dynamics and state of exploitation, has been reviewed by GARROD (1977). Areas 1 and 5 correspond reasonably well to individual stocks, but the large Labrador stock is caught in both Area 2 (mostly by large, distant water fleets) and in northern Newfoundland (mostly by local vessels), i.e. the data for Area 2 cod are only part of the catches from the Labrador stock. On the other hand, the data for Areas 3 and 4 include catches from several stocks—though the state of exploitation of these stocks has been broadly similar over the years.

Despite the superficial similarity of the events in these stocks, the underlying nature of the dynamics vary. A few stocks have been suffering from little more than 'growth

Table 5-9

Northwest Atlantic Ocean: Catches from selected species and areas in the year in which the greatest catches were taken, compared with present catches (Based on ICNAF Statistical Bulletins)

| Species | Area | Peak year | Peak catches | 1978 catches |
|---|---|---|---|---|
| Haddock | 3 | 1955 | 104 | 1 |
| Redfish | 3 | 1959 | 246 | 68 |
| Cod | 1 | 1962 | 451 | 38 |
| Haddock | 4 | 1965 | 85 | 33 |
| Haddock | 5 | 1965 | 155 | 27 |
| Silver hake | 5 | 1965 | 323 | 27 |
| American plaice | 3 | 1967 | 101 | 58 |
| Cod | 2 | 1968 | 451 | 34 |
| Cod | 3 | 1968 | 736 | 192 |
| Cod | 4 | 1968 | 247 | 164 |
| Herring | 5 | 1968 | 407 | 51 |
| Herring | 4 | 1969 | 459 | 217 |
| Mackerel | 6 | 1971 | 232 | 1 |
| Silver hake | 4 | 1973 | 299 | 48 |
| Mackerel | 5 | 1973 | 315 | 1 |
| Capelin | 2 | 1975 | 145 | 11 |
| Capelin | 3 | 1976 | 266 | 74 |

over-fishing'. These include many of the cod stocks, with the notable exception of West Greenland. In the 1960s and early 1970s these stocks attracted very heavy fishing from long-range fleets. Because of their mobility these fleets tended to concentrate on a stock when it was particularly abundant, for example when one or two outstanding year classes were present. Furthermore, the catches during the initial burst of intense fishing on the generally long-lived species of the region often represented to a large extent the removal of an accumulated stock of a number of year classes. For both these reasons the annual catches during the initial period were usually higher—often considerably higher—than the annual catches that could be maintained from the stock indefinitely, given average year classes, however well the stocks were managed.

Catches from several of these stocks have been lower than the potentially sustainable level in the most recent years for a number of quite different reasons. Once they had become heavily fished, it proved exceedingly difficult during the international open-access conditions prevailing prior to the extension of jurisdiction by the coastal state, for any effective action to be taken to reduce the overall amount of fishing. Management actions were restricted to controls on the mesh size, though, as in the northeast Atlantic Ocean, there has been concern that the effective mesh size may have been lower than that legally in force. Once they had extended jurisdiction the coastal states were in a position to control the amount of fishing to whatever level was held to be desirable. For the Canadian cod stocks this level was clearly felt to be such as to provide high catch rates for the coastal fishermen, i.e. below that giving the maximum sustained yield, and well below the level occurring when jurisdiction was extended. It was possible, however by allocating only small catches to foreign fishermen, to reduce the fishing effort quickly; it takes a much longer time for this reduction to attain its full effect on the stock. During the transition period, while the stocks are being rebuilt to the desired level, catches are bound to be low. The consequent drop has been particularly abrupt in the case of cod in Area 2 because most of the Canadian inshore catches from the stock concerned are taken along the northern coast of Newfoundland, outside Area 2.

In other stocks the role of over-fishing has been much less clear, and in some has probably been no more than marginal to the role of natural factors. The decline in the West Greenland stocks, due almost certainly to a fall in water temperatures, has already been noted. Another stock whose decline is probably due to natural causes is the haddock in Subarea 3. This supported major fisheries (principally by Canadia and Spanish trawlers) from 1945 onwards. The peak catches in 1955 were due to the extremely good year classes, but until 1956 there were other average to good year classes which maintained the stock. However, since the very successful 1955 year class and the good 1956 year class, all year classes have been poor. The cause of this is unknown, but the result has been the virtual disappearance of the fishery.

In other stocks there have been declines or collapses of recruitment which might also be due to natural factors, but for which the impact of excessive fishing can also be held responsible to a greater or lesser extent. One example is the Georges Bank haddock. On superficial examination the events in this stock provide a clear case of recruitment over-fishing. Prior to 1965 this stock had been exploited almost exclusively by the United States with some fishing by Canada. Though there had been concern that the stock was being over-fished on a yield-per-recruit basis (i.e. suffering from 'growth over-fishing'), recruitment, though variable, had held up and catches had remained around 50 000 tons or a little less for a long period. In 1965 the USSR entered the fishery, and in not much more than a 12-mo period took well over 100 000 tons. Total catches (including the USSR) were 155 000 tons in 1965 and 127 000 tons in 1966. Since then the stocks and the catches have steadily declined, catches reaching a minimum of only 5000 tons in 1974. The decline in the stocks resulted in ICNAF applying catch quotas to this stock for the 1970 season and onwards, the first stock to be so controlled in the region. The immediate cause of the decline was a failure of recruitment, no good year classes having entered the fishing since that of 1963. Recruitment has lately been rather better, and the stock (and catches from it) are slowly increasing.

While the match between increased fishing and reduced recruitment is strikingly good, the match between abundance of the spawning stock and year class is not particularly good, even though it would be expected to be better since it is closer to the direct cause and effect.

In fact, the initial impetus which disturbed what had been a reasonably stable situation came from recruitment, not from fishing; this was the occurrence of the outstanding 1963 year class. When this first reached commercial size as a 2-year-old fish around the middle of 1965, the haddock stocks on Georges Bank first became attractive to the long-range fleets. When they moved in the abundance was much above normal, and even when they moved out, in the second half of 1966, the abundance was not much below normal. What was abnormal was the run of extremely poor year classes from 1964 until well into the 1970s. Of these the 1964 year class came from a spawning stock that was similar to that which produced the outstanding year class, and it was not till about 1968 (when that year class had been fished out) that the poor recruitment could be unequivocally matched to a very poor adult stock. The exact cause of the poor year classes is therefore far from certain, but until it is better understood it would be sensible to be cautious and to maintain a substantial spawning stock.

Similar remarks apply to herring and, more especially, mackerel where also low recruitment has contributed to low catches. The recruitment of mackerel in particular has been highly variable, even in the absence of fishing. Also for these species the possibility of interspecific factors having a strong effect on recruitment has been examined (e.g. LETT and KOHLER 1976). The variance in estimates of some of the parameters, e.g. the quantity of fish larvae eaten by an adult of another species, which have to be incorporated in most of the models, make it difficult to confirm or reject them through direct observation, or to use them to make quantitative predictions. However, the estimates do give opportunities for some qualitative insight into the likely interactions between species.

These interactions, and the need to look at all the commercial species together, have been given particular attention in the southern part of the area, notably on Georges Bank. The need originally arose from the operational interactions between fisheries on different species. When total catch quotas began to be applied in the early 1970s, first for haddock and then for other demersal species, the incidental catches of the species being managed by vessels targeting on other species raised big problems. Some by-catch is inevitable, and prohibiting the landing of any incidentally caught fish would only have meant that they would have been discarded, almost entirely dead, with little conservation advantage. ICNAF, in common with many other regulatory bodies, used to allow a small percentage of protected species (often 10%) in the landings of vessels primarily fishing for other species. However, where the 'protected' species is scarce, and the other species common, even a small percentage of incidental catch can account for a large percentage of the allowable catch of the protected species. ICNAF, in addition to looking at the problems of setting Total Allowable Catches (TACs) for a set of individual species with interacting fisheries (e.g. FUKUDA, 1976) therefore looked at alternative strategies, including the control of the total catch or total amount of fishing. Another reason for doing this was that such a control would tend to put some limit on the exploitation of species which became the new target of the mobile fleets in the early years of such exploitation, until such time as direct assessments of the stock, and calculations of a TAC, could be made.

Though the first impetus to look at the total stocks came from operational reasons, it was also recognized that the biological interactions provided another and more basic reason for looking at the total biomass. Fortunately, unlike the situation in the North Sea, a data base existed on the results from trawl surveys (CLARK and BROWN, 1977) which was independent of the commercial fishery data and the latter's biasses due to changes in species composition. Though there are difficulties in using the survey data directly because of differential availability of the various species to the gear, it did enable estimates of total biomass of all commercial fishes to be obtained. Together with the statistics of total catches, these were incorporated in a SCHAEFER-type analysis (see Fig. 5-4 taken from BROWN and co-authors, 1976). On the basis of analyses of this type, indicating full exploitation of the total biomass, ICNAF introduced the so-called second tier system of TACs. This set a limit on the total catch of all species; when this was reached (or in practice when the total national catch reached the national allocation of the second tier TAC), all fishing had to stop, even though some or all of the species TACs (or the national allocation) had not been reached. This system was introduced in 1974, and represented one of the first attempts to manage a multi-species fishery in a manner that gave some recognition of the interaction between species (FAO, 1978). It may be noted that the second tier TAC was usually less than the sum of the species TACs (e.g. 924 000 tons in 1974 compared with a species total of 1 056 000 tons).

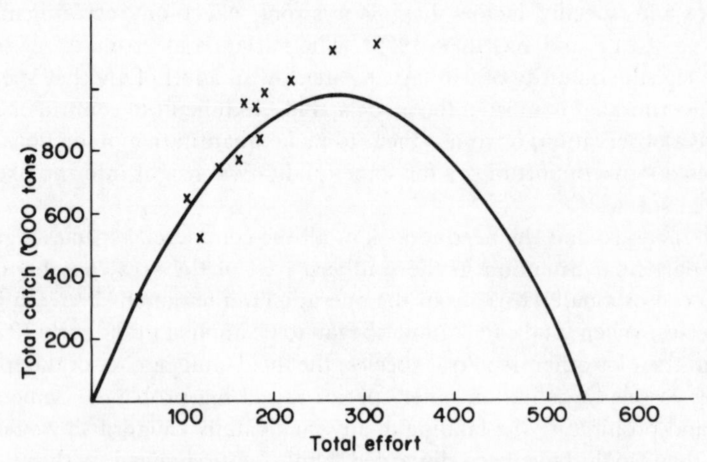


Fig. 5-4: Yield–effort curve for the total fish biomass in the southern part of the ICNAF area. (Points are actual observations, but the curve is fitted to the mean effort over 3 yr). (After BROWN and co-authors, 1976; reproduced by permission of the Northwest Atlantic Fisheries Organization)

*Future prospects*

The northwest Atlantic Ocean, with the establishment of ICNAF in 1949, saw one of the earliest attempts to manage large international fisheries under the old pattern of the Law of the Sea. ICNAF achieved some successes, notably in the compilation of statistics, co-ordination of scientific research, and in the consideration of some of the general problems of management, including the incorporation of economic considerations (e.g. Templeman and Gulland, 1965; ICNAF, 1968). While the success of the Commission in the new era of extended jurisdiction. These attempts have been most significant in the ultimately acceptable to the coastal states, it was by no means negligible. The implementation of mesh regulations in the 1960s, and particularly the setting of TACs, initially for Georges Bank haddock in 1974 and then for all the most important stocks, was as much as has been achieved by many national governments in respect of stocks under their sole control. The TACs usually corresponded to fairly high levels of fishing effort, though here again ICNAF made a procedural advance of quite general value by using the value of fishing mortality of $F_{0.1}$ (Gulland and Boerema, 1973) to provide an objective value which is closer to what is likely to be socially or economically desirable than that giving the maximum yield (or the maximum yield per recruit, which is more usually the one that can be calculated).

The northeast Atlantic is now seeing some of the major attempts to manage fisheries in the new era of extended jurisdiction. These attempts have been most significant in the United States, where the extension of jurisdiction was implemented by the comprehensive Fishery Conservation and Management Act of 1976. In addition to establishing a fishery conservation zone, out to 200 miles from the base line (in Title I), and establishing the legal mechanisms for control over any foreign fishing in the zone (in Title II) the Act, in Title III, sets out in considerable detail a national fishery management programme, concerned essentially with the control of fishing by US nationals. The main sections of this table set out the general objective—to achieve the optimum yield, defined earlier as the amount of fish that 'A, which will provide the greatest overall benefit to the Nation . . .' and 'B, which is prescribed as such on the basis of the maximum sustainable yield . . . as modified by any relevant social and economic factors'—established regional Councils with general responsibility for overseeing management, and set out what should be contained in fishery management plans for various fisheries.

It is too early yet to see exactly how this ambitious undertaking, which involves all the waters surrounding the United States from Alaska and the Pacific islands to New England and Gulf of Mexico, will work out. As in any attempt to design almost from scratch machinery to deal with a very complex situation in new circumstances, it is certain that there will have to be modifications in the light of experience. On the other hand, the major principles—that fishery management should be based on broad objectives encompassing more than the maximization of the physical yield, that all major interest groups should play a part in determining the management policy, and that the policy should be implemented through clearly enunciated plans—will surely stand. Some changes can be foreseen, at least in general terms. For example, the definitions of optimum quoted above do not seem to be entirely compatible, and it is probable that Definition A, reflecting broad national interests, may tend to be given greater weight than Definition B, which reflects the wording in the draft texts of the new Law of the Sea. Again, the detailed working of the regional management councils, the division of responsibilities between them and the US National Marine Fishery Service, and the preparation of management plans, has proved to be cumbersome, leading to slow reactions and adjustments of management measures to changing conditions in the sea, e.g. the occurrence of an unexpectedly good (or bad) year class.

So far as the other major coastal state in the region is concerned, there has been much less basic upheaval in the way Canadian fisheries are managed. Nevertheless the impacts on the fisheries have been much the same. The first obvious result was a great decrease in the amount of foreign fishing (Table 5-7). Such foreign fishing as remained was very much the exception—mostly catches from the Flemish Cap, and the extreme tip of the Grand Banks which lie beyond the 200-mile limit, or catches of species such as squid for which there is so far no established local fisheries, but some small catches from the stocks which the coastal states cannot fully utilize.

This was followed by a period of euphoria among the local fishermen, partly based on a feeling that all their troubles had been caused by foreign fishing, and had thus been removed, and partly on the sounder basis of direct experience of rising catches. It soon became clear that to achieve long-term success, fishery management had to include control of local fishing. This is proving much more difficult, and it is too early to see how successful the coastal states will be. By the minimum, but still essential, criterion of the maintenance of the biological productivity of the stocks the future looks good. The coastal states have sufficient authority, as well as the necessary scientific information, for there to be some confidence that the stocks now being re-built will in future be maintained at a productive level. What is not so clear is whether management will be successful in realizing 'the greatest overall benefit to the nation'—whether the nation is the United States, Canada, or Greenland. This is particularly the case if the costs of management (including the costs of research, administration, and enforcement) are taken into account.

### *Mediterranean Sea and Black Sea*

#### *The stocks*

Though the Mediterranean Sea and the Black Sea have long coastlines, especially along the northern shore of the Mediterranean Sea, the shelf is generally narrow. Only in the northern part of the Adriatic Sea and in the Black Sea are there moderately large areas of shelf favourable to commercial fisheries. The hydrographic conditions are also unfavourable to high levels of production. Being nearly completely enclosed, the Mediterranean lacks, in general, the strong currents and good vertical mixing that renews the nutrients in the surface waters. There is an inward flow at the surface through the Straits of Gibraltar (which is balanced by an outward flow at greater depths) which moving along the North African coasts makes these waters rather more productive than the European coasts, though this is in part balanced by the greater runoff of nutrients from the land in the north.

The Black Sea is even more isolated, with just a little exchange of water through the Bosphorus. The sea is highly stratified, with very low salinity water (16 to 18‰ S) overlying moderate salinity water (21 to 25‰ S) often deficient in oxygen and containing high concentrations of hydrogen sulfide.

The result is that the abundance of fish in the Mediterranean Sea is not high, with the shoaling pelagic fish being rather more abundant than the demersal fish. The species composition is similar to that of the adjacent part of the open Atlantic Ocean, though a few Indo-Pacific species have entered the eastern Mediterranean following the opening of the Suez Canal (BEN-TUVIA, 1966). Among the pelagic species, sardines and anchovy are most abundant throughout the region, though mackerel and horse mackerel are also important. In the eastern Mediterranean, sardinellas (both *Sardinella maderensis* and *S. aurita*) also occur. A large number of species occur in the demersal community, including a variety of cephalopods. No one species is particularly important throughout the region, though local fisheries may contain high percentages of one or two species. For example, hake is important in several parts of the northeastern part (around the Balearics and in the Gulf of Lyons).

Only a few species have been successful in the unusual conditions of the Black Sea, but some of those that have, occur in large quantities. The Black Sea anchovy stock is probably the most abundant stock in the whole region.

*The fisheries*

The countries bordering the northern Mediterranean Sea have a high demand for fish. Until recently this demand arose largely because meat was scarce and expensive, and fish was a cheap supply of animal protein. In addition to encouraging local fisheries this resulted in large imports of salt cod and herring from the richer waters further north. Now the demand is as much or more for fish as a part (and increasingly an expensive part) of a day or a holiday by the sea. This demand is for fresh fish, even though it may not be as fresh as the closeness of the restaurant to the sea might suggest. Despite the relative scarcity of fish, the fisheries of the Mediterranean Sea have therefore been well developed. Because of the high prices reached in the European countries the value of Mediterranean fish is among the highest of the regions of the world, even though the weight is relatively low (FAO, 1981).

The fisheries are mostly local, involving small boats making short trips, usually of a day. If it is worth making longer trips, it is generally—especially since the advent of freezing at sea—worth making really long trips into the Atlantic Ocean to the much richer ground off northwest Africa, or in the northwest Atlantic. The exception is in Sicily. From here the relatively wide and lightly exploited shelf areas of Tunisia and Libya are within reach of short voyages of a few days. Sicilian ports, particularly Mazara del Vallo, have thus developed the biggest fleet fishing within the Mediterranean Sea comprising hundreds of trawlers of up to 250 tons and 400 to 500 hp. Similarly, a proportion of the Spanish fleet, especially those based in Malaga and Almeira, fish along the Mediterranean coast of Morocco and Algeria. Elsewhere the typical Mediterranean vessels are small trawlers for demersal fish, or similar size vessels using surrounding nets for small pelagic fish, usually with lights. Among other types of gear mention should be made of two spectacular types used—though now to a much decreased extent—for large pelagic fish. Enormous traps—madragues—were used along the coasts of Spain, North Africa, and Sicily to catch large bluefin tuna as they passed along the coast on their spawning migrations. Also, in Sicily a special vessel with one extremely high mast with a crows-nest (to spot the fish), and an equally long bowsprit projecting well forward of the vessel (on which the harpooner stood) is used to harpoon swordfish when they lie quietly on the surface on calm warm days.

Summary statistics of catches in 1978, according to the main species groups and the major areas into which the Mediterranean Sea has been divided by the General Fishery Council for the Mediterranean (GFCM) are given in Table 5-10. Apart from some

Table 5-10

Mediterranean Sea: 1978 catches (thousand tons) by area and main species groups (Based on Statistical Bulletins of the GFCM)

| Catch content | Balearics‡ | Gulf of Lions | Sardinia | Adriatic | Ionian Sea¶ | Levant | Aegean | Black Sea | Total |
|---|---|---|---|---|---|---|---|---|---|
| Diadromous fish | 0·1 | 4·5 | 0·9 | 0·9 | 0·4 | + | — | 1·6 | 8·3 |
| Flatfish | 1·1 | 0·6 | 0·8 | 2·8 | 1·7 | 0·7 | 1·1 | 2·3 | 11·0 |
| Gadoids* | 11·4 | 2·2 | 5·7 | 4·3 | 7·8 | 0·2 | 2·4 | 5·7 | 39·6 |
| Breams | 15·2 | 1·6 | 5·2 | 2·9 | 11·2 | 3·8 | 6·1 | — | 45·9 |
| Other demersal fish | 9·2 | 6·5 | 6·4 | 8·3 | 17·4 | 5·7 | 11·2 | 3·0 | 67·7 |
| Mullet | 0·1 | 1·3 | 2·1 | 3·1 | 3·3 | 1·6 | 2·6 | 1·4 | 15·6 |
| Horse mackerel | 20·1 | 0·6 | 2·1 | 2·8 | 5·7 | 0·4 | 7·4 | 21·6 | 60·5 |
| Anchovy | 37·1 | 2·1 | 12·1 | 28·6 | 16·7 | — | 7·2 | 239·1 | 342·8 |
| Clupeoids† | 78·2 | 11·4 | 13·1 | 59·9 | 17·4 | 5·1 | 12·4 | 116·0 | 313·5 |
| Mackerel | 3·8 | 0·9 | 1·5 | 0·6 | 2·6 | 0·5 | 1·8 | + | 11·7 |
| Other pelagic fish | 0·1 | + | 0·6 | 0·2 | 0·3 | 0·7 | 0·3 | 3·6 | 5·9 |
| Tunas | | | | Not specified by area | | | | | 10·2 |
| Sharks and rays | 0·7 | 0·3 | 1·7 | 1·3 | 4·0 | 0·3 | — | 2·7 | 11·1 |
| Shrimp | 3·7 | + | 1·1 | 0·2 | 7·8 | 1·1 | 0·9 | 0·4 | 15·0 |
| Other crustacea | 1·5 | + | 2·3 | 4·7 | 3·7 | 0·2 | 0·7 | — | 13·0 |
| Cephalopods | 0·8 | 1·6 | 6·7 | 9·1 | 17·2 | 0·7 | 2·0 | + | 45·5 |
| Other molluscs | 3·8 | 5·5 | 3·1 | 11·1 | 4·3 | — | 0·6 | 8·2 | 36·6 |
| Unspecified fish | 19·4 | — | 9·4 | 52·7 | 24·9 | 4·9 | 11·8 | 8·8 | 131·9 |
| Total | 213·6 | 39·3 | 74·5 | 193·4 | 146·3 | 26·1 | 68·5 | 414·4 | 1200·9 |

*Mostly hake.
†Mostly sardine, but includes sardinella, and some anchovy not explicitly identified.
‡Includes north African waters as far east as the Algerian/Tunisian border.
§Waters north of a line from the Gargano peninsula to the Yugoslav/Albania border.
¶Includes north African waters from Cape Bon to 25° E.

fluctuations in the catches of anchovy in the Black Sea (which reached a peak of over 350 000 tons in 1978), the figures in the table are close to the catches in any year over the past decade.

Table 5-10 shows that the catches are not evenly divided between the regions, even allowing for the somewhat different sizes. Almost as much comes from the small area of the central and northern Adriatic Sea as from any other area except the Black Sea (this can be mainly explained by the relatively wide shelf area), whereas very little is caught in the eastern Mediterranean (Levant and Aegean areas). This distribution is also shown by the national catches. Over a quarter of the total catch (around 350 000 tons) is taken by Italy. This is closely followed, especially in years when the Black Sea anchovy catch is good, by catches in the Black Sea by the USSR. Other countries exceeding 100 000 tons are Spain and Turkey (almost wholly in the Black Sea).

*State of the stocks*

There is a long history of marine biological research in the Mediterranean Sea. In addition to the scientists from the coastal countries, scientists from northern Europe have felt the attraction of the Mediterranean sun and sea, and have worked at institutes like the Biological Station at Naples. However, relatively little attention has been paid to the quantitative study of the fish stocks. Recently, more attention has been given to resource assessment, particularly through GFCM and the activities of its Working Party on Resources Appraisal and Fishery Statistics (TROADEC, 1979). The reports of this working party (e.g. GFCM, 1978) and also of its subsidiary groups that have looked at individual areas in detail (e.g. GFCM, 1980) contain convenient summaries of the results, as well as entries to the literature describing national studies in more detail.

The basic material for making assessments has not been very good, particularly as regards statistics of catch and corresponding fishing effort. As shown in Table 5-10, even in 1978 much of the reported catches (over a quarter in the important Adriatic region) consists of unspecified fish, without definite indication of even whether they were demersal or pelagic. The situation is still worse as regards effort data. Figures on numbers of voyages made, or days at sea by vessels of different categories, with their corresponding catches, which are now supplied as a matter of routine in several other parts of the world, have until recently been lacking in most Mediterranean countries. Action is being taken in most countries to rectify this. Similarly, only a limited number of estimates of the vital parameters of growth and mortality (especially the division of total mortality between that due to fishing and to natural causes) are available. The best-known stocks are those in the western Mediterranean, where Spanish research workers have been particularly active, but even there the data are far from complete (see Tables 6.1 to 6.7 of GFCM, 1980, which show what estimates are available for the main species by subarea). Most assessments have therefore been done by application of simple production or Schaefer-type models, though some yield-per-recruit calculations have been made. Also, estimates of standing stock are available for some pelagic stocks from acoustic surveys, for demersal fish from trawl surveys (JUKIC and PICCINETTI, 1979), and for some adult stocks from surveys of eggs and larvae (PICCINETTI and co-authors, 1979).

These estimates give a reasonably consistent picture of the degree of exploitation decreasing from west to east, and from north to south. Exploitation of demersal stocks is also much heavier than that of pelagic stocks. For demersal fish off the coasts of Spain, France, and Italy the amount of fishing has passed the level giving the maximum sustained yield in the 1970s. In general, the total yield has decreased little, if at all (i.e. the stocks appear to be suffering only from 'growth over-fishing'), but an exception is the turbot in the Black Sea. Here there has been a serious decline in recruitment, and hence in total catch, which appears to be related to a reduction in adult stock, i.e. there is 'recruitment over-fishing' (GFCM, 1979). The other demersal stocks, as well as the pelagic stocks (except possibly anchovy) in the northwestern part of the Mediterranean Sea, are also at least moderately heavily fished, so not much increase in yield can be expected. The main opportunities for any increase comes from the pelagic stocks along the North African coast, and in the eastern Mediterranean (though the potential of the latter area is probably small).

Apart from the direct impact of fishing, at least one fishery seems to have been affected by other human activities. This is the Egyptian sardinella fishery (SHAHEEN, 1976). Prior to the construction of the Aswan High Dam there was a substantial fishery (mostly in the period September–November, towards the end of the period of annual Nile flood), which took some 7000 tons of sardinella annually. After 1965 this declined, reaching less than 500 tons in 1968. It is, however, not known whether this represents a real decrease in the stock, or merely the failure of the stock to move into the region off the Nile delta when this area ceased to be so attractive.

*Future prospects and management*

Given the state of exploitation of the stocks the future well-being of the fisheries depend on how well they are managed. For most of the demersal stocks this means a reduction in the amount of fishing (or at least a halt to the current increasing trend), and better protection of small fish, e.g. through increasing the mesh size in the trawls. Several countries have had for some time legislation setting a minimum mesh size—mostly at 40 mm—and others are introducing similar legislation, as a result of discussion and recommendations of GFCM. However, as these discussions have shown, legislation by itself does not improve matters unless it is properly enforced, and enforcement has been weak in most Mediterranean countries. This is partly due to the past lack of communication between fishermen, administration, and scientists, which also resulted in lack of research directed towards fishery problems. Thus, neither the administration nor the fishermen understand what benefits would arise from protecting the smaller fish or reducing the amount of fishing. Equally, the fishermen do not understand why certain measures have been introduced, nor do they accept them as being necessary. The difficulties of enforcement are enhanced by the fact that most of the fishing is carried out from a large number of small and scattered fishing ports, rather than the few industrial ports of northern Europe, where a few full-time enforcement officers can have some chance of controlling most of the total national catch.

This situation is now well appreciated by GFCM and its member countries. Recent recommendations of GFCM put as much weight on explaining to fishermen why a minimum mesh size is to their advantage, as to the formal steps of legislation and legal enforcement. It can be hoped that in a few years this will result in the general use of a larger mesh of at least 40 mm (possibly larger) and some increase in the weight (and still more in value) of demersal landings. As always this will then increase the incentive to add more vessels, so that the benefits will be dissipated unless there is control on the total fishing effort. Though this control is more difficult than that on mesh size, several Mediterranean countries have already gone far in adopting some form of effort control, and limitation of entry (PEARSE, 1980).

For the pelagic species the current need is, with some exceptions (e.g. sardines in parts of the western Mediterranean), for expansion of effort. Several countries have programmes for developing this, including campaigns to increase public demand for these types of fish. If successful, this would also help divert demand and hence effort from demersal species. Though attempts to change public taste are often unsuccessful, and this may well be the case in the European countries, the plans for developing the pelagic fisheries of the north African countries may be more successful. Optimistically, therefore, in the future the Mediterranean fisheries may achieve a small increase in total catch, due to better managed demersal stocks and more fully exploited pelagic stocks. This increase need involve little increase in total costs.

## *Northeast Pacific Ocean*

### *The stocks*

The region is divided into two dissimilar parts by the Aleutian Peninsula, and the chain of islands extending from it. To the north lies the wide shelf of the eastern Bering Sea—at over half a million square kilometres one of the largest expanses of shelf in the world. South of the Alaskan chain the shelf is narrow, lying off mountainous country, with very broken coastlines. The fish stocks reflect these differences, with the more strictly demersal fish being most abundant in the Bering Sea. In general the species composition is analogous to that in the North Atlantic Ocean: flatfishes, gadoids (notably Alaska pollock), and rockfishes (related to the redfishes) are most important among the demersal, herring and capelin among the pelagic fishes. The northeast Pacific Ocean, particularly its eastern side, differs from the North Atlantic Ocean in the abundance of salmon and marine mammals.

The Pacific Ocean, has, in the steelhead, a fish that in behaviour and in status among sports fishermen is very close to the Atlantic salmon. In general importance, though, this species is dwarfed by the five species of Pacific salmon (*Oncorhynchus*) (excluding the Masou salmon, *O. masou*, which is confined to the Asian side). These have, at least to the locally based fishermen, always been of much greater interest than pelagic or demersal fish. Much research has been carried out on salmon—summarized, *inter alia*, by LARKIN (1977a) and FOERSTER (1968). All salmon species are born in fresh water; they move to the sea when small and achieve nearly all their growth in the sea (mostly in the open Pacific Ocean, though some fish stay in coastal waters); when fully grown they finally return to their place of birth, where they spawn. Unlike Atlantic salmon, all Pacific salmon die after spawning. The species differ chiefly in the location and duration of their freshwater life. Pink salmon spawn at the mouth of the rivers or streams, and move to the sea as fry; sockeye salmon move from the spawning streams to lakes, where they may spend more than a year before migrating to the sea. Pink salmon also spend the least time at sea—two years—a period so well fixed that runs in odd and even years operate almost as independent stocks, usually with one or other being strongly dominant.

A large variety of marine mammals, from whales to sea otters, occur in the region (SCAMMON, 1874), some of them, notably the fur seal, in very large numbers. The larger baleen whales and the exploitation and management of these whales are to a large extent independent of the fisheries, but this is not so for some of the other mammals. Estimates of the consumption of finfish in the Bering Sea (1·3 million tons by pinnipeds, and 1·1 million tons by toothed whales, of which over 1 million tons is pollock) exceed the total catch by fishermen, except in a few exceptional years (LAEVASTU and FAVORITE, 1977).

Clearly this consumption, and any major change in it, must have some impact on the species consumed, and on the fisheries harvesting them.

*The fisheries*

Until the arrival of the Europeans, the human population along the Pacific coast of North America was small. Salmon and other fish were well known to the local Indians all along the coast. In the north, hunting whales and other marine mammals had been, and still remains, an important part of the Eskimo way of life—now raising difficult problems in the management of the bowhead whales. The total catches were small and had little impact on the stocks. This changed when the white man arrived. Among the first were fur hunters from Russia and from eastern North America. They found riches in the smaller marine mammals, particularly the fur seal and the sea otter. Both were depleted during the nineteenth century, the latter close to extinction. Since then marine mammals have ceased to be of more than minor direct economic importance, but their management is of scientific and general public interest, and as some of their populations recover their interactions with fisheries are receiving increasing attention.

Large-scale exploitation of fish started with salmon towards the end of the nineteenth century. Catching salmon is easy, so that without controls most of the run to a given river can be caught. The precise homing of the salmon and the comparatively close link between the amount of spawning stock and subsequent recruitment (RICKER, 1954) mean that the effects of taking too much from a run show up clearly 2 to 4 yr later (depending on the species concerned). The need for controls has therefore been accepted from early times, and soon after the access to large outside markets (at first principally in the form of canned fish, but now increasingly as frozen fish) encouraged large-scale fishing, the form of the salmon fisheries has been determined mostly by competition between fishermen and those setting regulations. The most efficient methods (traps, weirs, fish wheels) which operated in the rivers have long been prohibited. Current methods—purse-seines, gill-nets and trolling—are mostly used in the coastal waters in the approaches to the spawning rivers. The Japanese developed a large gill-net fishery for feeding in the open Pacific Ocean, which reached its peak in the 1940s. After the Second World War it was restricted to waters east of approximately 175° W; at the time this was expected to eliminate catches of fish born in North American streams. Later studies (e.g. ROYCE and co-authors, 1968) show that the oceanic movements of salmon are more wide-ranging than was believed, and the Japanese fishery did take numbers of American salmon, particularly from fish coming from Bristol Bay in Alaska. In addition to commercial fishing, sports fishing from central British Columbia southwards has considerable political influence. Especially in the states of Washington and Oregon, it takes a significant proportion of the total catch of the preferred species (particularly chinook). Finally, the catches by native Indians are still significant; they have achieved much more political importance as a result of legal decisions in the United States.

The other long-established fishery by the coastal states is the long-line fishery for halibut by the United States and Canada. This fish grows to a great size; together with its long life and high value this has made it very vulnerable to growth over-fishing (as opposed to the recruitment over-fishing of salmon). The need for management of the halibut fishery has been long recognized (e.g. THOMPSON, 1950). Because the halibut moves actively along the coast, and because most of the fishing, until the change in the general legal regime of the sea, was carried out beyond the limits of national jurisdiction, this has required joint action by the two countries concerned, and catches have been controlled by the International Pacific Halibut Commission since 1933.

The other demersal species supported some small local fishing, but large-scale catching did not start until the distant-water fleets became interested in the area. The pattern of 'pulse-fishing'—i.e. rapid development of fishing on one species, high catches for a few years, and then a switch to another species as the abundance of the first one declined—already noted in the northwest Atlantic Ocean has been particularly marked in this region. Earliest attention was given to yellowfin flounder and other flatfish in the Bering Sea, of which catches (almost entirely by Japan) reached over half a million tons in 1961, falling to under 100 000 tons in 1965. Attention of Japanese, and now also USSR, vessels then turned to Pacific Ocean perch in both the Bering Sea and the Gulf of Alaska; catches of this species approached half a million tons in 1965, but then also fell. Since 1966 the chief attention of the long-range fleets has been concentrated on Alaska pollock, mainly in the Bering Sea. Catches of this species passed one million tons in each year from 1974 to 1976, but following the extension of jurisdiction over fisheries, lower limits on total catch have been set by the coastal state. Long-distance fleets also fish, on a smaller scale, on hake along the Washington—Oregon coast.

Pelagic fish have not in general been high enough in price to attract local fishing, or abundant enough to attract distant-water fisheries. The exception is the herring in the Washington–British Columbia region. This supported a medium-sized fishery, mostly for meal, around Vancouver Island, but collapsed, apparently from heavy fishing, in the late 1960s. Following (and quite possibly as a result of) strict controls, the population has recovered (HOURSTON, 1980). At the same time a market was established for the export of roe to Japan. The very high prices for roe meant that the value of the small catches allowed was many times that of the larger catches taken earlier for meal and oil, leading, at least in the short run, to great satisfaction of the fishermen with the management programme.

A number of crustaceans are valuable enough to attract local fishing. The most important are those for the very large king crab and snow crab in the Bering Sea (though the fisheries on these were first developed by the Japanese), but there is also a highly localized and very intense shrimp fishery round Kodiak. Catches of Dungeness crab, mostly off Washington and Oregon, are also significant.

Statistics of the catches in 1979 are summarized in Table 5-11. This table shows that there is very little overlap in the species taken by the two coastal states and by the few countries that maintain long-range fleets in the area (Bulgaria and the German Democratic Republic fished briefly around 1975). Though the latter account for some two-thirds of the weight caught, these catches are of the bulk, lower-valued species, and in terms of value, the proportions are at least reversed.

*The state of the stocks*

For the North Pacific Ocean there is no organization like ICES. There is no forum for international review of the fish resources as a whole whose publications could provide ready access to current knowledge of the state of exploitation. However, a number of organizations do this for specific stocks or groups of stocks. The most general of these is the International North Pacific Fisheries Commission (INPFC). Under the so-called abstention principle of this Commission, drawn up in 1952, non-coastal countries, i.e. Japan, should abstain from fishing stocks fully exploited and properly managed—with

Table 5-11

Catches of main species and species groups in the northeast Pacific Ocean in 1979 by countries (thousand tons) (Based on FAO *Yearbook of Fishery Statistics*)

| | Canada | Japan | Korea | Poland | USSR | USA | Total |
|---|---|---|---|---|---|---|---|
| Salmons | 59·9 | — | — | — | — | 243·2 | 303·2* |
| Halibut† | 3·9 | — | — | — | — | 9·6 | 13·5 |
| Other flatfish | 6·3 | 117·4 | 1·7 | + | 53·9 | 24·1 | 203·5 |
| Pacific cod | 9·1 | 36·7 | 3·2 | + | 2·4 | 6·3 | 57·8 |
| Alaska pollock | 3·4 | 619·8 | 108·8 | 40·4 | 33·9 | 3·1 | 809·3 |
| Pacific hake | — | 3·6 | — | 23·0 | 101·1 | 13·9 | 141·6 |
| Pacific Ocean perch | 8·8 | 0·3 | — | + | 0·9 | 3·3 | 13·3 |
| Other demersal | 3·7 | 49·7 | 1·7 | + | 10·1 | 43·5 | 108·7 |
| Pacific herring | 43·5 | 0·4 | — | — | — | 29·8 | 73·6* |
| Dungeness crabs | 1·2 | — | — | — | — | 17·6 | 18·7* |
| King crabs | — | — | — | — | — | 70·1 | 70·1 |
| Snow crabs | — | — | — | — | — | 59·6 | 59·6 |
| Shrimps | — | — | — | — | — | 43·1 | 43·1 |
| Total‡ | 154·4 | 832·4 | 119·7 | 64·0 | 210·3 | 593·3 | 1974·1 |

*Includes small quantitites in eastern central Pacific (off California).
†Excludes incidental catches in trawl fisheries for other species.
‡Includes small quantitites of other species, especially molluscs.

the implication that other stocks could be exploited. This puts the onus on the coastal states to demonstrate, at INPFC sessions, that the stocks they did not want the Japanese to touch—salmon, halibut, and some herring—did in fact fulfil these conditions. The state of certain salmon stocks is also reviewed by the International Pacific Salmon Fisheries Commission, concerned with salmon spawning in the Fraser River, which enters the sea just north of the US/Canadian border. Its drainage basin, though largely in Canada, includes parts of the United States. The studies show that the salmon stocks are fully exploited, and are, by and large, being properly managed at least in biological terms. The state of the stocks, and the success of management, depends on the balance between escapement (number of adult salmon going upstream to spawn, having escaped from the fishery), numbers of juvenile salmon, run of adult salmon approaching the home streams, and catches. The stock-recruitment curve for salmon is dome-shaped (RICKER, 1954). The greatest recruitment, or run, is produced by a fairly high escapement. Under average conditions the greatest sustainable catch (run less the escapement necessary to reproduce the same run) would be taken by maintaining a slightly lower escapement, though one still giving a high run. In the past, many stocks have been fished down to below this level.

Salmon runs have also been affected by other human activities: dams can prevent adult fish moving upstream to spawn, and fry can be killed passing through turbines or in the tail-race; insecticide spraying may kill fry and juveniles; changes in land-use can damage or destroy spawning grounds, etc. The classic example of incidental damage to salmon runs was the rock slide at Hell's Gate on the Fraser River caused by railway construction in 1913. The overall effect of man's activities has been to reduce the total runs of salmon on the northeast Pacific, and catches have fallen accordingly. Catches were around 350 000 tons (1·0–1·5 × $10^8$ fish) between 1920 and 1940 (LARKIN, 1977), but fell to little more than half this in the 1960s, though since then there has been some increase (the 1979 catches, listed in Table 5-11, were the highest for many years).

So long as halibut was caught almost exclusively by the North American long-line fishery, for which it was the target species, its dynamics fitted well with the simple models. The stock, or stocks, had been clearly over-fished by the 1930s, and slowly rebuilt since then to yield somewhere around the maximum sustainable yield of approximately 40 000 tons (THOMPSON, 1950; CHAPMAN and co-authors, 1962). Since the 1960s there has been a steady decrease in total catch, which cannot be explained by changes in the fishery directed at halibut. This is almost certainly connected with the increase in trawling for other species in both Bering Sea and Gulf of Alaska. These fisheries, mostly by Japan and the USSR, certainly take a small proportion of small halibut in their catches, which in total weight exceed the halibut by much more than an order of magnitude. However, there are some doubts, reflected in the reports of INPFC and IPHC, as to the exact magnitude of this incidental catch, and thus as to the extent to which the decline in the directed halibut catch (from 39 000 tons in 1965 to 13 000 tons in 1979) is wholly due to the trawl fisheries.

These fisheries are certainly fully exploiting most of the other demersal fish. This can be seen, at least in a crude way, by the successive collapse of the fishery on yellowfin sole (550 000 tons in 1961, 95 000 tons in 1979) and Pacific Ocean perch (470 000 tons in 1965, 13 000 tons in 1979). The same can be demonstrated more elegantly by analysis of age data, catch rates in the commercial fishery, and—additional information that makes the study of the Bering Sea and Gulf of Alaska easier and more accurate than of other areas—data from surveys carried out before large-scale exploitation took place (ALVERSON and co-authors, 1964; MOISEEV, 1964; HITZ and RATHJEN, 1965). The status of the Alaska pollock is less obvious; there has not been the same dramatic fall in catch rates, the survey data is equivocal because many of the fish are well off the bottom and are not sampled quantitatively by the trawls used; also the Alaska pollock is much shorter lived, particularly than the Pacific Ocean perch, and therefore in some ways less vulnerable to heavy fishing. However, it seems fairly clear that it is close to being fully exploited and not much increase in sustained yield could be expected. The same is true of the other demersal fish.

The status of the stocks of pelagic fish has not been so well studied, with the exception of the herring around Vancouver Island. This stock (or stocks—the exact stock structure is not entirely clear) is probably one of the few examples of a pelagic stock that collapsed and then recovered after protection (MURPHY, 1977). No other significant catches of pelagic fish are taken, though it appears that some stocks are quite substantial.

One source of evidence of the abundance of pelagic stocks comes from the study of the consumption of these species by the higher predators (larger fish, marine mammals, birds). Estimates of food consumption have been a large element of studies of the interaction between species carried out in this region at least as intensively as anywhere else in the world (LAEVASTU and FAVORITE, 1977, and other reports from the same group). This interest in species interaction in part arose from the very different interests of coastal and non-coastal states. Another factor was the high public and political interest in marine mammals, and the concerns on the one hand that the increasing catches of fish would harm marine mammals, and on the other that the recovering stocks of marine mammals would have an impact on species on which they feed, and which also support important commercial fisheries.

This impact is clearest for the sea otter. After being nearly extinct at the beginning of this century, this attractive animal is now re-establishing itself along many parts of the North American coast as far south as California. It is an efficient predator on some of its preferred prey—abalones and sea urchins. Indeed, it is so efficient that sea otters do not leave enough to support an economically viable catch rate for commercial fishermen on these species, especially since they will take prey at a smaller size than is attractive (or even legal) for the fishermen. The picture is made complicated further because the prey of sea otters, at least the sea urchns, are themselves very efficient in grazing down their food, the large kelps. Without sea otters it seems that it is difficult for kelps to establish or maintain large growths, whereas with sea otters there can be dense kelp beds. These in turn provide a favourable habitat for juvenile fish which, when grown, may move offshore to support commercial fisheries. The balance between sea otters and fisheries is not simple, especially as different fishermen may be interested in abalones, sea urchins, and commercial finfish favoured by kelp beds.

*Future prospects and management*

The future of the fisheries in the northeast Pacific Ocean must to a large extent be determined by salmon. Salmon fishing moved comparatively quickly out of the early phase of uncontrolled exploitation—the choice was control or rapid disappearance of the stocks. The last 50 yr or more have seen the establishment of more and more complex controls over the fisheries. They have also seen the growth at a somewhat slower pace and less strictly, of controls over assorted activities in the drainage basins of the spawning streams, aimed at maintaining the biological productivity of the stocks. This has been successfully achieved, though the success has not been 100%—witness the decline in catches between 1940 and 1975. It is now recognized that this achievement is not enough, and attention is being given increasingly to two other aspects: controls that enable greater net economic benefits to be realized, and measures to increase the biological production.

The irrationality of the traditional pattern of salmon management, at least as viewed from the economist's point of view, has been pointed out by many economists, notably CRUTCHFIELD (e.g. CRUTCHFIELD and PONTECORVO, 1969). Until the early 1980s, entry to the salmon fisheries was open to all. Hence, the fishing capacity that could be deployed at the beginning of each season greatly exceeded—by an order of magnitude or more for some stocks—that necessary to harvest a desirable amount of catch. As a result, the biological objective could only be achieved by severe restraints on the fishermen's activities. These took the form of banning the more efficient types of gear, and restricting the times and places where the other gears could be used—often to only 1 or 2 d a week. This had the effect of raising the total costs of the fishery to approximately the value of the catch, while without these constraints they need be no more than perhaps a quarter or less. That is to say, an economic rent from the resource of perhaps three-quarters or more of the gross value of the catch (i.e. some hundreds of millions of dollars), which might have appeared as cheaper fish for the consumer, greater net income for some fishermen, or income to the central government, or to fishery agency (i.e. ultimately to the tax-payer), had been dissipated in building extra boats and gear, additional fuel costs, etc. However, economics are not the only criterion. If the economic rent is to be obtained, many people who would like to fish must be prevented from doing so, either by some form of limited entry, or by discouraging them through a high licence fee; the latter would funnel the rent out of the fishery and into the government. Any practical alternative has met strong opposition on social, political, or even, in the United States, constitutional grounds. Progress has therefore been slow, with some trend of increasing effectiveness from south to north—i.e. from where fishing is only marginal to the regional economy, and where there are vast numbers of amateur or part-time fishermen wishing to go salmon fishing at weekends or in their holidays as much as for a change from city jobs as for the money, to Alaska, where fishing is one of the keys to the state economy. Alaska has now an effective limit on entry, and licences to operate a vessel in one or other of the local salmon fisheries were in 1978 changing hands at some $25 000 (ADASIAK, 1979)—an indication of the extra returns, over and above the normal return on investment in the hardware of vessel and gear, that was expected by these fishermen. Licences in the British Columbia fishery have been changing hands at a similar rate (FRAZER, 1979). These schemes have gone some way to realizing the rent potentially available from the resource, which now go to the present fishermen in the form of increased capital value of their vessels, provided they have a licence. However, they have not gone far in reducing the excess capacity, and there is a long way still to go.

The possibilities of positive enhancement of fish stocks in mitigation of over-exploitation has a long, and not particularly happy, history. In the early years of the twentieth century, soon after it became recognized that a number of preferred species had been depleted in the North Sea and elsewhere, a number of stations were set up to hatch and release fry of cod and other species. It was soon clear that these efforts were not effective; if the fry were released soon after hatching, the few millions produced were a tiny fraction of the natural production; if they were held until they had reached a size at which the greatest incidence of mortality was over, the costs of doing so would exceed the likely benefits. It now seems that Pacific salmon are an exception to this, and that some forms of enhancement are practicable and economically attractive. At the simplest, this involves no more than improving the natural spawning success by providing—sometimes at no more cost than a few hours work with a bulldozer—the most favourable spawning conditions in terms of the nature of the flow, and of the stream bottom. Most efforts have gone into the development of hatcheries in which the fish are raised until a suitable size—usually that at which they would normally migrate to the sea. One reason for this success is undoubtedly the considerable expertise that has grown up over the years in the rearing of trout for stocking streams for anglers and, more recently, for direct marketing. Another reason may be that salmon smolts are adapted to a change in their whole way of life when they leave the rivers or lakes and move to the sea. The change from hatchery to natural conditions may not therefore be an excessive shock. In contrast, some experience of the release of hatchery-reared plaice in the Irish Sea is that for a few days the presence of these inexperienced young fish provides good feeding for the local predators, and that thereafter very few are left.

Whatever the reason, the percentage return of at least the more valuable species (chinook and coho) is sufficiently high so that the value of those fish that return when adult exceeds, often by a good margin, the costs of producing the small smolts. There is therefore a growing amount of salmon being raised by public agencies (state/provincial and federal) for release in order to increase the total supply. The problem with this is that hatchery-reared salmon are not immediately identifiable (apart from a fraction that may be tagged or marked by fin-clipping or other methods) so the benefits are not immediately identifiable. More seriously, the benefits cannot be directly realized by those responsible for the production of smolts. Attention is therefore being given to salmon ranching (THORPE, 1980). This takes advantage of the precise homing of salmon, and the fact that the place to which they return can be determined by imprinting them just before release and movement to the sea to ensure that fish released will return to a given place. This may be a spot or stream with no natural stocks. Here they can be harvested by those releasing them, and by no one else. Apart from the technical problems of raising fry and smolts in the most efficient way, these activities will probably in the long term raise the question of how much salmon the North Pacific Ocean can support. Until now the limits to production have been in freshwater, and the growth and natural mortality in the sea have been treated as being independent of density. Clearly there must be a limit to the amount of food available to salmon in the surface waters of the North Pacific gyre (where most of them seem to feed), vast though it is. Even if consumption by the current salmon stocks is still well short of this limit, there must come a time when it will be approached if the salmon enhancement programmes are successful, and expanded without limit.

Most of the other management problems in the northeast Pacific Ocean are relatively straightforward. Management of the halibut fishery has gone through similar stages to that of salmon, first restoring and assuring the biological production, and tackling the problem of also ensuring the economic well-being of the fishery (CRUTCHFIELD and ZELLNER, 1962). Complications that are being tackled include the impact of large-scale trawling for other species in the Bering Sea and the effect of the new Law of the Sea on balance in the fishery between the United States and Canada (COPES, 1981). Until the new fisheries jurisdiction, long-liners behaved very much like the halibut, moving up and down the coast (outside national territorial waters) more or less at will. This is now not possible. Canadian vessels cannot go into US waters and vice versa. This reduces the efficiency of both sets of vessels, as well as affecting the relative shares of the total catch taken by the two countries.

For some other fish stocks (e.g. cod, sablefish) and, more notably, some of the important crustacean fisheries, the effect of new legal situation has been, as elsewhere, to greatly reduce, if not to eliminate, foreign fishing, and to allow the local fishermen to increase their catches. A problem somewhat special to the northeast Pacific Ocean concerns the large stocks of low-valued species (flounder, Alaska pollock) in the Bering Sea, for which there is little sign of any locally based fishery developing on a scale commensurate with the size of the resource. This foreign fishing is being continued, though under control, and at a rate of exploitation that should maintain the stock at some optimum level. Licence fees are charged, at a few percent of the estimated value of the catch, and do not cover the full costs of the management and enforcement programme, including the necessary support activities, such as research. So far no attempt is being made to enforce a pattern of harvesting and licence fees that would maximize the net return to the coastal state, or indeed to estimate what such a pattern would be (presumably a high licence fee, combined with a low fishing effort, which would give catch rates high enough to cover the fees).

*Northwest Pacific Ocean*

*The stocks*

This region is the most productive in the world, and also one of the most diverse. Though conveniently discussed with the other northern temperate regions, it actually extends from the subarctic waters of the northwestern Bering Sea to the subtropical waters of the southern coast of China. Several distinct subregions can be distinguished: the northern region (Sea of Okhotsk and western Bering Sea); the waters around Japan, including the Sea of Japan; the Yellow Sea and East China Sea; and the northern part of the South China Sea.

The northern part is very similar to the eastern part of the North Pacific—except that, lacking the wide shelf area of the eastern Bering Sea, demersal species are not so important. Salmon spawn in streams from the Arctic circle south to northern Honshu (the main Japanese island), but for most species the chief centre of spawning is Kamchatka. Though the total catch is about the same as on the North American coast, and much research has been carried out in Japan and the USSR, the literature is much less well-known to the English-speaking world than the Canadian and US studies. A convenient entry to this literature is KASAHARA (1961). KASAHARA (1964) provides a good review of North Pacific resources in general. The main species of salmon are the same as on the eastern side, with the addition of masu (cherry) salmon—though chum are considerably more, and sockeye less, abundant. Among other species, Alaska pollock is by far the most important.

Around Japan, and also on the western continental side of the Sea of Japan, the continental shelf is narrow. The waters are productive, but relatively little of this production appears as demersal fish, and the main species are a variety of pelagic species. Among these, herring, saury, mackerel, sardine, anchovy, and jack mackerel are, or have been at some time, important. To this list should be added squid, which fills a very similar ecological niche. As in other areas these species have shown high variability, with a succession of species becoming dominant, and then apparently collapsing. Special note should be made of the sardine; catches of this species rose to a peak in the 1930s, collapsed, but now have returned to or even above the level of the peak years half a century ago.

In contrast, over the wide shelf areas of the Yellow Sea and East China Sea, demersal species become relatively more important. In some ways this region is comparable to the North Sea with several species each making up an important part of the total, and one or the other often being the main constituent of some local or seasonal fishery. The species are different, with yellow croakers, lizard fish, and others replacing cod and plaice. The presence of a good stock of prawns (*Penaeus japonicus*), which because of the exceptionally high price can support a directed fishery even when the proportion by weight in the catch is small, may perhaps be compared to sole, which plays a comparable role, particularly in the Dutch fishery. *P. japonicus* spawns and spends its early months in the Po Hai; then migrates through the Yellow Sea to the deep waters of the East China Sea. This pattern of movement raises several complicated problems of management.

The southern end of the region, in the northern part of the South China Sea, has most of the characteristics of other tropical and subtropical waters. There are a great number of species so that the fisheries are generally directed at a group of fish, rather than at any particular species. This is shown in the statistics of Hong Kong in which, as given in the

FAO Yearbook, only two of the 30 categories distinguished are individual species, and (except for the large category of various and unspecified fishes), no category makes up more than 10% of the total. Information on the fisheries and fish stocks in this part of the region is, except for Hong Kong, rather slight. Present catches make up some 10% to 15% of the total of the region. The problems of assessment and management of these fisheries—which are probably fully exploiting the demersal species in the shallower water, with possibilities for expansion on some of the pelagic species and, with less certainty, some of the demersal stocks in deeper water—are similar to those of the fisheries in the region to the south. Further discussion of the general problems is given in a later section (p. 938). The present section will concentrate on the more northerly part of the region.

*The fisheries*

Information on the fisheries in the region is more difficult to obtain than for any other region. The fisheries in the Democratic People's Republic of Korea were estimated by FAO in the 1979 *Yearbook of Fishery Statistics* to have taken 1·2 million tons in 1979, but that figure is based on an assumed division between seaweeds and fish (including molluscs and crustacea) in the officially announced figure of 1·6 million tons for 1976, and subsequent extrapolation to later years. The official figures themselves give no idea of what types of fish and other species were included in the total, or indeed why the total was so much bigger than estimates previously available outside Korea. All that can confidently be said is that catches by North Korea are large, and they are mainly taken in the waters around Korea, probably mostly on eastern, Sea of Japan, side.

Until a few years ago the same fog obscured any view from outside of the fisheries of the mainland provinces of China. Now the political changes there have opened events in China, including fisheries, to the outside view. Detailed studies of the fisheries are still scarce, even in China, but ZHU DE-SHAN (1980) gives a good general overview of the recent fisheries.

In the absence of a widely based multinational fishery body, to which the USSR is committed to providing detailed statistical information, the generally available statistics of the large Soviet fishery is of total catch in the region (around 3 million tons), with a breakdown by species (two-thirds being Alaska pollock), but with no details on the area of capture. However, the main obstacle to access by scientists in Western Europe or North America to information on Japanese or Russian (and to a lesser extent Korean) fisheries is the language. Reviews in English are available in KASAHARA (1961, 1964, 1972) and for the pelagic fisheries, in NAGASAKI (1973).

Japan has always been one of the world's most important fish-catching and fish-eating countries. The coastal fisheries around Japan have a long history of slow growth and development, which accelerated in the nineteenth century with the opening up of the country to Western ideas. This brought steam power and gears, like the bottom trawl (with single vessel, and pair-trawl) and long-line, and led to a rapid growth comparable to that in the North Atlantic Ocean at around the same time. The same pattern of full exploitation of local stocks leading to movement to more distant waters was repeated. Thus, by 1960 Japanese trawlers were moving into the Yellow Sea and South China Sea, initially to fish for porgies. As these were reduced, in the next decades the trawlers moved to less valuable species, such as croakers and lizard fish (KASAHARA, 1964). While this expansion is comparable with the expansion of English trawlers to Iceland and northern

Norway at about the same time, another expansion of Japanese fishermen to the north was more comparable to activities of European fishermen along the Newfoundland coast some centuries earlier. Along the coast of the Kamchatka Peninsula the rich runs of salmon attracted Japanese fishermen, who started building traps along the coast from the 1880s onwards. The legal rights to exercise this fishery, under which a limited number of positions (lots) along the coast for traps reserved for Japanese fisheries were established following the Russo-Japanese war in 1904 to 1905, and lasted up to and during the Second World War, ending in 1945. Since then the large Japanese salmon catch has been taken almost entirely by gill-nets, operated (at least in the pre-200 mile age) on the high seas.

The development of Japanese fisheries since 1945 has been governed by the fisheries law of 1949, which gave clear recognition to the limited size of the resources, and of the social interests of small-scale inshore fishermen. The fisheries are classified into three groups: coastal fishery (small boats, up to 10 tons, working a great variety of gears, with trips seldom lasting overnight); offshore fishery (boats from 10 up to 200 tons, mostly using seines of various types, and trawls, and making trips of around a week); and distant-water fishery (tuna long-liners, large trawlers, whaling, etc. mostly outside the northwest Pacific). The nature of the controls are discussed below, but the effects have been to channel new developments into the distant-water fisheries, and to a lesser extent the offshore fisheries, while maintaining the coastal fisheries. Of the 8 million tons caught in 1979 (Table 5-12) a little over 2 million tons (including most of the molluscs other than cephalopods) were taken in the coastal fisheries, a few hundred thousand tons (most of the salmon, and much of the Alaska pollock) in the distant-water fisheries, and the majority (around 5·5 million tons) in the offshore fishery (with pelagic species such as sardines, mackerel, and squid predominating).

Korean and Chinese fisheries have developed in much the same ways, though at different paces and rather later than the Japanese. Before the Second World War, Korea (under Japanese occupation) had a large fishing industry. Between 1937 and 1940 Korean catches of sardine averaged 1·1 million tons, compared with an average annual Japanese catch of just under 1 million tons. The division of the country and the Korean war disrupted fishing, but by the late 1960s the fisheries in South Korea were expanding rapidly, reaching over 700 000 tons in 1970 (twice the average of the 1950s), 1·2 million tons in 1972, and over 2 million tons in 1976. Some of this expansion has been into distant-water fishing (tuna long-lining throughout the world, trawling in the Atlantic Ocean), but there has been a steady development of all types of fishing in local waters. Table 5-12 shows that South Korea reports a more even and wider spread of species than any other country in the region. The catches are less dominated by the main pelagic species than those of Japan—in part reflecting the wider expanse of shelf on the west coast of Korea. Very little is known about the fisheries of North Korea, and doubts surround even the reported total. What is known, and comparison with information on the pre-war fisheries, suggests that fishery development has followed a similar course, though somewhat later than in the south.

Development of the fisheries in the mainland provinces of China, in the sense of the growth of motorization and the increased use of trawls, started in the 1950s. By the mid-1960s heavy trawling in the East China and Yellow Seas was leading to the decline of some of the preferred demersal species (yellow croaker and hairtail); more recently, greater attention has been paid to increasing effort on pelagic species, first with gill-nets

Table 5-12

Catches of major species (species groups) in the northwest Pacific in 1979 (thousand tons) (Based on FAO *Yearbook of Fishery Statistics*)

| | China | | | | | | |
|---|---|---|---|---|---|---|---|
| | Mainland | Taiwan | Hong Kong | Japan | North Korea | South Korea | USSR |
| Salmon | — | — | — | 131 | | — | 130 |
| Flounders | — | 3 | 1 | 182 | | 32 | 52 |
| Alaska pollock | — | — | — | 940 | ‡ | 188 | 2015 |
| Other gadoids | — | — | — | 47 | | 2 | 106 |
| Large yellow croaker | 83 | 3 | — | 4 | | 35 | — |
| Small yellow croaker | 36 | | | | | | |
| Other demersal* | 105 | 124 | 68 | 687 | ‡ | 73 | 40 |
| File fish | ‡§ | — | | (c) | ‡ | 230 | — |
| Saury | ‡ | — | — | 355 | ‡ | 17 | 69 |
| Various jacks† | ‡ | 59 | 13 | 381 | ‡ | 30 | 7 |
| Herring | 39 | — | — | 6 | | + | 73 |
| Sardine | ‡ | — | — | 1586 | ‡ | 47 | 368 |
| Anchovy | ‡ | — | — | 148 | ‡ | 162 | — |
| Other clupeoids | ‡ | 17 | 3 | 106 | | 4 | 2 |
| Tunas and *Scomberomorus* | 30 | 81 | 3 | 310 | | 12 | — |
| Hairtail | 437 | 15 | 3 | 28 | | 121 | — |
| Mackerels | 250 | 6 | | 1591 | | 120 | 194 |
| Shark and rays | ‡ | 42 | 2 | 43 | | 18 | — |
| Various and unspecified | 1073 | 164 | 60 | 329 | | 87 | 160 |
| Crabs and lobsters | 310 | 12 | 3 | 66 | | 27 | 18 |
| Shrimp | 131 | 86 | 16 | 51 | | 25 | 3 |
| Squids | 42 | 30 | 6 | 433 | ‡ | 82 | 31 |
| Cuttlefish and octopus | 90 | 17 | 2 | 64 | | 7 | — |
| Other molluscs | 283 | 38 | 3 | 636 | | 381 | 4 |
| Various | 26 | 33 | | 38 | | 15 | — |
| Total | 2938 | 731 | 183 | 8164 | 1264 | 1763 | 3268 |

*Group 33 of FAO classification, other than yellow croakers.
†Group 34 of FAO classification other than saury.
‡Significant quantities probably included in 'Various and unspecified', or in 'Total'.
§310 000 tons in 1978.

and later with purse seines, using light attraction. Despite these developments the bulk of the Chinese landings in the mainland provinces is taken with small vessels (below 60 hp and 30 t), though there are also over 1500 larger vessels (over 100 t, but most under 250 ts). In the other Chinese province, Taiwan, the local fisheries in the South China Sea have followed much the same pattern, being carried out by small to medium-sized vessels, using a variety of gear in the coastal zone, and mostly trawls (especially pair-trawls) and purse-seiners, or similar gear in the offshore areas. Taiwanese fishermen have also actively developed long-range fishing, notably long-lining for tuna in all oceans, and trawling off northwestern Australia and other parts of the Indian Ocean.

The Russian fisheries are an exception to this general picture of the fisheries being mainly carried out close to shore, expanding with the use of small to medium-sized vessels into the offshore regions. There is relatively little fishing by small vessels, and the main fishery is by medium to large trawlers for Alaska pollock (Table 5-12).

*State of the stocks*

In general, nearly all stocks in the region, like in most regions close to the major fishing nations, are heavily fished. The precise reaction of individual stocks to this heavy exploitation varies. In the north, the species have reacted according to standard single species models. The salmon, like the salmon on the American side of the North Pacific Ocean, has suffered mostly from 'recruitment' over-fishing, and—to a lesser extent—damage to or loss of spawning grounds. The fisheries reached their peak in the late 1930s, when average catches of pink salmon were some $1{\cdot}5 \times 10^8$ fish. These declined to about half this in the 1950s, but since then stocks, and catches, have recovered.

The various stocks of Alaska pollock are all now probably fully exploited. Catches rapidly expanded from less than half a million tons in 1960, to nearly 2 million tons in 1970, and only just under 4 million tons in each year from 1973 to 1976. Since then the reported catches have declined to a little over 3 million tons in 1979. Some of this decline may be an artefact, since it excludes any catches by North Korea, and these may have been increasing in the last few years. In addition, there have been real reductions in catches by the countries with better statistics as a result of actions taken by the coastal states. Because of the variations in the seasonal distribution of the fish, Japanese fishermen used to take pollock off the USSR, while Soviet fishermen have fished off Japan. Since the introduction of EEZs by both countries in 1977 the catches by fishermen of non-coastal states have been strictly controlled. While the immediate impact has been some fall in total catch, the stock is probably recovering and in due course catches can be increased again. The other species, including various species of flatfishes and rockfishes, also taken by trawl in the northern areas, are probably at least as heavily fished as Alaska pollock.

Further south, changes in species composition have been as important as declines in individual species in both demersal and pelagic fisheries. In the demersal fisheries in the East China Sea and Yellow Sea the catch rates of preferred species declined soon after large-scale trawling began. As so often, the data available is not sufficient to determine how much of the subsequent changes in the composition of catches are due to real changes in the relative proportions of different species in the sea, and how much to changes in the fishing practices of the fishermen. Some of the early changes, e.g. a decline of catches of porgies and a shift towards less valuable species, can be explained fairly directly by a direct reduction by fishing of the preferred species, explicable on a simple yield-per-recruit model, and a change in fishing tactics towards other species. Recent changes seem to have been more fundamental. Chinese catches of yellow croaker off the east coast have almost collapsed. Catches of small yellow croaker, for example, fell from 160 000 tons in 1957 to only 20 000 to 30 000 tons in the 1970s (ZHU DE-SHAN, 1980). At the same time, catches of file fish—which is a much lower valued fish—have risen greatly, not only in China but also elsewhere. This was first noticed in the mid-1960s in Japanese waters, in the late 1960s off southern Korea, and in the Yellow Sea and East China Sea in the 1970s (FAO, 1981). Such changes in demersal fish communities are not well understood, or even well documented, but an increase of a similar species (*Balistes carolensis*) has occurred off West Africa.

The changes in the pelagic stocks are rather better documented, and have been even more dramatic. The most dramatic events have been in the sardine stock. Until very recently the Japanese sardine served (with the Californian sardine and, later, the North Sea herring) as an example of stocks that had collapsed (MURPHY, 1977). Japanese catches fell from nearly 1 590 000 tons in 1936 to 9200 tons in 1965. A common impression was that this collapse was due to poor management, and that very careful management would be required if the stock were ever to recover. This recovery has now taken place. Catches dramatically improved from 1972 onwards, exceeding 1 million tons since 1976. A similar recovery has occurred off Korea.

The sardine population is not homogeneous, and probably consists of four subpopulations (KONDO, 1980); these have not changed in the same way. At the earlier peak of the population and of the fishery, the main Japanese fishery (as well as the equally large Korean fishery) was in the Japan Sea, on the Kyushu and Japan Sea subpopulations. The recent recovery has been largely due to an outstanding 1972 year class (though later year classes have also been good), which has been most apparent on the eastern side, mostly from the Pacific subpopulations. The appearance of the 1972 year class seems to have been preceded by a slow increase in abundance, and a spreading of the population and of the spawning areas, but the immediate trigger has been ascribed by KONDO to a shift in the Kuroshio current.

The other pelagic species, including squid, have all undergone similar, if not so extreme, changes. Thus, the catches in Hokkaido herring fishery which produced up to 800 000 tons in the period from 1900 to 1930, fell rapidly from 1930 onwards and, unlike the sardine, there has been little recovery (MOTODA and HIRANO, 1963; MURPHY, 1977). Conversely, catches of mackerels, particularly of the Pacific stock of chub mackerel *Scomber japonicus* increased rapidly in the 1960s, from less than half a million tons in 1965 to 1·4 million tons in 1970, staying at around that level ever since.

Though the trends in individual species are highly confusing, and not easy to relate, either to changes in fishing pressure, or (with some exceptions such as the apparent link between sardine recovery and changes in the path of the Kuroshio) to environmental factors, the trends in the total catch of pelagic species around Japan are simpler. NAGASAKI (1973) examined the period between 1958 and 1969, during which there had been big changes in species composition. The rank order of the five main species, according to weight in the Japanese landings—which in 1958 was in the order of saury, anchovy, horse mackerel, squid, and mackerel—had been exactly reversed in 1969; the sardine which was eighth in 1969 was in first place by 1976. Despite this, the total of all pelagic species changed very little; apart from an exceptionally low catch of 1 900 000 tons in 1964, the Japanese catch of these species varied no more than from 2 304 000 tons in 1960 to 2 758 000 tons in 1968 (a difference of only 20%). While there is undoubtedly some compensatory effects within the overall Japanese fishery—the various fleets can only catch so much, and the various markets can only absorb so much, so that if one species increases greatly, catches of the other species are almost certain, in the short run, to decline—it would seem that some compensatory mechanism exists also in the biological system.

In fact, the extreme steadiness noted in the period studied by NAGASAKI (1973) may be partly accidental. Since then the total of the main pelagic species has increased to rather over 4 million tons in 1979, probably representing a considerable increase in the effective amount of fishing.

An exact compensation, whether operational or biological, would be unlikely. The different species and different stocks of the same species, occur in different areas, are fished by somewhat different fishermen, and go to different markets, e.g. an important use of saury is for bait in the tuna long-line fishery, for which other species are second-rate substitutes. Most individual fishermen or fishery enterprises are therefore not in a position to switch indiscriminately from species to species, not worrying whether catching sardine or mackerel, so long as the weight caught is the same. Equally, the differences in the areas and seasons that are important for spawning, larval survival, or feeding are different for each species so there is no simple case of a constant basic biological production going indiscriminately into one species or another.

Nevertheless it does seem clear that the pelagic species do interact, even though we are unable to specify exactly how, and that the assessment of individual species cannot be done in a meaningful way without reference to what is happening to the other species. As a whole, the stocks are heavily fished; it can be stated fairly confidently that the total catch cannot be expanded much beyond the present level of a little under 6 million tons (excluding possible catches by North Korea); equally, catches are not likely to fall much below the present level. It is much more difficult to say anything definite about, say, sardine or herring. Sardine catches are probably near their peak, but might continue at the current level, or collapse back to the level of the 1950s. If sardine collapses, the 'gap' will probably be filled, but what species or combination of species will increase is unknown.

The most important demersal stocks in the southern part of the region are those in the Yellow Sea and East China Sea. With the improved information now becoming available from China—the major fishing nation—it should soon be possible to make reliable and fairly detailed statements about the state of this resource. Summary reports suggest that, except for file fish, the demersal resources are all fully exploited (ZHU DE SHAN, 1980). Until recently, the only detailed data readily available outside China concerned only the Japanese fishery. This has been directed only on those species reaching a good price on the Japanese market, and has been somewhat restricted in area. Several of these stocks had become heavily fished nearly half a century ago, as shown by the declining catch rates in the 1930s, and by the dramatic increase in catch rates as a result of very greatly reduced fishing during the Second World War, especially in 1944 and 1945 (KASAHARA, 1964). This recovery suggests that the impact of fishing was essentially on the post-recruits, with a reduction in survival of fishable sized fish, and hence in numbers and mean size. This effect can be rapidly reversed as soon as fishing pressure is relaxed. Of more concern are the indications that there are changes in the recruitment pattern, with the recruitment of both species of yellow croaker (and other valuable species) going down, and being replaced by file fish. Such changes may not be so easily reversible by reducing fishing.

*Management and future prospects*

The state of management in the region is highly variable (KASAHARA, 1977). Japan has perhaps the most advanced and sophisticated management system in the world, fully in keeping with her position as the world's leading fishing country. In other areas, e.g. in the offshore waters around the Korean Peninsula, the difficulties of communication between the countries concerned, even at the level of interchange of data, have inhibited any effective management.

Regulation of fisheries in Japan has a long history, which has been reviewed by ASADA (1973), with local measures dating back as far as the eighteenth century. The present system is mainly based on the fisheries law of 1949. This took account of several conflicting pressures present in any fishery—the social problems of the inshore fishermen, who often have little opportunity to diversify; the need to expand the national supply of fish (particularly important in the conditions of Japan in 1949); the need for economic efficiency (or at least to avoid gross inefficiencies) in all sections of the fishery; and the fact that all resources are limited, but some (especially those near Japan) are more limited than others.

The Japanese fisheries are divided so far as management controls are concerned into two main groups: fishing rights fisheries and licence fisheries. The former comprise the smaller-scale inshore fisheries. They are further divided into (i) common fishing rights, issued to local fishing cooperatives and permitting a defined group of fishermen to exploit a certain inshore area in common; (ii) set-net fishing rights; as the name implies this allows the operation of a set net in a prescribed position; and (iii) demarcated fishing rights, which allow the holders to engage in aquaculture in a given area. The effect of these controls has to maintain the status quo in these coastal fisheries, without the social tensions and economic losses that would otherwise occur, but without much opportunity for development.

The licensed fisheries are subdivided according to the specific fishery concerned. For the larger offshore and distant-water fisheries, e.g. tuna long-lining, factory-ship whaling, salmon gill-net fisheries, etc., the licences are issued nationally. For the smaller fisheries, e.g. small-scale trawling, purse-seining by vessels from 5 to 50 gross tons, etc., the licences are issued by the prefectural government. The numbers of licences are limited, according to the productivity of the resource. Though licences are only valid for a fixed period, and must then be renewed, preference in issuing licences is given to existing licence holders, so that licences, which are transferable, have a definite value. For example, in 1963 when the tuna long-line fishery was particularly profitable a licence was worth $1500 per ton. These controls have been especially successful in keeping the expansion of new fisheries under control. Though Japanese fishermen have been successful in developing new fisheries, especially on tuna, but also on trawl fish and cephalopods in many parts of the world, these fisheries have not gone through the periods of extreme boom and bust typical of other fisheries.

This Japanese system has worked well so long as only Japanese fishermen harvested the stock concerned, or at least the activities of other fishermen were negligible. This is the case for the local stocks around Japan, and also held true for a number of other stocks, even those a long way from Japan—e.g. most of the world's long-line tuna fisheries in the 1950s, as well as other stocks even today, such as squid off New Zealand. In the northeast Pacific Ocean many of the stocks exploited by Japan are fished also by other countries, either because of the movements of the fish, or because the Japanese fisheries have taken place adjacent to other countries, e.g. in the East China Sea or the Sea of Okhotsk. For such stocks the effect of independent Japanese action is becoming steadily less through the growth of fishing by other countries. This has been accelerated by the changes in the Law of the Sea, which is bringing many of the areas being fished by Japanese vessels under the control of some other state.

These controls, especially in northern waters, are mostly taking the form of limits on the total catch, e.g. of salmon in the gill-net fishery, or of Alaska pollock in the trawl fisheries, though very detailed other controls are applied in some fisheries. The management of the king-crab fishery off the west coast of Kamchatka involved the division of the whole area into a large number of strips, allocated to USSR, or Japan or closed to all fishery (KASAHARA, 1972). These internationally agreed catch limits and other controls are applied in conjunction with the national licence controls. Similar catch limits are being applied to foreign fisheries, e.g. by USSR for sardine, in the waters now falling within the Japanese EEZs.

Arrangements for the detailed management of the several stocks that migrate between different EEZs remain complicated because of the absence of any international fisheries body with comprehensive membership. Indeed, between some countries any communication, even at the informal or technical level, is extremely difficult. Nevertheless, in the apparently most difficult situations some agreements can be reached. For example, there was for some time agreement between Japan and China, at the level of the fishermen's association, at a time when formal links between the countries were most distant, regarding the number of Japanese trawlers that could fish in certain areas of the East China Sea, though admittedly this came after a period between 1951 and 1954 during which more than 150 Japanese vessels were seized in the East China Sea by Chinese patrol vessels (KASAHARA, 1972).

The other obstacle to effective management in the central area is that major scientific problems concerning the interaction between species remain unsolved. It is far from clear what action should be taken to restore herring in northern Japanese waters (possibly at the expense of sardine), or to reduce file fish in the East China Sea and to restore the yellow croakers, or indeed whether there is any action that could be effective in the absence of fundamental changes in environmental conditions. On the positive side, all the countries in the region have long fishing traditions and are perhaps more aware than others that fishery resources are limited, and unless conserved can be damaged, to the lasting loss of the fishermen. For example, China is giving the management and conservation of its marine resources one of the top priorities in its fishery policies. The immediate prospect is therefore of no great change in the fisheries in the region, with catches being maintained at around the current level, but without any significant improvement in the management of the resources.

### (c) Upwelling Areas

The upwelling areas include some of the richest fishing grounds in the world (CUSHING, 1969a) and attract considerable multi-disciplinary scientific interest (BOJE and TOMCZAK, 1980). RYTHER (1969) estimated that they produced around half the potential harvest of fish from the sea; though the exact figures are open to debate, and particularly how far a given upwelling region extends, the general point holds true. Upwelling is not a rare phenomenon, occurring seasonally in many coastal areas, as well as along the equator in the open ocean, but attention here is focused on the four major coastal upwelling areas—Peru/Chile, the California current, the southeast Atlantic (Benguela current), and off northwest Africa (Canary current). In each region the prevailing wind blows along the coast towards the equator, moving the surface water offshore and replacing it by cool, nutrient-rich water from the depths. The richness in the upwelling sea areas is usually matched by near desert conditions on shore. The physical aspects of upwelling have been described by SMITH (1968) and DIETRICH (1972).

The fish populations are broadly similar in all major upwelling areas. The continuous supply of nutrients gives rise to high production and standing stocks of phytoplankton and zooplankton, and thus to large populations of small pelagic species, some feeding directly (at least in part) on phytoplankton (e.g. anchovies), but most on zooplankton (sardines, mackerels, and jack mackerels). Perhaps surprisingly the larger predatory fish (e.g. tunas) are not noticeably abundant in these regions, but birds and marine mammals can be extremely abundant. Exploitation of guano deposits of Chile, Peru, and southwestern Africa were among the first commercial uses of the riches of the upwelling areas. Besides their abundance, the pelagic stocks of these areas are noticeable for their instability. The sudden collapses of the fisheries for Californian sardine, Peruvian anchovy, and Namibian pilchard are well known, as are the arguments as to the precise role of over-fishing and environmental causes in any specific case (e.g. WALSH, 1978). While fishing has had an important role in recent changes, evidence from scale deposits off California (SOUTAR and ISAACS, 1974) and off Peru show that changes were occurring well before the fisheries became significant. Changes have also occurred in distribution. In its heyday the Californian sardine occurred as far north as British Columbia (MURPHY, 1966); now its stronghold is off Mexico, and only small quantities occur as far north as California. Conversely, during the 1970s the Moroccan sardine has spread a long way south from Morocco as far as northern Senegal.

Demersal fish are much less important, particularly in the Peru and California systems. Partly this is because the shelves are generally narrow. There seems also to be a more fundamental reason, in that most of the production is recycled within the surface layers, only a small proportion reaching the bottom, mostly in the form of faecal pellets rather than the remains of phytoplankton and similar detritus which provides the basis of the rich benthos of areas like the North Sea (WALSH, 1975). The dominant demersal species in most regions are the hakes, which in fact get most of the food from zooplankton (when small) or fish in midwater.

Table 5-13 shows the main species in each upwelling area, together with some summary information on the catches—the catch in 1979, and also where the fishery has collapsed, the catch in the peak years, and the year in which that occurred. Also shown is the catch in the whole FAO statistical region concerned, except for the eastern central Atlantic Ocean, for which the figures are only for the northern part of the region. For the eastern central Pacific Ocean, the region extends well beyond the upwelling system of the California current to include the tropical regions off Central America, and the statistics include substantial quantities (perhaps 750 000 tons) of tropical species, notably thread herring in the Gulf of Panama. The other regions correspond well with the upwelling areas, at least as well as can be expected when attempting to match a fixed geographic grid to an ecological system which changes from season to season within a year, and also from year to year.

Some features of Table 5-13 may be noted here, though the fisheries are described in more detail below. First there is, as already mentioned, a broad similarity between regions; secondly, the handful of species listed dominate the catches (over 90% in the Peru current), unlike the typical tropical areas when the number of species is extremely large. In this respect the upwelling areas are more similar to temperate areas such as the North Sea. Finally, although catches of the Peruvian anchovy were very small in 1979, the total catches in the Peru current system were more than twice those of the Benguela current, and four times that in the other two systems. This is roughly consistent with the relative productivities of the four systems estimated by CUSHING (1969a), though his

Table 5-13

Main commercial species of fish in the principal upwelling areas with 1979 catches (thousand tons) In brackets: catches in the peak year, if markedly different from the present harvest (Complied from the sources indicated)

| | California | | Peru/Chile | | N.W. Africa† | | Benguala | |
|---|---|---|---|---|---|---|---|---|
| Hakes (*Merluccius*) | *M. productus* | 142* | *M. gayi*<br>(500, 1978) | 186 | *M. merluccius*<br>*M. senegalensis*<br>*M. cadenati*<br>(102, 1976) | 37 | *M. capensis*<br>*M. paradoxus*<br>(804, 1976)<br>*M. polli* | 454<br><br><br>16 |
| Anchovies (*Engraulis*) | *E. mordax* | 303 | *E. ringens*<br>(12 300, 1970) | 1413 | *E. encrasicholus*<br>unspecified | 2<br>44 | *E. capensis*<br>unspecified | 570<br>14 |
| Sardines (*Sardinops* or *Sardina*) | *S. caerulea*<br>(791, 1936) | 186 | *S. sagax* | 3347 | *S. pilchardus*<br>(983, 1976) | 387 | *S. ocellata*<br>(1632, 1968) | 96 |
| Sardinellas | | | | | *S. aurita*<br>*S. maderensis* | 203 | *S. maderensis* | 207 |
| Jack (horse) mackerels (*Trachurus*, *Caranx*) | *T. symmetricus* | 17 | *T. murphyii* | 1287 | *T. trachurus*<br>*T. trecae*<br>*C. rhoncus*<br>(492, 1977) | 247 | *T. capensis*<br>*T. trecae*<br>unspecified | 465<br>272<br>30 |
| Mackerels (*Scomber*) | *S. japonicus* | 36 | *S. japonicus* | 213 | *S. japonicus*<br>(170, 1977) | 112 | *S. japonicus* | 35 |
| Total, above species | | 684 | | 6446 | | 1032 | | 2124 |
| Regional total | | 1616‡ | | 6899 | | 1632‡ | | 2519 |
| Percent of specified species in total | | 42% | | 93% | | 63% | | 84% |

*Mostly caught in statistical Region 67 (northeast Pacific Ocean).
†Data from FAO (1981) for northern part of eastern central Atlantic Ocean; other data from *Yearbook of Fishery Statistics* (1979).
‡Excludes tuna and similar oceanic species, but includes catches of other species.

results showed the Benguela current to be roughly as rich as the Peru upwelling area. This difference may lie in the differences in species fished, and their trophic level.

### *Southeast Pacific Ocean*

The region considered here comprises the whole of the statistical region of the southeast Pacific. It thus includes the temperate waters of southern Chile, in addition to the upwelling areas off northern Chile and Peru. The fisheries of southern Chile are confined to a narrow continental shelf, and carried out a long way from major markets. They have features in common with those in the similar waters of the northeast Pacific Ocean; indeed, some of the species of Pacific salmon have been transplanted to Chile, though without marked success so far (JOYNER, 1980). Crustaceans (particularly squat lobster) and molluscs (mussels and clams) are at least as important as finfish.

#### *The fisheries*

The history of modern fishing in the region started with the development of the Peruvian anchoveta fishery in the late 1950s. Prior to that there had been fishing along the Peruvian and Chilean coasts, mostly on a small-scale artisanal level. The most important fisheries had been the Peruvian fishery for bonito and Chilean trawl fishery for hake, both averaging around 50 000 tons $yr^{-1}$. There was also an important guano industry, whose interests in a high supply of food for the birds conflicted with the interests of the growing anchoveta fishery (MURPHY, 1981).

The growth of the Peruvian anchoveta fishery was stimulated by the fact that following the collapse of the Californian sardine fishery, a lot of equipment for fish-meal factories was available very cheaply. It was further assisted by the simultaneous growth in animal feed technology, which provided an almost unlimited market for fish-meal and other high protein supplements to incorporate in feeding stuffs for broiler chicken and other intensively farmed animals.

Once the feasibility had been established, rapid growth was inevitable. No part of the process from catching the fish to marketing the meal required great experience or high technology. In the calm seas, and amid the vast schools of anchovy, anything that floated could, and did in the early days of the fishery, catch enough to make money. The outside investment was minimal; the earnings from a boat or plant in the first year or two were enough to provide for the construction of the next boat or the next fish-meal plant. The result was that, starting at a level of about 100 000 tons in 1955, the catches doubled each year for nearly a decade.

This could not last, and in the absence of market limits, the limits were set by the resource, enormous though it was. These limits were being approached by 1962, when the catches had exceeded 6 million tons, and the total catch by Peru had passed that of Japan, making it the biggest fishing country (by weight) in the world at that time. The fishery then entered a 10-yr period of quasi-stability. The Peruvian government was well aware, at least in principle, of the dangers of over-fishing, and using the advice of the Instituto del Mar del Peru (sometimes with the assistance of Panels including outside experts, IMARPE, 1970a), introduced an increasing array of controls (VALDEZ-ZAMUDIO, 1973). During this period the total catch edged slowly upwards, from just over 8 million tons in 1963/64 to 1965/66 to 10 million tons in 1967/68 to 1970/71. The catches in the calendar years varied rather more because of changes in the impact of the regulations within a season, and reached a peak of 12·3 million tons in 1970. This period also saw a great deal of rationalization of the industry. While the industry as a whole was grossly inefficient because of the excessive capacity—far too many boats and far too many fish-meal plants (IMARPE, 1970b)—the efficiency of individual vessels and plants had increased.

The vessels were equipped with echosounders, power blocks, and fish pumps, which greatly increased the speed with which they could find the fish, shoot, and haul the net and get the fish on board. Making the necessary corrections for these changes in order to use the catch and effort statistics to follow reliably the trends in stock abundance was an important early stage in the assessment of the anchoveta stocks (SAETERSDAL and co-authors, 1965). Ashore the use of stick-water plant and general improvements in the way the plants were operated meant that virtually all the protein in the fish was recovered, and the costs of the operation were kept to a minimum. Events in the Chilean fishery in the 1950s and 1960s followed much the same pattern but an order of magnitude lower. Soon after the Peruvian fishery became well-established, moves were made to develop a similar fishery in Chile. These plans were over-ambitious, since the potential was much lower, but the Chilean anchoveta fishery did grow to around 1 million—very respectable by most standards.

The third phase of the fishery started in 1972. In that year, recruitment from the spawning in the middle of 1971 was extremely poor. The possible reasons for this are discussed below. The effect on the fishery was that catches dropped drastically, and with subsequent recruitment also being poor, particularly in 1977, they have remained low ever since. The best year (1976) produced only a little over 4 million tons, and catches were only 800 000 tons in 1977 (Table 5-14). During this time the stocks were being carefully monitored, both by routine analysis of the catches and by acoustic surveys (JOHANNESSON and VILCHEZ, 1981), and the catch quotas were set on the basis of the results.

This phase of severe crisis, which was particularly severe in the north and centre of Peru—the more southern anchoveta stocks were less severely affected—merged into the present phase of new growth. With the decline of the anchoveta there was new interest in the other pelagic species, particularly since these had much better chances of being used for direct human consumption, either locally or for export, canned or frozen. A proportion of these species had always been included in the fishmeal landings, which though small and unrecorded could have added up to what by common standards would have been a substantial fishery. In the late 1970s, directed fishing for sardines, and to a lesser extent, for other species, grew rapidly. Though this was in part a reaction to the decline of the anchovy, it was not a simple diversion of effort. The other species, particularly the mackerels and jack mackerels, are more active than anchoveta, and cannot be caught by the small-meshed purse-seines used for anchoveta. Furthermore, if the fish are to be used for canning or other direct human consumption, they must be handled much better than the bulk catches destined for meal; this means usually a new boat, preferably equipped with refrigerated sea-water or other methods of keeping the catch fresh. Together with the need for new types of processing equipment ashore, the change in attention is requiring a lot of new investment. As in the case of the anchoveta boom some of this is coming from within the industry, but the problems of over-investment and over-capacity are looming ahead.

Table 5-14 reveals that there has also been a big increase in hake catches. Prior to the

Table 5-14

Catches of major species from the southeast Pacific Ocean (thousand tons) (Based on FAO *Yearbook of Fishery Statistics*

| Species | Country | 1970 | 1971 | 1972 | 1973 | 1974 | 1975 | 1976 | 1977 | 1978 | 1979 |
|---|---|---|---|---|---|---|---|---|---|---|---|
| Anchoveta | Chile | 783 | 961 | 378 | 192 | 383 | 270 | 434 | 19 | 229 | 51 |
| | Peru | 12·277 | 10·277 | 4·447 | 1·768 | 3·583 | 3·079 | 3·863 | 792 | 1·157 | 1·363 |
| Sardine | Chile | 68 | 175 | 132 | 187 | 399 | 232 | 327 | 621 | 738 | 1·619 |
| | Peru | + | 6 | 6 | 61 | 74 | 63 | 175 | 871 | 1·075 | 1·728 |
| Jack mackerel | Chile | 112 | 158 | 87 | 122 | 194 | 261 | 342 | 341 | 587 | 598 |
| | Peru | 5 | 9 | 19 | 43 | 127 | 38 | 54 | 505 | 463 | 151 |
| | USSR | — | — | 5 | — | — | — | — | — | 49 | 532 |
| Chub makerel | Chile | — | — | — | — | + | 15 | 53 | 141 | 183 | 89 |
| | Peru | 9 | 10 | 9 | 65 | 63 | 24 | 40 | 46 | 97 | 118 |
| Hake | Chile | 88 | 66 | 67 | 46 | 43 | 32 | 30 | 37 | 34 | 93 |
| | Cuba | — | 1 | — | 6 | 11 | 11 | 31 | 30 | 45 | + |
| | Peru | 17 | 26 | 13 | 132 | 109 | 85 | 93 | 107 | 421 | 93 |
| All species* | Chile | 1·200 | 1·487 | 792 | 664 | 1·128 | 899 | 1·379 | 1·319 | 1·929 | 2·633 |
| | Cuba | — | 1 | — | 10 | 25 | 24 | 44 | 43 | 55 | 19 |
| | Peru | 12·533 | 10·526 | 4·722 | 2·323 | 4·139 | 3·441 | 4·338 | 2·528 | 3·355 | 3·667 |
| | USSR | — | — | 35 | 39 | — | — | — | — | 54 | 546 |
| | Total | 13·733 | 12·016 | 5·550 | 3·037 | 5·293 | 4·364 | 5·759 | 3·895 | 4·384 | 6·876 |

*Excludes catches of tuna in the open ocean by long-range fleets, mostly long-liners.

1970s hake catches had been confined to Chile, but lately Peru has paid more attention to this species. These catches have been taken in the north of the country, and there is a big gap in the distribution of hake between these grounds and those fished by Chile. Since the hake is mostly in deep water, and caught by the bottom trawl, it is not readily taken by the local fleets. Instead, the fishery has been developed by large foreign vessels, either under licence or as part of joint venture agreements.

*State of the stocks*

A large number of scientists have studied the Peruvian anchoveta fishery, including many of the leaders in fish-population dynamics. The methods and approaches used have progressed from the simplest models, to more complex single-species analyses, to methods that attempt to understand the whole system. At each stage improvements were significant to explain some of the unexpected events of the past, only for further developments to show that the explanations were far from sufficient.

The first approaches were to fit some type of production (Schaefer-type) model to the catch and effort statistics. Between the 1960/61, the first for which good statistics were available, and 1965/66, catch-per-unit-effort, when corrected for changes in vessel efficiency, declined to about half. Such a decrease corresponds to a stock being exploited at around the level giving the MSY. Though the effort was thereafter controlled, it did tend to move up, and so also, albeit more slowly, did the total catch (contrary to the simple model). A number of modifications to the simplest form of the production model were then used; some of these were no more than mathematical refinements, but a more fundamental change in approach was used by SCHAEFER (1970). He included the predation by the guano birds as a component of 'fishing mortality', including the estimated consumption by the birds among the 'catch'. The significance of this is that during the development of the fishery there had been a decline in the bird population. With this adjustment the increase in total 'effort' and total 'catch' was less than otherwise calculated, and the increase in actual catch was in part due to a greater share of the total being taken by man rather than birds.

Up till 1971 these simple models provided a reasonable description of the fishery, and, in showing that the stocks of anchoveta were being fully exploited, provided some basis for management. However, even before the events of 1972 showed to everyone how incomplete the description was, some scientists were putting out warnings. At its second session in March 1971 the Panel of Experts of Population Dynamics (IMARPE, 1972) warned that

> 'the high effort has three bad results . . . (c) the instability of the population is increased and there is an increasing risk of the collapse of the entire stock, [and also] . . . under heavy fishing a succession of years of poor environmental conditions can so reduce the stock that it cannot easily recover when environmental conditions return to normal'.

The Panel also noted that several stocks seem to have reacted to impending crisis by what appears to be a final attempt to rectify matters with the production of a single outstanding year class, after which recruitment fails—the recent history of the Georges Bank haddock stock, discussed in a preceding section, in which the extremely good 1963 year class had been followed by an almost complete recruitment failure, was in the minds of several Panel members. The Panel therefore noted

> 'With this in mind the occurrence of the extremely good year-class which supported the fishery throughout 1970 may not be entirely comforting (IMARPE, 1972).

One of the Panel members, G. J. Paulik, went further in a detailed scenario (PAULIK, 1971) of how an 'unthinkable catastrophe . . . the destruction of the world's greatest single species fishery' could occur. The stages of what he called 'this apocalyptic vision' were: (i) heavy fishing pressure diminishes abundance in such a way that the bulk of the catch is taken from recruits before they spawned; (ii) adverse environmental conditions cause a near failure of the entering year class; (iii) weak upwelling currents concentrate the residual population in a narrow band about 10 km from the coast; (iv) within 2 months the entire population of 4·5 million tons is caught; (v) the government imposes an emergency closure when catch-per-unit drops to zero; (vi) the fishery is closed. But it is too late—only small, scattered schools of anchoveta survive.

This gloomy vision was fulfilled more quickly and more closely than the author probably expected. Spawning in the middle of 1971 (the southern winter) was a failure, and very few small fish recruited in the early months of 1972, during what had normally been the period of peak recruitment. At about the same time the environmental conditions were unusual. Starting roughly in the last months of 1971, surface temperatures were well above normal, and the fish were concentrated towards the coast. The fishery was closed in January and February as usual to protect the small fish—the peladilla—which are particularly abundant at this time. The catch rates in March and April were extremely good. Monthly quota of 1 800 000 tons were achieved in 13 and 20 fishing days, respectively. These catches were enough to remove most of the older fish in the stock; with the recruitment failure the stock dropped rapidly, to perhaps a third or less of its pre-1971 level. Recruitment in the next few years remained poor, at about half the average between 1961 and 1971 (CSIRKE, 1980) until 1977, when recruitment was very bad—possibly less than half that in 1972. Catches which had been controlled by the Peruvian government, in fair accord with the changes in stock, and which had recovered from a trough of 1·7 million tons in 1973 to 4·3 million tons in 1976, dropped sharply. They were below 1 million tons in 1977, and still only 0·7 to 1·4 million tons between 1978 and 1981.

The proximate cause for the collapse of the anchoveta fishery has undoubtedly been a failure in recruitment. There have been changes in the distribution and availability of the anchoveta, and in their growth, but these have had much less influence on the changes in abundance, and in catches, than the changes in recruitment. As in most fisheries direct evidence on natural mortality is slight, but there certainly has not been the large-scale mass mortalities, typical during 'red-tides' in some other areas (VALDIVIA, 1978). The doubts concern why the recruitment has been poor. Possible reasons are environmental effects, pure recruitment, over-fishing, and combinations of these factors. The most popular candidate among environmental effects has been the El Niño. The impact of the El Niño on the fishery and on Peru's economy is discussed extensively in GLANTZ and THOMPSON (1981). In this GLANTZ presents an interesting discussion of the value of being able to forecast an El Niño.

As commonly used, El Niño refers to the occurrence of above-average surface temperatures along the coast of Peru, often accompanied by rain, and unusual weather generally on land. These seem part of a complex phenomenon extending across much of the Pacific Ocean (WYRTKI and co-authors, 1976). There is probably nothing that is a typical El Niño, nor is there a precise black and white separation between 'normal' years and 'El Niño'. Nevertheless, some years—notably 1965, 1972, and 1976 among the most recent (WALSH and co-authors, 1980, Fig. 12)—can be picked out as having well above average surface temperatures, and in this sense being 'El Niño' years. The 1976 El Niño coincides neatly with the poor recruitment in 1977—or rather with the early weeks and months of life of the fish which recruited in 1977. It appears that conditions in the sea—abundance and quality of the plankton—during the 1976 El Niño were less favourable than usual for larval and early post-larval fish. The coincidence is less good in 1972. Though the abnormal temperature became apparent early in 1972, at about the same time as the poor recruitment to the fishery was occurring, it is probable that the size of recruitment had been determined some months earlier. Also, the 1965 El Niño had little obvious impact on the anchoveta stocks.

The effect of the abundance of adults on the recruitment of anchoveta was most carefully examined by CSIRKE (1980). He showed that while the low recruitment between 1973 and 1976 could be explained by the low abundance of the spawning stock in the previous years, this was not the case for the poor 1972 recruitment. It may be noted here that because it is not possible to use the scales on anchoveta to determine their age exactly, the recruit year classes of anchoveta are usually referred to by the year in which they enter the fishery, and not the year in which they were spawned. The spawning biomass in 1971 was not greatly below the average in earlier years. What was different was that the fish were much more concentrated, so that the density in the spawning areas was much higher than usual. CSIRKE noted that in the RICKER formulation for the stock-recruitment relation $R = aS \exp(-bS)$, the negative term in the exponential $(-bS)$ which described the interactions, including perhaps predation of adult anchoveta on the eggs, resulting in higher mortality was likely to be a function of the density of spawners, rather than abundance. He therefore suggested that it should be written in the form (with some change from his notation) $R = aS \exp(-bD)$, where $D$ is an index of density. For the anchoveta, independent indices of density (from catch-per-unit-effort data), and of biomass (from virtual population analysis, and acoustic surveys) could be obtained. With this model he obtained a much better explanation of changes in recruitment, including the poor recruitments in 1972, though the actual recruitment was even worse than the model predicted.

This discrepancy is presumably due to environmental effects, and it would be possible to identify one (or indeed several) environmental factors that correlate with the residual changes in recruitment, not accounted for the CSIRKE stock–recruit relation. At this stage the number of degrees of freedom available becomes small compared with the range of environmental factors that could be considered. It is therefore impossible to determine on statistical grounds whether a good correlation is significant and indicates a real cause-and-effect relationship, or is pure chance. What can be concluded is that the anchoveta can be fished down to the level where the adult stock is affected, and that the impact of this can be sharpened by environmental changes. The pattern of fishing on anchoveta since 1972 probably represented a quasi-equilibrium state. So long as there were no unfavourable factors, as great a harvest was being taken without further depletion of the stock, but this did not allow for rapid recovery.

Apart from the direct effect of fishing, and of the physical environment, the effects of other fish species should now be taken into account. Acoustic surveys off Peru (JOHANNESSON and VILCHEZ, 1981) have shown that the combined biomass of mackerel, horse mackerel, sardine, and saury ranged between 3 and 7 million tons in the period 1976 to 1979, compared with an anchoveta biomass which was no more than 4 million tons after 1976. Judging by the catches the relative importance of these species off Chile is even greater.

It is too early to assess their state of exploitation with much confidence, especially in view of the changes that have taken place in species composition of both stock and catch. Catches of these other species off Peru in 1979 were over 2 million tons. If the biomass were 6 million tons, this would imply a fishing mortality of two-sixths or 0·33—not exceptionally high, but still approaching the level at which other pelagic stocks have reacted severely to the effect of fishing. The impact is probably not equal on all the species. Sardines are probably already heavily fished, whereas there may be some room for expansion of catches on jack mackerel and, more surely, on saury.

The real question regarding the evaluation of the current stocks is not so much the quantities that can be taken from under current conditions (which is perhaps 1 million tons of anchovy, 3 million tons of sardines, and 2 million tons of other species), but how the species interact. In particular fishery managers wish to know the extent to which this harvest, of perhaps 6 million tons, can be considered as an alternative to the harvest of 8 to 10 million tons of anchovy, plus a handful of other species, which was taken under the conditions of the 1960s. Furthermore, if these are alternatives, to what extent can the manager control which of these conditions (or some other combination of species) occur? For example, if the greater gross weight of catch taken in the 1960s represents a preferable condition (and this is probably not the case, in view of the better use that can be made of sardines), can the anchoveta stock be restored to its former glory by deliberately over-fishing sardines and other species?

Answering this question may require wide-ranging studies. WALSH (1981) has looked at the whole production cycle in the system, and how it seems to have changed between 1966 to 1969, and 1976 to 1979. Primary production may not have changed much—though reductions in the grazing on herbivores by fish, and changes in the way the nutrients are recycled could alter this quite a lot—but other aspects such as the production of bacterioplankton and detrital carbon have changed in addition to fish production. It may be necessary to understand how these changes operate, and the influence of these various factors at all stages of the life of anchoveta and sardine in order to answer questions on what pattern of fishing (over-fishing one species, and careful protection of another) would result in the desired sustained mix of species in the catches. On the other hand, the determining factors may be quite few. For example, GULLAND (1971) found a good correlation between the survival of Californian sardine from spawning to recruitment, and the total abundance of spawning pelagic fish (sardines plus anchovy). If this correlation, which held over the period of dramatic changes in sardine stock from 1932 to 1957, in fact describes the result of real cause-and-effect mechanism, then achieving the right mix of species may be easy. All that has to be done to achieve say a high sardine biomass is to keep the spawning population of sardine high, and that of anchoveta low.

*Management*

Some of the countries of South America, including Peru and Chile, were among the first to embrace the idea of extended jurisdiction over fisheries. Off these countries 200-mile limits have been in force for a quarter of a century. Thus, they have been in a position for a long time to enforce regulations if they appeared necessary to the government without the need to involve the complex machinery of international agreements. The rather poor results show that having authority is only one of the prerequisites for success; failure to achieve management objectives have been due *inter alia* to taking too narrow a view of the problem (e.g. concentrating solely on biological aspects), or to incomplete scientific advice (particularly in relation to complicated recent changes in the biological system). As always, the tendency to apply unpalatable restrictions too late and too weakly has played a part, but compared with other fishery authorities the Peruvian government has usually acted firmly and without undue delay.

The development of management up to 1972 has been reviewed by VALDEZ-ZAMUDIO (1973). The rules set were based on biological assessments, and were aimed to keep the total catch within the recommended limits, and to protect the fish when they were small or in poor condition, so that the meal yield was small. In summary, the measures were as follows:

1965: Closed season during August.
1966: Closed season during June, July, and August.

1967: Closed season from mid-February to mid-March (the season when small fish were particularly abundant), and also from June to August. Fishing was also prohibited on Saturdays and Sundays.

1968: Closed season from mid-February to mid-March. A quota of 9·5 million tons was set for the period 1 September 1967 to 31 May 1968. In addition, the closed season for June, July, and August was continued.

1969: A quota of 8·2 million tons was fixed for the 1969/70 season. The other controls (closed seasons, and no fishing on Saturday and Sunday) were continued in this and following years.

1970: In addition to existing controls, fishing was stopped from a port where 30% or more of the landings were of fish less than 12 cm long.

1971: Vessels were only allowed to land fish up to a maximum of 70% of their hold capacity. Large boats (150 t capacity and upwards) were only allowed to fish on Mondays, Tuesdays, and Wednesdays. Smaller boats were allowed to fish also on Thursdays.

1972: After the collapse of the stock became apparent in the middle of the year, fishing was stopped. Since then the fishery has been opened and closed almost on a monthly basis, usually in the light of the latest information on the stocks from the Instituto del Mar del Peru.

From the biological side these regulations have been fairly consistent with what appeared to be needed in the light of current knowledge. The small fish were protected, so far as this can be done in a purse-seine fishery (for which mesh control is not a practical method) by closures at the beginning of the year, and other measures. The controls on the amount of fishing were also effective in keeping the fishing mortality well below what it might otherwise have been. Though the common pressures for increasing effective effort, that occur in nearly all managed fisheries, did allow the fishing mortality to increase above the target, the impact on the stock would not have been serious in many other stocks.

The other failure in the management scheme was not examining the economic aspects until they had become too serious for easy resolution. While the fishing effort (and fishing mortality) were being kept under control, the capacity was not. The number of boats and, more especially their size and efficiency, expanded continuously, as did the capacity of the processing plants ashore. The results are shown in Table 5-15. This table shows how the number of days open for fishing steadily declined from the time that regulations were first introduced. The decrease, at least in the first years, did not represent a real reduction in the amount of time that the individual vessel operated, since the closed periods were used for maintenance, etc. otherwise done during the open season. However, soon this slack was taken up, and from 1967 or so onwards the regulations meant that expensive catching and processing equipment were unused, and crews and factory staff idle, when they could have been productively occupied. The total costs of fishing were therefore much higher than they need be. This effect of a control on effort, without controls on capacity, are readily predictable on theoretical grounds and have been observed in many fisheries, including some of the first to be controlled under effective international agreement, e.g. Antarctic whales and Pacific halibut.

The extent of the over-capacity is shown by such figures as the daily catch of over 150 000 tons in November 1971. That would be equivalent to taking the entire British

Table 5-15

Changes in capacity, effort and catches in the Peruvian fishery. (Based on IMARPE, 1974

| Year | No. of boats | Total gross tonnage | Total catch (thousand tons) | No. of fishing days |
|---|---|---|---|---|
| 1959 | 414 | 16 342 | 1909 | 294 |
| 1960 | 667 | 21 949 | 2943 | 279 |
| 1961 | 756 | 43 261 | 4580 | 298 |
| 1962 | 1069 | 71 991 | 6275 | 294 |
| 1963 | 1655 | 127 670 | 6423 | 269 |
| 1964 | 1744 | 140 059 | 8863 | 297 |
| 1965 | 1623 | 138 080 | 7233 | 265 |
| 1966 | 1650 | 150 856 | 8530 | 190 |
| 1967 | 1569 | 154 727 | 9825 | 170 |
| 1968 | 1490 | 138 561 | 10 263 | 167 |
| 1969 | 1455 | 128 652 | 8960 | 162 |
| 1970 | 1499 | 179 698 | 12 277 | 180 |
| 1971 | 1399 | 199 114 | 10 282 | 89 |

catch in a week. The Panel of Experts on the economic effects of alternative regulatory measures, which met in 1970 and 1973 (IMARPE, 1970b, 1974), pointed out that there was enough processing capacity to turn the whole world's fish catch into fish-meal. At the prices then prevailing, the long-term benefits from a more rational capacity could have been some $30 million annually. While some degree of over-capacity above the economic optimum is probably desirable on social grounds to provide additional employment, the 1972 capacity was well above any such optimum. The need to keep, so far as practicable, large numbers of fishermen and factory hands in employment, was a considerable, though not insuperable, obstacle to rapid and drastic action to reduce catches when the crisis occurred in 1972. The management actions taken since then, which have included phased opening of the fishery to different groups of vessels and factories, have had as an objective equal to that of maintaining the resource, that of minimizing social distress.

The collapse of the anchoveta has removed any immediate chance of a complete rationalization of the anchoveta fishery, and of achieving greatly increased economic benefits. It has produced two new management problems. The first concerns the rapidly growing fishery on sardine and other species. Capacity ashore (in canning plants, which also have their own fish-meal plants for offal, etc.) and afloat (in boats with better nets, and the ability to keep the fish in better conditions) is growing rapidly. Action is needed to ensure that even if the stock is protected, the same story of over-capacity is not repeated.

The other problem is the possible interaction between species. Biological problems have already been mentioned. The economic problems are almost as intractable and equally challenging. Assuming that it is possible to know what to do to achieve any one of a variety of combinations of different species, what combination should be aimed for? It is reasonably certain, because anchoveta have a higher proportion of phytoplankton in their normal diet, that the greatest weight would be taken under conditions of almost pure anchoveta, i.e. the conditions of the 1960s where the sustainable yield off Peru might be around 8 million tons. It is unlikely that a system that included many sardines and mackerels could produce as much, perhaps no more than 3 to 5 million tons (including the production off Chile). However, anchoveta can, at present, be used only for fish-meal; it therefore fetches a low price, and virtually all the meal has to be exported, mostly to the markets of the richer developed countries. The anchoveta have, therefore, never made any significant direct contribution to feeding the people of Peru—though as a provider of employment and an earner of foreign exchange the fishery has been very important. Sardines can be used for direct consumption, and canned sardines fetch a much higher price on the international market than fish-meal. With the collapse of the fishery off southwestern Africa there is a gap in supply that the developing fisheries off Chile and Peru have been excellently placed to fill. Also, the Peruvians have developed a tuna-type of pack for sardines that has found a new market in the United States where tuna prices have risen beyond the level where tuna and tuna sandwiches are a cheap food.

As stated, the choice does not appear too difficult, but there are complicating factors. Anchoveta are easier and therefore cheaper to catch than sardine. Prices of fish-meal and canned sardines (or other products) on the world market are variable; the fish-meal price could rise substantially if something went wrong with the US soya-bean crop for 2 or 3 years. Conversely, the price of canned fish could drop if some of the stocks now depleted recovered. Intensive research is going on in many places to find better uses for the smaller pelagic species such as anchoveta, and a technological breakthrough could greatly raise the value of anchoveta.

Even this understates the manager's difficulties. At present, and probably for some time to come, no one knows what effect management will have on the species balance in the sea. The manager has to deal with probabilities of unknown size, e.g. that fishing sardine hard may increase the recovery rate of anchoveta. His concern that the level of the sardine quota desired by the industry could damage the stock may be relieved by the feeling that if it does it may also help the anchoveta. The best answer to some of these problems lies with the industry; if it is well equipped to deal with a variety of species, then the decline of one particular species may cause no great harm. Peru suffered more than Chile from the collapse of the anchoveta not only because its anchoveta fishery was much bigger, but because the Chilean industry moved more quickly onto the other species. At the same time this ability to switch from one species to another, if not matched by a similar mobility on the part of the management, can lead to a progressive depletion of all elements in the system, as was tending to happen in the northwest Atlantic Ocean in the late 1960s and early 1970s.

### *Eastern Central Pacific Ocean*

This region, as defined for the purposes of FAO statistics, extends from 40° 30′ N (the north of California, USA) to the Peru–Ecuador border (about 5° S). It therefore includes the upwelling region of the California current, and also the tropical areas along the central American coast from central Mexico southwards. In addition, the influence of the upwelling system of the Peru current sometimes extends strongly northwards into the waters off Ecuador. The area of continental shelf in the tropical zone is small, and there are relatively few islands with surrounding reefs. The truly tropical fisheries of this region are therefore of relatively minor importance; they are included here, and not discussed separately. The exceptions are the fisheries for tuna, for which this region is one of the world centres. Tuna and other fisheries are discussed in detail in a later section.

In the Californian current system the main species are typical of upwelling systems (Table 5-13). In addition to those listed in the table there are a variety of demersal species, among which rockfishes are most notable, though their abundance is much lower than that of the pelagic species. Pelagic species, e.g. central Pacific anchoveta in the Gulf of Panama, are also important in some parts of the more tropical areas further south, particularly where there is local upwelling, and in the northern extension of the Peruvian upwelling systems.

In the tropical areas generally the economically most important species are the varied species of penaeid shrimp.

*The fisheries*

Some half a century ago this region supported what was among the biggest fisheries in the world—that of the Californian sardine. At that time, the very large sardine stock—when combined with Californian drive and know-how, and particularly the large markets for canned sardine, and to a lesser extent fish-meal—allowed the fishery to grow rapidly. Even the economic crises of the early 1930s caused no more than a temporary reduction in growth rate, until catches reached a peak of nearly 800 000 tons in 1936 (Fig. 5-5). Though the economic pressures for expansion probably continued, and were

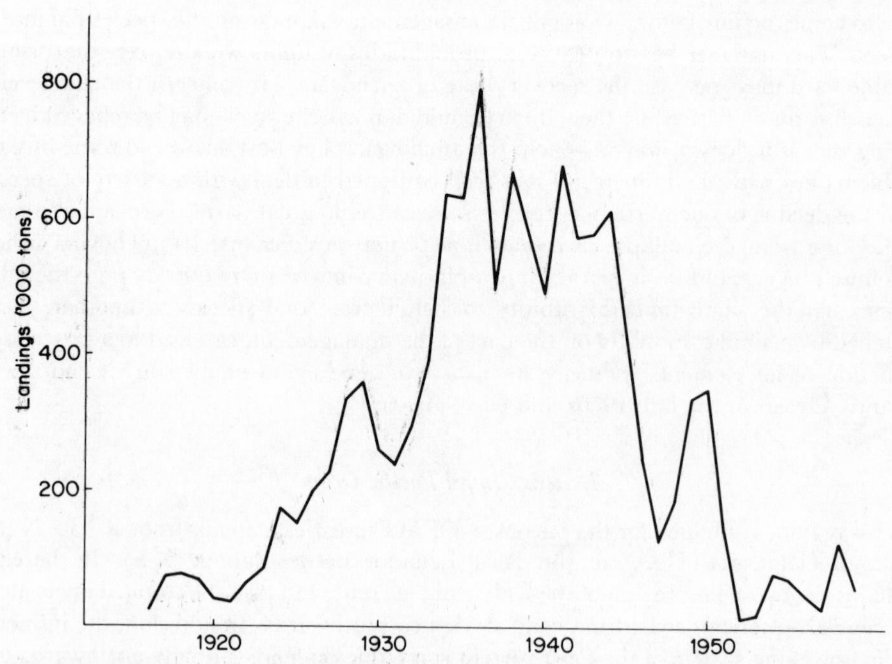


Fig. 5-5: Changes in the catches of sardines from the US Pacific coast. (After MURPHY, 1977; reproduced by permission of the California Cooperative Oceanic Fisheries Investigations, Vols. 8 & 9)

intensified during the Second World War, catches never again reached more than 700 000 tons, and after 1945 started a continuous and catastrophic decline. The history of this decline has been told many times, the most convenient references being by MURPHY (1966 and, in summary, 1977). Almost as striking as the decline in total catch, has been the shrinking southwards of the area of the stock (as measured by the location of the catches). In the early years as much as a quarter of the total was taken off British Columbia (Canada). These catches fell from 35 000 tons in the 1945/46 season to zero three seasons later; several seasons later the catches off Washington and Oregon (USA) had also disappeared. In the 1930s three-quarters of the Californian catch came from northern California; this fell to half or less by 1950, and by 1960 nearly all the catch was taken off southern California or further south.

The Californian pelagic industry was never based wholly on sardine; it used some of the other species as supplement or alternatives, especially as the sardine stock declined. However, none of the species ever supported catches approaching those of sardine at their peak. The mackerel fishery in fact behaved very similarly to that of the sardine, reaching a peak of 67 000 tons in 1935, and declining to only 3000 tons in 1953, though with some recovery thereafter (PARRISH, 1944; KRAMER, 1969; TROADEC and co-authors, 1980). Jack mackerel catches have recently been somewhat larger, but no more than 70 000 tons at their peak. The biggest resource is undoubtedly the anchovy. Attention was focused on this species by the extensive CalCOFI studies. These, among the most intensive studies of the fish stocks in any region of the world, were set up following the collapse of the sardine. They included regular and widespread surveys of fish eggs and larvae; the basic data from these surveys, and the various CalCOFI reports, provide an excellent source of information. Among other things, these surveys showed the presence of a very large and increasing stock of anchovy.

The development of an anchovy fishery in the United States has been slow. After the collapse of the sardine (and also mackerel) there was understandable concern that the same might happen to other species. Apart from its actual and potential value for fish-meal, the anchovy is also important to the large Californian sports fishery. Though statistics are not good, it is possible that the sports catch of many of the larger species—especially of striped bass, but also of many demersal fish—is much larger than the commercial catches. Anchovy is a popular bait for sports fishermen, who also believe that the presence of a substantial stock of anchovy and other food fish is important in maintaining a good stock of sports fish. The fishery managers in California have therefore adopted a cautious attitude, and allowed only a slow and carefully controlled growth in the fishery.

The main recent growth in pelagic fishery has therefore been in Mexico—though often with the participation of US commercial interests. Apart from a rapid growth in anchovy catches—up to a quarter of a million tons in 1979, three times the catch in 1976—the sardine stock has remained in healthy condition in the southern part of its range, and also now supports large catches. In contrast, the interests of the Californian fishing industry is now concentrated very largely on tuna, mostly to the south and west of the Californian current system, and to some extent outside the whole eastern Pacific Ocean, off the west coast of Africa, and in the western Pacific Ocean.

Pelagic fisheries are also important in the south of the region. The Gulf of Panama is a small pocket of relatively rich waters in the generally poor tropical area, and has supported a fishery for anchoveta *Cetengraulis mysticetus* and thread herring for a long period.

Total catches and the balance between species have fluctuated, with a peak of nearly 200 000 tons, mostly anchoveta, in 1977.

The central, tropical part of the region lacks the large varied fisheries on the mixed demersal communities of the shelf area, or of the reefs, that are typical of many other

Table 5-16

Selected statistics of catches in the eastern central Pacific Ocean (thousand tons) (Based on FAO *Yearbook of Fishery Statistics*)

| Catches | Country | 1938 | 1948 | 1961 | 1970 | 1975 | 1981 |
|---|---|---|---|---|---|---|---|
| Mackerel | USA | 36 | 18 | 20 | + | + | 39 |
| Jack mackerel | USA | | | 46 | 22 | 13 | 14 |
| Sardine | USA | 503 | 169 | 20 | + | — | + |
| | Mexico | † | † | 20 | 35 | 122 | 346 |
| Anchovy | USA | + | 5 | 4 | 87 | 150 | 57 |
| | Mexico | † | † | † | 5 | 60 | 367 |
| Anchoveta | Panama | † | † | † | 29 | 44 | 84 |
| Thread herring | Panama | | | | 7 | 20 | 26 |
| | Ecuador | † | † | † | 35 | 140 | 613 |
| Tunas | | † | † | † | 314 | 368 | 406 |
| Various demersal§ | | † | † | † | | 59 | 32 |
| Various and unspecified | Mexico | | | | 91 | 25 | 257 |
| | Ecuador | † | † | † | 30 | 40 | 26 |
| Shrimp | Mexico | † | † | † | 43 | 44 | 42 |
| | Panama | † | † | † | 7 | 7 | 16 |
| | Ecuador | † | † | † | 6 | 7 | 20 |
| Squid | USA | † | † | † | 11 | 8 | 21 |
| | Mexico | † | † | † | + | + | 10 |
| All species | USA | | | | 332 | 433 | 354 |
| | USA (excluding tuna) | † | † | † | | | 180 |
| | Mexico | | | | 228 | 332 | 1200 |
| | Panama | | | | 52 | 81 | 132 |
| | Ecuador | | | | 91 | 222 | 686 |
| | Japan* | | | | 102 | 75 | 86 |
| | Others | | | | 61 | 137 | 110 |
| Total | | † | † | † | 867 | 1·279 | 2569 |

*Almost entirely tunas.
†Data absent or incomplete.
‡Reported figures. The bulk of the catch was mostly mackerel.
§Species group 33 of FAO; includes croakers, snappers, etc.

tropical areas, particularly the South China Sea area. As Table 5-16 shows, the total catch of the large variety of demersal species included in the FAO category 33 (croackers, snappers, etc.) only amount to a few tens of thousands of tons. This figure should be increased by a large proportion of the varied and unspecified fishes shown in the statistics, and also, as far as catches, but not landings, are concerned, by the fish caught incidentally, and then discarded by the shrimp trawlers. Even with these adjustments the total is not large.

The important tropical fisheries are those for shrimp, which are carried out along much of the central American coast. Two main groups of fishermen are concerned; the industrial fishery using what are more or less standard 'Gulf of Mexico' trawlers (based largely on US designs and practice), which are used in many parts of the world. These work mainly on the offshore grounds on the older adult and subadult shrimp. Nearly all the catch is frozen and exported, largely to America, but to an increasing extent also to Japan. The other fishery is the subsistence fishery by inshore fishermen, often using simple traditional types of gear. These mainly exploit the juvenile shrimp in the coastal and lagoon areas. Because of the high value of export shrimp, the interests of the industrial fishery are given high priority in setting controls on the fishery—for which there is also a biological basis because of the differences in size caught. There has, therefore, been a decline in the catches of the Mexican inshore fishery, from around 2000 tons in the early 1950s to a few hundred tons in 1970 (McGOODWIN, 1979).

*The stocks*

In the absence of a local international fishery body there is no convenient source for reviews of the state of the stocks in the region, looked at as a whole. This is also true of the stocks in national waters, though there have been some summaries of knowledge at the national level (e.g. MACCALL and co-authors, 1976). The pelagic fisheries in the Californian current have also been reviewed by TROADEC and co-authors (1980) in a paper presented at the ICES symposium on assessment and management of pelagic stocks. The ICES report (SAVILLE, 1980) contains also several other valuable studies on the very important and difficult topic of the pelagic stocks (herring, sardine, etc.) which have proved very sensitive to over-fishing, and also very difficult to provide good scientific advice for.

While there is little formal international action, and nearly all work in the region is carried out at the national level, it is clear and increasingly recognized (e.g. ROEDEL, 1974) that the migration of stocks, especially between US and Mexican waters, requires that stock assessment (and, when this becomes necessary, management) be carried out in co-operation between both countries. So far as research is concerned this is now being done increasingly effectively, albeit mostly through arrangements less formal than the type of international commission set up in some areas. Co-operation in management, as discussed below, has not progressed so far.

The best studied stock has been the sardine. The results of these studies, up to 1963, by which time the stock and the fishery had almost fully collapsed, have been reviewed by MURPHY (1966). The most obvious changes, other than in distribution, between the 1930s, when a fishery was growing to its peak, and the 1950s, when it was collapsing, have been a decrease in the proportion of older fish and in the strength of the recruitment. Fish over 8 yr old were quite common in the 1930s, whereas in the catches of 1960, 6-yr-old fish made up only 0.1% of 2-yr-old fish. Omitting the exceptional year classes of

1932 and 1939, which both consisted of some $14 \times 10^9$ fish at Age 2, the average year class in the 1930s was around $7 \times 10^9$ fish, while the average in the 1950s was less than $2 \times 10^9$.

It is difficult to relate either of these changes directly to fishing because the recorded statistics of nominal fishing effort—number of boat months is the statistic used by MURPHY (1966)—may not be closely related to fishing mortality. The shrinking southwards of the area of distribution of stock and fishery would certainly be expected to increase the effectiveness of a unit of effort. Also, in pelagic fisheries the catchability would be expected to increase as the stock decreases, especially if the decrease is reflected in a reduction in the number of fish schools, rather than in their average size. These considerations make it likely that the increase in nominal effort during the growth of the fishery, and during its peak years, underestimated the rise in fishing mortality, and that after the collapse fishing mortality remained high (so long as fishing on sardine was permitted) even though nominal effort was low. Any purse-seine vessels fishing for anchovy or jack mackerel kept an eye open for sardine schools, and if one was sighted went after it. Only if sardine were caught would any sardine 'effort' be recorded. Despite these difficulties it is clear from a variety of analyses that fishing is responsible for the increase in mortality, and hence the decline, from a given recruitment, in total abundance, and in abundance of the spawning stock.

In addition to the common techniques of analysing age-composition and tagging, the study of the stocks in the Californian current has included, to an extent greater than anywhere else in the world, regular surveys of eggs and larvae (e.g. AHLSTROM, 1966). Apart from documenting the rise of the anchovy, these surveys allow estimates to be made of the abundance of the spawning stock of the main species. These estimates depend on the frequency with which the individual adult female spawns—about which there are some doubts. Does the number of large eggs in the ovary represent all, or one-half or one-quarter of all the eggs that she will produce during the spawning season? There are also doubts about the precision that can be achieved from such surveys, at least without great efforts (e.g. ENGLISH, 1964). Nevertheless, these surveys can be used to produce estimates of adult abundance (SMITH and RICHARDSON, 1977), and hence, with data on catches of adults, estimates of fishing mortality. These estimates for sardine are in broad agreement with those based on other methods.

The questions that have surrounded the assessment of sardine therefore concern the decrease in recruitment, and the change in distribution. Are these due to fishing or to changes in the environment? To what extent have either type of effect been modified or increased by the rise in the anchovy stock? In either case, changes in distribution and abundance are probably linked. Any unfavourable change in environment is likely to have had a greater impact on the fringes of the distribution, in this case towards the north, where it may be presumed that the sardine was already on the margin of its tolerance. Conversely, it is possible that a reduction in abundance caused by fishing will result in shrinkage towards the centre of population, especially in the case of a shoaling fish like the sardine.

These questions have long been the subject of arguments (MARR, 1960; MURPHY, 1966, 1977), and probably no one answer is fully correct. The hypothesis of a fishery-induced collapse is supported by the close relationship between adult stock and subsequent recruitment for sardine (e.g. Fig. 9 of MURPHY, 1966). A similar close correlation is found for mackerel (e.g. Fig. 2 of PARRISH, 1974). However, the strong serial correlation in the series of both spawning and recruitment mean that the underlying causal mechanism may be no more than that high recruitment in Year $X$ gives rise to high abundance in Years $X + 2$ and $X + 3$.

Similar considerations apply to the role of the anchovy in the changes in sardine stock. Studies of the frequency of scales of different species in the annual layers in the anaerobic bottom deposits off California (SOUTAR and ISAACS, 1974) have shown that the abundance of both species has varied greatly, even before fishing started. However, there are periods when the anchoveta was abundant and sardine rare, and others (though less frequent) when the reverse occurred. The changes cannot be represented by a simple switch between one or other species. The recent increase in anchovy occurred at about the time of the collapse of the sardine, though the precise timing (the big increase in anchovy did not occur until after 1950, by which time the collapse in sardine was almost complete) suggests that the anchovy has been expanding to fill a gap left by the sardine (AHLSTROM and RADOVICH, 1970; MURPHY, 1977). The presence of the large anchovy stock could well now be inhibiting any recovery of the sardine, and the large-scale fishing of anchovy has often been suggested as one mechanism for encouraging the rebuilding of the sardine stock.

The more southern pelagic stocks have not been assessed in detail. It is unlikely that the stocks in the Gulf of Panama are large by the standards of the major upwelling areas. Current catches are probably approaching, or already at, the limiting potential of the resource. The very high catches off Ecuador represent a big jump from previous catches. Though reported in the 1979 FAO *Yearbook* as thread herring, it is likely that the bulk of the catches were mackerel. Thread herring, in contrast, has become scarce in Ecuadorian waters. These changes are probably part of the widespread changes that have occurred off Peru and Chile, as discussed in the previous section. If so, the possibilities of maintaining catches at the 1979 level, or even increasing them, depend less on the pattern of fishing in Ecuadorian waters than on whether general conditions return to what might be considered 'normal', i.e. those prevailing before 1972.

Assessments of the shrimp fisheries have been more straightforward. Available units of effort, e.g. number of days at sea, provide a good measure of true fishing mortality. As early as the 1960s BOEREMA and OBARRIO (1962) found a clear relation between catch-per-unit-effort and effort in the Panama shrimp fishery, indicating that those stocks were already fully exploited. Many of the other shrimp stocks along the central American coast also became heavily fished at about the same time. Some increase of total catch has been possible by expanding onto some of the less valuable stocks, either of the less preferred species, or the smaller or more dispersed stocks of the prime species. The main opportunities for expansion in recent years has come from the deep-water shrimp, though these are considerably less abundant than those in shallow waters.

The two groups of shrimp fishermen take different sizes of shrimp. While small-scale inshore fishermen catch mainly juveniles, the offshore industrial fishery takes larger, or mostly adult shrimp. The small-scale fishery is therefore responsible for some degree of 'growth over-fishing', in that for the heavily fished stocks the weight caught from a year class may be rather greater if it is harvested only in the offshore fishery. The extent of this difference depends on the balance between growth and natural mortality. Neither of these is well known. Growth is certainly rapid, but available estimates of mortality in shrimp are correspondingly high. Both probably vary from stock to stock. The gain in weight is presumably not much, but because of the higher price obtained for large shrimp, there may be an appreciable increase in total gross value of the catch from harvesting offshore only.

*Management*

Direct management of the Californian sardine, one of the most serious points raised in the 1940s and 1950s, is no longer a problem. The complete collapse of the stock has made it easy for a complete prohibition on the commercial catching of sardine in Californian waters to be accepted, but has left a number of questions unanswered. The first concerns the continued and expanding fishing of sardine in Mexican waters. So far as Mexico is concerned this stock appears to be in good shape. Though there may not be much scope for a sustained increase in catches, and attempts to increase catches may cause damage to the stock, there is, on present evidence, little reason to restrict catches below the present level. On the other hand, a reduction of Mexican catches would lead to a greater abundance, and hence encourage an expansion of the stock northwards. The extent of any effect in the United States would depend on the environmental conditions, and also on the genetic structure of the stocks of sardine—at its peak the sardine population was not uniform, and the fish that were abundant off California may have been quite different genetically from those now off Mexico. Nevertheless, the chances of some beneficial effort of reduced sardine fishing off Mexico on future catches further north are not negligible. If all the resources were under the jurisdiction of a single authority, that authority would probably find it desirable to make such a reduction. This is, however, an academic point. No one would expect, particularly in view of the difference in economic well-being of the two regions, that Mexico would make sacrifices in its fishery for the benefit of Californian fishermen. Collaboration between Mexico and the United States, and the establishment of machinery for co-ordinated management as recommended in the Law of the Sea texts, would clarify these issues, but not make the choice any easier.

Another question is far from academic. This concerns the desirable pattern of harvesting anchovy in Californian waters. The scientific uncertainties have been discussed above. However, the fishery manager cannot afford to wait for certainty in his scientific advice, and must act on reasonable probabilities—in this case that the current anchovy stock can support higher catches, and that increasing the catches would increase (though perhaps very marginally) the probability of a return of the sardine. From the point of view of the commercial fishermen, and of maximizing the yield in weight of fish from Californian waters, there is little doubt that a considerable increase in anchovy catches should be allowed—and indeed encouraged. However, current regulations strictly control the total quantity that may be caught, and additional controls on where and when anchovy may be fished further reduce the amount that is actually harvested. These controls have been introduced as a result of pressure from sports fishermen who fear that too much anchovy fishing will reduce their supply of bait, and also (by reducing their food) reduce the abundance and size of the sports fish. If these fears are well grounded (and the evidence either way is not strong), then the management actions are reasonable. They would illustrate clearly that management objectives are diverse, and the balance between objectives varies from area to area. California is not short of food or employment, and the fish of its seas may well be more valuable as a source of relaxation for some tens of thousands of sports fishermen than for employment of a much smaller number of commercial fishermen.

Social and economic problems also lie at the heart of the other serious management problem of the region, that of shrimp. Shrimp fishing is highly profitable, and even when the stocks are fully exploited there is great incentive for additional boats to enter. Though this does not depress the total yield much, and the boats concerned may make a profit, or at least break even (at the cost of the vessels already in the fishery), the country as a whole loses because of the extra costs (especially of fuel and equipment, much of which may be imported). Thus, BOEREMA and OBARRIO (1962) strongly recommended a limit on the number of trawlers in the Panama shrimp fishery. A limit was in fact set, but because this gave high profits to the vessels allowed to fish, the resulting pressure for additional boats to enter was too strong for the government to resist completely and the numbers gradually crept up to a level where most of the potential economic rent has been dissipated, and there is little further incentive for expansion.

Similar over-capacity exists in most other shrimp fisheries in the region, including Mexico. Here also result serious social consequences (MCGOODWIN, 1979). Because of the decline of the stocks brought about by the large size of the industrial shrimp fleet, and the political power of the industrial fishery, partly as a major earner of foreign currency, controls have been set on all sections of the shrimp fishery, including the inshore fishery. In fact the open season for this fishery has been less (10–12 wk) than that for the industrial fishery (35–40 wk). As a result the catches in the inshore fishery of South Sinaloa studied by MCGOODWIN declined from around 2000 tons in the 1950s to only 400 tons in 1970. This has led to distress among the local fishing communities, who have little alternative opportunities. As in the case of California anchovy there is a conflict in management objectives between maximizing gross yield (by fishing anchovy, or reserving shrimp to the vessels catching them when relatively large) and social objectives (providing recreational facilities, or helping the poorer section of the community). The different decisions reached in the two fisheries show that there is no unique solution, and that the actual choice will be based as much on political pressures, as on the results of biological analyses.

## *Northwest African Waters*

### *The resources*

The area covered in this section consists of the northern coastal divisions of the FAO statistical area 34, the eastern central Atlantic Ocean, i.e. the area along the African coast between the Straits of Gibraltar and 9° N (about the boundary between Sierra Leone and Guinea). The whole area of the eastern Atlantic, from the Straits of Gibraltar to the Congo River, is the area of responsibility of the Fisheries Committee for the Eastern Central Atlantic (CECAF) which is a subsidiary body of FAO. It has broad terms of reference including the promotion of management of the resources (in which it is similar to many others, non-FAO, fishery commissions), co-ordination of research, and promotion of fishery development. This last involves providing mechanisms for identifying, supporting, and in some cases executing programmes of technical assistance to the developing countries of the region, particularly that provided by the United Nations Development Programme. The southern, tropical, part of the CECAF region, from 9° N to the Congo, is discussed in a later section. The boundary at 9° N corresponds roughly to the area of strong upwelling, at least for some part of the year. Seasonal changes, and changes from year to year, in the southern boundary of the upwelling system make it impossible to match this to a fixed boundary that can be used for collection of statistics. The boundary at 9° N is a practicable compromise; corresponding roughly to the southern boundary of the upwelling system during the northern summer, and has been used by FAO for some years, notably in the CECAF *Statistical Bulletin* which is now produced annually. The species complex is typical of upwelling areas (Table 5-13), with a predominance of small pelagic species, but there are some differences. The proportion of these species is lower than in the other areas where the pelagic fisheries are fully developed—possibly because there is no substantial fishery on anchovy, though they occur in good quantities—and there are large stocks of octopus and cuttlefish which support the economically most valuable fishery in the region. Squids possibly occur in numbers in the other upwelling areas, though they are not fished elsewhere in quantity, but there does not seem to be equivalent stocks of the other types of cephalopods. These stocks occur mostly off the Sahara coast, between southern Morocco and northern Senegal.

Of the typical upwelling species those in the north, off Morocco, are largely identical to those found off Portugal. However, though sardine, horse-mackerel, and hake are found in abundance north and south of the Straits of Gibraltar, there is little to suggest that they are the same stocks. Rather, the strong currents flowing through the Straits would tend to keep the stocks of essentially coastal species separate north and south of the Straits. There is however some movement of the more active oceanic species, e.g. of albacore and bluefin, from the CECAF region into the northeast Atlantic Ocean.

The distribution of the main species, and the separation (if known) between different stocks of the same species, has been described in a number of CECAF/FAO reports (TROADEC and GARCIA, 1979; CECAF, 1979a,b; BELVEZE and BRAVO DE LAGUNA 1980). Among the pelagic species the pilchard is restricted to the northern part of the region, but seems lately to have extended its limits southwards from about 26° N in 1966 to 17° N in 1977 though most of the population is found off Morocco. Further south it is replaced by the two species of *Sardinella*. Mackerel and horse mackerel are more widespread, occurring in most of the region. While they are active, two or three separate stocks have been identified for each.

The demersal and cephalopod stocks appear to be more localized. For the latter, four distinct fishing grounds have been identified (CECAF, 1979c, Figs. 1, 2, and 3) in the area from Cape Bojador (*ca.* 26° N) to just south of Cap-Vert in Senegal (*ca.* 14° N). The migration patterns of the various species is still being studied, but enough can already be deduced e.g. from the movements of the centres of fishing activity, that the pelagic species at least move appreciable distances up and down the coast. Including this movement of adults, and the movements of younger fish from the nursery areas (mostly inshore) to the feeding and spawning grounds, it is apparent that most stocks (the most important exception is the sardine stock off Morocco) will occur, and can be fished, off the coasts of two more coastal states during its life.

*The fisheries*

The richness of the resources, and their closeness to the markets and advanced fisheries of Europe, have long attracted fishing vessels from outside the region, first from south and west Europe, and later from eastern Europe and Japan and other Asian countries. Local fisheries have also always been important in the north (in Morocco) and in the south (from Senegal southwards). In the central part of the area—from southern Morocco to Mauritania and northern Senegal—the coastal lands are mostly desert and only a handful of local people live along the shore. Some of them along the Mauritanian coast do depend on fishing, but their small numbers and their primitive methods, mean that their total catches have been virtually negligible.

Among the earliest European fishermen to visit the region were the Breton lobster fishermen, in the first decade of this century, and soon after the English trawlers came to the waters off Morocco in pursuit of hake. The latter fishery did not last long, but was followed not long after by Spanish and Portuguese fishermen. They extended their local activities to a relatively short distance across the Straits of Gibraltar, fishing for both pelagic and demersal fish (including hake). However, these fisheries were relatively small, accounting, in 1958 for example, for only about 80 000 tons, some 14% of the total catches in the region.

The activities of distant-water vessels only began to dominate the fisheries in the region from about 1964 onwards, when much larger vessels from Japan (and later Korea), eastern Europe and Mediterranean countries began working the more productive waters from southern Morocco southwards. The development of these fisheries up to 1970 has been described by GULLAND and co-authors (1973); the later history is covered by the various CECAF publications, particularly the *Statistical Bulletins*. By 1970 catches by these vessels had reached 1·7 million tons, 66% of the total; 1970 marked almost the end of the period of rapid growth of the distant water fisheries, and also the year when they took the greatest share of the total catch. Since then the total by these fisheries grew slowly until 1977, when the exercise of extended jurisdiction began to lead to their reduction. By 1979 the total had fallen to 1·3 million tons (47% of the total), compared with nearly 2·3 million tons (60% of the total) in 1979.

The distant-water fisheries include a variety of very distinct fisheries. Apart from the tuna fisheries, described in another section, they included the cephalopod fishery; the trawl fishery for high-valued species (hake, sea bream, etc.); the trawl fishery for lower-value pelagic species (particularly mackerel and horse-mackerel); and the purse-seine fishery for sardinella and other pelagic species. Some of these fleets have arrangements with African countries for landing or transhipping their catches—to a growing extent as part of the condition for being allowed to fish in the EEZ of the country concerned—and in Nigeria these landings by Polish and Russian trawlers, especially of those species which are not particularly liked by European countries have at times been an important part of the total fish supply. Other vessels take their catches directly home, without touching at an intermediate port. A very large part of all the distant-water fleets use ports in the Canary Islands as base in the region for transhipment of catches, interchange of crews, purchase of supplies, and routine maintenance. Las Palmas has thus become one of the most varied and important fishery ports in the world.

The cephalopod fisheries have been mainly carried out by Spain and Japan which have the highest home markets for these species. Between them they have accounted for between 80% and 90% of the total catch. Other Mediterranean countries (Italy, Greece) have taken small catches, and lately catches by Korean vessels have increased. Total catches built up steadily to a peak of some 200 000 tons in 1974 and 1975 but have since decreased. Cephalopods attract a high price on both the Spanish and Japanese markets. Therefore the total value of the cephalopod fishery is probably greater than that of any of the other fisheries—some $200 to 300 million from 1975 to 1977—though the values of the catches by some countries, notably the USSR, are not well known (CHRISTY, 1979).

The trawl fisheries for a variety of high-valued demersal fish (hake, sea bream, etc.) were the first distant-water fisheries to develop on a large scale, and are still among the most important. Most of the non-African countries are involved, and the vessels from some countries, e.g. Italy and Greece, are engaged almost entirely in this fishery. The ships involved range from moderate sized vessels, transhipping their catches at local ports, to very large trawlers of several thousand tons, capable of very long round trips and bringing their catches directly back to their home port.

For the Russian fleets and others for which the price of fish is not important, or at least is not determined purely by market forces, trawling for high-valued but increasingly scarce demersal species has, to a growing extent, been supplemented or replaced by trawling (with mid-water and bottom trawl) for the more abundant mackerels and horse mackerels. These fisheries grew particularly strongly in the late 1960s and early 1970s. Catches of horse mackerels increased from 45 000 tons in 1966 to over 400 000 tons in 1970, varying thereafter, with peaks of over half a million tons in 1971 and 1973. Similarly, catches of mackerel grew from less than 40 000 tons in 1966 to nearly quarter of a million tons in 1970, though decreasing rather sharply thereafter.

The last, and shortest-lived, foreign-based fishery was that for fish-meal, using small seiners attached to converted whale factory ships. This was an extension of the fishery that first started off southwest Africa as a method of circumventing the local quota regulations. While this was technically and economically a success, South African authorities took only a few years to amend the regulation so as to bring this type of fishing under control. It then had no advantage over the local-based vessels and the factories of Walvis Bay. The same method of fishing did however have economic advantages off northwest Africa, particularly the desert sector between southern Morocco and central Mauritania. To be efficient, a fishery of this type relies on very large quantities of fish, and in the first year of operations (1970) over 300 000 tons of fish (mostly sardinella) were caught. Later catches were lower, partly because one of the large factory ships sank, and the factory ships ceased operation after 1977.

Fishing by coastal states is discontinuous, with a gap in the middle, off the desert coasts of the western Sahara, where there is very little local population to do any fishing. In the Sahara coastal division of CECAF, between 19° and 26°, the total reported annual catch in the past decade has averaged over 1 million, with a peak of 1·5 million tons in 1976 but the catches by the African countries as reported in the CECAF *Statistical Bulletin* have only been some 20 000 tons. In contrast, Morocco to the north, and Senegal and others to the south, have substantial local fisheries.

The Moroccan fisheries are more typical of those of southern Europe than of Africa, and are little different from the local fisheries of Spain and Portugal across the Straits of Gibraltar. The biggest fishery is that by small purse-seiners for sardine, mostly for canning. This long-established fishery has produced around 150 000 tons in most years, but rose to over 200 000 tons in the early 1970s. This rise seems to have been associated with a spread of sardine to the south of Morocco, and increased landings in the southern Morocco ports. In addition, there are smaller trawl and other fisheries for a variety of species (hake, sea bream, etc.), but these account for not more than a quarter of the catches of sardine.

In the south the catches of Senegal and the neighbouring countries are largely taken in the traditional canoe fisheries. The majority of these, which number several thousand (over 6400 in Senegal alone in 1974), are motorized, but none makes voyages of more than a few miles from shore. The fishing methods include small seines and gill-nets for sardinella and other pelagic species, and handlines for grouper and bluefish. Altogether, despite the lack of advanced technology (and thus without the cost of much imported material), catches are large and increasing. The catches of Senegal have risen from some 160 000 tons in 1970 to over 330 000 tons in 1978, mostly from the traditional fishery. The catches by the other coastal states (Gambia, Guinea, and Guinea-Bissau) have also increased, but are considerably smaller, less than 50 000 tons in all.

In addition to the traditional fisheries Senegal has an industrialized sector consisting of a small fleet of trawlers fishing for shrimp and high-valued demersal species, medium-sized purse-seiners fishing for sardinella, and larger purse-seiners fishing for tuna. The tuna fishery is very closely related with the tuna fishery by French flag vessels that in season also work out of Dakar.

*State of the stocks*

The main pelagic stocks do not seem, at least until very recently, to have undergone the dramatic change, such as the collapse of the Californian sardine, as have occurred in the other upwelling areas. This may be merely that the stock in this region have not been so heavily fished, or fished over such a long period, as those elsewhere. The pelagic stock that has had the largest history of moderately heavy fishing is the Moroccan sardine. While this stock is probably fairly fully exploited, and without much opportunity for substantial increase in total catch, the amount of fishing has been mainly controlled by economic factors on the demand side. The main use of the fish is for canning, for which there is a good, but inelastic market. There has therefore not been the pressure to increase fishing effort, and no need to apply the range of controls that were used (albeit apparently ineffectively) in Peru or southern Africa.

This somewhat peaceful picture may now be changing. In the last few years trigger fish (*Balistes* sp.) has been found in increasing numbers in some of the southern parts of the region, as well as in parts of the Gulf of Guinea where there is seasonal upwelling. Since there is so far no commercial fishing for *Balistes*, it is difficult to determine just to what extent *Balistes* has increased its range and abundance, but data from acoustic surveys show that in some areas it now dominates the pelagic community. Tough skin and strong spine render a commercial use difficult—even for fish-meal *Balistes* is unattractive. If the increase in its abundance causes, or will cause, a decrease in abundance of other, more valuable species, the long-term prospects could be gloomy.

The stock to show the most obvious sign of collapse is the relatively minor stock of *Sardinella aurita* off Ghana and the Ivory Coast. It inhabits the mainly tropical area to the south of the region considered here, in an area of seasonal upwelling, with ecological characteristics similar to those further north. Catches from this stock, which had been in the range of 50 000 to 100 000 tons for some time, dropped to a low level in 1972 to 1975; in 1978 catches recovered, but in 1979 there were signs that another collapse might occur.

The state of the larger stocks of sardinella, mackerel, and jack mackerel between Morocco and Senegal are less well known. Catches of mackerel dropped from a peak of 250 000 tons in 1970 to less than 50 000 tons in 1975 and 1976, and this stock has probably been too heavily fished. Probably both horse mackerel and sardinella could produce higher yields-per-recruit if fished harder, but the increase would not be great, and the effect of increased fishing on the recruitment and in the long-term stability of the stock, is unknown.

The situation of the demersal stocks is a little clearer, particularly for cephalopods

(CECAF, 1979c). Though other species, both demersal fish such as breams and flatfish, and pelagic species such as horse mackerel, are taken in the fishery, they are of only secondary interest, and much of the catches are discarded. Possibly for this reason the catch and effort data fit the simple production models well, with the catch-per-unit-effort falling off steadily as the effort increases. The correlation coefficient between total effort and c.-p.-u.-e. for the period 1966 to 1975 ranged from −0·67 for cuttlefish, to −0·97 for octopus. The recent effort has been at or beyond that giving the MSY for all three groups of species.

The bottom-living fish also are heavily fished, and there are suggestions that as fishing pressure has increased there has been a shift in the species composition away from the preferred species (the Poisson nobles of the French fishermen) towards the smaller and less valuable species.

*Future impacts and management*

The important issue in this region, and the issue that is different for most other regions, is not one of maintaining the biological productivity of the stock or the economic efficiency of the fishery, but one of allocation of the benefits from the resources between the coastal states and the distant-water fishing countries. At a time when the extension by countries of jurisdiction over the fish resources off their coasts has become generally recognized, a situation in which the local states take less than half the total catch is difficult to accept. The fact that the distant-water vessels come mostly from the large and rich developed countries make it even more difficult at a time when there is a growing recognition of the need for the rich to help the poor.

The balance between local and non-local fishing is not uniform. In the north, Morocco has a thriving local fishery and there is little foreign fishing, most of it by moderate-sized Spanish and Portuguese boats coming across the Straits of Gibraltar. From southern Morocco to northern Senegal there is little local fishing, but stocks are rich and non-local catches are high—probably 95% or more of the total. From Senegal southwards African fishing increases, and non-African fishing decreases, to perhaps no more than a third or a quarter of the whole.

The problems faced by Morocco (at least in respect of the northern and central waters), and also by Senegal and her southern neighbours, are a little different from those faced by Canada in respect of the foreign fisheries in the northwest Atlantic Ocean. These are to maintain the well-being of the existing artisanal fisheries (there is not so much difference between the canoe fisheries of Senegal and inshore fisheries of Newfoundland, the main one being the climate), to build up an industrial offshore fleet, and to control the activities of foreign vessels so that they take no more than can be taken from the stocks after the requirements of the local fishermen have been satisfied (in terms of both total catch and catch rate).

For countries with less well-established national fisheries—notably Mauritania—the short-to-medium term problems are different. Only in the long term can Mauritania expect to have a truly national fishery capable of taking more than a small fraction of the potential catches from the waters off its coast. Until then, unless the resources are left more or less untouched (and this would be to no one's advantage), most of the catches will have to be taken by foreign fishermen. The problem then becomes one of determining the proper conditions under which the foreign fishery should continue in order to provide high benefits to the coastal states on the short-to-medium term, while giving the best opportunities for the long-term development of a local fishery.

There are a variety of solutions to this problem. One is the establishment of joint ventures, usually between the local government and a foreign fishing company. The local involvement may be no more than sufficient share capital to ensure control, and a part of the profits, with the whole operational side being carried out by the foreign partner. Most arrangements however look to a greater local involvement, perhaps starting with local deck crews, and progressive takeover, until foreign involvement is no more than in some high-level managerial input, and in marketing the product. While joint ventures are fine in principle, they have often failed in practice, due to misunderstandings, or excessive greed on one side or the other.

An alternative is to allow fishing to continue, but on condition of payment to the coastal state. This can be a direct cash payment, of licence or access fees, based on the amount of fish caught, or of the number of vessels operating, or of payment in kind. The foreign country may for example provide training for local people in various aspects of fisheries, or build facilities ashore, e.g. for fishery research. Since the shore side of fishing is usually more profitable than the actual catching, another provision may be that a proportion of the catch is landed in the coastal state for processing. Like licence fees these payments in kind could be advantageous to both sides—the coastal state getting what it needs, and the foreign country providing what it is in the best position to supply. However, this is often not so in practice, the aid provided being not appropriate to the real needs of the coastal state.

In two important aspects the problems of the African states in determining and implementing an appropriate policy towards foreign fishing are harder than those of Canada. Most of the stocks of fish are found only in the Canadian EEZ (though there are important exceptions), and Canada has the capacity to enforce whatever management rules she determines should apply to foreign fishing vessels.

The sardine off Morocco (with the exception of the sardine in the extreme south, which may be a separate stock) remain throughout their life in Moroccan waters. The other important pelagic stocks, and also several other stocks, migrate up and down the coast, and these migrations will take them into the EEZs of two or more countries. Unless the countries concerned co-ordinate their policies, the actions by one country to control fishing in its zone can be nullified by increased fishing in another zone. The problems of the management of shared stocks are always complicated (e.g. GULLAND, 1980). They depend on the nature of the movement between national EEZs and the similarity of national objectives, and are increased by the presence of large foreign fleets. Without good collaboration between the coastal states, these fleets can shift their activities between one EEZ and another, seeking where the greatest catches can be taken, and where the conditions of access (amount of licence fees, etc.) are least burdensome. CECAF, as the responsible regional fishery body, and with the support of FAO, is therefore making considerable efforts to strengthen regional and subregional collaboration, and to assist the countries concerned with a particular stock to reach a common approach towards fishing by non-local vessels, and the establishment of a common management policy.

A common policy, however good in theory, will be useless in practice unless each coastal state can know what the foreign vessels are doing, and whether they are complying with the management measures, and unless—in the cases without compliance—effective control action can be taken. The question of monitoring control and surveillance is general, and is becoming recognized as one of the keys to effective management (FAO, 1981). The choice of management policy needs to take account of how a specific measure can be enforced, as well as the benefits that are expected if it is complied with fully. Measures that require frequent surveillance of the fishing grounds by expensive patrol vessels or aircraft may cost as much or more as the benefits that they are expected to produce (GULLAND, 1981).

These problems are more critical off northwest Africa than anywhere else in the world, except possibly the island states in the southwest Pacific Ocean. The resources are large, and potential benefits from good management high. For example, GRIFFIN and co-authors (1979) estimated that in the cephalopod fishery a licencing programme that reduced the fishing effort by 40%, and transferred only 15% of the increased net returns enjoyed by the vessels to the coastal states, would generate revenues to the coastal state of some \$27 million $yr^{-1}$. However, the coastal states do not have the capacity to operate a sophisticated system of surveillance such as those of Canada, UK, or USA, involving very expensive ships and aircraft, linked to a central control centre. These systems are necessary when the management system is complex, involving the detailed specifications of how much of each species can be caught by each group of vessels in each region. A simpler system may be sufficient with simpler management systems, e.g. a limit on the total number of vessels. The effectiveness of individual systems can be increased by co-ordination between adjacent countries. In CECAF moves are being made in this direction, e.g. by reviewing on a regional basis the problem of enforcing fishing regulations (JENNINGS, 1980), and holding consultations between countries. The process is bound to be long, but the ultimate prospects are good of attaining a regional approach to managing and controlling the fisheries in the region so that the stocks are healthy, and the fisheries are economically efficient, and that the African countries directly participate to the fullest extent practicable in these fisheries, and where this is not practicable, they enjoy their proper share of the benefits.

### *Southeast Atlantic Ocean*

The southeast Atlantic Ocean stretches from the mouth of the Congo River in the north; south and eastwards into what is strictly the Indian Ocean as far as 30° E. This eastwards extension makes fisheries sense in that the cool water fauna on the Agulhas Bank southwards from the tip of the African continent are similar to those in the Atlantic. It is only off Natal that the temperature rises, and the fish fauna and fisheries become similar to those of the Indian Ocean as a whole.

In the Atlantic Ocean, the boundary coincides with that of the International Commission for the Southeast Atlantic Fisheries (ICSEAF), though to the east the ICSEAF boundary extends further into the Indian Ocean off the coasts of South Africa and Mozambique as far east as 40° E. All the countries with significant fisheries in the southeast Atlantic belong to ICSEAF, and despite being politically a very mixed group (ranging from Cuba and the USSR to South Africa) have worked well together, especially on the scientific side. The various documents of ICSEAF—especially its statistical bulletins, the working papers presented to its Scientific Council, and the reports of the Council—provide very complete information on the fisheries of the region. A more convenient summary was given by NEWMAN (1977).

The region has been divided into six statistical areas, numbered by ICSEAF from north to south 1.1. to 1.6 in the Atlantic Ocean, and 2.1 and 2.2 to the east of the Cape of Good Hope with the boundaries at each 5° interval of latitude. These arbitrary divisions do not coincide neatly with political boundaries ashore (the waters off Angola cover divisions 1.1, 1.2, and part of 1.3; those off Namibia, parts of 1.3 and 1.5, and all of 1.4; those off South Africa, parts of 1.5 and 1.6, 2.1, and 2.2). Nor do they match what is known about stock separation. However, this knowledge is incomplete and was even less when the boundaries were established. The clearly empirical boundaries have enabled good series of detailed statistical data to be compiled.

*The stocks*

The fish stocks are typical of the major upwelling areas (table 5-13), with a variety of small pelagic species predominating. The three species of hake in the region are relatively commoner than in the other areas, perhaps because of the relative wider continental shelf, especially in the southern part of the area. Though even here the extent of the shelf does not approach the wide expanses of some of the northern temperate waters. Apart from the species shown in Table 5-13, the rock lobsters—particularly the west coast rock lobster *Jasus lalandii*—are, because of their high value, economically very important.

Of the three species of hake, *Merluccius polli* only occurs in the north (Divisions 1.1 and 1.2 of Angola). It is much less abundant than the others. The other two, *M. paradoxus* and *M. capensis*, are not easily distinguished, and are not separated in most statistics. *M. paradoxus* occurs in deeper water than *M. capensis*. The separation of stocks within the species is not clear, but the pattern of fishing and of changes in catch-per-unit-effort suggest a separation between a southern stock (or stocks) in Divisions 1.6 and 2, and a northern group in Divisions 1.3, 1.4, and 1.5. This separation has been used in much of the analysis of the state of stocks, as well as in setting management policies.

No other demersal stock approaches the hakes in importance. The larger catches are those of kingclip, panga, and the large-eyed dentex. Most of the kingclip is taken incidentally in the hake fisheries, and occurs throughout the region, but mostly in the southern half. The other species support directed fisheries, for panga in the southeast part, principally Area 2, and dentex in the centre (Divisions 1.3 and 1.4).

Of the pelagic species the clupeoids show three distinct areas of concentration—off South Africa (Division 1.6), off central Namibia (Division 1.4), and in the north (Divisions 1.1 and 1.2). In the first two areas the main species are sardine and anchovy, while in the third sardinellas are most important. The separation between the first two concentrations may be exaggerated by the fact that the fisheries on these species is mostly by local fleets, and their operations are constrained by the shortage of suitable ports. Nearly all the catches have been landed either around Cape Town or in Walvis Bay. However, the ecological information does support the separation, and tagging studies show that the cape and Walvis Bay stocks of sardine are separate (NEWMAN, 1970).

The other major species, mackerels and jack mackerels, are more active and probably move over longer distances. They do not present such clear areas of concentration; presumably, probable stock separation prevails. The Cunene horse mackerel *Trachurus trecae* does not occur in the southern part of the region, but the other species, *T. capensis* and mackerel, occur throughout the area, except in the far north (Division 1.1). This would be a large area to be inhabited by a single stock, but there are no clear indications of what divisions, if any, between stocks there are.

*The fisheries*

Both locally-based and long-range vessels are important in the region, and because of the political situation ashore the changes in the general legal regime of the sea have had much less impact than elsewhere. The total catches are divided roughly equally between the two groups of fisheries, the proportion changing as much with the changes in local catches (particularly the collapse of the Namibian pilchard stock) as with controls on the non-local fleets. Among the local fisheries it is convenient to distinguish the fisheries of Angola, the pelagic fishery based in Walvis Bay, and the fisheries in South Africa. Each of these have followed very different trends, especially in the last few years.

Though the weight of lobster caught is small, at most only a few percent of the total weight landed in the region, the high price results in its value being a significant part of the total—up to 20% of the total value of the South African production in some years. Its high value also means that the fishery developed early, at first largely for canning, at the end of the nineteenth century (GERTENBACH, 1963), but now virtually all the catch is frozen, much for export. Lobsters are caught, mostly in traps, by a large number of small boats operating on what are probably a number of independent stocks all along the South African coast, as well as the southern part of Namibia, where Luderitz is a major landing place. Catches reached a peak some years ago—of 13 100 tons in Division 1.5 in 1952 and 11 100 tons in Division 1.6 in 1960. They have since declined considerably. NEWMAN (1977) ascribes these changes, and minor peaks in later years, to the fishing out of accumulated stocks (including those made newly available by reductions in the minimum legal size).

Until 1964 virtually the only demersal fishing in the region was the South African trawl fishery for hake (and minor quantities of kingclip and other species) on the Cape grounds (south of 31° S and west of 20° E). This produced catches approaching 100 000 tons, with a value almost exactly the same as the pelagic fishery producing four times the weight. There was in addition a small fishery for mixed species in Angola, producing some 10 000 to 15 000 tons. In 1964 the distant-water fleets discovered the hake grounds further north, and catches steadily increased, reaching a peak of over 1 million tons in 1972. Since then they declined due to heavy fishing. The main distant-water countries are, roughly in descending order of recent catches, USSR, Spain, Poland, Japan, Bulgaria, Portugal, and Cuba, all of whom have taken more than 10 000 tons $yr^{-1}$. The peak catches by any one country reached over 600 000 tons for the USSR in 1972. South African vessels have also extended their activities, with larger vessels of similar catching and freezing capacities as at least the smaller distant-water vessels being added to the fleet. These vessels are now fishing on the more northern grounds first exploited by the distant-water vessels.

The pelagic fisheries based around Cape Town and Walvis Bay have developed in close parallel, with the same few companies being engaged in both. Initially the catches were mainly used for canning (pilchards, mackerel, horse mackerel). Production of fish-meal was merely a side-line using the waste from the canning plants. Later, fish-meal (and fish-oil) production became an activity in its own right, especially for anchovy and minor species (including some quantities of a lantern fish *Lampanyctus hectoris* of which 42 000 tons were landed in the exceptional year of 1973) which are not suitable for canning. The addition of reduction plants in the 1940s led to a rapid extension in the following years. By the early 1950s catches in both areas had passed a quarter of a million tons, with pilchard being the main species.

These fisheries are notable as being among the first to apply stringent management measures, described in detail by GERTENBACH (1963, 1973). A variety of measures were adopted, starting initially with the least disruptive, such as controls on the mesh size of the nets used (lampara nets in the early days, but soon the more efficient purse-seines were introduced). A little later, a closed season during the southern spring (September–October) when the fish were in poor condition was introduced, but very soon the authorities were tackling the problems of excessive fishing effort, and excess fishing capacity afloat and ashore. The details of the controls changed fairly frequently, and limited the number of processing plants, the amount of fish that each plant could handle, the number and capacity of the fishing vessels, and the total catch. For example, the total catch landed in Walvis Bay was limited to 250 000 tons in each year from 1955 to 1958, and the catches (at least as reported in the official statistics) differed by a maximum of 4400 tons from these quotas (GERTENBACH, 1963, Table 7). These regulations worked well, and resulted in sustained and moderately high level of catches, and (what was probably at least as attractive) very high profits to those participating in the industry, for some years. However, under the usual commercial pressures the catches allowed steadily increased. What now seems to have been close to a death-blow to the Namibian pilchard fishery was given by the entry of South African owned factory ships (converted Antarctic whale factory ships) operating just outside the 12-mile limit. Together with the locally-based fleet, these ships caught 1·4 million tons of pilchard in 1968. Later catches declined. The Namibian stock seems to have virtually collapsed, with only very small catches being taken (TROADEC and co-authors, 1980; BUTTERWORTH, 1981). The main attention has switched to anchovy, which now provides the majority (80% in 1979) of the total catches. Since this species can only be used for reduction, the value of the Namibian catch has decreased greatly.

The South African pelagic fishery has fared better, possibly because it has been able to spread its effort onto a wider range of species, including mackerel, horse mackerel, and red-eye herring. However, from a peak of over 100 000 tons in the early 1950s horse-mackerel have now almost disappeared from the catches of the purse-seine fishery. Since 1960 the catches of this fishery have remained around 400 000 tons, with a peak of 509 000 tons in 1967, and a low value of 325 000 tons in 1971—a remarkable degree of constancy for a fishery of this type.

Up to the time of independence, Angola also had an important industrial pelagic fishery for a variety of species (sardinella, anchovy, pilchard, horse-mackerel) with catches running up to 400 000 tons. Following independence most of the Portuguese settlers, who owned and operated nearly all the fishing vessels, as well as the plants ashore, left, often taking their vessels with them. Angola catches fell from nearly 400 000 tons in 1974 to only 74 000 tons in 1976. Since then catches have increased somewhat, but attention is being given to building up the small-scale fishery on a variety of species, mostly demersal, for direct human consumption, and catches of most pelagic species, except horse-mackerel, have been low.

Unlike the clupeoids, which have been caught almost entirely by the coastal states, using purse-seines or similar gear, mackerel and horse-mackerel are caught also by trawl, and have become increasingly the target of the distant-water fleets from eastern Europe. Catches of horse-mackerel increased greatly in the mid-1970s, probably due in part to a real increase in stock abundance, the total catches reaching a maximum, for the two species combined, of over 900 000 tons in 1978.

*State of the stocks*

The state of all the stocks in the region are kept under review by ICSEAF. Particular emphasis is paid to the hake because this is the most important species for most of the member countries, and it is for the hake fisheries that ICSEAF is best placed to introduce effective management. The results of these studies are conveniently available in the ICSEAF publications, particularly the annual proceedings (which includes the report of the Scientific Council) and the collection of the scientific papers. The latter are produced annually, and contain most of the original studies on which the reports of the Scientific Council are based, as well as the report of the Standing Committee on Stock Assessment.

The assessment of hake stocks have been carried out using production models (various methods of analysing catch and effort statistics) and age-structured models. These give results that agree well in showing that each of the main hake stocks are heavily fished. Fig. 5-6 shows the trends in catch and effort since 1965, as well as the empirical relation between catch-per-unit-effort and catch for the Cape hake in Divisions 1.3 and 1.4. The data are taken from Table 1 (p. 185) of ICSEAF (1980). The total catch figures are taken directly from published statistics. The catch-per-unit-effort data represent the number of tons caught per hour fishing of the Spanish fleet. From this the equivalent number of hours of the whole fleet for the year has been calculated and this used to give an average effort over the three years up to and including the year of observation. Following the method of GULLAND (1961) the observed c.-p.-u.-e. has been plotted against the mean effort to estimate the relation between c.-p.-u.-e. and effort under equilibrium conditions. This relation and the corresponding relation between total catch and effort is shown in the Fig. 5-6. The maximum catch occurs at 500 to 600 000 h $yr^{-1}$, a level that has been exceeded , often by a wide margin in every year from 1972 to 1977.

The empirical relation could have been drawn in different ways, and different theoretical relations could have been fitted objectively. Also, there can be reservations about the effort units used, since the efficiency of the average Spanish trawler may have changed (probably upwards) during the period. However, the main conclusion—that the hake stock is heavily fished— will be changed only in detail. It is indeed obvious from Fig. 5-6A, which shows that since 1970 the effort has greatly increased, but the catch (except for the peak in 1973) has stayed the same or fallen. The same conclusion has been reached for the hake stocks in the other Divisions. The analysis of age-composition data adds some important details. First, the calculations of yield-per-recruit give a relation between Y/R and fishing mortality that is very similar to the yield curve of Fig. 5-6B—though the detailed form depends on the value used of natural mortality which, as always, is poorly known. Calculations have also been made of the effect of changing the size at first capture. This shows that, even if the effect on the proportion of the catch that is discarded is not taken into account, the yield-per-recruit would be increased by an increase in mesh size used in the trawl fishery from the sizes commonly used in the early 1970s (this varied from fleet to fleet, but was often no more than 70 to 90 mm) up to at least 120 mm. Estimates were also produced of the strength of the recruitment in each year; these showed considerable year-to-year variation, but no clear trend, or relation to adult abundance.

To this extent the Cape hake stocks have shown themselves to be, in relation to stock assessment, well-behaved. So far there have been no surprises, or obvious departures from predictions based on the most simple models. This is perhaps surprising. Since hake are highly cannibalistic (MACPHERSON, 1980), PRENSKI (1980) for example found that between 85% and 92% (depending on depth of sampling) of all hake longer than 60 cm had smaller hake in their stomachs. This percentage falls off with decreasing size, but even among hakes as small as 40 cm , some had consumed smaller hake. Cannibalism must be a significant (possibly even the major) source of natural mortality among

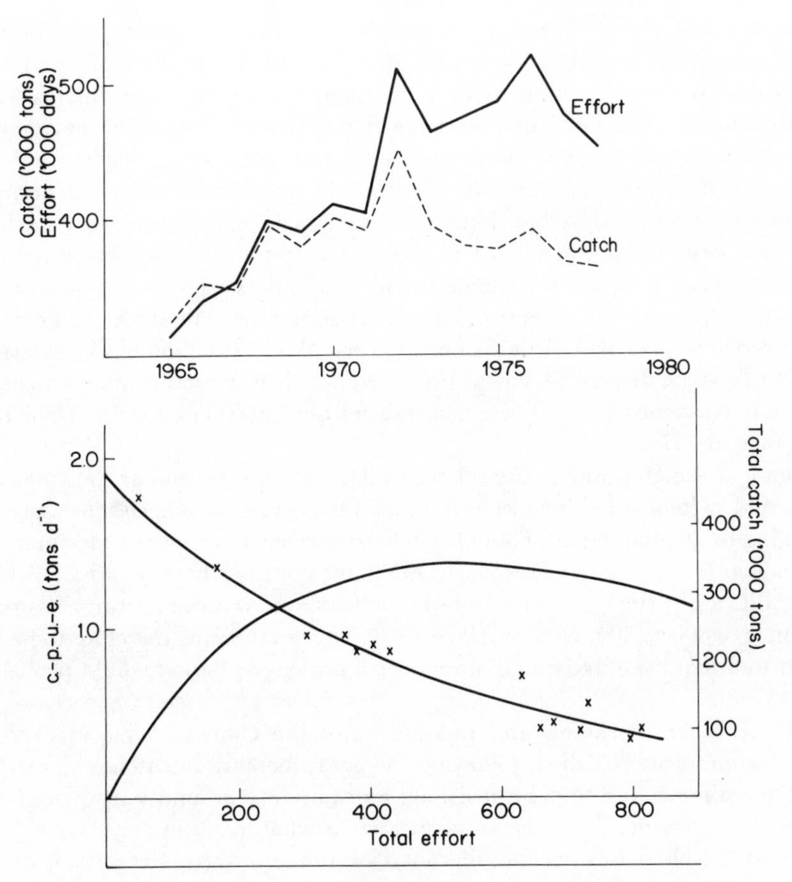


Fig. 5-6: (A) Trends in catch and effort in the fishery for Cape hake in Divisions 1·3 and 1·4 of the southeast Atlantic. (Based on data from the International Commission for the Southeast Atlantic Fisheries Report.) (B) Empirical relation between effort, catch-per-unit-effort, and total catch for the Cape hake. (Points are observed c.-p.-u.-e. plotted against average effort over two years.) (Based on data from the International Commission for the Southeast Atlantic Fisheries Report)

small hake, particularly pre-recruits, and immediately post-recruits. The pronounced reduction of larger hake due to heavy fishing might therefore have been expected to result in a higher effective recruitment. This does not seem to have been the case, at least not to the extent of having to modify assessments that ignore the possible effect of cannibalism.

Surprises have not been lacking in the pelagic fishery off Namibia. This fishery has had a history of sudden and unpredicted collapses that have occurred elsewhere, but for which there is no satisfactory quantitative explanation. In the case of the Namibian pilchard fishery, the management system, and the lags in it between changes in the fish stock and changes in the management measures, seems to have helped the final collapse (TROADEC and co-authors, 1980). There seems to be a high degree of variation in the stock, particularly in year-class strength, with a space between recent peaks of some 5 yr (it would not, on the present evidence, be correct to talk of a period). When the population increased, the reaction of the managers was to increase the catch quotas—but only after the change in abundance had been detected, and with some other inevitable delays in the system. Thus, the increased quotas became effective as the natural peak or abundance had passed, accelerating what would have been a natural decline. The system then reversed, with reduced quotas being applied only after the decline had been going on for some time, possibly when a recovery was already occurring (as seems to have been the case in 1972). The management system thus exaggerated the 'cycles'. The crash from the peak in 1975 seems to have been disastrous. The stock (and catches) are now at a very low level, with little sign of recovery. While attention has been switched to anchovy, this stock does not seem to produce enough to maintain total catches in the Namibian purse-seine fishery. These have fallen from 750 000 tons in 1973 to a little over 300 000 tons in 1979.

The state of the clupeoid stocks off South Africa is rather similar, but management controls, and switches of emphasis from species to species have been more successful in maintaining total production. Though pilchard catches have varied, declining to only 17 000 tons in 1974, they have recovered and have not since been less than 60 000 tons. Total production (principally of pilchard, anchovy, and red-eye herring) has remained in the range of 300 to 400 000 tons since 1968. This level seems therefore to be sustainable, but to attempt to take much more over a prolonged period would probably be a failure.

The Cape horse mackerel, and probably also the Cunene horse mackerel, have recently become heavily fished, following the great increase in catches since the early 1970s. The estimates of the present fishing mortality are around that giving the maximum yield-per-recruit. The interesting question is what the future levels of recruitment are likely to be. There has been a great increase in recruitment since 1975, which is in part responsible for the increased catches. What is not known is whether this is a natural event unconnected with events in other fisheries (e.g. the collapse of the Namibian pilchard stock), whether the high levels will be maintained in the future, or whether the recent high catches will make it more likely that future recruitment will be equal to, or even lower than, recruitment before 1975.

*Management*

Management in the southeast Atlantic Ocean has been largely out of step with the rest of the world. Prior to 1964 only one country was concerned, except in the north, and that country could and did set management controls without the need for international discussions that were so necessary in most other areas with large industrial fisheries. In 1983 this region is one of the last areas where, because of the political uncertainties ashore, large multinational fisheries (other than for tuna) are carried on outside national jurisdiction.

National management by the coastal states has concerned the lobster and shoaling pelagic fisheries. One of the small lobster fisheries presents what comes close to an ideal solution to many of the problems of fishery management. This is the isolated fishery around Tristan da Cunha (NEWMAN, 1977). UK authorities have given a concession to a single company, and the isolation of the island means that there is little likelihood of illegal fishing from outside. The company has therefore been able to operate in a way that maximizes its benefits, voluntarily limiting its catches to 900 tons, and avoiding catching the smaller lobsters.

In principle the same approach—limiting the amount caught each year, and protecting small fish—is used along the African coast. However, without sole ownership (and thus the incentive to take a long-term view of the problem) the short-term pressures to increase catches, and to be allowed to land small lobsters, have reduced the effectiveness of the measures. In addition, where there are large local markets, e.g. in hotels and restaurants, for illegally caught lobsters, the high value of lobster can make this a major problem. The local trade press reported in 1981 that this was approaching the level of organized crime of the Mafia type. The degree of illegal fishing can, however, be seen as a reflection of the success of management and the value of being able to fish legally. Application of stricter controls on the total catch, and on the minimum size (after what now seems to have been unwise reductions in the minimum size in 1964 in Division 1.5) are rebuilding the stocks.

The success in managing the shoal fisheries has been distinctly mixed. Elsewhere mixed success has usually meant that the biological productivity, and the level of the gross catch, has been maintained, but the economic and social problems of the fishery have intensified. Here it has been almost the reverse. From the early stages in the 1950s controls were aimed as much on controlling the inputs to the fisheries (number of processing plants, and of vessels) and hence the costs, as on the purely ecological aspects. As a result, the fisheries have in most years been highly profitable. This has raised its own problems as those outside the privileged group of companies holding existing licences attempted to enter the fishery. One result was the short-lived operations already noted of ex-whale factory ships of Namibia in 1966 to 1969. Another has been the continuing pressure to set quotas at least as high as scientific evidence allows. In the fluctuating environment off Namibia this raises special difficulties because it takes time for scientists to detect natural changes (reflected in the abundance of pilchard) and for authorities to react to the scientific findings. It seems that the duration of the time-lag involved, relative to the period of the natural changes, is just about the right length to put the system into the violent undamped oscillations that can be predicted from control theory. Whatever the cause, the result has been the collapse of the Namibian pilchard fishery.

The pattern of management of the hake fishery has been along the classic lines of management by an international Commission, and ICSEAF provides in many ways a better example of how such a body should operate than many of the older and better known bodies. After its establishment in 1969 its statistical and scientific groups quickly got down to work, and were soon pointing out the need for management. The Commission tackled the easier actions first, and in 1973 introduced a minimum mesh size of 110 mm, and also limited the amount of hake (30%) that could be taken in fisheries directed at other species.

As usual, these measures were incomplete and attention was turned to control of the amount of fishing. In international fisheries where allocation is important this can be done best by setting limits on the total catch, rather than controlling effort directly. This may not be the case in national fisheries. Since there has not been much discussions or agreement in the Commission on the general principles to be followed in setting the target amount of fishing (whether for example to aim at that which would give the MSY on a production curve, or at the fishing mortality giving the maximum yield-per-recruit,

Table 5-17

Catch quotas for Cape hake in 1980 to achieve various objectives (Data from ICSEAF Annual Proceedings)

| Objective | Catches during 1979 | |
|---|---|---|
| | Equal to quota | 80% of quota |
| $F_{0\cdot1}$ (high economic return) | 113 400 | 126 800 |
| $F_{max}$ (highest yield-per-recruit) | 171 000 | 191 100 |
| $F_{79}$ (maintain mortality at current level) | 304 600 | 272 000 |

or the $F_{0\cdot1}$ value (GULLAND and BOEREMA, 1973) or at some fraction of the current amount of fishing), the scientists were not in a position to recommend a single value of Total Allowable Catch.

Instead, they have usually given a range of values, corresponding to different possible policy objectives. Table 5-17 sets out the alternatives suggested by the Standing Committee on Stock Assessment at its 1979 meeting for quotas for hake in Divisions 1.3 and 1.4. Two sets of figures were given because at the time of the meeting it was not clear whether the full quota in 1979 would be taken, and this would influence the abundance of the stock at the beginning of 1980.

The proposed quotas vary greatly even though the long-term average yield under different objectives would not differ so much (with $F_{79}$ giving the lowest yield). This is because the current fishing mortality was far too high. Until the fish have adjusted to new, lower fishing mortalities, catches from stocks with these mortalities will be lower.

Naturally, an early reaction by administrators faced with a choice like that in Table 5-17, and the necessity for hard international and national bargaining if the lower figures were to be acted on, was to choose a TAC near the top of the range. It took only a few years of declining stocks, and progressively lower sets of alternative figures, for quotas—more or less along the lines predicted by the scientists—for this attitude to change. Now the figures chosen by the Commission are usually in the middle of the range. This provides the correct division of responsibility between scientists and administrators. It may be contrasted with situations in the North Sea where the scientists seem afraid (perhaps not without reason) to leave the responsibility for choosing management measures to the administrators, and only provide a single figure for the recommended TAC. Since there is no biological reason for preferring one level of the amount of fishing to another (so long as the existence of the stock is not threatened), and the scientists do not discuss non-biological objectives, this procedure is obvious nonsense, and can be defended only by the argument that any other system would be worse. The quotas agreed by ICSEAF are then divided between countries during discussions which formally take place outside the ICSEAF framework, being later reported to ICSEAF. The allocations in 1980 for the combined TACs for Divisions 1.3 and 1.4 (220 000 tons), and Division 1.5 (100 000 tons) were as follows (in thousand tons):

| | | | | | | | |
|---|---|---|---|---|---|---|---|
| Angola | 12·1 | German Dem. Rep. | 4·0 | Italy | 5·1 | Rumania | 4·0 |
| Bulgaria | 8·0 | Fed. Rep. Germany | 5·7 | Japan | 6·6 | S. Africa | 21·9 |
| Cuba | 16·6 | Iraq | 4·0 | Poland | 19·6 | Spain | 87·0 |
| France | 4·0 | Israel | 4·6 | Portugal | 9·4 | USSR | 107·4 |

Management of horse mackerel, the other important group of species fished by the distant-water fisheries has not advanced so far. This is partly because the fisheries themselves developed later, and partly because, with the natural increase in recruitment that has occurred it has been difficult to assess the effect of fishing accurately. Nevertheless the Commission did set TACs for 1981 at its 1980 session (though there was no agreement between countries on how these should be allocated).

*Future prospects*

The future prospects of any activity in southern Africa must be discussed within the framework of the political situation, and fisheries are no exception. The greatest fish resources lie off Namibia, where there is the greatest political uncertainty. Until Namibia achieves independence it cannot extend its jurisdiction over fisheries out to 200 miles (and certainly not in a way that would be recognized by the majority of the distant-water fishing countries). Angola too has internal problems, and its attitude to some of the countries with the largest long-range fleets off its coasts may well be influenced by the support it gets from these countries in other matters.

Once Namibia gains control over its resources it will probably be concerned not only with the problems already discussed (rebuilding the badly depleted pilchard stock, and moving the fishing mortality on hake more rapidly towards the lower and more economically attractive levels) but also with ensuring that Namibia itself gains more benefits from the fish resources. Though the Walvis Bay shoal fisheries are controlled largely by South African-based companies, so that the profits have not stayed in Namibia, the fisheries are an important source of employment both ashore and afloat. The distant-water fisheries at present contribute nothing, though in money terms they are more important. In an unpublished study, ROTHSCHILD has estimated that—at the prices current on the world market in 1980—the catches from distant-water trawl fishing was worth around $350 million. This is some two to three times the value of the pelagic fishery (depending on the state of the pilchard stock), and represents nearly $200 for each individual inhabitant of Namibia—a very considerable sum for an African country.

It is most unlikely that Namibia will itself have the manpower or other resources to engage directly in these fisheries for some time; when it does, emphasis may be placed on developing the shore side of the industry (processing and freezing) which is generally more profitable than catching the fish. Catching hake will therefore continue to be done by non-local vessels for a long time, but Namibia will aim at getting benefits from them in one form or other. This may be direct payment of licence or access fees (possibly substantial) or through other transfers, e.g. in training. In any case, since these fees or transfer must ultimately come from the net operating profit of the fishing fleets, one management objective will be to keep the fishing effort (and hence costs) reasonably low, so as to increase the amount of fees that can be paid. This may be a value of fishing mortality even less than $F_{0 \cdot 1}$, and will involve quite low catches for a period while stocks rebuild.

In the long run, a similar, relatively conservative management regime is likely to be adopted for most other fisheries in the region, particularly the shoal fisheries, where the attempts to maximize production from Namibian pilchard have been disastrous. That this stock will recover at some time seems, on the record of other pelagic stocks, highly probable, but the time-scale is less certain. It does not seem that a high abundance of Namibian pilchard is an unusual biological event—unlike the situation off California. A recovery time of a few years to a few decades seems most likely, but the timing may depend greatly also on the pattern of fishing. Selective fishing on anchovy or other potentially competing species could help. Any fishing on pilchard, especially too early a resumption of heavy fishing on a recoving stock, will certainly delay full recovery.

### (d) Tropical Areas

The tropical areas of the world have mostly escaped the attention of the large commercial fisheries from the developed world, and have thus also received little attention from fishery scientists concerned with stock assessment, and scientists and others concerned with management. However, these areas are rich in fish. In terms of the supply of food to hungry people and the provision of employment to poor people, these resources are among the most important in the world. They also supply some of the more significant management problems. However, some of the same factors that make them unattractive to large-scale industrial fishing, e.g. the large number of species involved, also make them difficult to study, and to provide good scientific advice for managers.

The tropical fisheries are at least as varied as those in temperate waters. For the purposes of discussion they can be grouped into the mixed artisanal inshore fisheries, typified by the coastal fisheries of southeast Asia; the industrial trawl fisheries for a mixture of species, including shrimp (in some places, e.g. the Gulf of Mexico, these fisheries are predominantly or exclusively for shrimp; elsewhere, e.g. the Gulf of Thailand, shrimp provide an extra incentive to a fishery that also lands large quantites of fish); the fisheries on pelagic fish, often with very simple gear, typified by the mackerel and oil sardine fisheries of southwest India; and the coral-reef fisheries, typified by those of the western Pacific Ocean.

#### *Southeast Asian Waters*

*The resources*

This region comprises the whole of FAO's statistical Area 71 western central Pacific Ocean, except for the scattered islands of the southwest Pacific, lying eastward of New Guinea. The production in the latter area is very small. Virtually all the 5 to 6 million tons of fish that puts the western central Pacific in the third or fourth place among the regions of the world (behind the northeast Atlantic and southwest Pacific and, in some years, the southeast Pacific) comes from southeast Asia.

One reason for this high production are the wide areas of shelf that occur in the south and western parts of the South China Sea, including the Gulf of Thailand and Java Sea. Moderate to high values of primary production are also promoted by the marked changes in current systems brought about by the patterns of monsoons. Apart from causing good mixing of the surface waters, there are local and seasonal upwellings, e.g. off the coast of Vietnam in June and July and seasonally in Banda Sea–Arafura Sea (WYRTKI, 1961).

The demersal stocks are rich in number and in variety. RITSRAGA (1976) lists 40 taxa caught in the Thai research vessel in the Gulf of Thailand. Nearly all of these are groups of species, at the genus or higher taxonomic level. The total number of species that occur regularly in the commercial landings run into the hundreds, and several dozen may occur in a single trawl haul. There are differences in the species composition from place

Table 5-18

Catches of small pelagic species in southeast Asian waters (Based on FAO *Yearbook of Fishery Statistics*)

| Species group | Indonesia† | Malaysia | Philippines | Thailand† | Total‡ |
|---|---|---|---|---|---|
| *Decapterus* spp. | 78·4 | 20·4 | 146·3 | 86·1 | 33·9 |
| Other jacks, etc.* | 147·5 | 63·8 | 90·2 | 62·9 | 393·4 |
| *Sardinella* spp. | 134·8 | — | 106·4 | 116·5 | 346·2 |
| *Stolephorus* spp. | 102·0 | 37·2 | 70·5 | — | 196·5 |
| Other clupeoids | 18·1 | 22·5 | 26·3 | 11·0 | |
| *Rastrelliger* spp. | 84·6 | 53·7 | 56·9 | 66·0 | 245·0 |
| Total | 565·4 | 197·6 | 496·6 | 342·5 | 1515·0 |

*Other species in FAO group 34.
†Includes catches in the Indian Ocean.
‡Includes catches by other countries, but excludes catches by Indonesia and Thailand in the Indian Ocean.

to place, particularly according to depth and the type of bottom, but there is a broad similarity of composition throughout the region. This ranges from the large predatory fish (sharks, groupers, etc.) through the intermediate predators and larger invertebrate feeders (snappers, rays, etc.) to small invertebrate and detritus feeders (which include the valuable penaeid shrimps). In general the abundance is greatest nearshore, and decreases in the deeper offshore waters. Not much is known about the movements of the fish, or the stock structure of demersal fish. Though the same species may be found over much of the region, limited tagging studies suggest that the individual demersal fish does not move far (CHOMJURAI and BUNNAG, 1970). For management purposes the fish can be treated as forming a number of independent stocks, without much interchange.

The major of small pelagic species (excluding tunas) (Table 5-18) are probably more active, though migrations are reasonably well known only for the *Rastrelliger* in the Gulf of Thailand. Tagging experiments (KUROGANE and co-authors, 1971) have shown considerable movements within the Gulf. Though most of these movements remain within the Thai jurisdiction, some fish migrate south and west into the waters off Malaysia. Much less is known about the movements of *Decapterus* and other carangids. Indeed, the location of their spawning grounds is unknown. Probably they migrate at least as far as *Rastrelliger*, so that there is some interchange of fish between adjacent countries, but there is a separation of stocks over longer distances.

Table 5-18 shows that there is considerable similarity in species composition in the catches by the four major countries in the region. The similarity in the actual stocks is probably even greater. The absence of *Sardinella* spp. in the catches reported by Malaysia, and of *Stolephorus* spp. in Thailand are probably either because a fishery on these small species has not yet developed, or because, while they are caught, they are not distinguished in the statistics.

*The fisheries*

The fisheries in southeast Asian waters are predominantly small-scale and artisanal, carried out by small vessels (often unpowered) close to the shore. The process of development has generally been one of evolution rather than revolution, with gradual increases in the efficiency of the gear, in the size of the boats, and the power of the engines (when fitted), as well as in the number of fishermen. Thus, a great variety of gears (traps, lines, surrounding nets, trawls, etc.) are used, especially in inshore areas. Most developed of the fairly traditional fisheries are those for pelagic species, but even here a variety of different types of surrounding nets (Thai purse-seine, luring purse-seine, Chinese purse-seine, etc.) are used.

The exception to this slow development has been the Thai trawl fishery. Its development since 1960 has been only slightly less dramatic than the Peruvian anchoveta fishery. Prior to 1960 some trawling had been done in the Gulf of Thailand by local pair-trawlers, and standard North Sea-type single boat trawls had been used by research and survey vessels. The latter was unsuccessful because the gear used had been too heavy for the soft muddy bottom of much of the area. Demonstration by a German aid project with a more suitable trawl proved highly successful, and was quickly learnt by Thai fishermen. Catches grew from less than 100 000 tons (mostly by pair-trawl) in 1960 to approaching 1 million tons by 1973, nearly all by otter-trawl. The expansion of the activities of the Thai trawl fleet after 1960 has been very similar to, though much more rapid than, the expansion of English trawl fleet between 1890 and 1950. As local grounds became heavily fished and abundance dropped, the fishermen moved further and further from the initial grounds in the northern Gulf of Thailand. In the South China Sea they fished further down the coast of Cambodia and Vietnam and off Malaysia. They have also moved to the west coast of Thailand, and in the Indian Ocean as far away as Bangladesh. This expansion has been accompanied by a gradual increase in average size and power of the vessels. The larger Thai trawlers are now some 30 m long, with engines of up to 1000 hp, i.e. similar to the North-Sea trawlers.

*State of the stocks*

The assessment of the state of stocks in southeast Asian waters faces much greater difficulties than in say the North Sea. Many more species are involved, there are fewer scientists and a shorter history of fishery research, and much of the basic data is poor or missing. In particular, the statistics, at least until recently, have been incomplete and often inaccurate. An exception to the lack of good basic data is the Gulf of Thailand demersal fishery. As part of the German assistance programme, a trawl survey of the Gulf was started in 1963, and has been carried out regularly from 1966 onwards. This provides data on changes in fish stocks, free from the complications involved in catch and effort data from the commercial fishery, that is equalled only by data from similar international surveys on Georges Bank in the northwest Atlantic Ocean.

The immediately striking feature of these data is the continued and severe fall in the total catch-per-hour trawling from nearly 250 kg in 1963 to barely one-fifth of this in 1975 (BOONYUBOL and HONGSKUL, 1978). This is very closely related to the increase in the amount of fishing by Thai trawlers, as measured by the total catch of demeral fish (Fig. 5–7A).

The normal procedure of calculating catch-per-unit-effort from independent statistics of total catch and total effort can for this fishery be reversed, and an index of the total effort obtained by dividing the survey c.-p.-u.-e. into the total catch. Relating this total effort to the survey c.-p.-u.-e. and applying the SCHAEFER model as though the demersal stocks could be treated as a simple biomass of undifferentiated fish gives a very good fit (Fig. 5-7B). The derived relation between total catch and total effort suggests that the stock is heavily fished, and that probably the amount of fishing has already passed the level giving the greatest sustained catch.

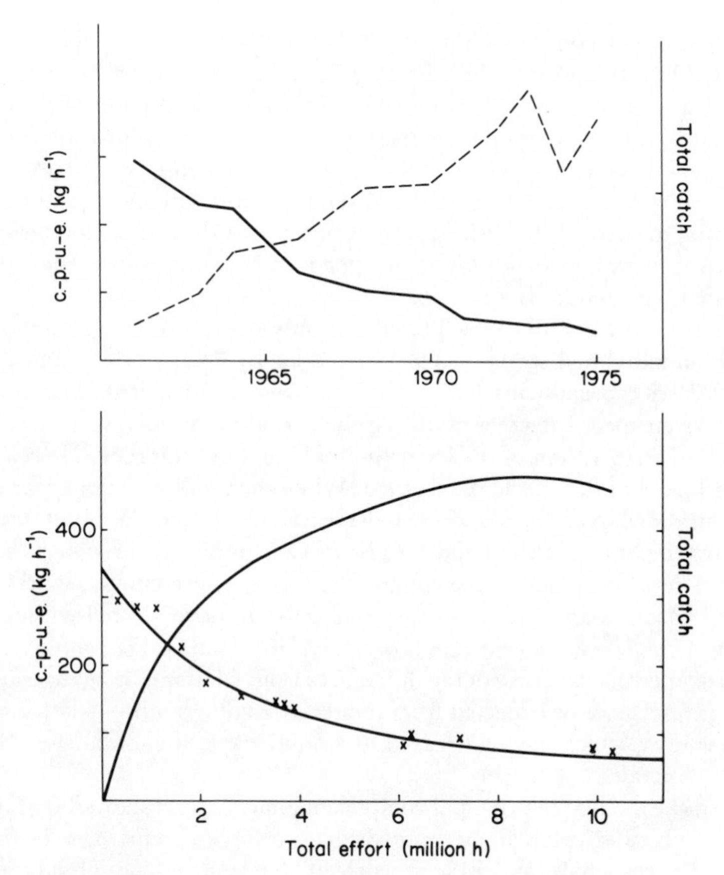


Fig. 5-7: (A) Trends in catch rates by research trawlers, and of total catch in the Gulf of Thailand. (Based on data provided by BOONYUBOL and HONGSKUL, 1978). (B) Relation between effort, c.-p.-u.-e and total catch for demersal fish in the Gulf of Thailand. (Based on data provided by BOONYUBOL and HONGSKUL, 1978)

Several objections can be raised against this. First, the Thai catch statistics do not distinguish where the fish were caught, and the figures include large quantities taken by Thai trawlers outside the survey area, off Vietnam and Cambodia. Thus, the total catch and the total effort have been less than the figures suggest. However, the decline in the stock is unchanged, as presumably would be the conclusion that it is heavily fished. The principal changes would be in the estimates of the potential yield from the survey area, and the level of fishing effort at which it would be obtained. The second, and more serious, theoretical objection is that a community of several hundred species should not be treated as a single mass, especially as there have been big changes in the relative frequency of the various species.

There are some weighty counters to these arguments. The most telling, in the context of the management of the fisheries, is that a proper analysis, taking account of the dynamics of each species (or at least all the major species) and their interactions, as well as obtaining reliable information on the location of Thai catches, would be prohibitively time consuming. The conclusion that the stocks are too heavily fished is so clear that the analysis would probably only alter the details, and not the general conclusion, and the important matter is to start achieving some control of the amount of fishing. Furthermore, the apparently simple-minded analysis may be more sophisticated than it seems. If the interactions between species act fairly quickly—as may be the case in short-lived species—what is observed in the sea will tend to represent the equilibrium position, corresponding to each level of fishing, taking account of the interaction between species. That is to say, the simple analysis of total catch may be more realistic than attempting to combine single-species analyses.

Nevertheless it is likely to be useful and certainly interesting to look more carefully at the changes in individual species, as has been done by PAULY (1979) and POPE (1979). PAULY (1979; see especially his Table 8 and Fig. 8) looked particularly at the proportion by which a given species decreased during the period of the survey, as measured by the coefficient in the regression of the logarithm of abundance on time. He related these to the feeding habits of the species concerned. While the relation is not exact, there was a clear tendency for larger species to have decreased most, and the small 'trash' fish and invertebrates (including shrimp and squids) to have decreased least or to have actually increased. There were some exceptions, and large predators (sharks, serranids, *Muraenesox*) which might from the general pattern have decreased most, actually changed by about an average amount—less than large zoobenthos feeders and medium-sized predators. Some of the differences from what might be expected on purely ecological grounds can be expected from the relative vulnerability to bottom trawls. For example few rays, which have decreased to around 1% of the initial stock, escape from the path of the trawl.

POPE used the method of principal component analysis, and showed that the variation in annual catch rates could be accounted for to a large extent (83% in the area as a whole, and between 55% and 81% in each of the nine individual subareas) by two components. The first of these represents the steady trend throughout the period, decreasing for most species, but increasing for squid and crabs. The second component reflects changes in species (*Scomberomorus* and snappers) which did not decline initially, but started to decline after a period.

These, and other studies, while still somewhat tentative, do show that changes in species composition are susceptible to analysis and understanding and hence to prediction. The significance to management is that the value of different species varies greatly. POPE (1979; his Table 4.21) quotes prices on the Thai auctions in 1973 ranging from 40 baht (20 baht = 1 US$) per kg for silver pomfret down to 1 baht for trigger fish (the large quantity of fish used for fish-meal would fetch an even lower price). In general high price is associated with large size, and with vulnerability to fishing, and an above average decrease in abundance (shrimp is an obvious exception). The highest gross value therefore probably occurs at a level of fishing effort lower than that producing the greatest weight.

For no other stock in the region is the data so good, nor has any been so carefully examined, as the Gulf of Thailand demersal stocks. For a number of the inshore stocks it has been possible to obtain sufficient statistical data on catch and effort (in terms, for example, of number of vessels or number of fishermen) to apply simple types of the production model. These have usually shown—for example in the Malacca Straits (SOUTH CHINA SEA PROGRAMME, 1976a; WANG and PATHANSALI, 1977) and in parts of the Philippines (SOUTH CHINA SEA PROGRAMME, 1976b, 1977)—that in the areas with better data—which are also those with the best developed fisheries—the inshore demersal stocks are heavily fished, with little possibility of increased local catches.

A similar approach has been successfully used for some of the pelagic fisheries (SCSP, 1978b). These have shown that some of the mackerel stocks are heavily fished (e.g. in the Gulf of Thailand and in the Malacca Straits), while most of the scad stocks seem to be lightly exploited, with reasonable opportunities for increasing the total catch.

Apart from direct assessments of individual stocks (or rather stocks in particular areas), attempts have been made, especially by regional or subregional working groups, to look at the resources of the region as a whole, and to estimate the potential or state of exploitation of stocks in less well studied areas by extrapolation from the better known stocks (AOYAMA, 1973; SCSP, 1976a, 1978a). Research vessels have now made surveys similar to those in the Gulf of Thailand, though for a shorter period, in many parts of the South China Sea and adjacent areas. These surveys show that there is a broad similarity, not only in species composition, but also in catch rates (when due account is taken of the current amount of fishing) over much of the region. The differences in density between different areas is no more than the differences within areas, and is less than the consistent difference between depth zones. In all areas catch rates decrease with increasing depth, and in depths greater than 50 m catches are nearly always small. This causes the larger vessels, which can fish anywhere, to concentrate their effort in shallow water (at least when local regulations, and the activities of local enforcement agencies allow them to do so). Since the numerous small-scale fishermen have little choice but to fish along the coast, the intensity of exploitation falls off with increasing depth and increasing distance from shore. This makes it difficult to conceive any simple and realistic summary statement of the state of the stocks. The participants at the workshop on demersal resources of the Sunda Shelf (SCSP, 1978a) felt that demersal catches (currently around 1·3 million tons) could be increased to some 3 million tons, but a great deal of this biological potential occurs in the larger areas of deeper water. The fish are undoubtedly there to be caught, but it is far from certain that the catches of an individual boat could be high enough to support an economically viable fishery. Shallower areas where there is the biological potential for increased demersal catches, and probably also the economic potential, include the east coast of the Malaysian Peninsula, and the north coast of Borneo. In contrast, the coastal waters of the Gulf of Thailand, the Malacca Straits, and the north coast of Java are among the areas which are fully exploited.

One difficulty in attempting a differentiation between deep and shallow waters, whether at 50 m depth or at some other depth, or at some fixed distance from the coast, is that the extent of movements of fish across such boundaries is unknown. There are differences in species composition, and some species are confined to shallow water, or to deep water, but many occur at all depths, or at a wide range of depths. For several of these species there is a graduation of size, with the fish in deeper water being, on the average, larger. This would suggest some movement offshore as the fish grow.

*Management*

In the Gulf of Thailand trawl fisheries, the administration faces problems familiar to those of many fishery administrations in developed countries—too many trawlers are chasing too few fish. These problems are likely to increase in the future as Thai fishermen lose access to fishing grounds off neighbouring countries. At present, many vessels—particularly the larger and more powerful ones—are fishing off foreign coasts. Although these countries have now declared some form of extended jurisdiction, this is not fully enforced in all countries, and also some non-local fishing is allowed until local fishing capacity is developed. In due course, few of any Thai trawlers are likely to obtain access to foreign waters. When all the other vessels return to the already congested waters of the Gulf of Thailand, or the small area off the Thai west coast facing the Indian Ocean, the problems of over-fishing and over-capacity in these waters may get even worse.

Reducing severe over-capacity in a fishery has proved to be very difficult, even in the rich countries around the North Atlantic and North Pacific Oceans. In developing countries, with generally weaker administrations and fewer opportunities for alternative employment for those engaged in the fisheries, the problems are even more severe. In the short run it may not be possible to do much, and the emphasis needs to be put more on preventing the growth of over-capacity in other fisheries in the region—the trawl fisheries in other countries, or the pelagic fisheries generally. For example, BOONPRAKOB (1976) has pointed out the need for measures, including limited licences, in the Thai mackerel fishery, which is now at about the optimum level, in order to avoid future problems.

The other serious management problem is the conflict between the groups of fishermen, particularly the small-scale fishermen using simple gears along the coast, and the more industrialized fisheries. The former can only exploit a limited strip of water, perhaps no further from the coast than one or two miles, or perhaps out to 10 or 12 miles if the weather is favourable. This leaves the resources further offshore unutilized, to the extent that they do not move into the coastal zone. For some time therefore emphasis in the development plans of Indonesia (ZACHMAN, 1973) and elsewhere was on the development of the industrial fisheries, especially as these were in the best position to supply high-valued species (particularly shrimp) for export. However, as these fisheries grew, the trawlers found that the best catches were close to the shore. Their presence in these waters has brought growing complaints from the small-scale fishermen, against the direct damage to their gear, and the reduction in the stocks. Most countries have therefore introduced legislation, where this did not already exist, prohibiting the use of trawling and other industrial types of fishing within specified distances from the shore, or in waters less than certain depths. In some cases these regulations are complex, with a variety of different distances or depths for different sizes of trawlers or types of gear.

A big problem with this type of regulation, even in its simplest form, is that it is difficult to enforce. Vessels can fish in the legal zone, and move inshore at night, in poor visibility, or whenever they feel the chances of detection and arrest are small. For this reason Indonesia has moved towards a complete ban of trawling in all waters. Another problem is that it is far from clear to what distance trawling should be prohibited to protect the inshore fishermen. This arises from the uncertainty about the movements and migrations of the fish, and the extent to which fish caught, say 20 miles off the coast by the industrial vessels, might have later moved into the coastal zone to be available to the small-scale fisheries. There are therefore difficult scientific and administrative problems to be solved before this approach can be fully effective. However, given the social conflicts and the growing emphasis on problems of the poorest section of the population (which in many countries includes the small-scale fishermen), it is clear that the management of fisheries in this, and similar regions, will be increasingly concerned with measures, like the prohibition of trawling in the coastal zone, which implicitly allocate some of the resources to these poorer groups of fishermen.

## *Indian Ocean*

### *The resources*

In fishery terms the Indian Ocean is the forgotten ocean. It has largely been forgotten by scientists because it lies a long way from most of the great centres of marine research. However, the International Indian Ocean Expedition did something to reduce this neglect. The Indian Ocean has been largely neglected by major fishing industries because it lacks the wide shelf areas, or the major permanent upwellings that produce rich fish stocks, and because it is far from the major markets. While the Indian Ocean accounts for over 20% of the total surface area of the World Ocean, it contains less than 13% of the shelf area, and supplies only a little over 6% of the total world marine catch.

A unique feature of the northern Indian Ocean is the impact of the monsoons on the surface currents, and hence on the general ecology of the region. Twice a year the main wind systems reverses from the southwest monsoon (in April–September), to the northeast monsoon (October–March). This in turn causes many of the surface currents, especially in the Arabian Sea, to reverse. Thus, during the southwest monsoon there is a strong and very swift current northwards along the Somalia coast, but during the northeast monsoon the flow is southwards. These currents produce upwellings at one season of the year or another along much of the northern Arabian Sea from Somalia to the west coast of India. Upwelling is particularly strong along the south coast of the Arabian Peninsula during the southwest monsoon. However, in few if any places does the upwelling persist throughout the year (CUSHING, 1969a).

The distribution of the resources is therefore uneven, with most of the potential occurring either in upwelling areas in the northern Arabian Sea, or in shelf areas around the Bay of Bengal. Table 5-19 shows that these areas account for some 80% of present catches from the Indian Ocean, though the apparent concentration may be exaggerated because these areas also have the highest human populations and the highest intensity of fishing.

In the Arabian Sea the main resources are the small pelagic fish typical of most upwelling areas, with oil sardine (*Sardinella*) being particularly important. In addition, various species of mesopelagic fish appear to be more common in the northern Arabian Sea than anywhere else in the world (GJØSAETER and KAWAGUCHI, 1980). Demersal fish are less important, and in some areas the presence of large permanent demersal stocks is prevented by the invasion of water with extremely low oxygen content at some stages of the monsoon.

In the Bay of Bengal the richest areas are the wider shelves in the north and northwest, where the productivity is increased by the great runoff from the land through the estuaries of the Ganges and the Irrawaddy. The main species are those adapted to muddy bottoms and somewhat brackish conditions. They include several, like *Hilsa* (shad) that run up the rivers to spawn.

Along the east African coast, from southern Somalia southwards, including the islands—which range in size from Madagascar downwards to the small atolls—the

Table 5-19

National catches in the Indian Ocean in 1979 (thousand tons) (Based on FAO *Yearbook of Fishery Statistics*, Vol. 48)

| East Africa and Islands | | Red Sea and Gulfs | | Northeast Arabian Sea | | Bay of Bengal | | Southeast | | Others† | |
|---|---|---|---|---|---|---|---|---|---|---|---|
| Somalia | 32·6 | U.A.E. | 64·4 | South Yemen | 51·6 | Sri Lanka | 148·8 | Indonesia | 134·1 | Japan | 40·4 |
| Tanzania | 84·6 | Ethiopia | 25·8 | Oman | 198·0 | India (E) | 405·0 | Australia | 65·1* | Korea | 57·6 |
| Others | 50·5 | Others | 100·0 | Pakistan | 259·7 | Bangladesh | 100·0 | | | Others | 42·0 |
| | | | | India (W) | 1090·8 | Burma | 412·8 | | | | |
| | | | | Maldives | 27·7 | Thailand | 184·6 | | | | |
| Regional totals (and % of whole ocean) 167·7 (4·7) | | 183·6 (5·2) | | 1626·7 (45·6) | | 1251·3 (35·1) | | 199·2 (5·6) | | 140·0 (3·9) | |
| Whole Indian Ocean 3568·5 | | | | | | | | | | | |

*Includes catches off southern Australia to 150° E.
†Catches from countries outside the Indian Ocean, mostly of tunas.

bottom drops off rapidly, and the shelf is mostly extremely narrow. The current systems do not favour high production, and only in a few favoured localities (e.g. between Zanzibar Island and the mainland, in Sofala Bay off Mozambique, and in northwest Madagascar) are there even moderate-sized stocks of small pelagic and other fish. Elsewhere the main resources are reef fish, which along the Kenya coast increase the attractiveness to tourists, but supply only small catches.

Similar conditions prevail in the Red Sea and, on the opposite side of the ocean, along most of the Indian Ocean coast of Indonesia and western Australia. The Persian Gulf, however, is shallow and has relatively more abundant (but still not particularly rich) stocks.

Crustaceans are exceptions to the general poverty of the Indian Ocean. They account for about 15% of the total world production (i.e. its correct share in terms of shelf area) of both lobsters and shrimps. The stocks of rock lobster *Panulirus cygnus* of western Australia forms one of the biggest resources of rock lobsters in the world (second only to the true lobsters of the North Atlantic Ocean), and smaller stocks of more tropical species are scattered around most of the Indian Ocean where suitable rough bottoms occur. The biggest shrimp resource is off the southwest coast of India, but other smaller, but still important stocks are found in several parts of the Bay of Bengal, in the Persian Gulf, off northwest Madagascar, and elsewhere.

*The fisheries*

The statistics of the fisheries in the Indian Ocean for 1979 are summarized in table 5-19 (in terms of national catches) and Table 5-20 (in terms of species and groups of species). In general, the distribution of catches follows that of the resources. The biggest catches are taken in the northeastern Arabian Sea and the Bay of Bengal, but the distribution of populations is also important—much more is taken off the Indian west coast, with a large population and a long tradition of fishing, than by the scattered population along the Arabian coast. On the other hand, the Indian population is even higher on the east coast, particularly in West Bengal, but here the catches are only half those on the west coast. Although Java is one of the most densely populated parts of the world, the catches along its south coast—as well as west and south of Sumatra and the rest of the Indian Ocean side of Indonesia—are small.

Catches by countries outside the Indian Ocean account for less than 4% of the total, and of this nearly all is tuna, caught by the long-line fleets of Japan, Korea, and China (the island of Taiwan). The other non-local vessels are mainly trawlers, now mostly operating under joint venture and similar agreements, working off the coasts of Somalia, Yemen, and Oman, though Taiwanese trawlers are also fishing off the Indian west coast. Among the other vessels in the region that fish more than a short distance from their home ports are the Thai trawlers. As the resources in the Gulf of Thailand became depleted some trawlers moved to the west coast of Thailand and spread out north and westward. Until the late 1970s they were taking substantial catches off Bangladesh and Burma. With the extension of jurisdiction by the coastal states concerned, the activities of these vessels has been reduced. Another minor but interesting medium-range fishery has been that of Mauritius. In a manner very similar to the traditional Portuguese dory fishing in the northwest Atlantic Ocean, motherships have carried a number of small dory-type boats to the banks between Mauritius and Seychelles, where they fish with hand-lines for larger predatory fish. Only a few vessels have ever been involved, and the catches have been small.

With these exceptions, the fisheries of the Indian Ocean have been with small vessels, often without engines, seldom if ever staying out for more than a few hours. A convenient description of the Indian fisheries is given by VIRABHADRA RAO (1973). The fisheries in most of the rest of the area are not much different. The species composition of the catches, as reported to FAO, is given in Table 5-20. This shows that nearly 30% of the catches are of unspecified fish. Some of these are the mixed species caught in the trawl and other fisheries, which are not distinguished on the market after the more valuable shrimp and large fish have been sorted out, but the high proportion is due to the fact that for several countries, including Bangladesh and Burma, with a total between them of half a million tons, no data at all on species breakdown are available. Indeed, for some of these countries, and some others for which nominal species data are given, there are considerable doubts about the precise value of the total catch. The statistics in the tables should therefore be treated with more than the usual caution.

Despite these reservations it is clear that in terms of volume, the most important fisheries are those for pelagic species—particularly the Indian Oil Sardine *Sardinella longiceps* which accounts for over 8% of the total catch, and possibly as much as 10% if the quantities reported as unspecified fish, or unspecified clupeoids are included. The centre of this fishery is the Kerala coast of southwest India. The fish live offshore during the first half of the year, but during August they begin to move inshore, starting with the older fish. During the last quarter of the year the population is mostly close to the shore, where they are caught with a variety of gears including large beach-seines, boat-seines, gill-nets and cast-nets. Because the fish are delicate, and are caught in large, and somewhat unpredictable quantities along open beaches without much processing or

Table 5-20

Catches of major species groups from the Indian Ocean in 1979 (thousand tons) (Based on FAO *Yearbook of Fishery Statistics*, Vol. 48)

| Demersal | | Small Pelagic | | Tunas, etc. | | Others | |
|---|---|---|---|---|---|---|---|
| Flatfish | 15·8 | Shads | 42·1 | *Scomberomorus* spp. | 50·9 | Shrimp | 257·1 |
| Cods | 2·4 | Oil sardine | 315·0 | Small tunas | 65·2 | Lobsters | 16·4 |
| Bombay duck | 125·5 | Anchovies | 165·8 | Skipjack | 32·6 | Other crustacea | 30·6 |
| Sharks and rays | 152·4 | Other clupeoids | 200·9 | Large tunas | 122·7 | Cephalopods | 26·5 |
| Others | 447·7 | Jacks etc. | 204·7 | Billfishes | 11·6 | Other molluscs | 29·4 |
| | | *Rastrelliger* spp. | 114·8 | | | Others | 1·5 |
| | | Other mackerels | 95·0 | | | | |
| Total | 743·8 | | 1138·3 | | 283·0 | | 361·5 |

| Totals: | | | |
|---|---|---|---|
| | Demersal fish | 743·8 | (20·8%) |
| | Small pelagic fish | 1138·3 | (31·9%) |
| | Tunas | 283·0 | (7·9%) |
| | Unspecified fish | 1041·9 | (29·2%) |
| | Others | 361·5 | (10·1%) |
| | Grand total: | 3568·5 | |

transport facilities, only a proportion is used for human consumption, traditionally a little as fresh fish and the rest cured. Other uses of the fish are for oil used in jute, leather, and soap industries, and as fertilizers in the plantations. New efforts are being made to use more for human consumption, especially canned.

The short season of peak landings has been an obstacle to the development of a consumption market. Acoustic surveys have shown that during the off season (roughly from February to July) the fish are not far offshore and are accessible to modern types of gear from mechanized boats. Moves have been made to encourage the development of an industrial-scale fishery, with vessels in the 30 to 50 ft range. However, this has led to clashes, violent at times, with the traditional fishermen, who see such a development as a threat to their livelihood.

The other pelagic fisheries follow broadly similar patterns. The traditional fisheries depend on the movements of the stock into the coastal zone where they can be taken with a variety of gears. This makes them highly vulnerable to changes in migratory behaviour, as well as to possible changes in actual abundance. The variations in sardine and mackerel were noted a century ago by DAY (1868). Reasonable statistics are available from 1925 onwards (RAJA, 1969), and these show that the catches of sardine dropped from some 20 000 tons in the 1930s to an average of no more than 500 tons in the years 1943 to 1949. In the 1960s and 1970s the variation in both the sardine and mackerel fisheries have been striking. Even so, there have been pronounced differences in individual year classes. In the data presented by BANERJI (1973) for the mackerel fishery at Karwar the year class entering the fishery as O group in the 1963 to 1964 season appeared to be some ten times as abundant as those in the two preceding years, and twice as abundant as those in any of the preceding 20 years except 1949 to 1950 and 1957 to 1958.

Attempts have been made to relate the year-to-year fluctuations in sardine year-class strength to environmental factors. Connections have been found with the patterns of the monsoon (RAJA, 1973). Bearing in mind the importance of the monsoon to the whole climate conditions around India, this is reasonable, but as in all such correlations the small numbers of pairs of observations and the wide choice of possible environmental parameters, make the statistical significance of the relation somewhat suspect. Even less is known about the long-term fluctuations, e.g. the 7-yr period of virtually zero catches in the 1940s. Possibly little will be known until such time as another collapse of this type occurs.

In terms of value (or at least economic value) the pelagic fisheries are much less important than those for shrimp. Penaeid shrimps find a ready and high-priced market, particularly in Japan and the United States. As a result a high proportion of the catches of the larger species are exported. While there have been complaints, particularly in the United States, about the quality of some of the shrimp imported from Indian Ocean countries, leading to some batches being condemned and a lower price being obtained than for shrimp from some other exporting countries, the value of shrimp exports from India alone in 1978 was $200 million.

In several of the countries round the Persian Gulf, and also in northwestern Madagascar, the shrimp fisheries were developed by foreign companies concerned in the international shrimp trade. These brought in large industrial shrimp trawlers, mostly from the Gulf of Mexico, usually with their own crews, or at least captains. The biggest shrimp fisheries, such as those of Pakistan and India, which catch 20 000 tons and over 200 000 tons, respectively, are of more local development. The main catches are taken offshore, where the larger individuals are found, by trawlers, but much is caught inshore. The latter areas, especially the backwaters and lagoons of southwest India, provide the nursery grounds for several of the larger *Penaeus* species, which move offshore as they grow, but they also contain smaller species e.g. of *Metapenaeus*, several of which remain in the backwaters. In these areas a variety of traditional gears are used, mostly with non-powered vessels—gill-nets, stake-nets, and some such as lift-nets which do not need a boat at all.

In contrast to the pelagic and shrimp fisheries, demersal fisheries are not important. Much of the bottom fish that are landed are caught while fishing for shrimp, though it is often not possible to draw clear dividing lines in the spectrum of trawl fisheries for shrimp only (with any fish that are caught not necessarily being released), for shrimp with fish being a valuable by-catch, for fish with shrimp as the by-catch, and purely for fish with few if any shrimp being caught.

Trawling is in fact an advanced gear, requiring an engine—preferably a fairly powerful engine. Apart from the shrimp fisheries (including trawl fisheries for mixed fish and shrimp) and the Thai fishery, trawling is an uncommon type of fishery in the Indian Ocean. Hand-lining, trap-fishing (especially in the rough bottom and coral reefs of East Africa, Seychelles and Mauritius), and gill-netting are the main methods used for catching demersal fish.

*State of resources and management*

While no comprehensive and detailed review of the state of the resources of the Indian Ocean and their potential has been made, a number of studies exist of smaller regions, particularly the waters around India (QASIM, 1973). In the western Indian Ocean the resources have been reviewed at two workshops organized by the FAO/UNDP Indian Ocean Programme in Karachi in January 1978 (ANON, 1978a), and Seychelles in October–November 1978 (GULLAND, 1979a) dealing with the resources north and south of the Equator, respectively.

South of the Equator the resources are not rich, and though catches are small a number of surveys with acoustic equipment, trawls, and other fishing gear have failed to reveal major unexploited resources. The largest of those that were found were pelagic stocks (mainly *Decapterus*) over the shelf area around the main islands of the Seychelles. The potential yield has been estimated as some 75 000 tons annually. In contrast some local stocks are heavily fished, notably around Mauritius, as well as the reef fishes along parts of the African coast. Management of these fisheries, at least in the sense of introducing total catch quotas and other sophisticated measures, is difficult. Though measures are needed, they will take simpler forms such as banning the more destructive type of fishing (e.g. dynamite), and encouraging those types of development which will shift the fishing activities away from the more heavily fished stocks.

North of the Equator, in the area from Somalia to Pakistan, the picture regarding the resources is brighter, and the general need for management less. After the International Indian Ocean Expedition showed that the general production in this region was high, more specific fishery surveys, notably by the Norwegian vessel *Dr. Fridtjof Nansen*, were directed to this area. These surveys have shown that an unusually high proportion of the primary production appears in the form of mesopelagic fish, such as lantern fish (GJØSAETER and KAWAGUCHI, 1980) rather than as sardines and other familiar types. Until technical and economic problems of harvesting and utilizing mesopelagic fish are solved, the prospects of increasing fish production from this area are less than the magnitude of total primary production would suggest. At the same time the stocks of oil sardine are substantial, and present catches of pelagic fish, as well as of demersal fish in some areas (e.g. off Somalia), could be increased. This does not mean that there are no management problems. The fact that the resources are relatively rich is well known, and in at least the central, Arabian, sector, capital is fairly readily available. Since the resources are far from limitless, and are far from uniformly spread along the coast, there are considerable dangers of over-capitalization and of building large fish-meal plants in inappropriate places. Resource management considerations are also needed to determine the arrangements under which foreign vessels are fishing for high-valued demersal fish and cephalopods off Yemen and Somalia. How large a catch can the resources withstand? Should catches be set at a rather lower level so as to increase the catch rates, and hence improve the ability of the foreign fishing interests to pay high licence fees?

Surveys have also been made in the Red Sea and the Gulfs. The former is similar to much of the East African coast. It has a narrow shelf with moderate to low productivity. The catches come from small-scale artisanal fishermen, and there are neither major opportunities for development nor are there problems calling for management action. The Gulfs feature a wide area of shallow water. Extensive surveys, particularly those of a co-operative regional programme organized by FAO (ANON, 1981c), have shown the existence of a substantial quantity of demersal fish and provided extensive details of their distribution. The potential yield from the whole area, including the Gulf of Oman, has been estimated at some 180 000 tons (SIVASUBRAMANIAM, 1981). Undoubtedly the biological potential exists for quite large catches (at least several tens of thousands of tons) to be taken. Unfortunately, the fish are scattered, and possible catch rates are not high. Costs in the Gulf area (except for fuel) are high, while there is only a moderate demand for fish. Under these economic conditions the prospects for developing a demersal fishery are not good. The prospects for fishing for small pelagic fish, of which the most abundant appear to be sardines, are similar. Estimates of biomass have been obtained by acoustic methods during the Gulf surveys (ANON, 1981b), and they are probably rather more abundant than demersal fish. They are on the average rather shorter lived, implying a higher natural mortality and a relatively higher potential yield. However, they are on the whole less valuable, so that the economic prospects are less attractive.

For the present, therefore, the main fishery in the Gulf area is likely to remain that for shrimp. This presents at least one peculiar feature, in that the large-scale industrial fishery developed first. Between the 1964/65 and the 1967/68 period industrial catches along the west coast of the Gulf, from Kuwait to Oman, increased from under 400 tons to over 9000 tons. This fishery was carried out by relatively large vessels, mostly of the Gulf of Mexico type owned by large fishing companies, often with interests in many countries. It is only later, from about 1973 onwards, that a so-called artisanal fishery has developed with motorized dhows. In the last few years their catches, especially around Bahrain, have made up a significant part (up to 30%) of the total landings. It soon became apparent that most of the stocks in the Gulf were becoming heavily fished (BOEREMA, 1969; ANON, 1977). The resulting need for management has become more apparent with a recruitment failure in 1979. This has led to several industrial companies ceasing operations, and the introduction of a closed season by several of the coastal states.

The stocks around India have been well studied. Vessels of what is now the Exploratory Fishery Project based in Bombay have carried out a very large number of hauls over a long period. Though these have concentrated on east-coast grounds, the data enable estimates to be made of the standing stocks of fish over most of the shelf waters around India down to some 200 m (JOSEPH, 1974; JOSEPH and co-authors, 1976). For the west coasts of India and Pakistan, the information available suggests that the stocks are becoming quite heavily fished. Estimates of the potential yield are in the range 400 000 to 600 000 tons annually. The actual catches of demersal fish, as reported in official statistics, have risen from an average of 240 000 tons in 1965 to 1969 to over 400 000 tons in 1977. The situation regarding pelagic fish, including sardines and mackerel, is somewhat similar. Estimates of biomass have been obtained from acoustic surveys. Converted into potential yield, these are higher, but not very much higher, than recent catches. A similar conclusion was reached by BANERJI (1973) on the basis of age-composition data, from which he deduced that the fishing mortality was about half that giving the maximum yield-per-recruit.

While formal assessment of the state of shrimp resources has not been made, or at least none is generally available, it appears probable that the stocks are being heavily fished. Shrimp catches rapidly increased in the late 1960s and early 1970s. These reached a peak of just under 250 000 tons in 1974, and have since declined to only around 180 000 tons in 1979. This decline has occurred despite the continuing high demand for shrimp, and the encouragement given by the Indian Government to what is an important export industry. The high price of fuel might have brought about some decrease in the fishing effort among the smaller mechanized boats. There have also been extensions of fishing efforts into areas that previously had not been significantly exploited, especially on the east coast.

For the shrimp stocks, therefore, it is likely that the Indian Government faces a management problem of the classical kind—too many boats chasing too few fish—involving the same social problems and pressures from established interests in attempting to reduce the excess capacity.

In the fisheries for demersal and small pelagic fish the problems are somewhat special or, at least in India, are met with particular force. Increased catches could be taken from these stocks. In principle, higher catches are highly desirable in a poor country short of protein. However, increasing catches from the demersal stocks will require greater emphasis on these stocks, lying offshore in deeper water, which are at present only lightly exploited. Here, larger mechanized vessels are required. Similarly, increased pelagic catches and greater catches during the off-season—when the distribution system is more able to handle greater quantities and to ensure that a higher proportion is used for human consumption—will also require larger and more powerful vessels, e.g. purse-seiners.

Since the stocks are already at least moderately heavily fished (the distinction with the demersal fishes of lightly exploited species offshore, and more heavily fished inshore is far from absolute), increased fishing mortality caused by the addition of the mechanized vessels will cause a fall in stock abundance, and a drop in the catches of artisanal vessels. The development of the industrial fishery is, therefore, strongly opposed by artisanal fishermen, even to the point of riots. There is no simple answer to this management problem. The Government of India has to make a choice between national objectives—increasing food production and improving the well-being of the rural poor—which are in this instance irreconcilable.

In the Bay of Bengal little is known about the state of the stocks off Bangladesh and Burma. Considerable doubt surrounds even the figures of total catches. Surveys by research vessels suggest that there are substantial stocks and that the total catches can be increased, especially by working further offshore, beyond the range of the present vessels. Off Thailand, however, and also off the west coast of Malaysia, fishing is already intense, particularly on demersal fish. The management problems are similar to those in the Gulf of Thailand—how to reduce the over-capacity of the trawl fleet, and, in the case of Thailand, how to deal with the exacerbation of these problems caused by the loss to the Thai trawl fleet of the distant-water grounds following the extension of jurisdiction to 200 miles by neighbouring coastal states.

Along the Indonesian coasts the management problems are somewhat similar to those off India. In general, the stocks are not so rich as to attract a high intensity of fishing, and are therefore not so heavily fished as to require urgent management action. One minor exception occurs at Cilicap in the centre of the south coast of Java. Here, the local stock of shrimp has attracted considerable attention, especially from small local vessels. It is heavily fished, and management is needed if the potential economic and social benefits from this fishery are to be obtained (VAN ZALINGE and NAAMIN, 1977). Another exception involves the Bali Strait (between Bali and Java) where there has for a long time been an important fishery for oil sardine by small local vessels. The stock is moderately heavily fished, but catches could probably be increased, especially by introducing larger and more powerful boats, capable of fishing outside the Bali Strait in the open Indian Ocean. There is then, again, the same choice as for the Indian pelagic fishery between increasing total fish supply and maintaining the living of fishermen.

Off western Australia, catches, except for crustacea, are very small, and the stocks as a whole are only lightly exploited. However, there are two important management problems. Off the northwest coast, roughly from around Broome into the Gulf of Carpentaria, the shelf is wide, and, at least by tropical standards, supports a moderately rich fish stock. This has been exploited for some time by long-range, but small, vessels—mostly pair-trawlers from Taiwan—fishing for the more valuable demersal species. Their intensity of fishing is probably only moderate, so that stocks as a whole are not fully exploited. At the same time their activities have reduced the abundance of stocks, and this reduction appears to be sufficient to make the development of a local Australian-based fishery on these stocks difficult. The catch rates are too low for such a fishery to be profitable under Australian conditions.

The major management problem concerns the large crustacean fisheries, especially that for the western rock lobster. This is extremely valuable, and the large export market is almost limitless in relation to the size of the stock. Once the fishery became properly established it rapidly expanded until the stocks were fully exploited, with annual yields around 8000 tons (BOWEN and CHITTLEBOROUGH, 1966). Since the resource was under the control of a single country, and a single state within that country (it may be noted here that reaching agreement on fishery management policies between states or provinces within a country with a federal or similar structure can often be as difficult as reaching agreement between countries), the decisions on the need for, and implementation of, management measures could be taken comparatively quickly. As usual, these first concentrated on the simplest measures to maintain the biological production starting with size limits (later supplemented by the use of escape-gaps in the pots so that the under-sized lobsters could escape, and did not undergo the risks involved by being brought on deck, sorted out from legal-sized animals, and dumped back over the side), and soon progressing to closed seasons. Later, emphasis was placed on reducing the overall fishing effort, and thus the costs of fishing. This has been achieved by limiting the number of boats that are licensed to engage in the fishery, and the number of pots that each vessel can carry (based on the length of the vessel). In turn, this resulted in a great increase in the value of a boat with a licence, compared with a similar, but unlicensed vessel, representing a considerable capital gain to the fishermen. Other, less directly measurable effects of a long period of progressively more detailed management can be seen in the general prosperity of the fishermen. While there are many problems in the management of the western Australian lobster fishery—the difficulties of enforcement when the animal can be easily caught, and easily sold (e.g. to local restaurants); the continuing attempts by fishermen to outwit the regulations (e.g. by using larger pots, thus increasing the fishing power of a vessel while staying with the fixed number of pots); and the fact that the benefits are realized mainly by the fishermen, not by the consumers, the taxpayers as a whole, or the management authority—the fishery does stand as one of the better examples of the benefits that can come from good management.

### *Gulf of Guinea*

This region covers the southern part of the eastern central Atlantic statistical region, from 9° N south to the Congo River. In broad terms it corresponds to the band of tropical water lying between the rich areas of permanent upwelling of northwest and southwest Africa. As pointed out by TROADEC and GARCIA (1979) in their detailed review of the resources of the area the region is, however, far from being homogeneous. In the northern winter most of the area is tropical, with surface temperatures exceeding 27 °C over the whole region. In the northern summer, however, the Benguela system in the south moves north and causes upwelling and surface temperatures to fall below 24 °C as far north as Pointe Noire and beyond almost to the Equator. There is also significant upwelling off the Ivory Coast and Ghana. This tends to raise the productivity of these areas above that normally expected in tropical areas. On the other hand, the purely tropical areas with high temperatures throughout the year and no upwelling, e.g. in the Bight of Biafra, are poor.

Another reason for the general low level of the resources in the region—and for the differences between different parts of the region—lies in the width of the continental shelf. Along most of the coast it is narrow, mostly between 10 and 30 miles wide, but in the extreme north, around the Bissagos Islands, it extends out to some 100 miles. A less wide shelf, but still above average for the region as a whole, occurs in the eastern part of Nigeria, from the delta of the Niger westward to the Cameroons.

The Gulf of Guinea, perhaps because it is the tropical area most accessible to marine scientists of Europe, has been well studied. In particular, two major international activities—the International Cooperative Investigations of the Tropical Atlantic (ICITA) and the Guinean Trawling Survey (GTS)—were carried out in the early 1960s (MONOD, 1967; WILLIAMS, 1968).

Among the demersal fish, LONGHURST (1969) and others have distinguished three main communities: (i) those of shallow water, above the thermocline, living on muddy bottoms, including the deltas of the Niger and other rivers; (ii) those of hard bottoms in shallow waters; and (iii) those in deep water below the thermocline. Within each class there are differences as one moves along the coast, but there are pronounced similarities throughout the region, and also considerable similarities with the communities on similar bottom types and depths throughout the tropics. On muddy bottoms, the main groups of fish are sciaenids, polynemids, ariids, and cynoglosids. The same groups, though with fewer sciaenids, are found in the western Atlantic Ocean and in the western Indian Ocean. In the western Pacific Ocean and eastern Indian Ocean, Leiognathidae becomes a very important element of the soft bottom community. As elsewhere, shrimps (in this region the pink shrimp *Penaeus duorarum*) are an economically very important component of this community. Shrimps are found in most locations throughout the region (GARCIA and LHOMME, 1979, in their Fig. 1, identify 14 separate grounds), with the biggest stock in the Niger Delta's Cameroons grounds. The Cameroons take their name from Rio dos Camaroas (shrimp river) of the early Portuguese navigators.

The fauna of the shallow hard-bottom areas is much poorer in number of species than in the western Atlantic, or in the Indo-Pacific region, and there are few coral reefs with their rich fauna. The main commercially used genera—*Lutjanus*, *Lethrinus*, and *Pagrus*—are, however, much the same as in other oceans.

In terms of abundance a clear trend emerges from the Guinean Trawling Survey. Densities, deduced from catch rates of the survey trawlers, and expressed as kg ha$^{-1}$ fell off from a maximum of 70 to 80 kg ha$^{-1}$ in the Bissagos area to a minimum of not much more than 10 kg in the south of Liberia. They increase again to some 20 to 30 kg off the Ivory Coast (presumably in relation to the seasonal upwelling) and then fall off before rising again to a maximum of some 40 kg off the Congo.

The typical pelagic fish of the region, and more particularly of the areas where there is some seasonal upwelling, is the round sardinella *Sardinella aurita*. A distinct, and locally important stock exists off Ghana and the Ivory Coast. The species is also found in large individual numbers in the north and south of the region in the extensions of the major upwelling areas. The flat sardinella *Sardinella eba* has a similar general distribution; but it is found further inshore. The other pelagic species of the major upwelling areas (mackerels, horse mackerels) are seldom found, except in small quantities. In the purely tropical waters, with no upwelling, the most important clupeoid is the bonga *Ethmalosa fimbriata*. This is essentially a coastal species, tolerant of low salinities, and found as much as 200 km upstream in large rivers (Senegal and Gambia) with still some, but minor, influx of saline water (BOELY and FREON, 1979). Being a coastal species it probably forms local stocks, without much migratory movement. This is borne out by tagging experiments (LONGHURST, 1960).

Mention should also be made of tunas. Yellowfin and skipjack support large fisheries, initially with pole and line, but now mostly with purse-seine, in the central Gulf of Guinea. These are discussed in a later section (p. 994).

*The fisheries*

Though the Gulf of Guinea countries are poor in sea fish, they are rich in people. Nigeria alone has some 55 million people, out of a total African population of some 350 million. West Africa, though not devoid of wild life, lacks the great herds of wild animals that are such a feature of the savannah lands of east Africa, and is not particularly favourable for raising cattle or other domestic animals. Therefore fish, both limnic and marine, are an important source of animal protein. Most of the coastal people have, therefore, a long tradition of fishing. Of these, the Fanti people of Ghana are among the best known. With thousands, or tens of thousands of fishermen engaged in several countries, the total catch of these traditional fishermen is large. Also, within the limitation of small open boats and, until comparatively recently, the absence of engines, the fisheries are well developed. Where the canoes have to work off the open beaches they are usually very well adapted—long, narrow and light—to passing safely through the heavy surf. Some of these canoes are large, up to 10 m (30 ft) in length, with crews of a dozen or more. Most of these larger canoes are now equipped with outboard engines. The commonest gears are gill-nets, hand lines, and various types of seine. In Dahomey there is in the lagoons a highly specialized type of fishing—the acadja fishery—which is somewhat intermediate between ordinary fishing of wild stocks and fish culture. It consists of fixing dense masses of brushwood in shallow water. These serve to aggregate the fish for easier harvesting but probably also, perhaps by providing shelter from predators, increase the total production.

Large though the catches by the traditional fisheries are, they have not been able to supply all the fish needed to satisfy local demand. In several countries there has been pressure to increase supply by developing an industrial fishery. Usually such pressure has come from governments, but on occasions from commercial interests engaged in the distribution and marketing of fish. The first such industrial fisheries were based on the use of small to medium-sized European vessels, typical of the home fisheries of England or France. Much of the landings in the Ivory Coast come from a fleet of moderately small (15 to 30 m long) trawlers and seiners. The smaller trawlers (less than 400 hp) work only in Ivorian waters, the larger trawlers (400 to 900 hp) and also the purse-seiners, work also as far off as Senegal. Other countries with local industrial-scale fisheries, mostly trawling, include Ghana, Nigeria, and Togo, but their catches are less than those taken by the Ivorian fleet (some 15 000 tons of trawl-caught fish, and up to 20 000 tons in some years from the purse-seiners). The waters off the Ivory Coast are among the richest in the region. To increase their catches several countries have turned to distant-water fishing in the more productive areas to the north or south.

Nigeria is engaged in distant-water fishing by proxy. Large trawlers from Poland and other countries, which have caught a mixture of species in the bottom-trawl fisheries, land a proportion of catches in Nigerian ports, usually of the species that are unfamiliar and therefore less appreciated in the markets of north and eastern Europe, but which fetch a good price in African markets. The distribution of these fish to inland markets is unusually simple and effective. A large solid block of frozen fish is taken directly from the trawler, placed on the back of a truck, and covered with sacks. After a day's drive it is nicely thawed out and ready for sale.

Ghana and Liberia have their own national fleets of large distant-water trawlers. In the case of Ghana, these are mostly government owned, directly or indirectly, but the Liberian vessels are owned by a company that was initially concerned with the operation of a cold chain and fish distribution, and which moved into the catching side of the industry to better assure the supply of fish.

Given the poor resources, and the strong local demand, there has been relatively little interest in the area from foreign vessels, or from foreign businesses concerned with the export of fish. The exception are tunas and shrimp. Abidjan and Tema are important bases and transhipment ports for the long-range tuna fleets (particularly France and Japan), but these African countries are also developing their own national tuna fleets, usually in association with the traditional tuna fishing (and tuna importing) countries.

On the Ivory Coast, emphasis has been placed on purse-seining, principally for yellowfin. The Ivorian fleet works closely with that of France and Senegal. In Ghana, emphasis has been on live-bait fishing for skipjack, in association with Japanese interests. Most of the countries with shrimp stocks export much of their catches. In most cases they use moderate-sized trawlers, typical of industrial shrimp fisheries elsewhere, and often in association with large multinational companies engaged in world-wide shrimp trading. In addition, large amounts of shrimps are also taken by artisanal fishermen in the coastal area, particularly when some stocks of juvenile shrimp migrate to sea through the narrow entrances of the coastal lagoons.

*State of stocks and management*

For a few of the industrial trawl fisheries, notably off the Ivory Coast and Nigeria, statistical data and biological information are sufficient to make assessments by standard methods, using either production or analytic models. These analyses, which have been usefully reviewed by DOMAIN (1979) and by the CECAF working party (ANON, 1979), show that even 20 yr ago some of the stocks were already quite heavily fished (LONGHURST, 1964). Apart from demonstrating that in the fisheries examined little increase in catch would be expected from fishing harder, the analyses often also examined the effects of changing the size of fish capture, and revealed that increases of catches could come from protecting the smaller fish.

For example, FONTANA (1979) presented yield isopleth diagrams (or strictly isopleths of yield-per-recruit) for selected species in the Congolese fishery, showing the yield as a function of the mesh size used (for sizes from the current 45 mm up to 77 mm), and the amount of fishing as a fraction of the present. Taking the five larger and most valuable species the greatest catch would be taken with a 77 mm mesh (or even larger), though with an amount of fishing greater than the present. If the smaller and less valuable *Brachydeuterus auritus* is also included the optimum mesh size falls to 70 mm (at very high rates of fishing) or around 60 to 65 mm for moderate rates. This illustrates an important theoretical dilemma for most multi-species fisheries, in that the optimum strategy depends on which combination of species is included.

The dilemma is, however, under present circumstances more theoretical than practical. Even when including the smaller species the optimum mesh size is considerably greater than that currently used by Congolese vessels (60 to 65 mm compared with 45 mm). The same conclusion that catches in the main demersal fisheries would be improved by the use of larger meshes—at least up to around 60 mm—has been reached by nearly all analyses, for example by the study of BAYAGBONA (1965) of the Nigerian croaker fishery. The mesh sizes currently in use in several of the coastal countries are shown in Table 5-21. The table lists the small sizes used in many trawl fisheries, and also indicates that the mesh size in use is not necessarily the same as that set in the regulations. Enforcement of mesh sizes, and indeed of all fishery regulations, is never easy. Even the countries around the North Sea, with strong administrations, have not been uniformly successful in implementing mesh regulations. The important factor seems to be as much achieving the understanding and co-operation of the fishermen as the police-type aspect of control and enforcement. Nevertheless, there are indications (ANON, 1979) that the Nigerian regulations introduced at the end of 1971 did result in some increase in the mesh size used, and that this was at least in part responsible for the increase in catch-per-unit-effort observed in 1973 and 1974.

Table 5-21

Mesh sizes (stretched mesh in the cod-ends of trawls and beach-seines of countries in the Gulf of Guinea (After ANON., 1979; modified)

| Country | Trawlers | Shrimp trawlers | Beach-seines |
|---|---|---|---|
| Ivory coast | 40–60 mm* | 33–40 mm | No data |
| Ghana | 35–50 mm†<br>20–26 mm‡ | No fishery | 20 mm |
| Togo | *ca.* 40 mm | No fishery | 20–50 mm |
| Nigeria | § | § | Little fishing |
| Cameroon | *ca.* 40 mm | *ca.* 40 mm | Little fishing |
| Gabon | Probably 40 mm | Probably 40 mm | Little fishing |
| Congo | 40–45 mm | No fishery | Little fishing |

*1968 regulations set a mesh size of 35 mm each side, equivalent to about 63 mm stretched mesh.
†Large trawlers; there is a regulation size of 70 mm.
‡Small trawlers.
§No direct information available; a regulation size of 76 mm for trawlers and 44 mm for shrimp trawlers were introduced in December 1971.

These assessments only covered those stocks for which there has been a well-established fishery, i.e. the shallow-water stocks off certain countries. Elsewhere the assessments have to be based on the results of the GTS and other surveys. These show quite clearly that in total there is a substantial biological potential in the demersal fish in deeper waters, as well as in some shallower areas. However, the fish are scattered, so that the catch rates are low and include quantities of low-valued fish, notably *Brachydeuterus*. Thus, the economic returns from fishing these stocks—even at their maximum unexplored abundance—is poor, and the prospects for development of these fisheries are slight.

Assessment of the potential is complicated by the changes that have occurred in the population of trigger fish *Balistes capriscus*. The recent outburst of *B. capriscus* has already been noted in relation to the upwelling areas to the north, where it appears to be becoming a dominant member of the pelagic community. However, at the time of the Guinean Trawling Survey it was noted as being only a regular, not particularly common member of the hard-bottom community, with catches of a few individuals per hour.

The increase of *Balistes capriscus* first became noticeable in Ghana in 1970. By 1972 it was dominating the catches in the inshore vessels and in the trawl surveys. The percentage in the catches in inshore trawlers rose from 27% in 1972, to 53% in 1974, and nearly 90% in 1976 (ANSA-EMMIN, 1979). Various reasons for the rise of *B. capriscus* have been discussed (DOMAIN, 1979). Though commonly classed as demersal, the young (8 to 12 cm) are pelagic, and the individuals up to 20 cm feed mostly on plankton (BECK, 1974). Only the larger fish appear to be fully demersal. Thus, it is not unreasonable to examine most of the other changes that have occurred off West Africa for possible relationships with the *B. capriscus* outburst. Possible reasons considered by DOMAIN (1979) are the decline of sardinella, the growth of shrimp fishing, and the decline in sparids. Probably it will be some time before the reason is properly established, but in the meantime the rise of *B. capriscus* and its interaction with other species will have to be borne in mind when setting management policies.

Among pelagic stocks the only one with good data, as well as the most important, is

Table 5-22

*Sardinella aurita.* Catches (tons) near Ghana and Ivory Coast (After BOELY and FREON, 1979; modified)

| | Ghana | | Ivory coast | |
|---|---|---|---|---|
| Year | Canoes | Purse-seines | Purse-seines | Total |
| 1963 | 5500 | 1960 | 500 | 7960 |
| 1964 | 22 250 | 7180 | 10 900 | 40 330 |
| 1965 | 2350 | 1550 | 4300 | 8200 |
| 1966 | 4200 | 5800 | 5774 | 15 774 |
| 1967 | 25 200 | 11 000 | 10 930 | 47 130 |
| 1968 | 2500 | 1800 | 3941 | 8241 |
| 1969 | 15 900 | 6600 | 7304 | 29 804 |
| 1970 | 14 700 | 4800 | 10 911 | 30 411 |
| 1971 | 27 490 | 3724 | 4614 | 35 830 |
| 1972 | 72 350 | 14 716 | 7676 | 94 742 |
| 1973 | 4701 | 615 | 502 | 5818 |
| 1974 | 1409 | 260 | 29 | 1698 |
| 1975 | 1930 | 131 | 0 | 2061 |

the stock of sardinella off Ghana and Ivory Coast. Table 5-22 shows the great fluctuations in catches that have taken place in the last 15 yr. After 1975 there has been a recovery, with the catches reaching over 50 000 tons in 1978. Thereafter they dropped in 1979, suggesting a further collapse of the population (FAO, 1981). While some of the variation in the catches might be ascribed to changes in fishing effort, or even in the system of collecting statistics, the close correlation between the catches by all three gears show that the main causes are fluctuations in the stock. As in other pelagic stocks the relative role of fishing and environmental factors (of which the intensity of upwelling is the most likely to have a significant impact, either on recruitment, or on the growth and mortality of post recruits) is unclear. It appears that the large catches in 1972 came from a strong year class, but included a large proportion of juveniles; they may have seriously reduced the spawning stock. Assuming this stock does recover, it seems that the catches should be very carefully controlled, though at the moment the mechanisms to do this have not been clearly established.

The other management problem concerns the shrimp fishery. As in the majority of shrimp fisheries, most of the stocks are fully exploited. For the industrial sector this raises the usual questions of determining the optimum amount of fishing, and possible adjustment of the sizes of shrimp caught, e.g. by mesh regulation. In Ivory Coast, and in some other countries, there are (as in India) problems of conflict between the industrial fisheries and the artisanal fisheries in the lagoon (GARCIA, 1981). Since the latter catch small shrimp before they migrate to sea, the total weight is probably less than if all the shrimp were taken in the offshore industrial fisheries on larger shrimp. The government, therefore, has to choose between different objectives (maximizing total catch or total value of the catch, or the social aim of maintaining the welfare of the small-scale fishermen). Probably some compromise will be achieved which maintains the current artisanal fishing, but discourages further increase.

## *Western Central Atlantic Ocean*

### *The resources*

This region includes the coasts of the United States, north to 35° N, and of South America, south to 5° N, which face the open Atlantic Ocean, as well as the Gulf of Mexico and the Caribbean Sea. The fishing grounds can be divided into two main classes—the shelf fisheries off the continental coasts, and the predominant reef fisheries around the Caribbean islands. The former can be further subdivided according to type of bottom, nature of coastline, etc. Particularly important are the areas of soft bottom in the northern Gulf of Mexico and off the Guianas, which support large shrimp stocks.

The waters of the region are not particularly productive, but the shelf, especially in the Gulf of Mexico, is wide so that the total fish potential is quite good. The largest resource is that of menhaden, of which there are two species: *Brevoortia tyrannus* along the US Atlantic coast, and *B. patronus* in the Gulf of Mexico. Both are essentially coastal species, and the juveniles grow up in the lagoons and estuaries. No other small pelagic species in the region approaches menhaden in volume of current catches, but a number of species are quite common in the Gulf of Mexico. Data from egg and larval surveys HOUDE, 1973) suggest that the combined biomass of round herring *Etrumeus teres*, thread herring *Opisthoma oglinum*, and other species might approach that of menhaden. Outside the Gulf of Mexico there is a moderate-sized stock of Spanish sardine *Sardinella anchovia* off Venezuela, where there is a small upwelling system. Elsewhere, small pelagic species do not appear to be common.

Among bottom-living animals the shrimp are by far the most important commercially. Shrimp occur in suitable areas of soft bottom throughout the region, but the most productive area is the Gulf of Mexico coast of the United States. Other important areas are on Campeche Bank, off Venezuela, off the Guianas, and along the Atlantic coast of the United States (ANON 1978). A number of species are involved in each fishery. In the US Gulf coast the most important are three shallow-water species: brown shrimp *Penaeus aztecus*, white shrimp *P. setiferus*, and pink shrimp *P. duorarum*. These all have similar life histories. They spawn offshore on the continental shelf and the planktonic larvae drift inshore. For details consult Volume III: KINNE (1977). The very young shrimp move into the estuaries where they spend their first few months of life, moving out to the open sea where they begin to approach maturity. They spawn when about 1 yr old. Few shrimp appear to live much longer than 1 yr.

In other grounds the species composition is different, though the situation is confused by the use of the same common English (or Spanish) names in many areas. For example, off the Guianas the fishery is also based mainly on brown, white, and pink shrimp, but the species are *Penaeus subtilis*, *P. schmitti*, and *P. notialis*, respectively. HOLTHUIS (1980) provides a useful guide to the taxonomic classification and common nomenclature for shrimp. In addition in most areas smaller species, such as sea bobs (*Xiphopenaeus* spp.) are found in the shallower offshore areas, and deep-water species (particularly royal red shrimp *Pleoticus robustus*) are found in depths beyond those typical of the *Penaeus* species. Associated with the shrimp is a rich fish fauna typical of soft bottoms. The actual fish catches by shrimp trawlers nearly always exceed that of shrimp, by factors ranging from 3 or 4 to 1, up to 10 or 20 to 1 (JUHL and DRUMMOND, 1977). A large number of species are involved. The commonest is the croaker *Micropogon undulatus*. Virtually all these fish are at present discarded at sea after catching.

On hard bottoms the fauna is very different. Shrimp are absent, but spiny lobsters occur in most parts of the region and form a valuable resource. Several of the fish are also valuable, notably snappers and groupers. They support important local fisheries. However, the hard-bottom fauna, including the reef resources around the islands is, taken as a whole, less than that of soft bottoms.

Apart from the oyster fishery of the middle Atlantic states of the United States, on the northern fringe of the region, and some trawling for scallops a little further south along the US Atlantic coast, molluscs are a small, or at least a neglected resource in most of the region. Conches are harvested throughout much of the Caribbean, but the total production is low. Cephalopods appear to provide a possible resource but are not fished much, and there is little quantitative information about them (Voss, 1960)

*The fisheries*

Summary statistics of landings in the western Central Atlantic Ocean in recent years, as presented in FAO's *Yearbook of Fishery Statistics*, are given in Table 5-23. Apart from the usual caution regarding any statistics, it must be stressed that there are two major omissions in this table if it is to be regarded as a record of the removals from the sea. First, it excludes any fish discarded at sea (not landed). As already noted, up to 90 to 95% of the catches of shrimp trawlers may be fish, but virtually all is discarded at sea. Discarding also occurs in some other fisheries; for example up to 40% of the catch in the trawl fishery for large croaker (Juhl, 1974). Altogether perhaps half a million tons of various demersal fish may be discarded. Secondly, along the United States coast sports fishing is very important. Surveys have provided estimates of about 3·5 million marine anglers in 1970, catching about 400 000 tons of fish (Klima, 1977). For several species (king mackerel *Scomberomorus* spp., jack and sea trout) the sports catch exceeded by a considerable margin the commercial landings in the United States.

The table shows the extent to which the total catches are dominated by the United States, with some 70% of the reported landings, and an even higher proportion if the discards and sports catch are taken into account. In terms of volume the biggest fishery is that for menhaden. The catches are taken in purse-seine, and are all used for the production of meal and oil. The fishery developed in the Atlantic Ocean, but in the 1960s the stocks there showed many of the classic signs of collapse (McHugh, 1969; Schaaf, 1975). The distribution shrank and the catches fell from around 600 000 tons (over 700 000 in 1956) to less than 200 000 in 1967. Attention then shifted to the Gulf menhaden. Though the Atlantic stock has not fully collapsed, the biggest catches now come from the Gulf. The meal and oil industry has also shown interest in the thread herring and other pelagic fish in the Gulf of Mexico. However, the schools seem to be small, so that the catch rates are not high enough to support an economically viable fishery. The only other significant pelagic fishery is the Venezuelan fishery for sardine, which takes some 30 000 tons annually, but small quantities are taken in several other areas, especially by Mexico and Cuba.

In terms of volume, the second most important fishery is that for oysters. The main production comes from the US Atlantic coast, especially Chesapeake Bay. Including production from north of 35° N the total US harvest is some 300 000 tons. This comes from both culture and dredging of wild stocks, but unlike most other areas, the proportion coming from culture (which had been high) is tending to drop, perhaps because of the very high labour costs in the United States. Some 30 000 tons is also produced in Mexico.

In economic terms, however, the dominant industry is shrimp. While the United

Table 5-23

Commerical landings (omitting discards and recreational catches) from the western central Atlantic Ocean. Upper: catches by major countries; lower: catches by major species groups (Based on FAO *Yearbook of Fishery Statistics*)

| | 1970 | | 1973 | | 1976 | | 1979 | |
|---|---|---|---|---|---|---|---|---|
| Countries | Thousand tons | % | Thousand tons | % | Thousand tons | % | Thousand tons | % |
| United States | 1016·2 | 71·8 | 908·5 | 65·4 | 1080·7 | 68·6 | 1281·5 | 72·0 |
| Mexico | 114·2 | 8·1 | 130·3 | 9·4 | 122·9 | 7·8 | 163·7 | 9·2 |
| Venezuela | 122·6 | 8·7 | 153·8 | 11·1 | 139·1 | 8·8 | 128·2 | 7·2 |
| Other mainland | 46·3 | 3·3 | 51·8 | 3·7 | 47·4 | 3·0 | 60·7 | 3·4 |
| Cuba | 61·8 | 4·4 | 72·8 | 5·2 | 79·2 | 5·0 | 66·1 | 3·7 |
| Other islands | 42·4 | 3·0 | 49·5 | 3·6 | 54·8 | 3·5 | 65·6 | 3·7 |
| Non-local fleets | 12·7 | 0·9 | 23·3 | 1·7 | 50·4 | 3·2 | 15·2 | 0·9 |
| Total | 1416·2 | | 1390·0 | | 1574·5 | | 1780·9 | |

| | 1970 | | 1973 | | 1976 | | 1979 | |
|---|---|---|---|---|---|---|---|---|
| Species | Thousand tons | % | Thousand tons | % | Thousand tons | % | Thousand tons | % |
| Various demersal | 151·7 | 10·7 | 183·0 | 13·2 | 184·8 | 11·7 | 141·6 | 8·0 |
| Menhadens | 598·0 | 42·2 | 525·5 | 37·8 | 627·3 | 39·9 | 896·8 | 50·4 |
| Other clupeoids | 75·9 | 5·4 | 79·1 | 5·7 | 79·5 | 5·1 | 59·5 | 3·3 |
| Other small pelagic | 44·7 | 3·2 | 55·5 | 4·0 | 51·2 | 3·2 | 50·2 | 2·8 |
| Tunas | 38·0 | 2·7 | 45·8 | 3·3 | 55·4 | 3·5 | 54·1 | 3·0 |
| Sharks and rays | 6·4 | 0·5 | 10·2 | 0·7 | 11·6 | 0·7 | 10·3 | 0·6 |
| Miscellaneous and unspecified fish | 66·4 | 4·7 | 90·5 | 6·5 | 87·9 | 5·6 | 119·8 | 6·7 |
| Shrimps | 168·0 | 11·9 | 153·4 | 11·0 | 168·3 | 10·7 | 171·5 | 9·6 |
| Other crustaceans | 49·5 | 3·5 | 59·0 | 4·2 | 56·7 | 3·6 | 70·3 | 3·9 |
| Cephalopods | 2·1 | 0·1 | 4·3 | 0·3 | 6·5 | 0·4 | 7·6 | 0·4 |
| Oysters | 177·0 | 12·5 | 145·0 | 10·4 | 192·2 | 12·2 | 159·9 | 9·0 |
| Other molluscs | 18·2 | 1·3 | 23·3 | 1·7 | 44·9 | 2·9 | 24·8 | 1·4 |
| Others | 11·2 | 0·8 | 15·0 | 1·1 | 7·9 | 0·5 | 8·2 | 0·5 |
| Total | 1416·2 | | 1390·0 | | 1574·5 | | 1780·9 | |

States is the biggest producer, significant catches (several thousand tons or more) are taken by several other countries, notably Mexico, Venezuela, Surinam, and Guyana.

In the United States the earliest shrimp fishery was carried out by haul-seines, but this had given way by 1920 to otter trawling. This has remained the standard method. However, the gear has been modified by the introduction of double-rigged trawl (two trawls, one towed from each of two booms, one each side of the vessel) around 1955, which has been generally adopted, and by the double-rigged twin trawl (i.e. two trawls on each side), which was developed in 1971 (BULLIS and FLOYD, 1972) and is used where little turning is involved (ROTHSCHILD and PARRACK, 1981). In parallel with the improvements in gear there have been changes in distribution and marketing (virtually all the commercial catch is frozen), and a spread in the area fished, as well as in the species caught, which was initially only the white shrimp.

As they moved further afield and into deeper water, the US ships also extended their activities into waters off foreign countries, notably Mexico. For example, the US fleet developed the fishery on the Campeche Bank in 1945, and even in 1965 was taking one-third of the total catch. With the changes in the Law of the Sea all the fishing grounds for shrimp fall within the jurisdiction of one or other coastal state, and this form of non-local fishing is being phased out. All US shrimp fishing off Mexico was due to finish in December 1979.

Another form of foreign shrimp fishing is still continuing in the Guianas. Foreign shrimp vessels (mainly Japanese and American) are based, with the agreement of the country concerned, in Guyana or French Guiana, landing their catches in these countries for processing and ultimate shipment to the main importing countries.

Apart from these shrimp fisheries, and a little movement between adjacent countries in search of lobster and high-valued fish, mostly by small vessels, non-local fishing in the region is minimal. Table 5-23 (upper) shows that in 1979 less than 1% of the total catch was taken by countries from outside the region, mostly tunas and billfish caught by long-line vessels from Japan and Korea. Catches by the Soviet Union, mostly from a trawl fishery on the Campeche Bank, reached a peak of nearly 70 000 tons in 1975, but no Russian fishing has been reported since 1977.

The rest of the fisheries in the region are essentially small-scale. There is in fact not much difference between the artisanal and subsistence fisheries of the Caribbean Islands and the coasts of Latin America, and the sports fisheries of the United States, particularly in the size of boats involved and the difficulties of getting accurate statistics. For the latter it is fully accepted that comprehensive data cannot be obtained. The figures available are based on interviews of a statistical sample of the total fishery; this can provide figures for confidence limits of the estimated total catch, as well as some idea of any bias. The figures for the catches of small-scale fishermen are often little more than guesses, with little information on their reliability. Attention is now being given to improving the system of collecting statistics in most countries of the region, especially via assistance through WECAF (FAO's Western Atlantic Fisheries Commission).

The chief difference is that the sports fishery is mainly carried out by hook and line, and concentrates on a few species preferred because of their good eating, or sporting qualities. The artisanal fishery is less selective and uses a wider range of gear, particularly traps in the reef areas. Small shrimp is a preferred bait in the sports fishery, and the commercial fishery for small bait shrimp, plus a recreational fishery by small boats with a variety of gears—trawls, cast nets, and dip nets—and other catches in inshore waters takes about a quarter of the total catch (ROTHSCHILD and PARRACK, 1981; their Table 1).

*State of the stocks*

The most comprehensive attempt to review the state of the resources in the region and their potential for increased harvest is that of KLIMA (1976); it was prepared for the first session of WECAF (Port of Spain, October 1975). Further reviews of selected resources on a regional basis have been made by WECAF working parties (ANON., 1978b).

National studies have also been conducted, especially in the United States, as part of the background work for preparing management plans for the regional management commissions set up under the US Fishery Management and Conservation Act.

For most shrimp fisheries the assessments were fairly simple and straightforward. Available indices of effort (mostly days at sea) and catch-per-unit-effort seem to provide reasonably satisfactory measures of the amount of fishing and abundance. Application of simple production models usually results in a good fit to observed points (e.g. ANON., 1978). In most cases the resulting curves suggest that the fishing effort is approaching, or has already reached, the level giving the MSY. One exception is the Colombian fishery (ANON., 1978: Fig. 8) where there has been a decline in c.-p.-u.-e., but the total yield is still clearly increasing. The general result is as expected. Except possibly for the main US Gulf of Mexico fishery, the market for shrimp is, so far as production from any individual fishery is concerned, virtually unlimited. Once a fishery has become established and proved itself to be economically viable, the only factor to stop unlimited growth is a decline in catch rates.

Analytic models have been less commonly used in shrimp assessment (GARCIA, 1981). Fairly good estimates of growth—usually in terms of daily or monthly increments— have been obtained for a number of stocks, including several in the Gulf of Mexico/Caribbean region. Estimates of mortality (total, or separated into natural and fishing) are scarce; they have been obtained from tagging or deduced from size-composition data. These estimates used in a RICKER or BEVERTON and HOLT type of model give results in terms of the shape of the yield-per-recruit curve. So far as this can be estimated, the current position of the fishery matches the results from production models fairly well. Until recently, the effect of fishing and the resulting reduction in adult stock on recruitment has not been considered very seriously. There has often been a hypothesis that because shrimp only live 1 yr, they can be treated as an annual crop, without having to worry about recruitment. There is little logical justification for this hypothesis, and there is growing evidence that the average recruitment does change with changes in adult stock, but as usual, any underlying relation is obscured by variation in recruitment due to other causes. However, so far in no shrimp stock in the region does it appear that the recruitment has been affected to any serious degree.

The menhaden stocks have been evaluated using production model analysis and, for the Atlantic stock, an analytic age-structured model (SCHAAF and HUNTSMAN, 1972; KLIMA, 1977). Though the fit to the production model is reasonably good, the observed change in catch-per-unit-effort is small. The effort units used (vessel-weeks, or vessel-ton-weeks) do not include any correction for searching time. The experience in other purse-seine fisheries is that without some correction of this type, the catch-per-unit-effort can seriously underestimate the decline in abundance, sometimes to the extent that the stock may be on the point of collapse before there is any noticeable change in catch-per-unit-effort. Though the analytic model has similar problems in obtaining a reliable index of fishing mortality, the effect on the assessment is less serious. KLIMA (1977) is therefore correct in assuming that the analytic model, in which a spawner–recruit relation has been incorporated, and which predicts for the Atlantic stock a MSY of some 400 000 tons from average recruitment, is more reliable than the production model, which predicts a MSY of 620 000 tons (these figures include catches outside the present region, in the northwest Atlantic Ocean statistical area). Both models agree that the fishing effort, particularly in the middle 1960s, was above the optimum level. For the Gulf menhaden, the estimated MSY from the production model is some 430 000 tons; this has been exceeded in some years, notably in 1978 and 1979, and it will be interesting to see whether these high catches were due simply to one or two good year classes, or to an increase in the amount of fishing, and, in the latter case, whether the stock can sustain it. Variations in year-class strength clearly account for much of the year-to-year changes in the Atlantic catches, and probably also for much of the divergence between the two models. NELSON and co-authors (1977) have examined the deviations from the Ricker-type stock-recruitment curve, and found that they could account for some 84% with an environmental model, in which the wind-induced transport of larvae into favourable (or less favourable) areas was the most important factor.

For no individual stock of reef fish are there sufficient data to make assessments by the traditional methods. However, MUNRO (1975, 1977) has used a comparative approach similar to that successfully used in assessing the productivity of lakes and other inland waters. If the productivity of different reef areas is similar, and, as is likely, there is little interchange of fish between areas, then areas with different intensities of exploitation can be used to derive a yield curve in the same way that periods with different intensities can be used to derive a production-model type of yield curve for a single fishery. Fig. 5-8 (from MUNRO, 1977) shows the results for different areas around Jamaica. There is a very close correlation between the catch-per-unit-effort (tons canoe$^{-1}$ yr$^{-1}$) and the intensity of fishing, as measured by the number of canoes km$^{-2}$. The data strongly indicate that the maximum yield (about 4 tons km$^{-2}$) is taken with an intensity of fishing of about 3 to 4 canoes km$^{-2}$, and that the intensity of fishing in several areas exceeds this figure. Since each of the gears used by the canoe fishery (traps, hand-lines, gill-nets, etc.) are selective to a greater or lesser extent, this conclusion strictly applies to the current mixture of gears (and hence the balance of fishing mortality exerted on different species), which is probably broadly similar around Jamaica. The maximum yield, and the density of canoes at which it occurs, would presumably be different with a different mixture of gears. However, the results seem valid under present conditions in Jamaica—and probably in many other parts of the Caribbean Sea. They certainly give useful guidance for fishery policy makers.

For the remaining resources in the region, which includes most of the demersal resources, the main source of information on the potential are the results of trawl and other surveys, particularly those carried out by the US National Marine Fishery Service, and the UNDP/FAO Caribbean Fishery Development Project. From these, estimates of standing stock, and hence, under certain assumptions, also of potential annual yield, have been obtained (KLIMA, 1977). For demersal fish these estimates are, in round numbers, 0·6 million tons off the southeast coast of the United States, 1·1 million in the Gulf of Mexico, 0·5 million in the Caribbean Sea, and 0·3 million tons off the northeastern coast of South America. The estimates for shoaling pelagic fish are more variable, but their potential is probably smaller than, but more than half, that of the demersal fish. In total, these estimates considerably exceed the figures of reported catch given in Table 5-23, suggesting that there are good prospects for increasing production. However, when allowance is made for discards in the shrimp fisheries and US sports catch, as well as errors in both estimates, the gap between potential and catch is less striking. The biological evidence for the possibilities of increased catches include many of the stocks, e.g. thread herring and other small pelagic fishes in the Gulf of Mexico, which are economically least attractive.

*Management*

The region is probably less heavily fished than any other region, except the southwest Pacific Ocean, but it does have its management problems. Several of these are similar to

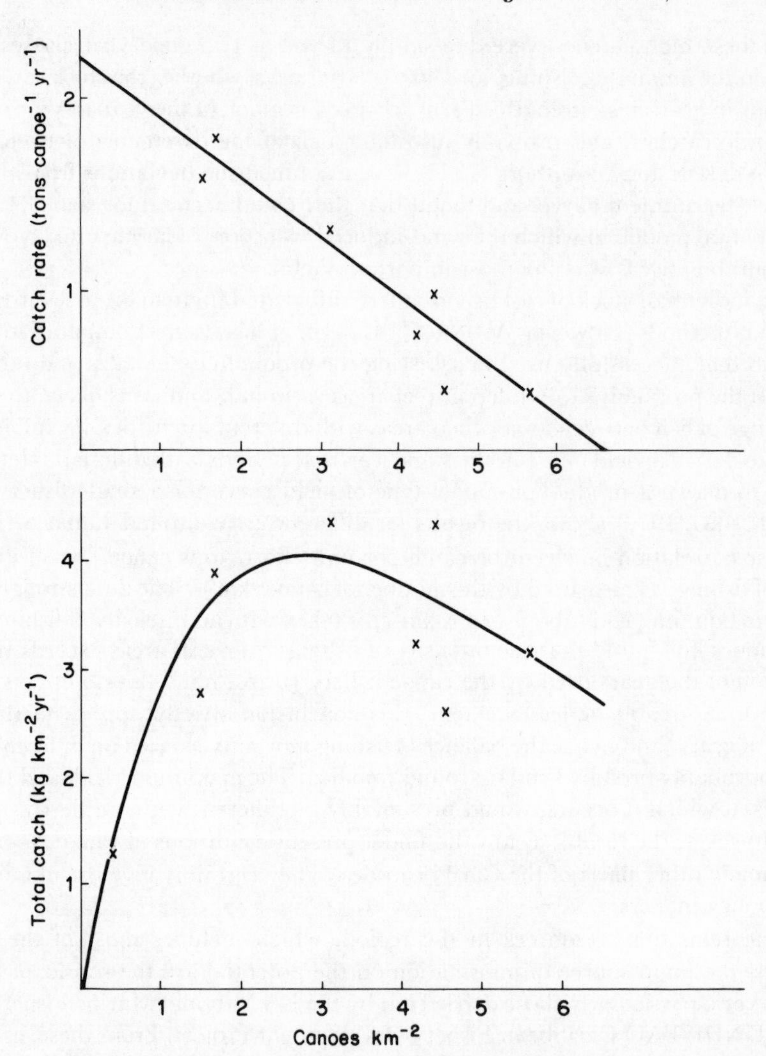


Fig. 5-8: Relation between the catch rates, total catch, and density of canoes in various parishes around Jamaica. (After MUNRO, 1977; reproduced by permission of FAO)

those in other regions, e.g. the biological problem of determining the optimum harvesting regime for a pelagic fish stock that may be unstable (menhaden), or the social problems of intensive fishing on local stocks when the fishermen concerned have little alternative employment (Jamaica, and probably other islands). The shrimp fishery has problems that, if not quite unique to the region, are presented in their most acute form in the Gulf of Mexico. In fact there are two problems, one that is real, but which has attracted until recently relatively little attention, and one that has attracted much attention, but which, while a challenge, may not be a problem of fisheries management.

The challenge arises from the large quantity of discards. At a time when large numbers of people, including millions in the countries around the Caribbean Sea and Gulf of

Mexico, are hungry and short of protein, it is obviously disturbing that half a million tons of fish, much of it of good quality, is thrown back into the sea. There is a clear challenge to devise ways in which this fish could be made available to those who need it. The problem is a wide one, involving questions of North/South exchange, e.g. how fish brought ashore in a Louisiana port can help feed a peasant in the Mexican highlands, as well as technological questions of vessel design, and of processing methods. Questions of resource management, i.e. of adjusting fishery policy so as to make better use (however defined) of the resources, may not arise. They would arise if the demersal fish presently harvested is caught in a directed finfish fishery, or if the incidental catch could be brought ashore by the shrimp trawlers without disrupting their operations. The typical shrimp trawlers fishing out of the Gulf of Mexico ports of the United States are designed to achieve a high economic efficiency in their operations, with small crews and limited carrying capacity. Bringing fish ashore would add considerably to their costs—the catch would need more people to handle, and trips might have to be shorter. The result is that under US conditions a combined fish–shrimp fishery is not economically viable. The situation may be different in other parts of the region where labour costs are lower and the demand (or at least the need) for fish is higher. Efforts are therefore being made to increase the landings of fish by shrimp trawlers in countries like Mexico or Guyana (the latter requires that foreign shrimp vessels include a proportion of fish in all their shrimp landings). It may be a mistake therefore to consider that the discarded fish are necessarily wasted, in the sense that with a little thought or care they could add to the diet of hungry people—at least they may be no more wasted than the vast majority of fish in the sea that are never caught. The fallacy lies in assuming that the most difficult or costly part of bringing a fish from the sea to the table is getting it out of the sea. The history of fisheries show that improvements in handling, processing, and using fish (freezing at sea, cold chains ashore, use of fish-meal in chicken rearing), have been just as important as improved catching methods in stimulating the growth of catches.

More interesting scientific and resource management questions arise if there is an actual or potential fishery for the discarded species.

The additional mortality caused by the shrimp fishery will reduce the fish abundance (especially if, as is likely, the small-meshed shrimp trawlers take a lot of small fish), and hence the catches of those fishing for fin-fish, possibly to the point that a fin-fish fishery would not be economically viable. On this argument anything to reduce the quantity of discards, e.g. by using a larger mesh, reducing the overall fishing intensity, or concentrating shrimp fishery in areas where the discard ratio is low, should be beneficial.

The conclusion is less clear if possible species interactions are taken into account, and it is accepted that, whatever improvements are made in marketing fin-fish, they will mostly still have a much lower price than shrimp. Some shrimp may feed on detritus, and the large volume of dead fish dumped by the shrimp trawlers could be a welcome additional source of food. A more important effect may be the reduction in abundance of several species of the larger species of fish that are predators on shrimp. PAULY (1979) noted that in the Gulf of Thailand shrimp were one of the species that has shown the least decrease since large-scale trawling started—even though they are one of the species that is most actively sought by fishermen. He has further suggested (pers. comm.) that the recruitment of shrimp can be negatively correlated with the abundance of the larger species of fish. On this argument, while better use of the by-catch would be desirable, measures to reduce the magnitude of the by-catch might be counter productive.

The real problems concern the shrimp fisheries themselves. These are increasingly exhibiting all the symptoms of an unmanaged fishery—over-capacity, falling catch rates, conflicts between different sectors of the fishery, etc. These problems have tended to be ignored until quite recently, partly because the harmful impacts have been reduced by the shifts onto other species of shrimp, or (in the case, particularly, of the United States) onto more distant grounds, and by the rising price of shrimp. Another reason has been the myth that short-lived shrimp cannot be over-fished. These factors are changing. The U.S. fleet has lost access to many grounds off other countries. Prices are less buoyant while fuel costs have greatly increased. It is also clear that with their rapid growth during the first few months in which they are exposed to the fishery, shrimp can suffer from 'growth over-fishing'. Hence the conflict between inshore and offshore fishery. There is less clear evidence of the risk of 'recruitment over-fishing', but there are suggestions that fishing could affect recruitment. If so, in the absence of management there could be a progressive, and possibly in due course catastrophic, fall in recruitment and hence in total catch. It is clear therefore that management questions, especially methods of controlling the total amount of fishing, e.g. by licence limitation, will receive increasing attention from all those (scientists, administrators, and fishermen) concerned with the shrimp industry.

## *Pacific Islands*

### *Resources and fisheries*

The Pacific Islands fall mostly within the western central Pacific statistical region. However, they are so distinct in nearly all ways from most of the shelf areas around the continental mainland of southeast Asia that they are worth being looked at separately. Also included are the islands of the eastern central Pacific Ocean, and the tropical and subtropical islands to the north and south. The region therefore covers an enormous area—some 90° of longitude from Palau (130° E) in the west to the Marquesas (140° W) in the east. A comprehensive review of the resources of the region has been made by UCHIDA (1978). More detailed information, particularly on skipjack, is available in the publications of the South Pacific Commission (SPC).

Despite the vast extent of the region the land area is minute. The total area of the countries of the South Pacific Commission (to which most of the island states in the region belong) is, omitting Papua New Guinea, only some 90 000 km$^2$. The human population is correspondingly small, that of the SPC countries (again with the exception of Papua New Guinea) being less than 2 million. Beyond the fringing coral reefs, the sea bottom around most islands drops rapidly, so that the shelf area, though greater than the land area in most cases, is still modest. However, most islands have claimed, in one form or other, 200-mile zones around their shores. These are of vast extent—of the order of 25 to 30 million km$^2$, i.e. several hundred times or more greater than the land area of most of the countries. In other words, the SPC countries have on the average some 15 km$^2$ of EEZ for each inhabitant.

The resources of the region are therefore sharply divided between those of the open ocean, which, while not particularly dense, cover a vast area and are quite large, and those around the islands and coral reefs, which are limited. The principal ocean resources are the tunas. These are discussed in a later section, but there are special problems in the management of tuna in this region, where much of the catch is taken within 200 miles of one or other island state, but the opportunities for many of these countries to engage directly in the commercial tuna fishery are limited. KEARNEY (1979) estimates that a quarter of a million tons of tuna are caught within the 200-mile zones of SPC countries, but only 90 000 tons are landed locally. Of this, nearly all is accounted for by joint venture operations in Papua New Guinea and the Solomons, or from transhipments of fish caught by foreign long-liners.

The tuna catches by wholly local fisheries are by comparison very small—only a few thousand tons—but this is quite significant in a total catch that is only about 100 000 tons. The methods vary from island to island, but most catches are taken by some variety of trolling. While the major tuna species (particularly skipjack and yellowfin) are important in most of the troll fisheries, several other species are taken, particularly various billfishes, dolphin fish *Coryphaena hippurus*, various species of *Scomberomorus*, and jacks (*Carangids*).

The typical fisheries of the region are those on the shallower parts of the reefs, and in the atolls. A great variety of gears are used. In his review UCHIDA (1978, p. 20) notes that for the Trust Territory of the Pacific

> 'the main fishing gears include handline, trolling line, pole and line, beach-seine, surround net, gill net, dip net, cast net, scareline, spear, fixed traps of rock or wire, and bottom trap'.

A similar list could be made for many other groups of islands. The most interesting point is that the two dominant gears of modern fishing—trawls and purse-seine—are missing, with very minor exceptions, throughout the Pacific islands. The bottom is nearly everywhere too rough for bottom trawling, and apart from tunas there are no large shoals of pelagic fish to make purse-seining, or mid-water trawling, worthwhile. Efforts are being made by US and Japanese fishermen to develop purse-seining for tuna in the region, but catches so far are small.

The vessels used and species caught in the local fisheries are equally varied. The boats are small, often without motors, but the Pacific is the home of some of the greatest small-boat ocean travellers in the world, and many of the canoes are highly seaworthy. On the other hand, in some other areas the canoes used are certainly not safe to go beyond the central atoll. Most of the fish, particularly those caught in traps or on hand-lines, are the typical reef species—snappers, parrot fish (*Siganus* spp.), etc. Among other species, flying fish are caught in many areas. Beyond the reef there are good stocks of larger bottom fish (snappers and groupers) in deeper water, down to several hundred metres. These are generally lightly fished if they are fished at all, but efforts are being made to encourage the greater utilization of these resources, particularly with lines (CROSSLAND and GRANDPERRIN, 1980).

A somewhat unique resource is that of pelagic armourhead *Pentaceros richardsoni* and alfonsin *Beryx splendens* on the sea mounts to the northwest of Hawaii. This resource was found by Russian trawlers in 1967 on sea mounts, which come to 100 to 400 m from the surface, far from land in a region of otherwise very deep ocean. The density of stocks when they were first found were very high (catches of tens of tons were taken wihin a few minutes trawling), and both Russian and Japanese took good catches (some tens of thousands of tons) in the early 1970s. Though it is highly likely that both the Russians and the Japanese have looked for similar resources elsewhere in the Pacific Ocean, it does not seem that they have found them.

Apart from fish, a variety of invertebrates are harvested. Lobsters (*Panulirus* spp.) support commercial fisheries in Papua New Guinea (with annual catches up to a few hundred tons in some years), and in Hawaii, and occur in sufficient numbers to support at least local fisheries in a number of other areas. Papua New Guinea has also a significant fishery for shallow-water shrimp (*Penaeus* spp.), but elsewhere in the region conditions are not favourable for these species, though deep-water shrimp (*Heterocarpus* spp.) have been found at depths of 300 to 400 m around Hawaii, and probably occur elsewhere. Molluscs are widely harvested for pearl shell, and for consumption, but the quantities are nowhere large. Sea cucumbers also occur widely, and in some places exports of bêche-de-mer to the Unites States and Japan have been important.

In general, the people of the Pacific islands are great eaters of fish and all marine products, but fish poisoning (ciguatera) is a problem. This is the curse of fisheries throughout most Pacific islands, as well as in the Caribbean (HALSTEAD, 1959; RANDALL, 1980). It can produce a wide range of symptoms—from tingling about the lips, nausea, and skin rash, to paralysis, convulsions, and not infrequently, death. The causes are not well understood, nor is it easy to determine which fish are toxic. Ciguatera can occur from a number of species, through larger predatory fish, e.g. *Lutjanus bohar* are particularly often involved, and it has occurred at time in most of the tropical Pacific islands. Its damaging effect is not only immediate poisoning, but the extent to which fishing is discouraged on species which for most of the time are perfectly good to eat.

Despite the small populations, and the wide seas around them, most of the Pacific Islands are net importers of fish. KEARNEY (1981) notes that some $20 million worth of fish products were imported into the SPC countries annually (excluding a number of countries for which data were not available), most of it canned mackerel and similar species. The countries concerned are therefore anxious to encourage greater production from local fisheries in order to reduce the drain on foreign currencies as well as to increase local employment and the general food supply.

*State of stocks and management*

The basic data on which to assess the state of the stocks (except for tuna) are poor or completely lacking. In all but one or two fisheries virtually all the catches are taken by subsistence or artisanal fisheries and only a small part of the total goes through any regular market. Statistics of the total catch, even when they are available, are generally unreliable. There is little or no data on detailed species composition or on fishing effort. Biological information, e.g. on growth and mortality, or on the size composition of the catches is equally lacking in most areas, though studies have been carried out around Hawaii and in a few other places.

Apart from tuna (among which it appears that the stocks of larger fish taken by the long-line fisheries are being fully exploited), no formal stock assessments are available. However, it is likely that many of the shallow-water reef stocks are being heavily fished. UCHIDA (1978, pp. 12–20)) in his comments on individual countries noted that:

> 'illegal fishing methods have seriously depleted the stocks of fish in the lagoon [Cook Islands]; . . . a severe depletion of the inshore resources [Trust Territory]; . . . a serious decline in the lagoon stock [Wallis and Futuna Islands]; the reef and lagoon fauna are undoubtedly overfished [Western Samoa]'.

The fact that reef and lagoon fish stocks are limited, and that their exploitation needs to be carefully controlled was well appreciated by many of the native peoples of the Pacific coasts. Their traditions, in addition to containing information on the biology of the more important species, and how they can best be caught, also set down patterns of exploitation, e.g. allocation of certain fishing grounds to specific villages, and determination of how many canoes should fish, which for a long time ensured that the resources were harvested on a rational and sustainable manner (JOHANNES, 1977).

This pattern largely broke down with the coming of the European. At first this did not matter, since the change towards a cash economy, and the drift to the towns caused fishing pressure to drop in most places. Now the rise in population, the increased demand for fish coupled with mechanization of boats, and the introduction of non-traditional gears is increasing the pressure on the stocks again. Management of these small-scale fisheries, in the sense of management as applied to the industrial fisheries of the North Atlantic, is highly unlikely to succeed. As UCHIDA (1978) has noted, even the simplest and least controversial types of management—the prohibition of the most obviously damaging types of gear, such as dynamite—have proved in several places not to be enforceable by governmental decree. The prospects are remote of successful government controls on the total catch, or the number of canoes, except close to the main towns, and in respect of the larger mechanized vessels. The solution may be to go back to the traditional pattern, in which management and control of fishing is a local village matter. While the use of a small-meshed net, or dynamite, is seen as merely outwitting the impersonal rules set by a distant government, it will continue, and the efforts of local fishery officers may be able to do little to stop it. On the other hand, if the pattern of fishing is decided locally, and the pressure of the local community is brought to bear on anyone breaking the rules, then there is a good chance they will be followed. This will require big changes in approach, both in the basic administration of fisheries, and in the way in which the stocks are studied and advice provided by the scientists.

The other management problem in the region—that of tuna—fits more neatly into the conventional pattern. The problem is not so much that of the overall management of the tuna fisheries—a world-wide problem that nowhere has been satisfactorily solved—as that of the island states obtaining a greater share of the benefits from much the biggest fishery in the region. The greatest benefits come from direct engagement in the fishery, particularly if this is not restricted to catching the fish, but also includes processing (in the case of tunas, principally canning). Some countries, notably Papua New Guinea and the Solomon Islands, have formed joint ventures with Japanese or other interests, and catches by these locally-based fleets are significant (40 000 tons of skipjack were landed in Papua New Guinea in 1974). However, the variations in the availability of tunas in a relatively small area represented by the EEZ of a single country is high; the PNG skipjack catch fell to less than 20 000 tons in 1975. A tuna fishery will therefore be more effective if it can change its areas of operation from one year to the next (as well as within a year) according to the pattern of movement of the fish. This the long-range fleets can do, but it is less easy for locally-based vessels. Processing presents even more serious difficulties, particularly for the smaller countries in the region. To operate on an economic scale a canning plant needs to handle at least several thousand tons of fish per year; this is on a scale beyond the reach of many of the South Pacific countries. At the other extreme, the water needed for a moderate-sized canning plant is more than the total fresh water supply of one or two of the smallest countries.

The countries are therefore on the lookout for obtaining benefits from licensing or similar controls of foreign vessels. Here geography favours them. In the southwest Pacific Ocean, the islands are well scattered, but are not too far apart. As a result, the 200-mile zone around each country run together, so that nearly all the ocean from the Marianas, Papua New Guinea, and New Caledonia in the west to the Marquesas and French Polynesia in the east lie within one or another of the 200-mile zones. If all the countries concerned can act together, they can set control on the operation of the long-range tuna fleets, which cannot effectively be avoided by fishing outside 200 miles. This will require close co-operation in deciding all the various aspects of management, e.g. how many vessels should be allowed to fish, on what conditions, how should the benefits (licence fees, etc.) be divided, how can the countries be assured that the foreign fleets are complying with the regulations, etc. The countries have therefore moved to set up the necessary co-ordinating mechanisms, establishing within the South Pacific Forum (which unlike the South Pacific Commission excludes the metropolitan powers) a Forum Fishery Agency. While it is too early to see how these policies will turn out in detail, the long-term prospect would be one in which those countries in the region which are well placed to do so, will develop their own tuna catching, and (to a more limited extent) processing industries. The other countries in the region will get their benefits from the tuna resources less directly, i.e. through controls on the long-range fleets, which will probably continue to take a substantial part of the total catch in the region.

### (e) Southern Temperate Waters

Geography has not favoured the growth of temperate water fisheries in the southern hemisphere. The amount of shelf in the temperate zone, say between 35° S and 60° S, is only a fraction of the corresponding areas in the northern hemisphere, and the hydrographic conditions are generally less favourable to high biological production. Ashore the human populations are small, and are all among the world's greatest producers and consumers of meat. The demand for fish for local consumption is therefore small, and it is only with the greater demand for fish in the northern countries that significant attention has been paid to fishery in these waters, either from local interests looking at export markets, or from the fishing fleets of northern countries.

#### *Southwest Atlantic Ocean*

*The resources*

The Patagonian shelf, from the River Plate southward to the tip of South America, provides the richest fishing grounds of the southern hemisphere, other than the upwelling areas of the Peruvian and Benguela currents. The shelf is extremely wide. Off central Argentina about 45° S, it is so wide that part of the shelf and the upper part of the slope lies more than 200 miles from the Argentine base line, and is thus outside its Exclusive Economic Zone. The main demersal stock of the northern part of the shelf is the Argentine hake *Merluccius hubbsi* which occurs in large numbers from the River Plate south to about latitude 50° S. It exhibits clear migrations both north and south, and from deep water in the winter into shallow water in spring and summer.

On the southern part of the shelf the Argentine hake is replaced by other gadoids, particularly the southern blue whiting, or polaca *Micromesistius australis* and the grenadier or merluza de cola *Macruronus magellanicus*. Both these are common on the shelf south of about 45° S, including the Burdwood Bank which extends eastwards from the top of Tierra del Fuego towards South Georgia. The two species have similar distributions, except that the blue whiting is found in rather deeper water from 150 m down to 300 m than the grenadier (30–250 m).

North of the River Plate the temperate species start to give way to warmer water species. The statistical region of the southwest Atlantic Ocean extends as far north as 5° N, and therefore includes some strictly tropical waters, but their resources are much less important than those of southern temperate waters. Of the tropical resources, the most important are shrimp. Off the mouth of the Amazon the shelf is wide, and supports a soft-bottom community that is a continuation of that off the Guianas to the northwest, and very similar to that in the northern Gulf of Mexico, and elsewhere. Economically the penaeid shrimp is the most important, but in weight various fish species, especially croaker, are more significant. South of the Amazon the shelf narrows, and around the eastern bulge of Brazil, south to about 15° S, the demersal resources are poor. South of 15° S the shelf widens again, and there is a moderate density of warmer water species, (croakers and weakfishes, as well as shrimps).

The best known, and almost certainly the largest, stock of pelagic fishes is that of anchovy *Engraulis anchovia* which occur from the River Plate area south to about 47° S. There seem to be two main spawning groups, one between the mouth of the Plate and Mar del Plata, and the other off the northern part of Patagonia (*ca.* 42 to 45° S). Mackerel are also fished in the same region, but the abundance of mackerel, as well as of other small-to-medium-sized pelagic fish (horse mackerel, bonito, etc.) is much less than that of anchovy. Another large pelagic stock is probably that of the Falkland sprat or herring *Clupea fuegensis*. Not much is known about this stock, but research vessels carrying out resource surveys on the southern Patagonian shelf have found substantial quantities from about 48° S southward, and especially off the coast of Tierra del Fuego.

In warmer waters the most important stock is that of sardinella off central Brazil between about 23° S and 26° S. Otherwise the warmer waters north of the River Plate seem to be poor in pelagic species.

Among invertebrates the stocks of shrimp in the north and south of Brazil have already been mentioned. Lobsters are found along the Brazilian coast, particularly in the northeast, in fair numbers. South of the Plate crustaceans are not important, but there are good mollusc resources. Both mussels and scallops are found on the northern Patagonian shelf, and there is a good stock of squid, which appear to migrate along the coast, from about 45° S in summer to around 35° N in the winter.

*The fisheries*

Fishing has not been an important occupation in the southwest Atlantic Ocean. The southern part of Argentina from Bahia Blanca southwards, where the largest fish stocks live, is not densely populated, and meat is cheap and plentiful. There has therefore been little or no incentive to develop a local fishery industry. Economic and social conditions are more favourable to the growth of fisheries in Brazil, despite the poorer resources (Table 5-24). There is a larger human population, and meat is not so plentiful. The lack of any big concentrations of fish and the long coastline have resulted in the Brazilian fisheries being highly diverse with a large artisanal sector. Taking a size of 20 gross tons as the dividing line between the artisanal and industrial sectors—a size higher than used

Table 5-24

Fishery catches. Upper: National catches (thousand tons) in the southwest Atlantic Ocean in selected years (Based on FAO *Yearbook of Fishery Statistics*). Lower: Catches of major species (thousand tons) in the Southwest Atlantic Ocean in selected years (Based on FAO *Yearbook of Fishery Statistics* and ANON., 1974)

| Country | 1961 | 1965 | 1967* | 1970 | 1975 | 1978 | 1981 |
|---|---|---|---|---|---|---|---|
| Brazil | 215 | 278 | 328 | 433 | 579 | 633 | 720 |
| Uruguay | 9 | 16 | 11 | 13 | 26 | 73 | 145 |
| Argentina | 88 | 192 | 228 | 186 | 196 | 504 | 350 |
| Japan† | No data | 22 | 4 | 14 | 1 | 21 | 22 |
| USSR | — | — | 678 | 10 | 9 | — | 17 |
| Poland | — | — | — | — | — | 21 | 73 |
| Germany (FRG) | — | — | 2 | — | — | 14 | — |
| Total‡ | 318 | 508 | 1253 | 664 | 820 | 1281 | 1339 |

| Species | 1961 | 1965 | 1967 | 1970 | 1975 | 1978 | 1981 |
|---|---|---|---|---|---|---|---|
| Hake | 41 | 104 | 604 | 118 | 126 | 417 | 333 |
| Croaker | 13 | 18 | 24 | 36 | 64 | 94 | 118 |
| Sardinella | No data | 51 | 80 | 89 | 139 | 196 | 223 |
| Anchovy | 11 | 17 | 13 | 13 | 19 | 16 | 13 |
| Mackerel | 10 | 11 | 13 | 11 | 10 | 7 | 11 |
| Shrimps | 25 | 39 | 35 | 38 | 44 | 47 | 82 |
| Squids | 0·3 | 0·7 | 3 | 2 | 5 | 75 | 54 |
| Others | 218 | 267 | 481 | 357 | 367 | 426 | 505 |
| Total | 318 | 508 | 1253 | 664 | 820 | 1281 | 1339 |

*Year of maximum distant-water catch.
†Mostly tunas, except in 1978 and 1979.
‡Includes catches by other countries.

elsewhere—there are some 50 000 boats in the small-scale sector, with only about 1 in 8 motorized. These use the variety of gears typical of such fisheries throughout the world.

The Brazilian industrial sector is also varied, with a total of some thousand vessels, many old and poorly equipped. The three main groups of vessels are the trawlers in the north, which principally fish shrimp and some catfish and are relatively the newest vessels; the trawl fleet in the south, fishing for a mixture of shrimp and the more highly valued demersal fish such as croakers; and the purse-seine fleet in the central south, which fishes for sardinella, and has the highest proportion of old vessels. There is also an industrial-scale fishery for lobsters in the northeast, and a number of line-fishing vessels, collecting high-valued species (snappers, etc.) in the north (principally for export) and in the south (for local consumption).

As Table 5-24 shows, the Uruguayan fishery was until very recently extremely small, with catches of a few thousand tons taken by small inshore vessels. In the last few years the government has given priority to the development of fisheries. This has led to the growth of a deep-sea trawling fleet, which works mostly in the joint Argentine–Uruguay zone off the River Plate, catching hake and croaker. Most of this catch is frozen and exported.

Special mention should be made of the Uruguayan harvest of fur seals. The seals of Lobos Island have been harvested for a very long time, but for an equally long time—at least since the early nineteenth century—the catches have been controlled. The amount of catch allowed to be taken has been determined purely arbitrarily (in the early period no scientific theory existed about the sustainable yield, or the level of the total allowable catch). Nevertheless, the success of the management of this seal stock, which was the subject of somewhat envious comment some 150 yr ago from an English sealer who had witnessed the decline of the uncontrolled South Georgia fur seal stock, illustrates that the most critical factor in nearly all resources management is the willingness and ability of the government to impose effective controls (FAO, 1978b, p. 118).

The Argentine fishery is essentially industrial; the small-scale and artisanal sector are unimportant. Most of the fleet are trawlers, fishing for hake, and to a lesser extent other demersal species, from the River Plate area southwards, particularly off Mar del Plata. Until recently, most vessels were of the traditional type, making short trips and bringing the fish back on ice for sale fresh, or for freezing ashore. Recently, a number of larger vessels, up to 110 m in length, have joined the fleet, which freeze the catch at sea. These vessels can make longer trips and can work further south. Much of the catch is frozen for export to Europe and North America. Apart from hake some of the trawlers are turning their attention to squid. Catches of squid by Argentine vessels have been very low, tending to increase slowly, but were only 2000 tons in 1977; they have leapt to nearly 90 000 tons in 1979.

Activities by foreign long-range vessels have been highly sporadic. Before 1965 the only foreign fishing in the region was a little long-lining for tuna in the oceanic area, principally by Japan. In 1965 Russian survey vessels found the rich trawl resources of the Patagonian shelf, and catches of hake by Russia rapidly built up to over half a million tons in 1967. Then Argentina, in common with several other South American countries, anticipating what has now become normal international practice, declared a 200-mile limit, which had the effect, if not the exact form, of a 200-mile Exclusive Economic Zone. Although Russia did not at the time recognize this action, it withdrew its fishing vessels to avoid what might have become a difficult political incident. Foreign catches of hake and of other fish on the Patagonian shelf fell to virtually zero by 1969, although some fishing continued to be done on the Antarctic grounds, principally around South Georgia, to the southeast. Long-range fishing has remained negligible until the last few years. Now, with the principle of a 200-mile zone generally recognized, and a wish by Argentina to utilize its fish resources more fully, the way is open for long-range fleets to return, but under the control of the coastal state. In part, these arrangements have involved foreign vessels from Japan and West Germany being transferred to the Argentinian flag (and their catches help account for the jump in Argentina catches since 1975). In addition, some long-range vessels have found that the hake, and other species, can be caught on the part of the shelf and upper slope which lie beyond 200 miles off central Argentina around 45° S.

*State of stock and management*

The most comprehensive review of the state of stocks in the region is contained in the report of a working group of CARPAS (Comision Asesor de Pesca para el Atlantico Sudoccidental, the FAO regional fishery body for the southwest Atlantic) which met in Montevideo and Rio de Janeiro in 1971 and 1974 (ANON, 1974). Since then a number of national studies have been made, but these have not been published in a generally available form.

Off Brazil a number of trawl surveys have been made from which estimates of demersal biomass have been obtained. Comparison of these figures with current catches, and other analyses, e.g. from size data, are in general agreement in showing a moderately high level of exploitation. As might be expected from the differences in patterns of human population (high in the south, low in the north) and fish density (fairly high in the north, lowest in the centre), there is a trend in the level of exploitation. The stocks in the south of Brazil are probably fully exploited, with little opportunities for increase in catch, while in the north, apart from shrimp and the most valuable species, catches can probably be increased significantly.

The Brazilian pelagic stocks have not attracted much fishing, apart from the sardinella, and even this probably still offers some opportunity for increasing catches.

With the exception of shrimp, and possibly lobsters, there do not seem to be urgent management problems. At the same time the fisheries, taken as a whole, are approaching the stage at which further increases in fishing effort could make some controls desirable. Also, the nature of most Brazilian fisheries—on a moderate scale, scattered along the coast—is such that the application of restrictive measures could be difficult. It is far better to consider future management problems now, and discourage further development (which is much easier than trying to restrict or reduce the excess fishing capacity once development has occurred), except when it is quite clear that considerable increases of catch are possible. This is well appreciated by some Brazilian scientists, and their attitude to the sardinella fishery is a good example of a careful and conservative approach. It seems that the stock is not fully exploited, and some increase in catch would be quite possible. The present fleet does not in purely technical terms operate very efficiently, though by supplying a well-defined local market it is economically viable, and the fishery is stable. By upgrading the technical operations of the fleet and opening new markets (e.g. for bulk catches for fish meal) it is highly probable that: (i) the fishery could be developed; (ii) for a short period catches could be increased; (iii) at least for the beginning of this period the fisheries will be economically highly successful. However, it is very doubtful if the stock could withstand a very much greater intensity of fishing. After the short boom period it seems very likely, looking at the history of other pelagic stocks, that there could be a collapse. Since strong mechanisms for controlling the fishery during a period of rapid development do not exist, it is preferable to leave the fishery in its present viable state, even though the theoretical possibility may exist for a moderate sustained increase in catch.*

The more abundant stocks from the River Plate southwards, were, until recently, only lightly fished. A number of surveys have been made over the shelf area, mostly by countries interested in the possibility of establishing distant-water fisheries. The more recent of these surveys have been carried out in collaboration with Argentinian scientists. The detailed results of some of the later surveys are not yet generally available, while there are some differences between earlier surveys. Surveys by the *Cruz del Sur* (operated by the FAO/UNDP project in Argentina) and the West German research vessel *Walter Herwig* in 1971 gave estimates of the biomass of hake of some 3 to 3·5

*I am grateful to my colleague, J. Csirke, for drawing my attention to this situation.

million tons, while the survey by the Polish vessel *Professor Siedlecki* yielded much higher estimates of 6·3 million tons (ANON., 1974). These differences can be ascribed, in unknown proportion, to differences in: (i) season and area in which the surveys were carried out; (ii) type of gear used; (iii) possibly also the way in which the survey was designed and carried out. Probably their main significance lies in illustrating the difficulties of estimating biomass quantitatively from trawl surveys.

At the time of the surveys the hake stock was not heavily fished, and if the estimates above can be used as estimates of the biomass of the unexploited stock, and a natural mortality of 0·3 is used, applying the GULLAND formula ($Y = 0{\cdot}5\ M\ B_0$), estimates of potential yield of 450 000 to 945 000 tons are obtained. Higher values of $M$ (0·4 and 0·6) were used by the CARPAS working group, and the total mortality estimated from samples of commercial landings at Mar del Plata, and from the research catches by the *Walter Herwig* were 0·746 and 0·75, respectively. However, such values of $M$ seem rather high for a fish which appears in significant numbers to age 10 yr and more. Using these larger values would give estimates of potential yield of from 1 to 2 million tons.

Until very recently the amount of sustained fishing has not been enough to check these figures directly. Some confirmation does come from the exceptional year of 1967, in which over 600 000 tons were caught, mostly by Russian vessels. The slight fall in the catch-per-unit-effort by the Argentinian fleet (5·55 kg $hp^{-1}\ h^{-1}$ in 1966 to 4·43 and 3·86 in 1967 and 1968, respectively) does not suggest that the long-range fleets took a high proportion of the total stock, though this conclusion must be tempered by the observation that the two fleets did not work in the same area. Simple algebra would suggest that if a catch of some 0·6 million tons reduces the stock by (5·55–3·86)/5·55 = 30% the initial stock was 2 million tons. This is subject to many reservations, but does give some confirmation to the direct estimates of biomass, more especially to the lower values.

The rise in catches in 1978 and 1979 is now bringing the catch close to the lower estimate of the sustainable yield. Therefore, as pointed out by FAO (1981a), the management of the hake stock needs attention. Although hake spend most of their time in the EEZ of Argentina, this will require international discussions on two aspects. In the north, hake move into the River Plate and into Uruguayan waters, including the common fishing area established by Argentina and Uruguay in the northern part of the Argentinian EEZ and most of the Uruguayan EEZ. The arrangements for managing this shared stock, including the principles on which the benefits from the fishery can be divided, are being actively discussed between Argentinian and Uruguayan scientists and administrators. Further south, around 43° S, the shelf widens sufficiently for the hake to be available, at some times of the year, to long-range trawlers fishing beyond 200 miles. The countries concerned must be brought into any scheme for managing the hake (limits on the total catch, control of mesh size, etc.).

The other major stocks, especially those on the southern part of the Patagonian shelf, are virtually unexploited. Together, the potential of these stocks (anchovy, sprat, and various demersal stocks) is large—perhaps 1 or 2 million tons. They are among the few resources with reasonable chances for the development of an economically viable fishery once the initial problems e.g. of building up local port facilities, are dealt with and the investments in catching and processing plants made. Given the imbalance in the world as a whole between growing demand for fish and virtually constant supply, these fisheries may develop soon. They will then need management.

*Southwest Pacific Ocean*

In this region, as discussed here, are included the temperate waters from the southwestern tip of Australia to Tasmania, which FAO includes in the eastern Indian Ocean statistical region. The region therefore includes two large shelf areas—the Great Australian Bight and the Campbell Plateau to the south of New Zealand—as well as the generally rather narrow shelf areas around the two main New Zealand islands, and the southeast coast of Australia, as far north as 28° 9′ S. The Great Australian Bight lies a little north of the zone of strong westerly winds and the West Wind Drift. The currents within the Bight are not strong and there is little mixing of water masses. Primary production seems low (though this is probably the least studied of any shelf area of comparable extent) and fish stocks are not abundant. A number of fishing trials have been made in the Bight, both by Australian and by foreign vessels. However, bearing in mind that they come from an unexploited stock, the catch rates have not been high (KESTEVEN and STARK, 1967). Similarly, productivity and catch rates on much of the Campbell Plateau have only been moderate. This may be because of the great depth of much of this shelf. Except around Auckland and Campbell Islands, most of the Campbell Plateau—as well as the Chatham Rise which extends from the South Island of New Zealand eastwards to the Chatham Islands—is deeper; in places considerably deeper than 200 m. This must reduce thorough mixing of the water column and hence tend to diminish basic productivity.

The productivity of the fish resources in the southwest Pacific Ocean is therefore relatively low, although there are some important stocks. In economic terms, the most important have been crustaceans, particularly the stocks of rock lobster off south Australia and Tasmania, and around the Chatham Islands. Shrimp are less important, being scarce around New Zealand, and the main Australian shrimp resources are outside the areas discussed here, in the warmer waters off northern and western Australia. However, stocks of deep-water prawn have been found off southeast Australia (ANON., 1978c).

Molluscs are also important, at least locally. Oysters of different species occur along the New South Wales coast, where cultivation is important, and in the Foveaux Straits, in the south of the South Island of New Zealand, where catches come from natural stocks. The region is one of the more productive parts of the world for abalones (particularly in southeastern Australia). Scallop resources, off Tasmania and elsewhere, are also reasonably good. Other molluscs are widespread, and the physical conditions for mollusc cultivation in New Zealand appear good (WAUGH, 1969a, b). In addition to typical molluscs, cephalopods are common in the area, with the arrow squid *Nototodarus sloani* and broad squid *Sepioteuthis bilineata* being the best known in New Zealand waters. A major Japanese fishery for squid has been established off the northwest coast of the South Island of New Zealand. Trial fishery by Japanese vessels for squid off Tasmania gave fairly good catches, but not sufficiently good for the establishment of a commercial fishery.

Among finfish, most attention has been paid to the demersal fish. In the colder and deeper waters to the south of New Zealand several of the species, e.g. southern poutassou *Micromesistius australis* and grenadiers (*Coryphaenoides* spp.) being very similar to those of comparable waters in the northern waters. The hake *Merluccius australis* also occurs, but does not dominate the system to anything like the extent that other species of hake do elsewhere. In the warmer and shallower waters the demersal fauna is more specific to the region, though having some similarities to other regions, particularly to the waters off southern Africa. Off southeast Australia the more important species are the flathead *Neoplatycephalus* spp. and morwong *Cheilodactylus macropterus* in the shallower water. Gemfish *Rexea solandri* are important in the deeper water out to the edge of the shelf. Many other species, including some species of flatfish, also occur in significant quantities. Around the main islands of New Zealand the main demersal fisheries have been around the North Island. Golden snapper *Chrysophrys aurata* is the single most important species in the catch, but other snappers, as well as gurnard, elephant fish, red cod, several flatfishes, and sharks are also caught in quantity. Jack mackerel and barracouta, although more usually considered pelagic species, also frequently occur in the trawl catches.

Apart from these incidental catches, and the catches of tunas (especially southern bluefin tuna), pelagic fish have not received much attention. Representatives of many of the typical pelagic genera occur in the region. Sardine *Sardinops neopilchardus* occur off southeast Australia and New Zealand (BLACKBURN, 1960, TURNBRIDGE, 1969). They have also been taken by Russian trawlers during exploratory surveys in the Great Australian Bight. Sprat *Sprattus antipodum* occur in most of the New Zealand waters; sprat eggs are common off Otago (in the central east coast of the South Island; ROBERTSON, 1980), but sprat have not attracted economic interest. Southern anchovy *Engraulis australis* are also widespread through New Zealand, though they live more northerly than sprat, which is absent from the northern part of the North Island (ROBERTSON, 1980).

The Jack mackerels *Trachurus declivus* and *T. novaezelandiae* may be the commonest of the pelagic species. They were often seen in aerial surveys around Tasmania and in Bass Strait (HYND and ROBINS, 1967), and have been caught in quantity by local and foreign vessels around New Zealand. Blue mackerel *Scomber australasicus* occur around most of the North Island of New Zealand. Here they are one of the species commonly sighted in aerial surveys. They also occur in the northern part of South Island.

In addition to these genera, typical of coastal pelagic fisheries in temperate waters throughout the world, the so-called Australian 'salmon' is an important coastal pelagic fish. There are two species of which the Australian salmon—or kahawai in New Zealand—*Arripis trutta* is the best known. The most important stock of this species migrates between eastern Australia (where it spends its juvenile life) and western Australia.

In addition, widespread oceanic species—such as saury *Scomberesox saurus*—and various species of lantern fish are also present. Not much detail is known about their occurrence in this region.

Finally, mention should be made of the introduction of salmon from the North Pacific. One species (the chinook, or quinnat, *Oncorhynchus tshawytscha*) has become established in the major rivers of the South Island (EGGLESTON and WAUGH, 1974). Compared with other regions, the fish in the region appear to be long lived. Snapper, tarakiki (*Morwong*), and sharks live to over 40 yr, and jack mackerel, gurnard, and other medium-to-small-sized fish live from 25 to 40 yr. To some extent the difference from temperate waters may be an artefact: few fish in areas that, like the North Sea, have been heavily fished for a long time, had a chance to live to high age. To the extent that the difference is real, it could affect the ability to take large sustained catches from the region, since the rates of production (and also potential yield) to standing stock is roughly proportional to turnover rate, thus inversely proportional to average life-span.

*The fisheries*

The first European settlers in Australia and New Zealand concerned themselves with opening up the land. They neither realized the richness of the sea, nor was there a market for any fish caught. Little attention was paid to fishing until towards the end of the nineteenth century. Then the growth of cities provided a market, and success of trawling in the North Sea, from whose coasts many of the settlers came, turned some attention to fishing. A number of North-Sea trawlers were brought out as trial commercial ventures, and to carry out exploratory fishing. These activities also prompted the beginning of fishery research in these southern waters. However, this faced a severe setback with the loss of the Australian survey vessel with all hands just before the First World War.

The larger North Sea type of trawler did not prove successful, though steam trawlers continued to work off the southern New South Wales coast. Smaller vessels, usually crewed by the owner and perhaps 2 or 3 others, became the typical fishing vessel in both Australia and New Zealand. Demersal fishing, with trawls and later Danish seines, as well as gill-nets and lines, have remained important, especially in southern New South Wales and western Victoria, as well as in the northwest part of the North Island of New Zealand; but other fisheries have also developed. These include the rich lobster fisheries of South Australia, Tasmania, and New Zealand (especially around the Chatham Islands); dredge fisheries for oysters (in southern New Zealand) and scallops (in Tasmania and South Australia); beach-seining for 'salmon' in western Australia; and pole-and-line fishing for bluefin tuna (mostly off New South Wales and on the same stock, but at different seasons, off South Australia). The variety of the local fisheries is suggested by the lack of any dominant species or groups in the statistics summarized in Tables 5-25 and 5-26.

Until recently few boats were longer than 15 to 20 m, but some larger vessels are now being added to the fleets of Australia and New Zealand, mostly for trawling in deeper water and purse-seining.

Fishing by larger vessels has been carried out almost wholly by foreign fleets. The first to work in the region were the long-line tuna vessels from Japan, for which southern bluefin tuna and albacore have been the most important species, and Japanese long-liners fishing for snapper off New Zealand. The latter were active in the 1960s, catching several thousand tons of fish. Their activities were disrupted around 1970 when New Zealand extended its limits from 3 to 12 miles, but after finding grounds beyond 12 miles, and rebuilding the fishery in the middle 1970s, this fishery was finally phased out after the establishment of a 200-mile zone by New Zealand in April 1978.

Trawling by foreign vessels around New Zealand started later. Again the first to carry out commercial-scale fisheries were the Japanese, around 1967. USSR exploratory vessels arrived a little earlier, but the first large-scale catches were taken in 1971. The Russian vessels have tended to work further south on the Campbell Plateau for grenadiers and poutassou (Table 5-26). In contrast to the Japanese, who have worked further north and in shallower waters, particularly for barracouta and jack mackerel, trawlers from Korea, and, for a short period, Taiwan, have also worked off New Zealand. The patterns of their operations seem to be similar to those of the Japanese vessels. A commercial trawler from West Germany, the *Wesermünde*, carried out a survey of the southern waters as a joint-venture enterprise with New Zealand partners. This survey has produced valuable scientific results (KERSTAN and SAHRHAGE, 1980), but the results during the commercially orientated part of the survey do not seem to have been sufficiently attractive to lead to continuing operations on a commercial scale.

Table 5-25

Landings (thousand tons) of major species and species groups from the southwest Pacific Ocean. (Based on FAO *Yearbook of Fishery Statistics*)

| Species | 1970 | 1972 | 1974 | 1976 | 1978 | 1981 |
|---|---|---|---|---|---|---|
| Flatfishes | 2·3 | 1·3 | 2·1 | 2·4 | 3·9 | 4·5 |
| Southern Poutassou (USSR) | — | 25·8 | 42·2 | 15·9 | 17·5 | 6·6 |
| Grenadiers (USSR) | — | 7·3 | 13·7 | 41·7 | 9·7 | 3·3 |
| Total gadoids | 1·0 | 37·2 | 62·9 | 95·9 | 41·7 | 24·1 |
| Snapper (New Zealand) | 12·8 | 13·2 | 13·9 | 14·4 | 17·7 | 11·9 |
| Total redfishes, basses, etc. | 34·8 | 48·5 | 63·6 | 57·8 | 83·7 | 131·1 |
| Jacks, etc. | 9·7 | 30·8 | 32·8 | 47·8 | 40·0 | 29·0 |
| Clupeoids | 0·2 | 0·1 | 0·2 | 0·2 | 0·3 | 0·2 |
| Tunas (Australia)* | 8·4 | 10·2 | 9·7 | 10·6 | 12·3 | 18·2 |
| Tunas (New Zealand) | 0·1 | 0·5 | 1·6 | 4·5 | 10·5 | 5·3 |
| Total, tunas and billfishes | 59·2 | 82·1 | 71·8 | 52·7 | 48·4 | 65·0 |
| Snoek (Japan) | 16·2 | 17·0 | 18·2 | 10·2 | 10·3 | 14·5 |
| Total mackerels, etc. | 17·0 | 18·5 | 22·3 | 14·7 | 18·4 | 30·8 |
| Other and unspecified fishes | 11·3 | 23·2 | 22·4 | 22·9 | 49·6 | 22·4 |
| Rock lobster (New Zealand) | 6·6 | 4·6 | 4·6 | 3·7 | 3·8 | 4·5 |
| Total crustaceans | 8·7 | 7·4 | 7·0 | 6·5 | 6·6 | 7·8 |
| Squids (Japan) | — | 0·1 | 19·6 | 19·6 | 29·0 | 45·2 |
| Total cephalopods | — | 1·0 | 19·7 | 19·7 | 36·8 | 60·6 |
| Oysters | 18·4 | 18·5 | 20·0 | 20·2 | 19·2 | 17·5 |
| Scallops | 0·6 | 2·1 | 3·4 | 6·1 | 2·4 | 2·1 |
| Other molluscs | 2·5 | 4·6 | 6·9 | 7·9 | 1·9 | 3·6 |
| Total† | 143·5 | 263·8 | 340·8 | 356·3 | 354·3 | 399·8 |

*Includes tunas caught outside FAO statistical area 81.
†Includes categories not specified above.

Table 5–26

Landings (thousand tons) by countries fishing in the southwest Pacific Ocean (Based on FAO *Yearbook of Fishery Statistics*)

| Country | 1970 | 1972 | 1974 | 1976 | 1978 | 1981 |
|---|---|---|---|---|---|---|
| Australia | 29·6 | 31·4 | 32·1 | 31·1 | 34·5 | 40·8 |
| New Zealand | 59·2 | 58·2 | 68·6 | 76·4 | 99·5 | 107·9 |
| Japan | 40·9 | 69·6 | 90·5 | 134·3 | 82·3 | 139·4 |
| Korea | — | 40·5 | 44·3 | 25·2 | 54·2 | 31·1 |
| Taiwan | 13·8 | 10·4 | 16·5 | 11·4 | 11·5 | 10·7 |
| USSR | — | 53·7 | 88·8 | 78·0 | 72·2 | 62·3 |
| Total | 143·5 | 263·8 | 340·8 | 356·3 | 354·3 | 399·8 |
| % Distant water | 38·1 | 66·0 | 70·5 | 69·9 | 62·1 | 62·8 |

The other foreign fishery in New Zealand waters is that of the Japanese for squid. At the peak of this fishery, around 1975, some 150 squid fishing boats were working. The main grounds were to the northwest of the South Island, but good catches have also been taken off the middle of the east coast of South Island. All these fisheries are now controlled by New Zealand, following the establishment of the 200-mile limit. Though continued fishing is allowed, except (e.g. in the case of snapper) where New Zealand fishermen are already heavily exploiting the stocks, foreign catches have tended to fall since 1978.

In contrast to these successful and varied foreign operations off New Zealand, which, together with the long-line tuna fisheries in the open areas, have accounted in recent years from some two-thirds of the total catch in the region, no foreign fishing on a commercial scale is carried out around Australia, except long-lining for tuna, and some fishing (e.g. trawling on the northwest shelf) outside the temperate waters considered here. This is not for want of trying, both before and after the introduction of the 200-mile Australian fishing zone. Since that introduction, activities of foreign vessels have been restricted to resources not exploited by Australian fishermen, but these are a large number. Apart from the resources of the northern Australian shelf in the Gulf of Carpentaria and northwest Australia (not considered in this section), these neglected resources include virtually the whole of the resources in the Great Australian Bight, as well as squid in the other parts of the temperate waters of Australia. Vessels (mostly trawlers, but also vessels equipped for squid jigging) from a number of countries including Russia, Japan, Korea, and United Kingdom, either independently or in joint ventures with Australian interests, have worked in these waters, but none has found the operation economically attractive.

*State of resources and management*

Until recently there have been few attempts to make a comprehensive review of the state of resources in the region. Assessments that were made concerned a few stocks of particular scientific or economic interest, especially those for which there had been concern regarding possible over-fishing. With the extension of jurisdiction the coastal states have found it necessary to review the resources over which they have acquired control, but these studies are at present mostly internal reports for the governments concerned.

The available information suggests that, with some exceptions, the stocks in the region are not heavily fished. The exceptions are highly valued species (rock lobster, scallops, shrimps, snapper), those near the main markets (demersal fish off New South Wales), or animals that are particularly easy to capture (Australian 'salmon', abalone). Over 30 yr ago it was shown that the tiger flathead *Neoplatycephalus macrodon*, the favoured species in the demersal fishery of southeast Australia, was heavily fished (FAIRBRIDGE, 1952, HOUSTON, 1955). Since then the fishery has reacted in a typical manner, turning its attention to other species and gradually expanding its operations into deeper water. Probably most of the demersal species around southeast Australia are now heavily fished.

Equally heavily fished are several other stocks of more valuable crustaceans (shrimps, rock lobsters) and molluscs (scallops, abalones) in southeast Australia. These and the trawl fisheries are now being strictly managed. In addition to a range of measures specific to the individual stock—minimum-size limits for many species, protection of berried female rock lobsters, bag limits, restriction of dredge size for scallops, closure of nursery areas of shrimp, etc.—entry into nearly all the fisheries is now strictly limited. The need to co-ordinate action in respect of several stocks between different states, and also between the states and the central Commonwealth governments, has probably tended to slow movement towards rational management. Australia, together with Japan, Canada, and the United States, is a country that has moved further towards an effective control on the inputs into fishing, and hence towards limiting excessive costs, and realizing greater net benefits.

A scheme of limiting licences was introduced into New Zealand even earlier—in the 1930s—but this was based mainly on the need to protect fishermen in the light of the very limited local market available at that time. It was not aimed at bringing the harvesting capacity into line with the capacity of the resources. As a result, it discouraged the expansion of the New Zealand fisheries at a time when nearly all resources were most lightly exploited. This licensing restriction was abolished in 1964. Though some expansion has taken place since then (Table 5-25, 5-26), the impact of the purely New Zealand fishery is fairly small except for some demersal species, e.g. snapper, found mostly in the shallower inshore parts of the shelf. The main management problems around New Zealand concern the stocks exploited by foreign vessels. In relation to the size of the New Zealand resource, the available capacity of the fleets of Russia and other countries is almost unlimited. Without controls the exploitation would almost inevitably follow the pattern of rapid growth followed by an equally rapid collapse, either partial or complete. At the same time, in the spirit of the text of the draft convention on the Law of the Sea, the New Zealand Government has wished to allow foreign access to resources which clearly cannot, in the immediate future, be fully utilized by New Zealand fishermen—particularly since granting of such access can be used as a card in other negotiations, e.g. for access of New Zealand products to foreign markets. However, although some surveys have been done, the information on most stocks is not sufficient to make detailed assessments with sufficient precision. A conservative approach has therefore been used, setting quotas on the basis of a Safe Biological Yield rather below the first rough estimate of sustainable yield (WAUGH, 1979). This slowly had the effect of controlling, and in fact slightly reducing, the catches by non-local vessels (Table 5-26).

The prospects are now for an expansion of New Zealand fishing, but one that will not, at least for a long time, completely replace foreign fishing. This will continue. Indeed, if the expansion of local fishing is slow, and the accumulation of information shows that an expansion of total fishing is possible—and is greater than can be taken up by local fishing—there may be, at least temporarily, an expansion of non-local catches, especially on those stocks (e.g. grenadiers and blue whiting on the Campbell Plateau) which are least attractive, economically and operationally, to local fleets.

This forecast applies only to stocks sufficiently attractive to either local or non-local fleets to ensure at least a moderate degree of exploitation. These include mostly demersal stocks, around both Australia and New Zealand, with the exception of those in the Great Australian Bight; squid around New Zealand, but not on present evidence those off Australia; 'salmon' of Australia (though not the 'kawahai' of New Zealand which is the same species); and most crustaceans, but only a small proportion of the pelagic species. These species—sardines, anchovy, mackerel, etc.—are not uncommon (e.g. ANON., 1978c,d) and the total potential yield could be as high as some hundreds of thousands of tons. For example, NOSOV (1978) gives an estimate of the biomass of jack mackerel off

New Zealand of some 800 000 tons. However, the species are generally of low value, and may not be easily caught in sufficient quantities to be worth fishing for fish-meal (for example, the average school size of jack mackerel off Tasmania seems to be too small to support a purse-seine fishery for fish-meal). It may, therefore, be a long time before all these species are exploited at a level approaching their ecological potential.

## (f) Other Resources

### *Southern Ocean*

*Introduction*

The Southern Ocean—the waters surrounding the Antarctic Continent—is the most isolated and least visited part of the World Ocean. However, it is far from being the least productive, or least interesting, and has indeed recently been receiving a good deal of international attention. Its southern boundary is the Antarctic Continent or, for many practical purposes, the northern limit of the permanent pack-ice—though a surprising amount of biological activity goes on even under the pack. The northern boundary is less clearly defined but is commonly taken as the Antarctic Convergence. The recently signed Convention on the Conservation of Antarctic Marine Living Resources uses as a northern boundary of its area of competence an approximation to the Convergence which runs along lines of latitude and longitude—mostly along 60° S in the Pacific sector, but as far north as 45° S in parts of the Indian Ocean sector.

Though inhospitable, the Southern Ocean is rich and full of life—of life in forms commercially attractive to man. Despite the distances, and the cold and rough weather, these riches have attracted sealers and whalers for centuries, and much of our knowledge of the region has come from their activities. During most of the present century the particular interest has been in baleen whales. Recently, the conservation and management of whales has attracted wide attention. The problems and history of whale management are discussed in the following sections. For convenience, problems of whaling elsewhere will be discussed in this section, in addition to the specific problems of whaling in the Antarctic.

*The resources*

Antarctic waters are notable for the very short food chain. Often there is only one link between primary production and the largest animals. This results in high overall efficiency of transfer to the animals that can be conveniently harvested by man and represents the main factor responsible for the apparent richness of Antarctic waters. The primary production is not particularly high, except locally (EL-SAYED, 1970). The most productive areas are along the coasts and ice edge, and especially to the east of the Drake Passage. Here the funnelling of the circular eastward flow of water around the continent, driven by prevailing westerly winds through the narrow passage between the tip of South America and the Antarctic Peninsula causes a thorough mixing of the water. This region has been the most actively studied, and this concentration of observations in the most productive area can have strengthened the impression of the Antarctic being particularly productive.

The key element in the Antarctic system, both as food provider for other animals and to an increasing extent as a resource in its own right, is the Antarctic krill *Euphausia superba*. Other species of euphausids (notably *E. crystallarophias* close to the ice edge), as well as other crustaceans and herbivores, may account for a substantial part of the total consumption of phytoplankton, as well as being an important food for some of the 'higher' animals. For example, copepods are the major food of sei whales (NEMOTO, 1968). However, no other single species is as important as the Antarctic krill, and for some animals, e.g. blue whales, it is by far the dominant food source. Krill has been the main object of many of the studies of the marine community, from the time of the Discovery work between the wars onwards. The report by MARR (1962) on the Discovery work remains the classic study of krill, and the major source of information—though this will presumably cease to be true when the extensive activities of the BIOMASS programme (organized by SCAR, the Scientific Committee on Antarctic Research in collaboration with SCOR, ACMRR and IABO) and especially the multi-ship FIBEX (First International BIOMASS Expedition) carried out during the 1980/81 Antarctic summer, become available.

The ecology of krill, and also of the other resources of the Southern Ocean relevant to harvesting and management, has been most conveniently summarized by EVERSON (1977). Krill is distributed throughout most of the area south of the Convergence, but is particularly abundant in the Atlantic sector, in the arc from the Antarctic Peninsula to South Georgia; it is also abundant towards the ice edge. Adult and subadult krill (which may live up to at least 2 yr) form large swarms. These are sometimes clearly visible on the water surface, but also occur at depth. The FIBEX surveys, using acoustic methods, have detected at least one swarm of vast size, estimated to contain several million tons of krill.

All classes of higher animals feed on krill. The least important are probably fish. Only around the Subantarctic Islands—noticeably South Georgia and Kerguelen—and the northern part of the Antarctic Peninsula are there areas of continental shelf which are clear of ice for most of the year and which can support substantial fish stocks; in some other parts of the Antarctic, such as the Ross Sea, the permanent ice may extend well beyond the continental shelf.

Little is known about the cephalopods of the Antarctic, but large quantities are consumed by the higher predators, particularly sperm whales and the larger birds. This indirect evidence indicates that squids are abundant and wide spread throughout the Southern Ocean.

Birds and mammals are particularly abundant. Birds are not considered a harvestable resource—though half serious examinations have been made of the potential value of penguins as a source of oil. However, several groups of mammals have been in the past, are now, or could become the object of important fisheries. Best known are the baleen whales (in order of decreasing size, and also roughly in order of exploitation, blue, fin, humpback, sei, and minke whales). All have similar life-cycles and ecological characteristics (for useful reviews see MACKINTOSH, 1965; and GAMBELL, 1976). They feed actively in Antarctic waters during the southern summer, migrating to the warm subtropical waters to breed in the southern winter. There are no marked differences between sexes, though the females tend to be slightly larger than males, and the female produces a single young (very rarely twins) roughly every 2 or 3 yr once becoming mature. Within the Antarctic there are differences in distribution (though with considerable overlap). Apart from the humpback whales, which have a very patchy distribution both in summer and winter (when they seem to congregate around a few preferred islands, e.g. Tonga), the other species are distributed fairly evenly (though with greater densities in the South Atlantic sector, and low densities in the Pacific) in zones around the Antarctic continent. Minke whales occur nearest the ice edge, with the blue, fin, and sei whales being found increasingly further north. These latitudinal differences have caused the whaling fleets to adjust their areas of operation as their main target species has changed from blue to fin to sei and finally to minke whales. The differences also have implications on the degree to which changes in abundance of one whale species can affect other species.

The biology of the sperm whale is different. There are also pronounced differences between the sexes. Males grow to a much larger size than females, and only the large solitary males migrate into the Antarctic and to the higher latitudes in the North Pacific and North Atlantic Ocean where (as elsewhere) their food is predominantly squids. Females, as well as young males, and adult males with harems remain in warmer waters throughout the year.

The true Antarctic seals living on the Antarctic continent, or the surrounding ice, have never been subject to significant exploitation. They form a potential resource for oil and meat, as well as fur. The commonest—the crabeater seal—is also the most abundant seal in the world, with a population estimated to approach 15 million individuals (little less than 3 million tons in weight). The crabeater—as might possibly be expected on a wide interpretation of its name—feeds predominantly on krill, eating only a small proportion of squid and fish. At the other extreme, the leopard seal is an active predator on penguin—though it also takes fish, squid, and krill. The number of leopard seals, as of the other truly Antarctic seals—Weddel and Ross seals—are between 1 and 2 orders of magnitude less than crabeater seals.

The Subantarctic seals—fur seals and elephant seals—were subject to heavy fishing from which they are still recovering. They are found principally on the Subantarctic Islands, notably the South Georgia group. They breed ashore in large colonies, which makes them vulnerable to exploitation. It also makes it relatively easy to study, and the biology of South Georgia seals and their recent increases in numbers is particularly well known (e.g. LAWS, 1977). The elephant seals—as their name implies, the largest of the seals with an average weight of half a ton (adult males may be bigger)—are predominantly fish and squid eaters, but fur seals have a lot of krill in their diet.

*The fisheries*

The Antarctic Ocean has a history of a succession of short-lived fisheries exploiting stocks well beyond the natural capacity. The result was that each fishery lasted only long enough to deplete the stock concerned to a level at which harvesting was no longer economic.

The first fisheries were those of the subantarctic seals, starting with the fur seals in the early nineteenth century. The high prices of fur meant that the stocks were reduced to a very low level before sealing became uneconomic. Attention of the sealers then turned to elephant seals, but these were only used for oil and were much less profitable. Thus, sealing them became uneconomic and ceased before they had been reduced to a critically low level. However, there was a brief resurgence of sealing as an activity incidental to shore-based whaling (also for oil) at South Georgia in the early twentieth century.

The latest boom-and-bust fishery has been that for various species of demersal fish around South Georgia and some other island groups. These fisheries developed around

1970 as about the last phase in the spread throughout the world of distant-water trawling by the USSR (and to a lesser extent by Poland and other eastern European countries). Catches in the Atlantic sector (mostly around South Georgia) reached over 400 000 tons in the 1970/71 season, nearly all notothenids, but immediately fell to virtually nothing in 1972/73. After 1976 attention of the fisheries was turned to icefish (mostly *Champsocephalus gunneri*) with catches reaching 200 000 tons in 1977/78, and since then decreasing.

During most of the present century, however, the interest of Southern Ocean fisheries has been in whales. Antarctic whaling can be considered to be nearly in the 'final' stage following a series of booms and busts on different whale stocks that started centuries ago.

The earliest whalers were the Basques. It is not known whether early Basque whaling was responsible for the extinction of the grey whale in the Atlantic Ocean (the species now only occurs in the Pacific Ocean). However, it is known that by the thirteenth century the Basques had a thriving industry on right whales. As local stocks declined they discovered how to process whales at sea, and at the end of the fourteenth century were working as far afield as Newfoundland. The basic pattern of whaling—the pursuit of right whales and, later, sperm whales from open boats, followed by the processing of the whales for oil and 'whalebone' alongside the mother vessel—remained unchanged until 1860, though whaling spread to other countries, notably Britain, Holland, and the United States. The whalers worked further and further afield as local stocks declined. In the heyday of New England whaling the Yankee fleets operated in all oceans of the world, to an extent matched only by the Japanese tuna long-line vessels more than a century later. The right whale fisheries collapsed to virtually nothing by the end of the nineteenth century, when most stocks had been reduced to very low levels. The sperm whale fishery collapsed rather earlier, though the reasons seem to have been as much the price drop for sperm oil following the discovery of petroleum, and the losses in ships during the American civil war, as the decline in the stocks.

Until 1860 only the slower and less active right whales and sperm whales could be hunted with open boats, but with the invention of the harpoon gun by Svend Foyn the more active rorquals (blue and fin whales, etc.) which sank after dying, could also be hunted successfully. This technique was first adopted by Norwegian whalers in the North Atlantic Ocean; as these stocks declined, attention turned to the Antarctic. The first whaling in the Antarctic was done from shore stations at South Georgia in the early twentieth century. While this was successful, the great age of Antarctic whaling came with the development of the factory ship with a stern ramp which enabled the whales to be hauled on board for processing. This opened up the whole Antarctic Ocean to whaling, and the fishery grew rapidly, reaching a peak in 1930/31 when over 40 000 whales, including nearly 30 000 blue whales, were caught. These catches depressed the market, and catches in 1931/32, and to a lesser extent for the rest of the pre-war years, were less, primarily for economic reasons.

Up to 1936 mainly blue whales were caught but as these declined the industry shifted to the slightly smaller fin whales. These provided the mainstay of the industry until 1964. Then attention again shifted, first to sei whales, and then in the last few years to minke whales, the smallest of the baleen whales. The other change in whaling practice in the last few decades has been the shift in emphasis to the use of meat for direct human consumption. Until about 1950 the only significant product was oil, mainly from the blubber, but starting with the Japanese fleets more attention was paid to producing frozen meat, principally for the Japanese market. While this has meant a more efficient use of the catch, it has also increased considerably the value of the individual whale, and this lowered the limit at which decreasing whale density makes whaling uneconomic.

Outside the Antarctic Ocean the use of factory ships to catch baleen whales has been for more than 30 yr restricted to the Pacific Ocean where the industry has gone through much the same phases of successive concentration on, and depletion of, the different species. Elsewhere, hunting of baleen whales has been restricted to shore stations, which with their restricted range (the distance from which a catcher vessel can tow back a dead whale in the short time in which it remains fresh) have not developed to such an intensity as the pelagic fisheries. Some of these shore stations (e.g. in Australia and South Africa) have harvested the Antarctic stocks during their northern migration to warmer waters. These have suffered from the depletion caused (mostly) by the Antarctic operations and are now generally closed down. Others, e.g. in the North Atlantic Ocean have been operating on what appear to be local stocks.

After the decline of the Yankee whaling in the middle of the nineteenth century, sperm whales have not been the main object of a major whale fishery. However, since the 1950s they have provided alternatives or substitutes for baleen whales, especially for the pelagic factories. A factory ship returning to Japan or the USSR from a poor Antarctic season would often attempt to make up some of its losses by catching sperm whales in tropical waters. On the scale on which these factories operated these catches, plus catches of shore stations, have had serious effects on several sperm-whale stocks.

The last fishery in the Southern Ocean to develop—indeed it is still only in the early developmental stage and the course of development is still far from clear (MITCHELL and SANDBROOK, 1980)—is that for krill. The declining catch rates of most of the better known species of fish, and the extension of fishing limits by many coastal states, together with the estimates of very large potential yields, persuaded several countries with large distant-water fleets to look at the possibilities of harvesting krill.

Technically, catching krill is not a major problem (EDDIE, 1977). The large swarms, including those well below the surface, can be readily located with sonar or echo sounders and caught in large quantities with small-meshed mid-water trawls. The problems arise once the krill comes on board. It is extremely delicate and must be processed quickly into some frozen or dried form. The meat in the tails is nutritious and tasty (generally similar to its relatives, the larger shrimps and prawns), but there is very little on each animal. Potentially krill could be used for direct human consumption. However, it is not easy to develop a method of processing which yields a product that will attract a large demand at an attractive price. Japan has developed a small market for boiled and frozen whole krill, and the USSR a form of krill-flavoured cheese, but neither has so far developed into a large market.

Krill is also potentially suitable for producing a high-protein meal for feeding chickens and other farm animals as an alternative or supplement to fish-meal or soya-bean meal. It might be particularly suitable for feeding salmon and trout, or lobster, in the growing business of culturing these high-valued species. However, there are technical problems (including the high content of fluorine in the shell), and very serious economic problems. Meal is only useful if it is cheap, but it is very expensive to operate vessels in the distant and inhospitable waters of the Antarctic.

The immediate prospects for a take-off of large-scale (multi-million ton) krill fishery are now small. Most countries, except the USSR have slackened off their efforts on developing a krill fishery. The USSR has maintained its interests, and its catches rose to as much as 300 000 tons in 1978/79, but even this is quite small compared with the total USSR catch, and with the likely potential from krill.

*State of stocks and management*

So far as the Southern Oceans are concerned, resource assessment and management mean almost wholly the assessment and management of whales, and refer to the activities of the International Whaling Commission. The performance of the IWC has been patchy, but not really as bad as its public reputation suggests. The history of management of the Antarctic whales can be divided into four periods: (i) the period before the formation of the IWC in 1946; (ii) the early years of the IWC, up to about 1960; (iii) the years of severe crisis from 1960 to about 1970; and (iv) the recent period.

During the pre-war years whaling proceeded unrestrictedly, but not wholly thoughtlessly. The whalers themselves had seen the depletion of the whale stocks in northern waters, and the disappearance of the once-thriving industries, and were well aware that the Antarctic stocks though vast were not inexhaustible. However, while at that time the whaling industry was dominated by Norwegian interests no one country could apply unilateral restrictions—at least not without letting the other countries gain most of the benefits. Reaching the necessary formal agreements took time, with a series of conferences and meetings which culminated in the signing of the IWC Convention in Washington in December 1946. For the first time this opened the way to the setting of internationally agreed limits on the number of whales taken. It was one of the first international agreements on the take of any wild animal, and as such represented a considerable step forward in the conservation of stocks of wild animals. Smaller steps—prohibiting the killing of right and grey whales, which were already at low levels, and the setting of size limits for the major commercial species—had in fact been taken at earlier, pre-war, meetings. Another important action which was taken in the pre-war period was the instigation by the United Kingdom of a major research programme—the Discovery Investigations—in the Southern Ocean in support of the whaling industry. With the benefit of hindsight it might be felt that insufficient emphasis was placed on studying the population dynamics of whales and perhaps too much on more basic research such as studies on plankton or currents. Thus, the results were of relatively little direct help in providing scientific advice to the IWC in its crisis years. However, the long-term value of the research, especially into the distribution and ecology of krill, is now becoming apparent.

The IWC, as established in 1946, had a number of defects. These became steadily more apparent in the period up to 1960. The more serious defects included: (i) the absence of arrangements for allocation of the overall quota; (ii) the fact that in the absence of action the annual quota (set at 16 000 Blue Whale Units in 1946) remained unchanged from year to year; and (iii) the absence of a permanent mechanism for provision of scientific advice, backed up by adequate national research. That the catch limits were expressed as a global figure in Blue Whale Units (1 blue whale = 2 fin whales = 6 sei whales) rather than for each species or each stock separately has also been pointed out as a weakness in the initial IWC arrangements. Undoubtedly, separate controls for each stock are desirable, but the use of the BWU did not seriously affect the impact of the IWC's policies on individual stocks. Indeed, by facilitating the switch between species, it probably had a beneficial effect in the 1950s and 1960s, at a time when the species were exploited at very different intensities, but there were no good estimates of the state of individual stocks. Certainly, if the catch of blue whales (rather than the number of BWUs) had been maintained at the same level during this period, the blue whale stock would have been even nearer to extinction in 1960 than was actually the case.

In the absence of allocation, catching was allowed until the total catch (reported weekly by radio, and as the limit was approached, daily, to the Bureau of International Whaling Statistics in Norway) was reached. The season was then closed. The result was that each country, and each company within the country, increased its catching capacity (particularly the number and power of catcher vessels attached to each factory ship) in order to maximize its share of the total. As a result, despite the falling stocks, the limit was reached progressively earlier each season, and the length of the open season, which had been 121 d in 1946/47 fell to only 58 d in 1955/56. With expensive capital equipment being fully used for less than 20% of the year, the costs of whaling were much higher than they need have been. Until this cause of inefficiency was removed the industry had great difficulty in agreeing to reductions in the quota. Agreements on national allocations of the total quota were reached in 1962, though not without considerable argument, including the temporary withdrawal of Norway and the Netherlands from the commission. This and further allocations within countries with several fleets eliminated the need for each expedition to scramble for its share of the quota, and has greatly improved the economic performance of the industry.

The shortage of research at the national level into the quantitative population dynamics of whales, the lack of good formal arrangements for providing the commission with scientific advice, and the fact the quota remained at the same level from one year to the next unless specific action was taken, all interacted to result in the great crisis of the early 1960s. In 1946 the blue whales had been depleted somewhat below their most productive level, though they were still reasonably abundant, and the other species were all in a very healthy state. The quota set by the commission was not far from the combined sustainable yield of all species, and certainly as close as could be expected given the information then available.

The quota was, however, too high, so that the stocks declined—at an accelerating rate as the sustainable yield also decreased. This would not have been too serious if the decrease in stocks had been properly measured and the necessary adjustments made to the quotas. While the decline in the whale stocks was commonly recognized, not least by those in the whaling industry, there was little agreement on the extent of the decline, or on the implications of the decline on the policies of the commission and the adjustments that should be made to the quotas. The majority of scientists (though with a few influential exceptions) did advise that the quotas were too high, and should be reduced, but did not explain the consequences if this were not done. The reply of the industry was that, while appreciating the scientists wishes, they felt that it would be nice if the catches could be maintained or even increased. Since much of the attention of the commissioners was diverted to the negotiations on the allocation of the quota, little was done.

In 1960 a Committee of Three (later Four) scientists was set up. While not (at that time) whale experts, the scientists were experienced in the dynamics of fish populations and the provision of management advice. These aspects of applied science had at that time advanced much further in relation to fish (notably in the North Sea) than to whales. However, some of the earliest work (e.g. HJORT and co-authors, 1933) was done in relation to whales, and several of the simpler fishery models (e.g. the SCHAEFER model and its relations) are more applicable to whale than to fish stocks.

When the Committee of Four made its report (CHAPMAN, 1964) the situation was put clearly before the commission. They were enabled to understand the principles underlying the concept of the sustainable yield, and the trade-offs that needed to be made between high catches now, and the assurance of continued catches in the future. They were also given quantitative assessments of the current status of the major stocks. The blue whales (except for the small stock of so-called pigmy blue whales in the southern part of the Indian Ocean) had been so depleted that the sustainable yield was neglible, and any continued catches could seriously endanger the continued existence of the stock. The commission shortly agreed to ban all catching of blue whales in the Antarctic (the humpback whales had been given protection slightly earlier), and these bans have continued ever since.

The fin whale stocks, though depleted well below their most productive level, were still abundant enough to support a reasonable sustainable yield. However, the yield comprised only about 7000 whales, compared with catches of over 25 000 whales taken every year between 1954 and 1962. The first task of the commission when it received the Committee of Four's report was to bring the catches down to less than the current sustainable yield and thus to ensure that the stocks did not decrease further. This required something like a 75% cut in the catches, a figure not easily accepted by any industry. Achieving this therefore took several years of hard argument. This argument was reduced, and the faith in the reliability of the scientists increased, by the events of the 1963/64 season. At the preceding commission meeting the proposals to achieve a drastic reduction in the quota had been unsuccessful. During the discussions the scientists pointed out that the existing fleets, even if they fished without restrictions, would— with the fall in stocks—only catch some 14 000 fin whales, and that any larger quota (or its equivalent in BWUs) would be ineffectual. However, a larger quota was set, the fleet fished unrestrictedly and only caught 13 870 whales—very close to the prediction. Agreement was then reached in 1965 to set quotas at or below the estimated sustainable yield for each stock, thus abandoning the BWU, and this has been done since. Thus, each stock should, if the estimates are correct, at worst not be decreasing further, and at best should be increasing. Though it is very difficult to monitor the slow changes that are expected to be occurring in the stocks, it is probable that indeed the stocks are recovering.

The agreements reached around 1965 did therefore end the most serious crises in the commission and determine that whales could be managed rationally. They did not end the arguments. These have indeed increased, partly because several countries without a whaling industry have joined the commission. While this has undoubtedly encouraged the commission to take a wider and longer-term view of its responsibilities, it has also led to an unfortunate split between whaling countries (chiefly interested in maintaining the short-to-medium-term interests of their industries) and non-whaling countries (whose voice in the commission is in some cases dominated by the views of these conservation groups chiefly interested in stopping all whaling).

The main arguments concerned the rate at which depleted stocks should be rebuilt to their most productive level, and particularly whether some small catches should be allowed from stocks which are below this level, thus slowing down their recovery. The commission attempted to resolve these arguments by establishing procedures, particularly the New Management Procedure, according to which management actions taken followed directly from the conclusions of the scientific committee. These procedures in fact tended to resolve the arguments in favour of the more conservationist views, by setting catches at zero if the stock was more than marginally below its most productive level.

The main difficulty, however, was that the data available to the scientists, whose activities were now organized within a strong scientific committee, were seldom conclusive, and could be interpreted in various ways according to national interests. The division between conservation and commercial interests was therefore carried into the scientific committee. In the early years of this past period, while the interests of the whaling industry still had a strong voice in the commission, the tendency was to take the rather more optimistic interpretations or sets of parameters, leading to higher quotas. This tendency is now reversed, partly because the conservation interests are now stronger, and partly because it has become appreciated that when the stocks are depleted the situation is asymmetric, such that errors in taking too optimistic a view of the state of the stocks have far more serious consequences than taking too pessimistic a view. The result has been that the catch quotas set by the commission have been steadily falling not only in the Antarctic, but elsewhere, even though during this period most stocks have probably been increasing.

The Antarctic minke whales present special problems in applying the New Management Procedure. They were virtually unexploited until the 1970s, but had been indirectly affected by the catching of other species. They have similar distributions and food to the blue whales (and to a lesser extent the other large baleen whales), and have responded in the same way to the reduction of the stocks of these whales (presumably as a result of better food supplies), most clearly in reducing the age at which the females first become mature. This means that at the time that exploitation of minke whales began, they were increasing. Despite catches in the last few years, their present abundance is greater than that which would under the conditions occurring in, say 1900 (i.e. before any exploitation took place) have provided the greatest net production. However, the abundance is probably less than that which would, under current conditions of the other species, give the greatest net production. The choice (apart from the short-term interests of maintaining at least a small Antarctic whaling industry) is therefore between assuming that conditions will stay as they are now, with few larger whales, in which case the minke whale stock should be further increased, and no catches taken for some time, and assuming that in due course the other stocks will recover, in which case the minke whales should be reduced if they are to be kept at the most productive level (which would probably also help the recovery of the other whales).

The scientific uncertainties concern the models employed, and the data used, in applying these models. The models are discussed in several reports, including ALLEN and CHAPMAN (1977) and ALLEN (1980). They have advanced from the simple SCHAEFER model used by the Committee of Four (CHAPMAN, 1964) to quite complicated models, including, for sperm whales, allowance for the social structure of the population, such that the males do not reach social maturity, and become able to look after a harem, until some time after reaching sexual maturity. The earlier models were based largely on fishery experience, whereas the newer models take more account of whale biology. However, there has been little inconsistency between the results used by different models, and there have not been the gross departures between expectation and observation that have occurred in some fisheries by, for example using yield-per-recruit models when recruitment has changed. The biggest difference is that whereas the SCHAEFER models predict an MSY at 50% of the initial unexploited population, the newer models, taking account of the detailed reaction of the stocks to exploitation (principally reduction in age at first maturity, and increased frequency of pregnancy), suggest that it occurs at higher levels, possibly as much as 60 to 70% of the initial stock. This change has brought several stocks into the range where, according to the policies of the commission, there should be no catch. It is another reason why the quotas actually set have been falling at a time when the stocks as a whole have probably been increasing.

The quality of the original data is a more serious problem. Examination of the ear plugs from a baleen whale can determine its age, and also the age at which it matured. A reasonable amount of sampling of the commercial catches can give useful estimate of age composition, mortality rates, and change in age at maturity, though in some fisheries the numbers involved, even if the total catch is sampled, are so small that there is considerable variance on the estimates of mortality.

Other parameters are more difficult to measure, particularly abundance, or relative abundance. A few stocks live in so restricted an area, at least at some times of the year, that they can be counted directly, from the shore or from the air. A long series of counts have been made of grey whales as they migrate past Point Loma, California. This has given a clear picture of how this stock has rebuilt. More recently counts have been made of bowhead whales passing Point Barrow, Alaska. Humpback and right whales have sufficiently clear but variable characteristics so that individuals can be recognized in the field, or at least from photographs (KATONA and WHITEHEAD, 1981). Some of their populations are so small that something like a complete count can be made, or the population size be estimated from the frequency with which known individuals are sighted again.

These methods cannot be applied to the large oceanic stocks, particularly in the Antarctic. For these, and for most of the stocks still being exploited by coastal fisheries, the most useful and most widely used measure of abundance remains the catch-per-unit-effort, usually calculated as number of whales caught per day's work by a catcher vessel. This suffers from several of the problems common to other units of fishing effort, though these can to a large extent be minimized by making corrections for the size and efficiency of the vessel, etc. and by taking account only of the time spent looking for whales, and eliminating the time lost by bad weather, pursuing and killing other whales, and bringing them to the port or factory ship. However, once commercial whaling is stopped these data are no longer available, and it becomes extremely difficult to determine directly how the population is changing. Some data for the Antarctic stocks are being obtained from sightings from special survey cruises, but these are subject to high variance. Over a short term they are practically useless for detecting the small increases, of a few percent per year, that should be expected to be occurring in the protected stocks.

With these reservations, the present status of the major groups of whale stocks is summarized in Table 5-27. Of these stocks two are in danger of extinction: bowheads in the North Pacific Ocean—the Atlantic stock may be extinct—and grey whales in the western Pacific Ocean. Although the number of bowhead whales is very low, they do not seem to have increased their reproductive rate in response to exploitation. Only few calves are seen, and the gross reproductive rate of the population seems to be little, if

Table 5-27

Estimates of current abundance of major whale stocks. High: abundance exceeds 80% of initial level; moderate: 20 to 80%; low: less than 20% (After ALLEN, 1980; modified, by permission of University of Washington)

| Species | Current abundance | | |
|---|---|---|---|
| | Southern Hemisphere | North Pacific | North Atlantic |
| Blue | Low | Low | Low |
| Fin | Moderate | Moderate | Moderate |
| Sei | Moderate | Moderate | Moderate |
| Humpback | Low | Low | Low to moderate |
| Minke | High | High | Moderate |
| Bryde | High | High | Moderate |
| Bowhead | — | Low | Very low |
| Right | Low | Low | Low |
| Grey | — | Moderate | — |
| Sperm (males) | Moderate | Moderate | Moderate |
| Sperm (females) | High | High | High |

any, higher than the rate of natural mortality (unless this is much lower among bowheads than it is among other whales). Despite its critical state, continued catching (though only of a limited number) by the local Eskimo population of Alaska is allowed and there are great political difficulties in stopping or reducing this aboriginal take.

One other stock is believed, at least by many scientists, to be decreasing. This is the sperm whale stock in the North Pacific Ocean. The number of large breeding males has been so reduced that the reproductive rate of the females has been affected. Thus, the number of births is less than the natural deaths, and this will continue for some years until the more numerous young males have increased in size and experience to bring the reproductive rate back to normal. This situation is strictly temporary, due to the odd structure of the current population, whose long-term prospects are quite healthy. All other stocks of whales are probably now increasing. However, only in some exceptional circumstances, e.g. when accurate counts can be made, is it possible to demonstrate this increase unequivocally.

The increase in some seals is much more obvious. None of the Antarctic seals is now being harvested, and many stocks are either recovering from over-exploitation in the nineteenth or early twentieth centuries, or benefiting from the lessened competition from the baleen whales. The most dramatic increase has been in the fur seals at South Georgia (PAYNE, 1977). Reduced close to extinction, this population numbered only about 100 individuals in the 1930s, but recovered to 15 000 by 1957, and to 350 000 in 1976. The future management of seals in the Antarctic seem assured by the Convention for the Conservation of Antarctic Seals. This was signed in 1972 at a time when commercial exploitation of the southern seals (particularly crabeater) seemed a strong possibility. It was a milestone in conservation in that the management arrangements were made before the industry started. Since commercial sealing did not in fact start, little action has been taken under the Convention, though provisional catch limits have been set for each species of seal.

The demersal fish are the group of Southern Ocean stocks that currently are most urgently in need of improved management. The knowledge of these stocks is not good because some of the critical statistical data on catches and fishing effort are not available. The BIOMASS Fish Biology Working Group (SCAR, 1980) reviewed the available data, and taking account of the biology of the main species, concluded that the stocks were probably very heavily fished.

*Ecosystem management*

The greatest attention in the last few years has been focused less on the state of the individual species, or groups of related species, than on the need to consider the ecosystem as a whole. Particular concern has been expressed about the possible impact of large-scale catching of krill on the whale stocks, and their ability to recover to their original levels. Part of the favourable changes that have been observed in the population parameters of the baleen whales (earlier maturity, higher pregnancy rates), as well as the observed increases in seals and penguins, are due to increased abundance of krill following the reduced consumption by whales. Heavy fishing on krill will reduce its abundance, and hence, presumably, alter the parameters of the whale stocks, leading to smaller rate of recovery and a reduction in level of abundance at which recovery ceases.

There has been considerable discussion on how much krill catching would have to take place to affect the recovery of the whales to a significant extent. The simple answer is that there are too many uncertainties to make definite assessments: how much of the changes in whale parameters can be ascribed purely to changes in krill abundance, and how much to other factors; where and when would the krill catches be taken; how might consumption by other predators change? Computer models by GREEN (1977) and by BEDDINGTON on behalf of IUCN (quoted by MITCHELL and SANDBROOK, 1980) give some insight into the nature of the impact of a possible harvest on whales, but conclusive evidence will probably arise only when a krill fishery develops.

It is therefore important that management mechanisms should be set up before large-scale krill fishery starts, so that the development of any fishery is kept under control and catches do not increase so fast that the impact on whales or other species cannot be detected and appropriate action taken before the impact becomes too serious. Often it seems to be assumed that krill fishing should be controlled so that there should be no significant negative impact on whales. However, krill might be able to make a significant contribution to the world food supply, and one that would be in tonnage much greater than could be taken from whales, since the sustainable yield of whales is only a few percent of the total stock, and is very much smaller than the consumption of krill by that stock. In that case it may be felt desirable to accept that a large krill harvest will involve a lower abundance of whales, but this should be a well-founded decision, and not happen by accident and neglect.

The need for controls has been accepted by the countries interested in the Antarctic and in possible krill harvesting. The result was the signing of a convention on the Conservation of Antarctic Marine Living Resources in Canberra in 1980. This is one of the first attempts to manage the resource of a large ocean area as an ecological unit, and as such represents a major step forward from the single-species approach, such as the whaling convention. In particular, the harvesting of one species, e.g. krill, has to be controlled in a manner that takes account of the impact on other species, e.g. whales. However, while this is fine as expressed in principle, it remains to be seen how it can be applied in practice.

## *Open Ocean*

*The resources*

The area considered here, i.e. beyond the range of the coastal and shelf resources considered in the regional sections, is vast. It covers some 80 to 90% of the entire World Ocean. However, it is not productive—though the poverty of some of the ocean areas such as the Sargasso Sea has been exaggerated. The production is also scattered. To become available in a form suitable for harvesting by man it has to be concentrated by passing through more trophic levels than in the coastal areas, so that ultimately a smaller proportion of the original primary production remains. RYTHER (1969) estimated that less than 1% of the potential world fish catch comes from the open ocean (with 90% of the surface area), compared with 50% each from coastal upwelling areas and other shelf areas. Another index of the low current valuation of the open-ocean fish resources is the lack of attention given to them in the discussions on the new Law of the Sea, which do little more than call upon interested countries to co-operate in the conservation and management of stocks beyond 200 miles. It may be noted here that, except off Newfoundland and Argentina, the 200-mile limit goes well beyond the limits of the coastal resources. The resources discussed in this section include much of the area within 200 miles, especially around oceanic islands and off coasts with narrow shelves.

Though the density of resoures is low, the large area means that in total they are far from negligible, and there have been important fisheries in the open ocean for a long time. The oldest is that for whales, especially sperm whales. The maps prepared by TOWNSEND (1935) of the distribution of catches by New England vessels in the nineteenth century show not only how widespread their activities were in all the oceans, but also how they concentrated in the areas of high production, e.g. along the equatorial upwelling zone. Even today, TOWNSEND's maps give one of the best overviews of the location of the most productive areas in the open ocean.

A number of groups of animals have been mentioned as potential objectives of exploitation in the open ocean, notably the squids on which the sperm whales feed, and the lantern fish and other small mesopelagic fish. Information on these resources is beginning to be put together (e.g. GJØSAETER and KAWAGUCHI, 1980) showing that in some areas, e.g. the northwest Indian Ocean, significant catches of mesopelagic fish can be taken. However, there is little sign of an early development of a commercial fishery on these species. Similarly, some other fish—such as saury and dolphin fish (*Coryphaenus* spp.)—occur in the open ocean, but are not harvested, except when they occur close to shore. For the present, with the decline of whaling, the important oceanic resources are tunas.

The species caught in the tuna fishery can be placed in three groups: (i) large tunas (*Thunnus* spp.) and skipjack; (ii) other small tunas (*Sarda*, *Euthynnus*, *Auxis*); and (iii) billfishes. In addition, sharks are often caught while long-lining for tuna.

The larger tuna and skipjack are cosmopolitan species occurring in the warmer waters of all oceans. An exception are the bluefin tunas in the southern and northern hemispheres which have been distinguished as separate species, *Thunnus thynnus* and *T. maccoyii*. These differ in their preferred temperatures. The bluefin prefers colder water, feeding as far from the Equator as northern Norway and the subantarctic waters of the Indian Ocean, though spawning in warmer waters, including the Mediterranean Sea, Gulf of Mexico, and off northwest Australia. The albacore also feeds in temperate waters, and the bigeye, skipjack, and yellowfin are progressively confined to the warmer tropical waters. They are very active. Having negative buoyancy they must keep swimming to stay at the desired depth. They therefore require water with a high oxygen content, and can find the dissipation of heat a problem at the upper end of their temperature range.

All tunas are classified as highly migratory species in terms of the draft convention of the Law of the Sea, but the extent of migration varies, being greatest for the more temperate species (MATHER, 1969; NAKAMURA, 1969). Albacore make regular migrations across the North Pacific Ocean between Japan and North America (OTSU and UCHIDA, 1963), and southern bluefin tagged off Australia have moved past southern Africa onto the Atlantic Ocean. On the other hand, though large numbers of yellowfin have been tagged in the eastern tropical Pacific Ocean and elsewhere, none has shown very long movements. Most fish stay within a circle of 500 miles around the tagging location, with only a few moving as much as 1000 to 1200 miles.

The other characteristics show similar gradations. The bluefin is both the largest (the northern bluefin grows up to nearly 3 m and 500 kg) and the longest lived (up to 20 yr). Other large tunas grow up to 1.5 m and 50 kg. The skipjack attains only some 90 cm (20 kg). Life-span and natural mortality rate are not well known, but skipjack probably only rarely live as long as 4 yr, with a natural mortality of 50% $yr^{-1}$ or more.

The most important and most widespread of the billfishes is the swordfish. It is the only billfish to support a directed commercial fishery. The others are caught incidentally to tuna fishing, or by sports fishermen. In Cairns, in northeast Australia, and some other places with a reputation for record-breaking billfish, sport fishing for billfish has become a significant element in the local economy. Largest of the billfishes is the black marlin *Makaira indica*; it grows up to 4 m in length and to over half a ton in weight (the world-record fish caught off Cabo Blanco, Peru, weighed just over 700 kg). The blue marlin is slightly smaller, and the striped marlin and sailfish reach only 50–100 kg. Like large tunas and skipjack they are found throughout the warmer oceans.

The smaller tunas are mostly coastal, and so far as known are fairly local, not making long migrations. The various species of bonito (*Sarda* spp.) live in somewhat cooler waters (e.g. Mediterranean Sea and coastal upwelling areas), than the more tropical frigate mackerel (*Auxis* spp.) and little tuna (*Euthynnus* spp.).

*The fisheries*

The tuna fisheries can be divided into the long-line fisheries for larger fish and the surface fisheries for schooling fish. The long-line fishery is carried out almost entirely by fishermen from eastern Asia (Japan, Korea, Taiwan). Their vessels cover all oceans, except the parts too cold for tuna. A typical vessel will shoot over 100 km of long-line each day, with perhaps 2000 hooks. The gear in itself is not selective, and in some areas several species of tuna and billfish may be caught in the same set, but the choice of location and the depth at which the line is set does influence the catch. The recent practice of setting the line very deep (200 m or more) has increased the catches of bigeye tuna, but lowered those of yellowfin. Virtually all the fish caught are large, 100 cm and upwards in length.

Mention should also be made of two traditional fisheries for large fish: harpooning of swordfish while they are basking on the surface on calm days, and the Mediterranean trap fisheries for large, mature bluefin. The latter is highly spectacular. It used to be very important to local fishermen. Neither now contributes much to the world catch.

The surface fisheries which catch the fish while they are at or close to the surface when they form schools, are of three main types: live bait, purse-seine, and trolling. Trolling—towing baited hooks from a small number of lines at slow speed through places where it is known or hoped tuna are feeding—is widespread, especially with small boats, but does not support much large-scale fishing. The largest industrial fisheries using troll gear are those for albacore off the American west coast (where the same vessels also troll for salmon), and in the Bay of Biscay. They catch only a few thousand tons.

The large-scale commercial fishery, which began with the development of the California canning industry around 1907, was based initially on the live-bait (or pole-and-line) fishery. This involves bringing a school of fish into a feeding frenzy by putting small live-bait fish (which are carried in tanks on deck) into the water. The tuna can then be easily caught with a baitless and barbless hook attached by a short line to a pole, with which the fisherman jerks the fish back over his shoulder onto the deck. This method is very labour intensive, and in the Californian industry was rapidly replaced by the purse-seine, following the invention of the power-block in the late 1950's. The Japanese skipjack fishery is still mainly a bait-boat fishery. By steadily improving the ability to carry live bait for long periods with few deaths, Japanese bait-boat fishermen have been able to extend their operations from their home waters into the South Pacific Ocean as far south as New Guinea and as far east as the International Date Line. Most bait fisheries, however, depend on a day-to-day supply of fresh bait, and their development in some other potentially productive areas is held back because suitable baitfish cannot be caught.

Both types of surface fishing depend on being able to locate the schools of tuna. This is mainly done visually, either by seeing the disturbance on the surface caused by feeding tuna, or indirectly by sighting the flocks of birds feeding on the small fish disturbed by the tuna. American fishermen have found that schools of yellowfin tuna are associated with porpoises. By setting their nets around the porpoises, sometimes after herding them with speedboats, they can catch the tuna that are underneath. If all goes well the porpoises escape after the net is pursed, but on occasion they become entangled in the nets and drown. In the first years after the technique was developed the total numbers drowned were very large, and some porpoise stocks were significantly reduced. The most seriously affected was stock of eastern spinner dolphin which was reduced from between 400 000 and 200 000 individuals in 1960 to around 30 000 in 1975. Later, especially after the problem gained public attention, modifications were made to the gear and to the way it is operated. This has greatly reduced porpoise mortality.

Table 5-28 lists recent catches of the major groups of tuna, according to area, and also gives the catch by the main fishing countries. Because of its size, and difference in the fisheries on each side, the statistics for the Pacific Ocean are separated into eastern and western Pacific, following the division used in FAO statistics—though these do not (and probably could not) correspond exactly to divisions between stocks of any tuna species. The values in the table should be treated with the caution needed for most statistics. The data for the larger species pertaining to the industrial countries are reasonably accurate. However, among the smaller fish and among the developing countries there can be errors in the totals, particularly with regard to species identifications. The entry 'small tunas' in the table probably includes some quantities of small yellowfin, skipjack, etc.

The table witnesses the dominance of the western Pacific Ocean (almost exactly half the total world catch) and, within that region, the importance of the Japanese skipjack fishery. The balance in the distribution between oceans of the catches of the larger tuna

Table 5-28

Summmary statistics of the world catch of tunas in 1979 (thousand tons) (Based on FAO *Yearbook of Fishery Statistics*)

| Area | Bluefin* | Yellowfin | Albacore | Bigeye | Skipjack | Billfishes | Small tunas | Total |
|---|---|---|---|---|---|---|---|---|
| Atlantic Ocean | | | | | | | | |
| France | 5·3 | 37·7 | 7·2 | 2·9 | 14·7 | — | — | 67·8 |
| Japan | 1·8 | 2·9 | 0·8 | 11·0 | 12·4 | 1·3 | — | 30·2 |
| Spain | 1·3 | 41·5 | 30·0 | 3·0 | 19·8 | 1·8 | 2·1 | 99·5 |
| Total | 13·3 | 144·2 | 74·3 | 33·1 | 87·7 | 19·4 | 32·4 | 404·4 |
| Indian Ocean | | | | | | | | |
| Australia | 10·9 | + | — | — | + | — | — | 10·9 |
| Japan | 21·8 | 18·0 | 1·6 | 3·8 | 1·0 | 3·9 | — | 50·1 |
| Korea | + | 17·7 | 1·8 | 21·2 | 0·1 | 3·6 | 0·1 | 44·5 |
| Maldives | — | 4·3 | — | — | 17·9 | — | — | 22·2 |
| Total | 32·9 | 60·6 | 12·2 | 31·6 | 32·7 | 11·6 | 65·2 | 246·8 |
| Western Pacific | | | | | | | | |
| Indonesia | — | 20·6 | — | — | 30·2 | — | 44·9 | 95·7 |
| Japan | 14·5 | 84·3 | 68·6 | 31·5 | 317·3 | 27·3 | 47·5 | 591·0 |
| Korea | + | 5·5 | 10·1 | 8·1 | 6·3 | 3·9 | 0·2 | 34·1 |
| Philippines | — | 49·2 | — | — | 45·1 | 4·7 | 103·1 | 202·1 |
| Total | 14·5 | 199·5 | 92·7 | 43·3 | 468·4 | 44·7 | 212·6 | 1075·7 |
| Eastern Pacific | | | | | | | | |
| Japan | + | 12·1 | 12·0 | 71·6 | 2·5 | 12·3 | — | 110·5 |
| USA | 5·5 | 120·5 | 7·0 | 1·3 | 79·7 | 0·2 | — | 214·0 |
| Total | 5·6 | 179·8 | 22·0 | 78·4 | 109·9 | 18·2 | 44·3 | 458·2 |
| Total of all oceans | 66·3 | 584·0 | 201·2 | 186·4 | 689·7 | 94·1 | 355·0 | 2185·7 |

*Indian Ocean figures includes catches of southern bluefin in the Pacific and Atlantic Oceans. Total includes both species.

is fairly equal—the low catches of albacore in the Indian Ocean can be ascribed at least in part to the absence of suitable conditions for this cooler water species, except in the southern Indian Ocean. This balance reflects the world-wide activities of the long-line fleets. In contrast, the surface fisheries either for the major commercial species that enter world trade (especially skipjack and yellowfin), or for the other small species that are mostly consumed locally are much more uneven. The main centres for the smaller tuna are Japan and other countries of South and East Asia (i.e. in the Indian Ocean and western Pacific Ocean). Skipjack catches are concentrated in the western Pacific Ocean.

*State of the stocks*

The stock with the longest history of assessment is the yellowfin in the eastern Pacific Ocean. This stock was the subject of the classic studies by SCHAEFER (1954, 1957) in which he developed the general theory of applying production-type models to fisheries. These studies were also among the earliest products of the Inter-American Tropical Tuna Commission, which was established in 1949 as an early reaction to fears that the stocks might be becoming depleted. However, its establishment was also in part a reaction by the United States to moves by some Latin American countries to extend their jurisdiction over fisheries.

I-ATTC has continued to collect detailed statistics directly from the fisheries, to assess the stocks, and to carry out other research. The assessments, the results of which are made available in the Bulletins of the Commission and in reports of its annual meeting, are still made mainly by variations in the production models, particularly the General Production Model (GENPROD) developed by Commission staff (PELLA and TOMLINSON, 1969). The results continue to show that the yellowfin stocks are heavily fished, though there has been a shift upwards in the estimate of MSY, which is now a little over 150 000 tons for the stock (or stocks) within the Commission's regulatory area. The reason for the difference between this and the figure of around half this value obtained from SCHAEFER's early analyses include the fact that there has been a considerable expansion of the fishery westward, so that a larger stock (or more stocks) is being exploited, and that the shift from bait-boats to purse-seine resulted in a shift upwards in the average size (especially when fishing on porpoises), which will increase the yield-per-recruit. It is also possible that the methods of fitting the production model tend to give an estimate of MSY that is not very great from the largest observed yield, even when the true MSY has not yet been approached. In the Atlantic Ocean similar analyses have been made for the yellowfin fishery, giving estimates of MSY of around 150 000 tons (see annual reports of the Standing Committee on Research and Statistics of the International Commission for the Conservation of Atlantic Tuna, ICCAT).

The long-line fisheries, in which for every species and in every ocean there has been a big fall in the catch-per-unit-effort (number of fish caught per 100 hooks) soon after the fisheries developed, can also be analysed by production models. The fit to the points is usually very good, particularly when allowance is made for the degree to which the fleets change their concentration on one or other species by changes in the area fished (HONMA, 1974) suggesting strongly that the stocks are fully exploited (ANON., 1980). However, long-lines are highly selective taking only the largest fish (above the critical size beyond which natural mortality exceeds growth). Also, there are doubts about the extent to which the decline in hook rate measures a real decline in stock abundance, rather than a reduction in that section of the stock which is vulnerable to long-lines.

Nevertheless, there is no doubt that the data do show that little if any increase in long-line catch can be achieved by increasing the long-line effort.

Application of analytic or age-structured models to tuna-stock assessment is at present difficult because of the problems of age determination. However, because of the size selectivity of all of the gears involved in catching tuna (mostly as a function of the behaviour of the fish, rather than the physical characteristics of the gear itself), the use of such models is highly desirable, particularly in assessing the changes in the proportion taken by different gears. The assessments made with analytic models on the larger species tend to confirm the results obtained with the simpler models, i.e. that no increase in sustained catch can be expected from fishing harder on older fish (i.e. with long-lines). Except where there is already a large surface fishery (e.g. on yellowfin in the eastern Pacific and eastern Atlantic Ocean) catches could be increased by fishing harder on the medium-sized fish. In total though the potential increase in catch from the larger species of tuna is not large.

The situation regarding skipjack is less clear. Application of production models has not proved enlightening; in the eastern Pacific Ocean and elsewhere the skipjack is a second-choice species. The directed effort on skipjack is high only when yellowfin is scarce, or skipjack unusually abundant. This leads to a positive correlation between observed effort and catch-per-unit-effort, which tends to obscure any negative correlation that might arise from the effect of fishing. The continued geographic expansion of the Japanese home-based skipjack fishery also makes analysis difficult for this fishery because the stock being exploited is continuously changing. The expansion does suggest that the stocks nearer Japan have become fully exploited. If it is conservatively assumed that this is true for the western Pacific as a whole, and that the same ratio of world catches to catches in the western Pacific which applies to the larger tunas ($2{\cdot}97 \pm 1$) also applies to skipjack, the present world catches can possibly be increased to some 1·4 million tons, doubling present catches.

Billfishes are caught in the long-line fisheries with the larger tunas, and have roughly similar biological parameters. It might be expected that they are being exploited to the same extent. This seems to be broadly true. Most species in the Pacific Ocean appear to be heavily fished (SHOMURA, 1980), as are some of those in the Indian Ocean (ANON., 1980), though there has been no great decline in the catch rate of swordfish, and increased catches of this species from the Indian Ocean may be possible.

Least is known about the state of exploitation and potential of the smaller tunas. These are mainly caught in local fisheries, for which the available statistics and other data are not suitable for applying the normal models. They are also difficult to survey quantitatively and are not caught in trawls or other gears which (with certain assumptions) can give direct estimates of numbers or weight-per-unit-area. Nor, being mostly in the surface waters, can they be surveyed with the usual acoustic methods. Surveys from aircraft have given interesting results, but the numbers seen probably depend too much on weather conditions and the behaviour of the fish to be useful in making quantitative assessments.

*Management*

Because their movements take them across national boundaries and beyond the limits of jurisdiction (even in the age of the 200-mile limit) onto the high seas, management of tunas requires co-ordinated international action. This need has been recognized in the draft text of the new Law of the Sea Convention, where tunas receive special treatment as highly migratory species (though some tunas migrate over shorter distances than, for example, the herring that migrate between Iceland and Norway). Slightly different institutional arrangements have been reached in different oceans. In the Atlantic Ocean, ICCAT has followed the pattern of most other commissions (most closely ICNAF) in relying on member countries to collect statistics and carry out research. The Commission itself provides a central compilation of statistics, a forum for the review of the scientific knowledge of the stocks, and of the needs of management, and, as appropriate, makes recommendations on agreed management measures. In the eastern Pacific, I-ATTC maintains its own staff for collecting statistics from the fishery and carrying out research. The staff makes recommendations to the Commission on management measures, such as the magnitude of the total allowable catch. Other aspects of management, particularly explicit or implicit agreements on how this total shall be shared between different countries, are discussed between governments outside the framework of the Commission. In the western Pacific and Indian Ocean, management of tuna is the responsibility of two subsidiary bodies of FAO—the Indo-Pacific Fisheries Commission and the Indian Ocean Fisheries Commission. These bodies have very broad terms of reference, concerned with all aspects of fishery management and develoment. To deal with the special problems of tuna they have set up special Committees on Tuna Management, which have normally met jointly.

While considerable attention has been paid to the mechanisms for possible management actions, less has so far, with a few exceptions, been done to implement actual measures. There are good reasons for this. First, no tuna stocks have appeared to be in such a serious state that some action had to be taken. Secondly, effective management has depended on the agreement of all those engaged, actually or potentially, in the fishery. For example, in the Indian Ocean it is generally agreed that the effort in the long-line fishery is too high, but the coastal developing countries having not been keen on accepting limits on the amount of fishing in the long-line fishery, at least not until such time as they have developed fisheries of their own, or until they can be assured that the form of the controls will not be such as to make if difficult for them to do so.

The exception has been in the eastern Pacific Ocean. Here, until recently, the fishery has been dominated by the United States, so that a regulation could be effective, provided it was acceptable to the United States. I-ATTC has been setting catch limits for many years that have held the amount of fishing within the regulatory area of the Commission to about the level estimated to give the MSY. Because of the way the controls are applied, which include exceptions for countries with small vessels or with small total catches, the effective restrictions have applied almost solely to US vessels. The controls have been successful in preventing a dangerous increase in fishing mortality. They have not prevented a large, and economically nonsensical increase in fishing capacity, such that the quota is reached—or at least the period of open fishing is ended—by the time each vessel in the fleet has made one trip. The greatest practical value of the regulations may have been that they have forced the fleet to work outside the regulatory area once the period of open fishing is ended, thus bringing stocks well to the westward of the regional fishing grounds under exploitation, and increasing the total yield of yellowfin.

## (3) General Problems of Management

### (a) The Nature of the Problem

In the previous sections, which reviewed the fisheries in each region of the world, numerous examples were mentioned of fisheries that have declined through heavy fishing, and of the measures that have been taken to arrest or prevent such declines. The present section examines in more detail the general principles that underlie these management actions. The section concentrates on fish stocks and their dynamics, as far as these are important in management. Special attention is paid to aspects that make it difficult to manage fisheries successfully. After considering the general background, the following points are reviewed: scientific input to management; stock and recruitment; interaction between fisheries; changes in the Law of the Sea; benefits from fisheries; new approaches to fishery management; and shared stocks.

The collapse of some fisheries and the serious, if less catastrophic, declines—especially in the economic performance—of other fisheries, have underlined the need for improvements in the way fisheries have been managed. In the sense used here fishery management is considered as encompassing all decisions that affect the influence of man on fish stocks in the ocean and the benefits that he receives from those stocks. Taking account of the variety of influences and benefits, this is a very wide field. However, so far as most major marine fish stocks are concerned, the direct influence of fishing is of much greater importance than the other influences (pollution of different types, land reclamation, etc.) and it is only the effect of fishing that is considered in detail here.

Among the potential benefits those that can be measured in simple biological terms, especially the weight caught, have traditionally received greatest attention. This is not surprising, given that those concerned with investigating the problems of over-fishing were nearly all biologists, but even in the period just before and after the Second World War when problems of fishery management were first receiving substantial attention, the biologists were pointing out that the benefits were not solely biological. In particular, GRAHAM (1943) gives a description of the English fishing industry between the wars which, with the details changed, stands unequalled as a picture of an industry cursed with uncontrolled expansion onto limited fish stocks. He shows clearly the pressures on fishermen to continue fishing even when the stocks are heavily fished, and frames his Great Law of Fishing, namely that all uncontrolled fishing becomes unprofitable.

Since then fishery management has received growing attention from economists and others. The literature involved has now grown too large to be easily summarized here. Important early economic studies are those of GORDON (1954) and CHRISTY and SCOTT (1965). ANDERSON (1977) provides a useful review of the subject from the point of view of the United States, and the problems of extended jurisdiction. A mathematical analysis of the interaction of biological and economic factors has been presented by CLARK (1976). A recent report by a working party of FAO's Advisory Committee on Marine Resources Research (FAO, 1980) has given one of the wider views of the problems of management. These studies have made it clear that maintaining a high yield or an abundant stock are only two of many objectives of fishery management. Admittedly if the manager fails to keep the fish resources in a productive state it is difficult to achieve other objectives, but large total catches do not in themselves guarantee that the fishery is being well managed.

From the point of view of society as a whole—and those concerned with fishery management are ultimately responsible to general society—the size of the catch from a particular fish stock is not of much interest. Of greater concern are such matters as the total supply of fish (or of animal protein or food in general), the net economic returns from the fishery, and the social well-being of the fishermen and fishing communities. In the absence of good management it can easily happen—and there are good theoretical reasons for expecting that in the long run it almost inevitably will happen—that the amount of fishing and its costs will rise to a point where these objectives fail to be met. The inputs of capital and manpower in the fishery will rise to an excessive level, with the result that the production will be less than if the inputs were used elsewhere, and the fishermen have only a poor living. The economic wastage is likely to be particularly unwelcome when, as is often the case, the fishing communities are in isolated or economically backward parts of the country.

Fishery management must therefore take account of the economic and social conditions of the fishery and the fishing industry, as well as the biological condition of the fish stock, but may have to take an even wider viewpoint. The concern in the past few years over whales and other marine mammals have shown that for some of these stocks, at least, the opinions and wishes of the public at large can have a weight greater than that of those directly involved.

We can now consider why, despite the great importance of proper management, success has so far seldom been achieved. One reason, which will be examined in more detail later in this section, is the difficulty in determining with sufficient precision what measures are needed. It is not too difficult to realize that too much fishing can harm the stock and hence, in due course, the fishery. But how much is too much? It is equally true that too little fishing means that less is caught than might be taken, and fishermen are most unlikely to accept restrictions unless there is reasonable evidence that these are necessary.

The other main reason for failure to manage successfully is that fish stocks are common property, or no one's property. The result is the 'tragedy of the commons' (HARDIN, 1968). There is no advantage to any one individual fisherman alone restricting his activities since the benefits would go to others—rather it is to his advantage to fish more, and so ensure that he has a larger slice of the cake, even if the size of the whole cake is reduced.

To be effective, management measures have to be accepted virtually universally by all those in the fishery either through voluntary agreement or by the imposition of authoritative controls. Given the individualistic nature of most fishermen, acceptance of voluntary controls—even within a purely national fishery—is unlikely. Given the differences in outlook and interests between different countries agreement on international fisheries is even more difficult—a fact that is illustrated by the generally very slow progress that has been made in managing international fisheries. Despite the clear benefits that might be obtained, and the great efforts of scientists and others concerned, the management measures applied in international fisheries have usually been ineffective, or such as to require from the fishermen little short-term sacrifice or interference with their normal activities (as in the case of regulation of mesh size in many North Atlantic trawl fisheries). The cases such as whales, where there has been a clear choice between accepting management measures and a complete collapse of the fishery, provide most of the few examples of severe measures being applied in a multinational fishery, and even in these cases the measures have too often been too lenient, or applied too late.

Successful management therefore almost inevitably presupposes the existence of some authority able to impose the measures that are necessary to ensure the long-term well-being of the resource and of the fishery on it. Without such an authority the pressures of short-term interests, the differences in the interests of the various participants, the contrast between individual interests and the interests of the fishery as a whole, and often, doubts about the exact outcome of specific measures, all combine to make any measures adopted weak and largely ineffective. Up to the present such authority did not exist for most of the major marine fisheries. These were carried out, at least partially, in waters beyond national jurisdiction where there was no direct authority.

As the need for management became apparent, some authority was given to various international commissions, charged with the management or conservation of resources in a particular area, or of some particular species. The authority given to these commissions was generally limited. Partly this was because of the unwillingness of governments to grant authority to a supra-national body in any matter, and partly because even if the original participants in a fishery might grant authority, there would still be no control over fishermen coming from other countries. While the use of 'flags of convenience' has been limited in fisheries—the operations for a couple of years nearly 20 yr ago of a whaling fleet owned by ONASSIS is probably the most significant example—it is quite easy to picture the large-scale use of such ways of circumventing regulations if fishery commissions were given strong authority by a limited number of countries. The most successful commissions have therefore been those where the number of actual participants have been few, and there has been little attraction for other participants to join. The best example is one of the oldest of such bodies—that controlling the harvest of fur seals in the North Pacific Ocean; it has existed in one form or another since 1911. The arrangements whereby harvesting is limited to a controlled take on the US and USSR islands on which the seals breed—Canada and Japan abstain from any catch in return for some of the proceeds from the land catches—has permitted the stocks to be rebuilt and maintained at a productive level after being reduced to a very low level at the turn of the century.

This situation of lack of authority is now changing. The draft texts for a new regime of the sea now emerging from the long discussion at the United Nations Conference on the Law of the Sea will grant coastal states jurisdiction over fisheries in an extended economic zone up to 200 miles from their coasts. Though the draft convention has not yet been ratified, most coastal states have taken action to introduce national legislation along these or very similar lines. These actions will mean that something like 99% of the world fish catches—essentially everything except a proportion of the tuna catch—will come from waters within national control. The implications of this change are discussed in a later part of this section.

## (b) Scientific Input to Management

The scientific input to management decisions is concerned with advising on the likely effect of different decisions. If, for example, a closed season of 2 mo is applied, by how much will the catches outside the closed season be increased, or, if an extra 20 boats enter the fishery, how will this affect the total catch and the catches of those already fishing? As the pressure on the stocks increase and management controls become more complex, the questions asked of the scientists become more detailed. However, the basic question is generally the same: what is the relation between the amount of fishing and the sustained catch, and more specifically how will the total catch change as a result of a decrease in the amount of fishing due to the imposition of some possible control?

It is not intended here to give a comprehensive review of the state of the art of fish population dynamics (and so far as management advice is concerned it has to be as much an art as a science). This can be obtained from the large and growing literature, from the classic works of BEVERTON and HOLT (1957) and RICKER (particularly RICKER, 1958) onwards. A convenient entry to this literature is provided by GULLAND (1977), especially the chapter by RICKER giving a review of the history of the development of the subject. Here the description of the standard techniques will be limited to providing a general understanding of the procedures used in giving scientific advice, the data requirements, and the assumptions made when using different procedures. The greater part of this section will be concerned with the consequence of the failures of these assumptions, and the basic scientific questions that have to be resolved if more precise procedures are to be developed.

'What is the effect of changing (particularly reducing) the amount of fishing?' The simplest method of answering this commonest question that fishery managers ask is purely empirical. Provided statistical records over some period of years are available, the catch in each year can be plotted against the amount of fishing in that year, and the relation determined by any appropriate statistical procedure.

Because the relation is usually curvilinear, fitting may not be easy. Instead the catch-per-unit-effort (or the catch divided by the amount of fishing) can be plotted against the amount of fishing. The result is often found to be approximately a straight line, which can easily be fitted. For this linear relation between c.-p.-u.-e. and effort the relation between total catch and effort (which will be parabolic) can be readily determined.

A procedure which is effectively identical to this empirical one can be arrived at by considering simple theories of population growth (SCHAEFER, 1954). These suppose that the rate of growth of a population is determined by some intrinsic factor ($r$), its current biomass $B$, and the difference between that biomass and the carrying capacity ($B_{max}$) of the environment in which it lives, i.e.

$$dB/dt = r\,B\,(B_{max} - B). \tag{1}$$

If the catch taken is maintained equal to this growth rate, the population will be kept at the same level, i.e. the catch is sustainable indefinitely.

The problems in this simple procedure are both practical and theoretical. The practical ones are that it is not easy to measure the actual abundance, $B$, or the fishing effort (in terms of its true impact on the stock). Though methods exist to determine the abundance of fish from various survey methods, e.g. using acoustic methods, trawling, plankton surveys of eggs or larvae (SAVILLE, 1977; ULLTANG, 1977), these are nearly always expensive and seldom very precise (i.e. usually have a high variance), and not infrequently are inaccurate in that they can have significant but unknown bias.

The practical approach has therefore commonly been to use statistics on the amount of fishing from the records of all or part of the commercial fishery to provide indices of the amount of fishing and from the catch-per-unit-effort of the abundance of the stock. The difficulty here is that even if the abundance of the stock of a given species does not change, the average catch per day of fishing can change, e.g. because of increased efficiency of the average vessel, or changes in preference between one species and another. More seriously, in a stock that is declining, the c.-p.-u.-e. may be maintained at around the same level by the fleet concentrating on the remaining patches of high density. Considerable attention has therefore been paid by those concerned with providing advice on management to finding methods of interpreting or adjusting commercial catch and effort statistics so that they provide as good indices as possible of the abundance and the amount of fishing (FAO, 1976).

The theoretical shortcomings of this very simplified approach are numerous. It does not take account of the other events (e.g. changes in climate, or impact of fishing on other stocks of fish that may compete with or feed on the species of interest) that can affect stock abundance. No account is taken of the composition of the stock. The same biomass of fish will have a different capacity for immediate increase, depending on whether it is composed of predominantly young or old fish. Some effects of abundance—e.g. on recruitment of young fish—may not exert immediate impact on the growth of the population biomass but may be delayed for a period, e.g. for the time between birth or spawning and the age at which the young fish enter the fishery. Even if there is a direct relation between biomass and its rate of change, this relation may well not be linear. On top of all this there can be a great deal of apparently random variation about the relation describing 'average' or 'steady state' conditions.

Some of these shortcomings can be dealt with by modification to the simple model that does not alter its essential character. For example, a non-linear regression of c.-p.-u.-e. on effort can be used, and some effects of lags can be dealt with by relating the change in population in one period to the abundance of the population at some earlier period.

These adjustments are, however, not generally sufficient, and an approach is needed that gives a better insight into the dynamics of the population. This is provided by the analytic models developed particularly by BEVERTON and HOLT (1957) and RICKER (1958). These consider the fate of each brood of fish from the time they recruit into the fishery until the last survivor has been eliminated by capture or natural death. During this period the numbers are steadily reduced by natural causes (usually represented by a fixed coefficient, $M$), and by fishing at a rate that depends on the amount of fishing (or fishing effort, $f$), while the individuals grow. The biomass at any moment can then be expressed in terms of the numbers originally present at the time of entry into the fishery (the recruits, $R$), and the parameters of mortality and growth. The catch rate at any time is given by the product of the current biomass and the fishing mortality ($F$) so that the total catch from the brood, or cohort, can be obtained by summing the catches over its life.

In the past some ingenuity went into devising expressions for growth and other factors so that the necessary calculations could be carried out algebraically without too much difficulty. With computers this is no longer necessary. If the life-span can be divided into sufficiently short periods so that within each period the change in total numbers, or in the weight of the individual fish, is small, then the expression for numbers and weight caught are simple. The calculation of total catch is then a matter of repeating these simple steps for each interval, which is a simple computer routine. This enables any pattern for growth or of natural mortality to be used, which can include allowance for seasonal variations.

Whichever approach is used, the outputs from the analysis, on which management decisions can be based, will be, in simplified terms, a series of curves relating important characteristics of the fishery to the amount of fishing. Typical results are shown in Figs.

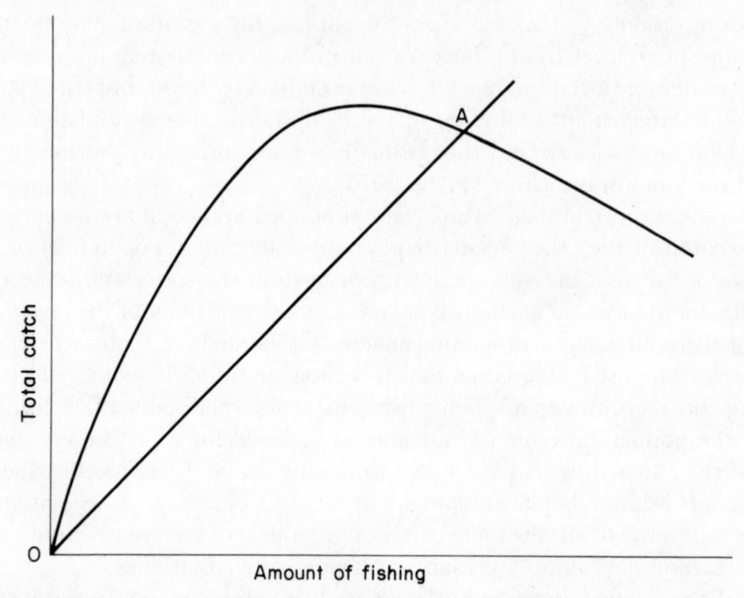


Fig. 5-9: General relation between the amount of fishery and total catch. Where the curve cuts the line on which costs of fishing and value of catch are equal is the equilibrium position under open access. (Original)

5–9, 5–10, and 5–11. The first of these shows the total catch. Initially this increases steadily with increasing amount of fishing, but then flattens out, reaches a maximum, and then declines. Apart from showing the loss in total catch that can result from too much fishing, this diagram illustrates the economic forces that, in the absence of any controls, result in over-fishing. If, as is a reasonable simplifying assumption, the costs of

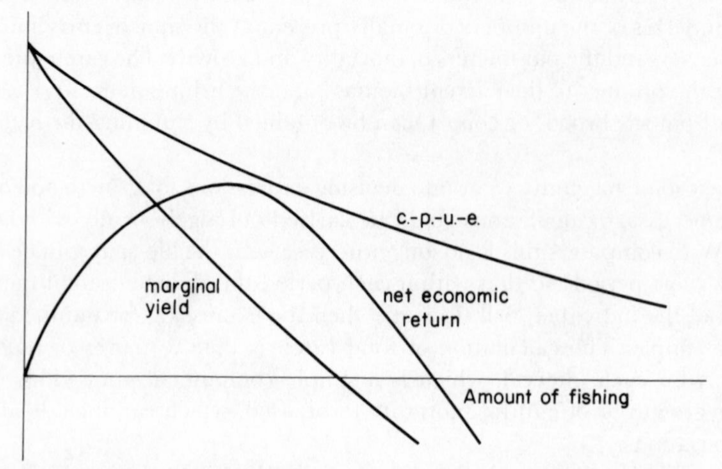


Fig. 5-10: Relation between the amount of fishing, and the catch-per-unit-effort (upper line), the marginal yield (lower line), and the net economic return. (Original)

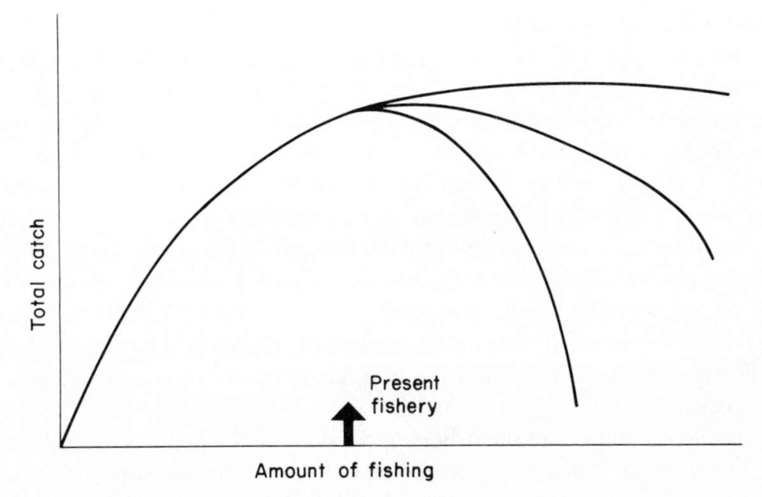


Fig. 5-11: Uncertainties in stock assessment. Possible changes in total catch as a result of increases in total effort. The broad arrow indicates the present position of the fishery. (Original)

fishing are taken to be proportional to the amount of fishing, and the value of the catch to be proportional to the weight caught—moderate departures from this assumption do not affect the argument much—then the line OA in Fig. 5-9 represents the line on which costs and values are equal. If the yield curve lies above this line, then the value of the catch will exceed the costs of harvesting it, and there will be incentives for the fishermen to increase their activities, and for new fishermen to enter the fishery. On the other hand, if the yield curve lies below the line—as it does for high levels of fishing—then costs will exceed returns. There will be little incentive to enter the fishery, and the amount of fishing will tend to drop as vessels are scrapped and fishermen retire. The result is that the fishery will tend, in the long term, to an equilibrium at the point where the line intersects the yield curve.

While the magnitude of the total catch is an important characteristic of the fishery it is by no means the only one, or even necessarily the most important one, to be considered when determining management policies. To the individual fisherman the most important is his catch rate—the amount he will take in, for example, a day's fishing. This will be roughly proportional to the abundance of the stock. Government planners are likely to have a number of factors in mind, such as employment or earning of foreign exchange, and among those the net economic return to the fishery, i.e. the difference between the value of the catch and the costs of fishing is likely to be particularly important to them. Another measure of essentially the same point is the marginal yield, i.e. the amount by which the total catch will change as a result of a small increase in the total amount of fishing.

These measures of the health of the fishery are plotted in Fig. 5-10. The catch-per-unit-effort decreases continuously from a high level when fishing is negligible. This is obvious—the more people that are fishing the less is there for each person—but it does show that for a fisherman any additional fishing is undesirable, and there is no such thing as an optimum level of fishing other than the situation in which he is the only one fishing.

The marginal yield also decreases continuously, but more rapidly than the c.-p.-u.-e. Except at very low levels of the amount of fishing it is therefore less than the c.-p.-u.-e. This has great practical significance. From the point of view of the fishery as a whole it is the marginal yield (i.e. the amount by which the total catch will be increased) that determines whether the addition of an extra vessel to the fleet is worthwhile, but from the point of view of the owner of that vessel it is the c.-p.-u.-e. (i.e. the amount of fish caught by the vessel) that is important. Because the c.-p.-u.-e. is larger, continued expansion in the fleet is possible and makes economic sense to the individuals responsible for that expansion, long after the additions, if any, to the total catch are negligible. In fact, as Fig. 5-10 illustrates, the marginal yield can become negative. This means that after a point—the level of maximum sustained yield—additional fishing actually reduces the total yield, even though the individual vessels may be catching enough to cover their costs.

The net economic returns to the fishery as a whole can be deduced from Fig. 5-9 as the difference between the yield curve and the straight line of equal costs and values. This is plotted in Fig. 5-10. Like the yield in weight, the net economic return at first increases with an increasing amount of fishing, but then flattens out (more rapidly than total yield), reaches a maximum, and finally decreases, reaching zero at the level of effort that corresponds to the long-term equilibrium in Fig. 5-9. The point of maximum economic yield (MEY) will occur at a lower level of fishing effort than the MSY. It is often suggested as an alternative to MSY, the weaknesses of which have often been pointed out (e.g. LARKIN, 1977b). It has some advantages over MSY but it must be recognized that the objectives in managing fisheries are, once the first step of maintaining the fish stock in a reasonably productive state has been satisfied, so varied that no single position on the yield curve will meet all objectives.

The curves shown in these figures would go a long way to satisfying the need for advice if they could be fully relied upon. More details are naturally required by those concerned with the management of fisheries, on both the administrative and industrial side, but these need not concern us here. What does concern us is that with the information and methods of analysis currently available, it is by no means certain that the curves produced in respect of a given fish stock do in fact describe the events that will happen to that fish stock under all levels of exploitation, especially above those so far experienced in practice. The typical situation is shown in Fig. 5-11. The broad arrow indicates the current position of the fishery, the effort having been less in some earlier periods. Three possible curves at higher levels of effort have been drawn. One predicts a catastrophic collapse if fishing increases much above its present levels; another that if fishing increases, there will not be much increase in total catch, but that effort would have to increase a lot before there were any significant drop in catch; the third predicts that increased fishing would increase the total yield, if only slightly, and by an amount that is proportionally much less than the increase in fishing effort.

To those attempting to set a policy for the utilization and management of the stock, and particularly for deciding whether increases in the amount of fishing should be permitted, the differences are enormous. Unfortunately, it can easily happen that all three curves are consistent with the available information, and that—given the variation that usually occurs—the methods of analysis currently in use offer no clear way of determining which represents the true reaction of the stock to heavier fishing. So far as the simple production models of SCHAEFER and others are concerned the difference may be no more than one of the value of a parameter used in fitting the data. Specifically, the different curves may represent different values of the parameter $M$ in the general production model derived by PELLA and TOMLINSON (1969) as a development of the simple SCHAEFER model. Since this parameter, and indeed all those involved in this type of model, is largely empirical, with little or no biological meaning, there is no reason to accept one value rather than another as being more reasonable, and there is, for this model, little additional information (other than new observations at the higher levels of fishing effort) that might be collected to resolve the question.

The situation is different, and in the long run more hopeful, as regards the analytic model. The difference between the three curves, and the uncertainty about the effect of increased fishing, concerns the part of the life history of the fish that is missing from the simple forms of this model. These follow events in a brood of fish from the time they recruit into the fishery onwards and can include calculations of the number and weight of mature fish. A little study of the fecundity of the species can determine the number of eggs produced, but thereafter there is a blank period until the young fish hatched from these eggs have in their turn reached the age at which they recruit.

The factors affecting the numbers of recruits may be ignored, or the numbers may be assumed to be constant, in which case the results of using the model may be best presented in terms of yield-per-recruit. However, this is an incomplete description of the behaviour of the fish stock, and consideration should be given to how the number of recruits can be affected by the abundance of the parent stock—which in its turn will be affected by the amount of fishing. The fecundity of most fish, with some exceptions such as sharks, is very high. This high fecundity is accompanied, under average conditions, by a high mortality between spawning and recruitment, mostly, it is believed, in the first few weeks of life. Relatively small changes in this mortality can be sufficient to balance or more than balance big changes in the abundance of the adult stock, or in the number of eggs produced. Thus increasing the amount of fishing may, over a wide range, have little effect on the numbers of recruits, though a point may be reached where recruitment begins to decline slowly or very rapidly. Given sufficient knowledge of the relation between stock and recruitment, it is possible to incorporate this into the analytic type of model, and thus construct a complete description of the dynamics of the fish stock. Some initial stock will give rise to a certain recruitment, which, given the pattern of fishing, will result in a certain abundance of adult fish; this in turn will give rise to a certain recruitment, and so on. In this way it would, in principle, be possible to determine the expected history of a stock of fish, given the pattern of fishing and the initial conditions. Unfortunately, our current knowledge of the relation between stock and recruitment is not sufficiently good in most cases to distinguish between the three situations outlined above—heavy fishing resulting in no change in recruitment, in a slow decline, or in a rather abrupt decline; this in turn corresponds to the three types of curves shown in Fig. 5-11. Thus, the improvement in our understanding of the stock–recruitment relation is a matter of high priority if better management decisions are to be taken.

Another scientific problem that has to be tackled to improve management arises not so much from any difficulties in determining the curves in Figs. 5-9, 5-10, and 5-11, but from the inadequacies of those figures as descriptions of present-day fisheries. They treat the relation between the amount of fishing and the output (catch) as a simple one-to-one relationship between two simple one-dimensional variables. This approach may be reasonably satisfactory for a single species, though even there fishing needs to be considered in two dimensions (as a function of the amount of fishing and the ages of fish which are caught). It is wholly inappropriate when, as is becoming increasingly necessary, it is the exploitation of the whole ecosystem, rather than single species, that has to be considered. It is now to be recognized that the abundance of a given species of fish, and the catches from it, depend not only on the amount of fishing directed towards that species, but also on the fisheries directed towards related species—especially those that feed on the first species, or are preyed on by it, or compete with it for food or space. This problem of multi-species fisheries, and the preceding one of stock and recruitment, are the two general scientific problems that are of particular interest—and difficulty—to those providing scientific advice on management. They are discussed in more detail in the following sections.

### (c) Stock and Recruitment

The serious failures in fisheries management, when the fisheries have collapsed, have come about through failures to maintain recruitment. In some cases these failures have operational or political causes, rather than being due to scientific difficulties. In the salmon fisheries of the North Pacific Ocean it is, within reason, possible to determine the escapement to a particular stream required, on average, to produce a given recruitment. However, it may be difficult in practice to achieve that escapement because fish are mixed with fish going to other streams, and because of the pressure from the immediate interests of the fishery industry, especially in years when the total run is low.

The latter problems were even greater in the case of whales. Though there was a period in early whaling days when the International Whaling Commission did not have adequate scientific advice—in this context quantitative advice on the size of adult stock required to maintain recruitment—the main reasons for the low level of whale stocks and the even lower level of whale catches were the slowness of the IWC during the 1960s to react to the advice, and the very long time required for the stocks to recover even when appropriate action was taken.

In some cases the practical problems are insoluble. BRANDER (1981) has described the disappearance of the common skate from the Irish Sea. Because of its size and behaviour the skate is vulnerable to trawls and several other fishing gears soon after birth and for a long period before it reaches maturity. Because of its low fecundity it cannot increase its net recruitment rate (recruits per adult) sufficiently to balance this fishing mortality, and the stock is steadily declining. This decline could only be checked if the general level of fishing in the Irish Sea were substantially reduced. However it is unlikely that a major fishery will be restricted for the sake of a single species with no popular appeal.

Other cases of recruitment failure—and the scientifically more challenging cases—are where the scientists were unable to give clear advice on the risks of recruitment failure until the collapse had occurred. Indeed, in some cases (Californian sardine, Peruvian anchoveta) the relative roles of too heavy fishing (and too low an adult stock) and environmental factors (e.g. El Niño) are still unclear. The immediate source of the problem is that the simple and direct approach of plotting the number of recruits against the abundance of the adult stock is nearly always unproductive. Only one point is available for each year of what is often a short series of observations. The points are very scattered and the nature of the underlying relation is unclear. These problems have been tackled in various ways (for useful entries and reviews see particularly RICKER, 1954; CUSHING, 1977; and the report of the ICES Symposium edited by PARRISH, 1973): (i) developing appropriate forms for the stock–recruitment relation; (ii) looking for patterns and similarities within groups of fish (clupeoids seem particularly susceptible to sudden collapse), thus in effect greatly increasing the number of observations; (iii) looking for relations between recruitment and environmental factors, thus reducing the scatter; and (iv) attempting to obtain a better and broader understanding of the dynamics of the population during the period (probably the first few weeks of life) when recruitment is determined.

Most observations of a theoretical curve relating abundance of adult to the average number of resulting recruits have, from the work of RICKER (1954) onwards, started with the assumption that the mortality of young fish at any age $t$ increase linearly with density, i.e.

$$M_t = M + bD. \tag{2}$$

This equation can take two forms, depending on whether the density is taken as the current density of young fish or as the density at some earlier period, e.g. of spawners. The practical difference is that the former can never result in lower recruitment from a larger spawning, while the latter can result in some intermediate stock giving the greatest recruitment. These have been distinguished by HARRIS (1975) as density dependence and stock dependence respectively. The two forms of the equation that gives the recruitment $R$ as a function of the adult stock are

$$R = \frac{P}{a + cP} \tag{3}$$

or

$$R = AP\mathrm{e}^{-BP}. \tag{4}$$

The latter equation can be written in a number of different ways, using different notations; an alternative form which has been used particularly for Pacific salmon, is

$$R = P \exp\left(\frac{P_{\mathrm{r}} - P}{P_{\mathrm{m}}}\right), \tag{5}$$

where $P_{\mathrm{m}}$ = adult population giving the maximum recruitment; and $P_{\mathrm{r}}$ = replacement population, the population level at which $P = R$. For fish like Pacific salmon which spawn once and then die, and in which recruitment is normally measured as the number returning to the parent river to spawn, the replacement population represents the stable level to which the abundance will tend to return or the absence of exploitation.

Equations (4) or (5) have been more generally used than Equation (3). They tend to fit the data better. Also, the theoretical basis is wider and thus somewhat better. There are certainly some stocks where the density of adults has a clear relation to the mortality of eggs or young fish. Later spawners in an abundant run of salmon can disturb the eggs laid by earlier spawners. Filter-feeding pelagic fish, e.g. anchovies or herring, may be significant predators on eggs or larvae. Even when the density-dependent effects are between young fish, e.g. competition for food, they may be expected to occur with some lag, i.e. the amount of larval fish present at one time, and hence the amount of food they eat, will affect the amount of food available for later larvae. This will in turn affect their growth rate, and hence either directly or indirectly (e.g. by increasing the time during which they are vulnerable to a particular size of predator) their mortality rate.

Both equations are in fact over-simplifications of the true relation, and more complicated formulae would be needed to give a full description of the relation of average recruitment to adult stock. Each contains only two basic parameters—essentially the scales on the two axes—and this is not sufficient to describe the range of forms observed in practice. In terms of Equation (4) or (5), there is a difference in the behaviour when the stocks are greater than those giving the greatest recruitment. For some, recruitment decreases rapidly so that a stock twice the optimum would give poor recruitment. For others, large stocks give average recruitment not much less than the maximum. Comparing data from a large number of stocks CUSHING and HARRIS (1973) found that there were some systematic similarities within related groups of fish. Flatfishes tend to have rather flat curves, often indistinguishable from constant recruitment over the range of population sizes observed in practice (*cf*. BEVERTON, 1962). Salmon on the other hand tended to have pronounced maxima, with recruitment clearly lower at high abundance.

Given the variability and the small number of points in most sets of stock and recruitment data, increasing the number of parameters—and hence the degrees of freedom in fitting a theoretical curve to observations—is likely to be an unrewarding exercise. A fit may be obtained, but it will probably be little better than that obtained from the simpler expression, and neither the improvement nor the estimated values of the parameters are likely to have much practical or statistical significance.

A more adequate approach is to examine the departures from the average relation. This may be approached by adding a random error term, usually multiplicative (i.e. good environmental conditions are more likely to increase the recruitment expected under conditions by say 20%, than by a certain number of recruits). In other words, the recruitment of young fish born in year $t$ will be given by

$$R_t = AP_t \, e^{-BP_t} \cdot x_t. \qquad (6)$$

Or, since later analysis is easier in terms of additional factors

$$\log R_t = \log A + \log P_t - BP_t + \log x_t. \qquad (7)$$

This formulation does not account for all the possible effects of environmental factors (i.e. those not directly related to the abundance of the fish stock). In addition to changing the magnitude of the recruitment, these could also change the shape of the stock–recruit curve, i.e. alter the point at which it has a maximum. This is equivalent to writing an additional random or environmental term in the exponent. Equation (6) becomes

$$R_t = AP_t e^{-By_t P_t} \cdot x_t. \qquad (8)$$

There is a long history of attempts to relate recruitment to environmental factors, either using the above equations in some form or other, or by direct correlation, without taking account of the adult stock. This interest has arisen as much from the desire to be able to forecast recruitment (and hence catches) a year or two in advance as from the need to elucidate the stock–recruitment relation. Several environmental factors have been considered in the search for a correlation: (i) the strength and direction of winds that could move the larvae towards favourable areas; (ii) temperature; (iii) amount of river runoff; (iv) feeding conditions for adults or larvae etc. Given that there is also a considerable choice of the precise time at which the chosen parameter should be measured, there are a very large number of possible values that could be related to recruitment. It is therefore not surprising that, over short series of years of observation, some striking correlations have been observed. The real statistical significance of these correlations is at best unclear. Most correlations have failed to be confirmed by later observations. The exceptions have been those stocks where one environmental factor is clearly dominant. For example, at West Greenland the cod is at the northern extreme of its range, and below-average temperatures are associated with low recruitment.

A better approach is to look more closely at the mechanisms that can cover departures from average. So far as the exponent in Equation (4) (i.e. the value of $y_t$ in Equation (8) is concerned, CSIRKE (1980) has pointed out that the impact on survival should be more closely related to density than to abundance. If the distribution changes from year to year, this should be reflected in the value of $y_t$. For the Peruvian anchoveta stock he obtained independent estimates of density (from catch-per-unit-effort) and of abundance (from acoustic surveys), and hence an estimate of the concentration. Using this in a variation of Equation (8) he found that he could account for much of the variation in year-class strength about the average relation between adult stock and recruitment.

Another aspect of the distribution of adults and of spawning products is that survival of larval fish in the sea may be an unusual event, and fish populations may have to spread their eggs widely so that some find favourable conditions. In the last few years a considerable attention has been paid to food requirements of larval fish (Volume III: KINNE, 1977). Much of the work has investigated the possible culture of commercial species, and has concentrated on finding a suitable food source that itself can be cultured. Other studies have looked at the natural foods and the concentrations that are necessary to give good growth and survival. Many of the studies, especially those carried out in the California current system, were reviewed at a workshop held in Lima in 1980 (SHARP, 1981).

The young larvae depend critically on an adequate food supply. If they have to undergo more than 1 or 2 d of starvation or near-starvation conditions their condition becomes non-reversible and they will not feed even if food now becomes available. The density of food organisms that are required to ensure feeding and good growth is high; it is considerably above the average in most sea areas, if the average is taken over large distances (kilometres horizontally, and tens of metres vertically). It seems that good survival can occur only in places where the density is much above average. In the Californian current these conditions occur in horizontal layers in which the zooplankton is concentrated. Such layers are most strongly formed when the weather has been calm for some time. High winds break up the vertical structure so that there are no food patches available even for the more fortunate fish larvae. Strong winds are therefore correlated with poor year classes (LASKER, 1978). The concentration may also be in time. In temperate waters the chronological outburst of phytoplankton, and later of zooplankton, is very sharp; the period of peak densities lasting only a couple of weeks. The success of a year class can therefore depend, as suggested by CUSHING (1969b) on the degree to which the fish are successful in timing their spawning so that there is a match between the peak demand of food by larval fish, especially in the apparently critical early feeding period and the peak in abundance of the favoured food organisms.

This, and similar information, particularly concerning the year-to-year changes in food abundance, are going a long way towards providing the answer why recruitment in one year is better than in another. However, much remains to be done if the more important questions of the manager are to be answered: 'Would the recent poor year class have been quite so bad if the spawning stock had been larger?' 'How big a spawning stock should be maintained in order to minimize the risk of poor recruitment?' 'What would the expected future benefits (from higher recruitment and higher catches) be from forgoing some of the present benefits by catching less and thus building up a larger spawning stock?'

Answering these questions may involve shifting the basic stock and recruitment problem from the whole population to the individual patch of favourable conditions. Will such a patch produce a fixed quantity of recruits, provided only that some minimum number of eggs or larvae reach it (equivalent to constant recruitment), or will a fixed population survive, or will there be a more complex relation, with the proportion surviving decreasing with increased initial abundance of eggs and larvae?

The first of these alternatives gives the simplest relation between total recruitment and adult stocks. Recruitment depends on the extent to which spawning is spread (in space and time) so as to maximize the chances of finding good conditions. It depends on the geographical and temporal extent of spawning, but not on the total abundance of spawners. This makes life easier for the manager. The danger sign of a possible fall in recruitment is a significant shrinking in the extent of spawning, which can be fairly easily monitored. However, it is almost as difficult to see how the mortality of larvae within a suitable patch can increase so neatly with increasing initial abundance in order to maintain a constant contribution to the total recruitment, as it is to see how a similar mechanism works for the population as a whole.

Therefore, on theoretical grounds the first alternative (and also the second, which would result in recruitment being, for given environmental conditions, proportional to adult stock) can be rejected. This would seem, to leave the problem as before, though on a smaller scale. There are, however, better hopes of studying the effect of increased density of larval or juvenile fish on the dynamics of food organisms in a small patch, and during the small time throughout which that patch is coherent, than over the whole spawning area and the whole spawning period.

For the present, though, the manager's questions can only be given qualitative answers. Only for some groups of fish (e.g. salmon) and here only when they are very abundant (i.e. excluding most stocks that are being fished) will reducing the spawning stock increase recruitment; for most, some, if only small, decrease is quite likely. Therefore, management should avoid further decreases in spawning stock unless there are clear benefits in other ways. How far the spawning stock can be reduced, for example in order to maximize yield-per-recruit, without a serious or catastrophic fall in recruitment remains uncertain, but there are clues. A shrinking in the extent of spawning in space or time is a danger signal. Some species groups, notably herring and sardine, seem more vulnerable than say flatfishes. More attention should be given to maintaining the spawning stock of the former groups.

### (d) Interaction Between Fisheries

The basic theories of stock assessment consider the dynamics of a single species, and the basic approaches of most management systems consider a single fishery (usually as though it was exploiting a single stock). Neither give much attention to the other species or other fisheries in the same region. In contrast to this simple approach of management and assessment theory, in practice in nearly all parts of the world many species are exploited, usually with several different fisheries in the region each exploiting a slightly different mix of species. The success of the management of any one fishery depends on taking adequate account of the events in the other fisheries and the other species, and the degree of the dependence is increasing as each stock becomes more heavily exploited.

The nature of the interaction can vary. The simplest case concerns direct interactions between fisheries, so that a fishery directed principally at one species incidentally catches species which are the main target of some other fishery. This 'by-catch' problem has been most serious in the North Atlantic Ocean, where each fishery tends to be directed at a single species, or a small number of species, out of the total range of potentially valuable species. Examples are the incidental catches of small haddock, whiting, and other demersal species in the fisheries directed at sand eel, herring, or Norway pout, usually with smaller mesh sizes than those used in the fisheries for larger species for human consumption. In the northwest Atlantic Ocean the problem involved the trawl fisheries directed at different species for direct human consumption. Here the by-catch of scarce and heavily-fished stocks, e.g. haddock, in fisheries directed at more abundant species, e.g. silver hake, may cause the total catch of the former species to exceed, possibly seriously, the target set by the management authority.

The scientific approach to these problems is generally straightforward. Provided there is adequate data on the quantities and sizes of fish caught in each fishery, the single-species models can be used to determine, for example, the effect on the directed haddock fishery in the North Sea of the by-catch of haddock in the Norway pout fishery. In the northwest Atlantic Ocean it is possible to determine how big the allowable catch in the directed haddock fishery should be so that, with the incidental catches in the silver hake and other fisheries, the total catch of haddock reaches the desired level.

The subsequent management actions are far from straightforward. Biological analyses show that it is impossible to maximize catches (or any other type of benefit) from both species simultaneously. For example, full exploitation of Norway pout (mainly taken by Scandinavian fishermen) would involve a large drop in the directed haddock fishery (in which Scottish fishermen are the main participants), as well as in the total haddock catch. In the northwest Atlantic Ocean the quotas in the directed fisheries can be adjusted, e.g. by linear programming, so that the total catch of each species comes closest to the desired figure, but it is generally impossible to get precisely the correct total for each species, and even if this is approached, the allowable catches in the directed fisheries for the scarcer or more valuable species (e.g. haddock) may be unacceptably low for the fishermen concerned (FUKUDA, 1976).

The reverse situation occurs in those areas, e.g. the Gulf of Thailand, on which the major fishing is directed more or less indiscriminately at a large number of species. The scientific problems of gaining an understanding of the dynamics of the system that is even approximately complete, are horrendous. On the other hand, the management questions addressed to the biologist are simple, in many cases involving not more than determining whether the amount of fishing should be increased or decreased. As the discussions in relation to the Gulf of Thailand show, a clear answer to this question can emerge from a very simple analysis, such as applying a production model to the statistics of catch and effort of the fishery as a whole, without distinguishing between species. While this answer will not be complete in every detail, it does seem (e.g. ANON., in press) that the results are reliable, and only likely to change in minor detail if more precise analysis could be carried out. The problems in these fisheries are then those of practical management, and of finding a way to reduce the excess effort in, for example, the Gulf of Thailand trawl fishery.

The situations that offer serious challenge to the scientist and the manager are where different fisheries exploit different species which interact biologically, so that management of one fishery will have effects on the other fisheries. These multi-species fisheries have attracted much attention (e.g. FAO, 1978, MAY and co-authors, 1979, ANON., in press). The direct method of approaching this problem is the construction of a multi-species model which will take account of all the biological interactions, and so predict the input of action in any one fishery on all the other stocks and the fisheries on them. Models of this type have been constructed for the North Sea (ANDERSEN and URSIN, 1977), and the Bering Sea (LAEVASTU and FAVORITE, 1978a,b).

The difficulties of these comprehensive models is that even when only a moderate number of species are considered, the number of interrelations, and hence the number of parameters that have to be estimated, becomes extremely large. Until recently, the computational problems alone would have made the use of such models impossible. Even now the demands on computer capacity are high, but a more serious disadvantage are the uncertainties in most of the parameters. In principle, many of these—e.g. the amount of 2-yr-old herring eaten each year by an average 3-yr-old cod, or by the entire North Sea stock of cod—can be directly observed. Considerable efforts are now being made in the North Sea to collect food-consumption data of this type. However, given the great seasonal and spatial variation in the feeding of most species, the sampling error, even with the great effort put into the collection of samples, remains large. There are also procedural problems (e.g. how long does a herring remain identifiable in the stomach of a cod?). In practice, therefore, the values of the parameters may have to be chosen somewhat empirically, and adjusted so as to fit the model to the observed history of the fishery. This leaves some doubt as to the true value of many of the parameters, and the predictive reliability of these models.

Even the most detailed models so far constructed are far from comprehensive, usually concentrating on the feeding interactions between fish of a commercial size, i.e. cod eating herring, rather than on the possible interactions at earlier stages, e.g. the possible influence on the survival of cod eggs and larvae (and hence recruitment to the commercial sizes of cod) of predation by herring or mackerel. A comprehensive and quantitative model and a comprehensive and simultaneous look at all possible interactions in a fishery region such as the North Sea is thus at this stage of the development of fishery science a practical impossibility. The examination, therefore, needs to be done in stages, looking in turn at the main types of interaction, e.g. between predator and prey; or competition at the same trophic level.

The best-known case, at least in theory of the predator–prey situation, is that of krill and whales in the Antarctic. This is also one of the simplest in that for blue whales at least there is essentially only one prey species. Following MAY and co-authors (1979) the basic features of the dynamics of the two species can be described in terms of an expansion of the single-species production model, using a simple Lotka-Volterra form of predation, by the equations

$$\frac{\mathrm{d}N_1}{\mathrm{d}t} = r_1 N_1 \left[1 - \frac{N_1}{K}\right] - aN_1N_2 - F_1N_1, \qquad (9)$$

$$\frac{dN_2}{dt} = r_2N_2\left\{1 - \frac{N_2}{bN_1}\right\} - F_2N_2 , \tag{10}$$

where $N_1$= prey population (krill); $K$= carrying capacity; $aN_2$= rate of predation on it. For the predator population (whales) the carrying capacity is proportional to the abundance of prey. The two stocks have intrinsic rates of increase $r_1$, $r_2$, and are harvested with fishing mortalities $F_1$, $F_2$.

The greatest yield of whales can be taken only if the krill harvest is zero, and the krill stock is kept high. Conversely, the greatest krill catch could be taken by first driving the whale stocks to extinction or close to it—though this does not mean that by allowing the collapse of the Antarctic whale stocks in the 1960s the IWC was really showing remarkable prescience of the needs of some future krill industry. It is also possible to determine the pattern of fishing which would produce the greatest total yield, provided some relative weighting (e.g. of price) is used to cover the yield of krill and whales. This will usually require taking the maximum possible krill or the maximum possible whales, depending on the relative weightings given to the catches of the two species, the intrinsic growth rates of the two populations, and the conversion efficiency from krill to whales. Only if these are balanced within a narrow range will the maximum total yield be taken with a combination of both krill and whale harvesting. However, determination of some theoretical maximum is not a realistic management problem. The practical questions that those responsible for management of Antarctic resources need to know are, for example, 'by how much would the potential harvest of whales be reduced by taking a given harvest of krill?' This can be deduced from Equations (9) and (10), and the resulting expressions for the yield of the two species. It is then a matter for the managers to decide whether the trade-off is worthwhile.

In the Antarctic this trade-off is still a hypothetical question, since a krill fishery on a scale large enough to interact significantly with the whales has not yet been developed. If it does, there may well be no whaling industry surviving, and the practical question will be the degree to which a given krill harvest will affect the abundance of whales, and their rate of recovery. In some other areas the questions of trade-offs between fisheries on predators and prey are real ones. In the northwest Atlantic Ocean the large catches of capelin in the mid-1970s have been followed by a decline in the capelin stock, and a fear that the several species of fish, particularly cod, and marine mammals that feed on capelin, are being adversely affected. The questions are whether, if capelin stocks do recover, further large-scale catching of capelin should be allowed, and if so, should the allowable catches be set at a conservative level so as to ensure a plentiful supply of food for the other species. In the North Sea, calculations show that if fishing mortality on cod were reduced to rebuild the stock so as to maximize the yield-per-recruit of cod, this would add greatly to the predation of cod on the small fish caught in the fish-meal industry. The question here is rather the reverse of that in the western Atlantic case; it must be determined whether an attempt should be made to maximize the catch of the predator (ignoring for the present any possible effects on cod recruitment of the abundance of adult cod or of other stocks), or whether something less than optimum utilization of cod has to be accepted in the interests of those fishing for the raw material for making fish-meal.

The management decisions are likely to be difficult to resolve in any case because in both regions different groups of fishermen are interested in the predator and the prey.

They are made more difficult because the biological information needed for making an accurate estimate of the extent of the interaction is lacking, particularly with regard to the effect of changes of prey density on the predator. Unlike the blue whale, cod (and most other fish) consume numerous different types of prey. If there are fewer capelin around Newfoundland, or even no capelin at all, the cod there might be a little more hungry, but they would switch to other fish or invertebrates. The net effect on their growth, or natural mortality rate, and hence on the commercial catches may be very small, and almost certainly quite different from what might be calculated from the loss of capelin.

The determination of the impact of the predator on the prey species and fisheries on the prey species raises fewer problems—a sand eel eaten by a cod is no longer available to be taken in the fish-meal industry—but there are still some. It is not known accurately just how many sand eels, and of what size, are eaten by cod; nor is it known how the fish consumption and species performances of individual cod might change following changes in the abundance of cod or sand eel. A more fundamental problem is to what extent the sand eels taken by cod are typical of the sand eel population as a whole, or that part of it taken in the fish-meal fishery. It has been suspected, particularly with reference to the possibly harmful effects of seals and other marine mammals on commercial fish stocks, that the predators may preferentially capture the sick or weakly prey, i.e. those which would be more likely than the typical fish to die from other causes of natural mortality if the predator had not come along. If this is the case simple calculations of how much is eaten by the predators, and what proportion of that would otherwise have been harvested, will probably over-state the impact of the predator, and hence the impact of any change in the way the fishery on the predator is managed, on the prey and the fishery on it. At the same time the simple models do indicate the direction of the input, and the sort of trade-offs that are involved if the interests of fisheries on predator and on prey are to be reconciled.

*Competing Species*

Assessment of interactions between competing species, i.e. species at the same trophic level, is difficult because the mechanisms of the interaction or competition are not always clear. In some cases it may not even be clear that any interaction exists. It is generally accepted that Antarctic minke whales are increasing (see reports of the IWC Scientific Committee) and that the reason for this increase is the depletion of blue whales, and to a lesser extent that of the other baleen whales. It is less generally accepted that off California (USA) the decline of the sardine and the rise of the anchovy is more than a coincidence. If it is accepted that the anchovy has benefited from the absence of sardine, there still may be argument about the responsibility of the anchovy in the sardine collapse. SISSENWINE and co-authors (1979), in a comprehensive analysis of the species caught in the groundfish surveys off New England, found no more statistically significant correlations between trends in different species than might be expected by pure chance. Some of these differences in scientific confidence may be related to the nature of the interactions and the amount of variability in the systems. The least variable seem to be the whales. Here changes in abundance of any one species, and its consumption of krill, is reflected in changes in krill abundance (though so far this has not been observed directly) and in the vital parameters of each whale stock, including age at first maturity and pregnancy rate which can be readily observed (at least as long as there are commercial catches from which samples can be collected).

The most variable system is that of shaling pelagic species. As noted above, these stocks have been the most susceptible to collapse as a result of 'recruitment over-fishing'. In several cases, sardine–anchovy in California, anchovy–sardine/horse mackerel in Peru, herring–sprat in the North Sea, the collapse of one species has roughly coincided with the rise of another. In any one case it is impossible to show that the coincidence is no more than chance. However, the collapses have too often followed a period of very heavy fishing to be due solely to chance or environmental effects and also the rise of a competing species has now occurred too often to be based on chance alone. Nevertheless, a collapse has not always been followed by a corresponding increase. No species—at least none capable of supporting a commercial fishery—has replaced the Atlanto-Scandian herring stock between Iceland and western Norway, and the anchovy off Namibia has so far shown little sign of benefiting from the decline of the sardine there.

The failure to observe any replacement of herring in the Norwegian Sea may be due to an absence of a suitable replacement (capelin might be a possibility, but it is a rather more northern species). The failure of the Namibian anchovy to increase in the few years since the sardine declined may, on the other hand, be only a chance effect. The decline of the anchovy may have increased the chances of a good anchovy recruitment, but while the anchovy parent stock is not itself high, good recruitment will depend also on environmental conditions being favourable too. That is to say, there may be two fairly stable situations, of abundant sardine or abundant anchovy, but the switch between them may involve an interim period when the abundance of both species is low.

The difficulties of establishing any interaction in demersal fish may also be merely a matter of the degree of noise in the system. Even with good survey data, or good data from the commercial fisheries, any estimates of the annual abundance of individual species are subject to considerable variance. The actual abundance also varies much from year to year for causes other than competition. Furthermore, the impact of the interactions may be less than in pelagic stocks, where it seems often to be a case of all anchovy or all sardine. With the other sources of variation, the chances of establishing a meaningful correlation showing that a large decrease in haddock causes say a small increase in cod, are slight over a short period of observations and by statistical techniques alone.

Despite the greater scientific difficulties, the administrative problems of managing competing stocks are usually less than those of managing two fisheries, one based on a predator the other on its prey. Because the fish are usually similar, they can normally be caught in the same way and by the same fishermen. Instead of adding to the complications of the work of the manager, the existence of a competing species and its reaction to the depletion of a preferred species can relieve him of some of the problems caused by failure to manage an individual species successfully. Thus, the existence of the minke whale stock in the Antarctic could have allowed the continuation of whaling for a few years after the depletion of the larger baleen whales, but the fact that the minke whales have increased allows continued whaling for a much longer period (possibly until some of the other species have recovered), or on a larger scale. The increase of smaller species (sand eel, sprat, etc.) in the North Sea has ensured the continuation of the fish-meal fishery which at one time depended to a large extent on small herring. However, those fishing for herring for direct human consumption (fresh, kippered, or preserved in various ways) have not found an adequate substitute, and the shore-based industries have turned to alternative suppliers of herring from Canada or elsewhere. The opposite has happened in Chile and Peru where the replacement of anchovy by sardine has enabled what was entirely a fish-meal fishery to diversify into frozen and canned products, with a considerable increase in the value of the product from a ton of fish.

The first need of the manager is therefore for information covering not only the possible biological changes but also the practical implications: what species might replace, partly or wholly, those now being fished? Can they be easily caught by the present fishing vessels? Can they be readily processed and sold at a good price? If the answers to these questions are favourable, i.e. if species apparently competing with species currently the main target of the fishery are themselves commercially attractive, the need for careful management is less urgent. On the other hand, if the competing species are not attractive, then it may be very important not to mismanage the present fishery and to stimulate a change in species composition that would be harmful and probably difficult to reverse. For example, the increase of *Balistes capriscus* off West Africa, and similar file fish in the east China Sea are dangerous signs. The fisheries in these regions may need careful management.

### *Other Interactions*

Some presumed interactions do not fit neatly into the compartments of predator/prey relations or of competition, particularly if attention is focused only on fish of commercial size. In the North Sea, the decline of herring and mackerel has coincided with a succession of unusually strong year classes of many demersal species, including cod and others which, at the commercial size, are predators on herring and would if anything be expected to have suffered as a result of the decline in herring. The increase in sand eel on both sides of the Atlantic Ocean has been ascribed by SHERMAN and co-authors (1981) to a decline in herring and mackerel with which it competes as a predator on zooplankton, but it may be due to other changes.

By considering the whole life-cycle of each species their interactions may be reduced to cases of direct competition, or of predation. For example, it may be that mackerel or herring are, as zooplankton feeders, sufficiently important predators on eggs or larvae of cod, plaice, or other demersal fish that reduction of mackerel or herring will allow improved survival and better recruitment. However, the match in space and time of the main concentrations of eggs and larvae of cod and of feeding herring and mackerel are not close enough to give much support to this particular hypothesis.

In any case, the range of trophic levels and niches, in the usuage of terrestrial ecologists, occupied by many species during their life is so large and gives rise to an even greater range of possible interactions between species that it would be impossible to predict which, if any, of possible interactions would be important. The manager is, therefore, reduced to reacting to events and to adjusting his policies to the interactions only after the changes have occurred. In the North Sea, the potential benefits to the cod or plaice fisheries were not known, and could not in practice have been known, before the herring and mackerel stocks declined, and the policies adopted during the years of crisis in the pelagic fisheries could not have been modified to take account of these benefits. The situation is now different, and some modifications in a policy towards the recovering pelagic stocks would be reasonable. It might, for example, be desirable to aim at only a partial rebuilding of the herring or mackerel stocks, in the hope that the demersal recruitment would remain at least close to its present level. The attraction of this policy, compared with one of rebuilding the pelagic stocks to the level giving the greatest catch, is increased by the highly elastic demand for pelagic species. When the supply is low the North Sea catches are used for various high-priced products, but at the levels of production in the 1960s much of it went for fish-meal. The monetary value of a North Sea herring catch of 1 million tons is, therefore, little greater than that of 100 000 to 200 000 tons—the value to the fishermen may in fact be less.

### *Implication for Management*

Species interactions make matters more complicated for all concerned with management. For those concerned with the theory of management the interactions may be beneficial in getting rid of the assumption that objectives of management consists of maximizing some easily definable quantity—the weight of the catch in the case of the often criticized Maximum Sustainable Yield. If species interact, it is impossible to achieve the MSY, or any other maxima, from both stocks simultaneously. It might be possible to derive some function, e.g. the weighted sum of the catches from the different species, which could be maximized. However, it is quite likely that to achieve such maximization would involve a fishery policy, e.g. elimination of whales to maximize krill harvests, unacceptable in practice. The loss to the management theory may not have much immediate impact on management practice, which has seldom if ever been based on maximizing anything, other than perhaps the administrators' immediate peace of mind. However, the loss may help in directing the attention of those who have been discussing the advantages and disadvantages of MSY or other theoretical objectives towards the ways in which the current practice of fishery management can be changed in a direction to increase net benefits.

The area of management which is probably being most affected by the existence of species interactions is the link between ecologists and administrators or decision makers. When managing a single species in isolation, the ecological advice has often been expressed as direct prediction—to achieve a certain objective (e.g. fishing at fishing mortalities giving maximum yield-per-recruit): the total catch during the following season should be $x$ thousand tons. While both scientists and administrators are well aware that absolute accuracy cannot be achieved, and that a catch of $x$ thousand tons would result in a mortality that might be rather above or below the target value, it has been convenient for both sides to ignore the uncertainty—the administrator, because this enabled him to concentrate on other problems such as the allocation of the catch between different groups of fishermen; the biologist, because it was often felt that the presentation of alternatives would encourage the taking of higher catches, with resulting damage to the stock. The uncertainties within the framework of advice as presented were in any case not great, concerning matters such as the strength of recently recruited year classes, or the values of the most recent fishing mortality to be applied in cohort analysis. They effected the details of management policy, but not the broad objectives.

The possibilities of species interactions raise more serious uncertainties: should the North Sea cod be rebuilt if this will harm the fish-meal fisheries? Should managers be greatly concerned about the risk of collapse of an anchovy stock in Peru or California if the result may be the increase or recovery of a more valuable sardine stock? There are at present no clear scientific answers to questions the administrators will ask on these matters. How much will be the loss to the fish-meal industry? What will be the size of the sardine catch in 5 yr if the anchovy stocks collapse? The scientists should not be expected to be able to produce such answers in the immediate future. Equally, the scientists cannot continue to produce advice based on single-species analyses that ignores possible species interactions. Future advice must be more comprehensive, if less certain. Though the degree of uncertainty needs to be no larger than that associated with, for example, long-range weather forecasts, or economic projections.

### (e) Changes in the Law of Sea

Future prospects for managing marine fisheries have been changed by the new pattern of the legal regime arising from the deliberations at the Third United Nations Conference on the Law of the Sea. This conference, which has lasted several years, has resulted in the signature of a convention in 1982. After that there can be expected a further period before such a convention has been ratified by sufficient countries to take full legal effect. Nevertheless, so far as fisheries are concerned the changes due to UNCLOS are clear, and are to a large extent already in effect.

Apart from rather general statements calling for conservation and management of marine resources, and for collaboration between countries concerned with the same stock, the most substantive part of the UNCLOS is the authority given to coastal states to establish exclusive economic zones (EEZs) to a distance of 200 nautical miles from the base line.

Virtually all coastal states have now introduced national legislation along these lines; indeed, several Latin American countries, notably Chile, Ecuador, and Peru, have had such national zones for some quarter of a century. Apart from understandable problems in defining the precise location of the boundaries between adjacent EEZs (in which the hopes for the presence of oil or other minerals under the seabed have often played as important a part as fisheries), and questions of the status of the so-called highly migratory species (mostly tunas) within national EEZs, these national measures are generally recognized.

The most immediate change brought about by UNCLOS and the establishment of EEZs has been in the balance between coastal states and local fisheries, on the one hand, and long-distance fishing countries and fisheries by large vessels, on the other.

There has often been the feeling that this dichotomy is equivalent to that between rich and poor or, to use United Nations jargon, between the developed and the developing countries. This is some way from being the case. Operating distant-water fisheries does not demand highly advanced technology, at least not as compared with say setting up an aircraft factory or building cars. It may require technology that is more advanced than that of the typical inshore fisheries, but not a technology that is out of the range of the typical developing country. What it does demand is a moderate degree of capital investment in vessels, and more particularly moderately skilled crews willing to work long hours under uncomfortable conditions. Furthermore, distant-water fishing can be carried out independently of existing local fisheries; it offers opportunities for increasing the production of food, either for local consumption or for export, without requiring the disruption of prevailing social patterns, which has been an obstacle in developing local fisheries (and also local agriculture) in some countries. Thus, several of the more advanced among the so-called developing countries have established long-range fishing fleets. Notable among these are the long-line tuna fleet of the Republic of Korea and the trawler fleets of Korea and Cuba.

It is at least equally untrue that distant-water fishing is done mainly off developing countries. In fact the requirements for successful long-distance fishing—high density of fish (i.e. high catch rates) and preferably only one or two species (to simplify processing and marketing)—are more easily satisfied in temperate waters, adjacent to developed countries, than in tropical zones where most developing countries are located.

This is shown more clearly in Tables 5-3 and 5-29, which present data on catches by non-local fisheries in 1972—a typical year in the period before the impact of the new regime of the sea became apparent. In that year, some 16 million tons—one-quarter of the total weight caught by all fisheries and probably significantly more than one-quarter of the total value—were taken by vessels of one country operating in waters off another country. Most of this total was taken by ships from developed countries, but not an inconsiderable minority, over 1 million tons, was taken by developing countries. Looking at it in another way, in terms of the location of the catches, two-thirds (over 11 million tons) were taken off developed countries, and one-third off developing countries.

The location of these catches is also shown in Table 5-29. The catches have been tabulated according to the major FAO statistical areas. This shows that some non-local fishing is done in all regions, but in some regions these catches are small and to a large extent composed of tuna. The tabulations include all catches in a region not taken by

Table 5-29

Total catches, and catches by non-local vessels in 1972, one of the last years before the general extension of limits (thousand tons) (After GULLAND, 1979b; modified; reproduced by permission of the author)

| Area | Total catch | Catches by non-local fisheries | |
|---|---|---|---|
| | | All species | Tuna |
| Northwest Atlantic | 4327 | 2292 | 10 |
| Northeast Atlantic | 10 699 | 3667 | 1 |
| West central Atlantic | 1488 | 143 | 5 |
| East central Atlantic | 3111 | 1930 | 180 |
| Mediterranean, Black Sea | 1165 | 40 | — |
| Southwest Atlantic | 805 | 24 | 12 |
| Southeast Atlantic | 3013 | 1771 | 29 |
| West Indian Ocean | 1809 | 201 | 67 |
| East Indian Ocean | 821 | 88 | 36 |
| Northwest Pacific | 14 531 | 2936 | — |
| Northeast Pacific | 2774 | 2254 | — |
| West central Pacific | 4770 | 479 | 114 |
| East central Pacific | 982 | 287 | 274 |
| Southwest Pacific | 275 | 199 | 100 |
| Southeast Pacific | 5539 | 48 | 13 |
| Total | 56 109 | 16 359 | 841 |

locally-based fleets, and catches taken more than 200 miles from any coast or base line. On the other hand, there are six regions where catches by non-local vessels exceeded a million tons. In four of these areas (northwest Atlantic, northeast Pacific, eastern central Atlantic, and southeast Atlantic) non-local catches exceeded local catches—by a large margin in the northeast Pacific Ocean. In another region—the southwest Pacific Ocean—non-local catches also accounted for well over half the total, though the absolute magnitude of the catch was small. A larger series of data on local and non-local catches has been presented by GULLAND (1973).

It is also possible to examine the data in terms of the countries concerned, i.e. to identify 'winners' and 'losers'. These are in some ways misleading terms, especially as regards the concept of 'losers'. As used here, they denote those countries which operate long-distance fleets and whose fishermen catch a quantity of fish off the coasts of foreign countries that is greater than the amount (if any) of fish taken by foreign fishermen off their own coasts. Certainly the freedom of operation of their fishermen will become more circumscribed, but these need not necessarily be wholly to their disadvantage. Local fisheries often serve important social purposes, e.g. bringing employment to isolated communities with little alternative economic activity. This cannot be easily equated with simple measures of economic efficiency and is, therefore, in some ways less vulnerable to the harmful economic effects of open-access fishing. Long-distance fishing, on the other hand, is essentially an economically-directed activity, employing large quantities of capital, labour, and other resources (e.g. money for fuel) which could readily be used in other ways. Distant-water fishing is, therefore, sensitive to misallocation of these resources that is the inevitable long-term result of open-access fishing. Put in more concrete terms, a company operating a distant-water fleet that used, prior to the establishment of EEZs, to take 10 000 tons of fish with 10 trawlers, may be very happy with controls imposed by the coastal state that only allow it to take 9000 tons, if at the same time the activities of all other fishermen are restricted so that this catch can be taken with only 5 trawlers.

There is little such uncertainty about the 'winners'. The extension of jurisdiction should undoubtedly bring them benefits, if not without some costs. Under open access the only way a coastal state could benefit from rich resoures off its coasts was by direct participation in the fishery itself, in competition with all others. Now it is possible, as discussed below, for coastal states to reap considerable benefits without necessarily engaging in actually catching fish (which in many countries is the least profitable part of the whole fish business).

These 'winners' are mainly the coastal states in the region identified above, where distant-water fishing is most important. These can be listed geographically as follows (GULLAND, 1979c):

*Northwest Atlantic Ocean:* Canada, Greenland, USA. These fisheries include the classical cod fisheries on the Grand Bank of Newfoundland, whose history goes back almost to the time of Columbus (in fact the tip of the Grand Bank extends just beyond the 200-mile limit), and the more recently developed long-range fisheries off Labrador and Greenland. Lately, cod has been only one of a large number of species pursued by long-distance fleets. Silver hake, herring, and mackerel have also been taken by these fleets, especially off the United States.

*Northeast Atlantic Ocean:* Iceland, Norway, UK. The situation in this region is confused because so many of the advanced fishing countries are located here, so that many non-local fisheries do not involve very long voyages from the home port. There is a general tendency for the richer grounds to lie in the northern and western parts of the region, and for the movements of the fleets to follow this pattern. Thus, Iceland and Norway gain control of the cod fisheries (in which English vessels used to take a large part), while in turn the United Kingdom gains control of a variety of stocks (hake, whiting, flatfish, etc. and also crustaceans) fished by Dutch, French, and Spanish fishermen. The regional authority and responsibilities of the European Economic Community adds further complexity to the management of fisheries in this region, but this will not be discussed here.

*Eastern Central Atlantic Ocean:* Mauritania, Morocco, Senegal. The pattern of winds and currents that allow high production in the sea also leads to deserts on land. These countries, therefore, have only small populations adjacent to the fish stocks. Non-local fleets take a leading share in all the major fisheries—purse-seining and trawling for small pelagic fish like sardinella and mackerels, and trawling for hake, sea bream, and other demersal fish, and also for cephalopods. The latter (squid, cuttlefish, octopus) are, because of the high prices for the product, among the most valuable in the region.

*Southeast Atlantic Ocean:* Angola, Namibia, South Africa. The ecological conditions both on land and at sea are very similar to those in northwest Africa, and so are, to a large extent, the fisheries. The difference is that there is a well-developed local fishery for shoaling pelagic fish (mainly pilchard), based in Walvis Bay. This has badly depleted the pilchard stock. The prime targets for the long-range fleet has, therefore, been hake and more recently massbanker.

*Northwest Pacific Ocean:* USSR, Japan, China. This region is unusual in that two of the major 'winners', with significant foreign fishing off their coasts, are themselves among the largest operators of long-range fleets. The explanation is that the USSR and Japan, while similar in being big producers and consumers of fish, have different preferences. Japanese fishermen, therefore fish off the USSR for high-priced fish (particularly salmon) and for Alaska pollock to supply the specialized surimi (minced fish) market. Russian fishermen come to Japanese waters for pelagic fish (sardine, mackerel) that do not fetch a good price in Japan.

*Northeast Pacific Ocean:* USA, Canada. The coastal states have long-established and highly efficient (too efficient in several cases) fisheries, but principally because of high labour costs these have mainly been for the high-priced fish (mainly salmon and halibut). It has been left to the distant-water fishing countries (mainly USSR and Japan) to develop the bulk fisheries for the lower priced fish (flatfish other than halibut, ocean perch, Alaska pollock).

So much for the major winners. The losers are most easily discussed in terms of the type of fisheries, which roughly fall into two classes: the locally-based and the world-wide. The first include many of the most successful groups of fishermen who, finding local resources inadequate for their demands, move into adjacent waters. The typical example of this fishery is probably the English trawl fishery. Trawling in its modern form began when the fishermen of the sailing smack fleet advanced from using steam power merely through tugs to get them in and out of the harbour to powering the fishing vessel itself. The English North Sea steam trawlers proved so successful (in conjunction with the use of ice and of railways to get the fish quickly from the ports to the big cities) and grew in numbers so much that by the end of the nineteenth century they had fully exploited the more attractive species, notably cod and plaice. As catch rates dropped in the North Sea, the more enterprising fishermen looked further afield for other stocks of the same species—particularly plaice—since it fetched a better price and lasted longer on ice. Early in the century they had found the plaice grounds off southwest Iceland and off the north Russian coast (the 'White Sea' grounds of the English fishermen). The fisheries in these and other grounds (Faroes, West Greenland, Bear Island, Spitsbergen) grew, until by the middle of the century the majority of the British fish supplies came from these middle- and distant-water grounds.

Fisheries can expand in this way over surprising distances, even with simple technology. At about the same time that the English fishermen developed the steam trawler, Breton fishermen found that they could keep lobsters alive on board for a long time, and that the railway gave them access to a vast market. Before long, Breton fishermen in small sailing boats were catching lobsters, of various species, from northwest Scotland to Mauritania, and fully exploiting these stocks.

The same pattern has been repeated—with modifications in scale, type of gear, and species sought—in all parts of the world. The English trawl fishery has been mirrored almost precisely by the trawl fishery in Thailand. Around 1960, Thai fishermen, helped by a small technical assistance team from Germany, started using the single-boat otter trawl, which proved very effective. As the number of trawlers grew, falling catch rates (which have been very well monitored by regular surveys by Thai research ships, TIEWS and co-authors, 1967) forced the Thai fishermen to go further and further to find good catches. Before changes in the pattern of jurisdiction made it difficult for them, Thai fishermen were working in most of the area from northeast India to Borneo.

These are only the more obvious examples. A list of all the countries whose fishermen did some fishing off the coasts of another country would include nearly all countries with fishing vesels big enough to stay at sea for more than a day trip. The main exceptions are geographically isolated countries like Australia or New Zealand. Often there is a movement of vessels in both directions, due perhaps to local preferences for different species of fish, or to taking advantage of seasonal concentrations of fish in one area or another.

The other group of 'losers' are the long-distance fleets, which commonly include a variety of support ships to bring supplies to the fleets and to take frozen or canned fish back home, in addition to the ships doing the actual fishing. The biggest fleets are those of the Soviet Union, but other eastern European countries have large fleets, as well as some western European countries and Japan. A distinction can be made here between the two groups. For the first, distant-water fleets have been developed largely as a supply of animal protein, more or less regardless of species, and are needed to compensate for failures to develop agricultural production. These fleets concentrate mostly on the high volume of relatively low-value species (e.g. mackerel and horse mackerel). For the second group, the distant-water fisheries have grown to maintain or increase the supply of preferred species of fish for which the home waters can no longer meet the demand. These fleets therefore tend to concentrate on high-valued species such as tunas, octopus, and hake.

Under the new regime of the sea, all these groups of 'losers' have had to come to some agreement with the coastal states in order to continue their activities, even at a reduced level. According to the draft Law of the Sea texts, the coastal state should determine the allowable catch that can be taken from a stock, how much of this will be taken by its own fishermen, and then make the 'surplus' (i.e. the total allowable catch minus coastal state capacity) available to other states, under conditions that mainly concern the conservation of the stock (sizes and quantities of fish caught, etc.). A good description of the practical operation has been given by LARKINS (1980) for the Gulf of Alaska.

There are many reasons why this apparently simple formula will not work in practice. The allowable catch is not a quantity that is uniquely determined by the ecological characteristics of the stock, but will depend on the objectives of those responsible for the stock (high gross catches, or high catch rates; employment for a large number of fishermen, or good living for a smaller number, etc.). In any case, it is difficult to calculate without a large research programme, which many countries lack. A coastal state is, therefore, likely to find it much easier to reverse the logical steps, i.e. to arrive at a figure of allowable catch by adding together the expected catch by its fishermen, plus any catch it may allow to other fishermen. In arriving at the latter figure any realistic administrator in a coastal state will take note of the fact that if there is an established local fishery on a given stock, any additional fishing (e.g. by a foreign fleet) on this stock will reduce the catches of the fleet—a fact that is not taken note of in the simple concept of 'surplus'. Coastal countries have adopted a variety of methods to control the activities of foreign vessels and to gain advantages for themselves from these operations. The method adopted is usually dependent on the extent to which the coastal state wishes to start or expand its own harvesting of the stock concerned, and the nature of the obstacles for doing so. Many developing countries are anxious to develop their own fisheries, but cannot do so because of lack of capital or technical expertise. Both of these can often be obtained from a distant-water fishing company. A common way of doing this has been the formation of joint ventures, i.e. companies combining interests in the coastal state (often the government, or a quasi-governmental organization) and in the distant-water fishing country (usually a large fishing company). Normally the coastal state will retain control (e.g. by a 51% holding of the voting shares), though the other partner will provide most or all of the capital input. Under favourable circumstances this can be a mutually satisfactory arrangement. The interests of the second group of major distant-water countries (e.g. Japan) and to a large extent also their fish trading companies are to a decreasing degree concerned with the hard, uncomfortable, and dirty business of catching fish, and, much more, with selling the product. If the coastal state can be provided with the capital and know-how to replace the large, expensive, long-range vessels with a locally-based fleet, so much the better, especially if the joint-venture activities include supervision of the initial processing so that the product (e.g. frozen fillets) is of a quality to match the demand of the consumer in the distant-water state.

The development of a locally-based fishery can also be pursued in other ways. In many fisheries the shore-based side—processing the catch in canning or freezing plants—is more attractive. It gives more employment or higher profits than the catching side. In such cases a condition of access by foreign vessels may be that a certain proportion of the catch is landed (often at predetermined and artifically low prices) at local ports for processing. In other situations getting the benefit may involve no more than charging a fee for fishing, based on the quantity caught or on the number and size of vessels allowed to operate, or a combination of both. Direct payment of licence fees can occur at opposite ends of the spectrum of economic development of the coastal state. At one end the coastal state (e.g. Australia or the United States) may be so well off, with so many alternative (and more profitable or comfortable) employment for people in the coastal towns or villages, that it is only foreign vessels with foreign crews that can be attracted to go fishing other than for the economically most valuable species. At the other end are a few countries that have yet to develop an infrastructure to support more than a small artisanal fishery.

The situation for the medium-range vessels, by fishermen in one country fishing in the EEZ of some adjacent country, is somewhat different. As already noted, this type of fishing is often a two-way affair. Vessels from one country will fish off the coast of a neighbouring country for some preferred species, or at some favourable season, while vessels from the second country will fish off the first for other species or at other seasons. The arrangements for fishing in each other's waters are then likely to be done in a comprehensive package, concerning all the stocks rather than on a stock-by-stock basis. Often these arrangements will be reached through some general regional grouping. In the northeast Atlantic Ocean, the main grouping—the European Economic Community—has special responsibilities for fisheries, at least in the Atlantic and North Sea, but to a less extent in the Mediterranean Sea. These responsibilities and the slow development of a Common Fisheries Policy have, at least in the short term, added considerably to the complications of fishery management in this area.

Further consideration of these international aspects of fishery management is outside the scope of this chapter. The new regime of the sea and the general establishment of EEZs by coastal states have altered the whole framework within which fisheries can be managed. It will now be possible to give more attention to passing social and economic objectives, and to getting more and greater variety of benefits from well-managed fisheries. What these are, and how they may be achieved, is considered in the following sections.

### (f) Benefits from Fisheries

Discussion of the future of fisheries, and of how improvements might be achieved by making better use of scientific knowledge of fish stocks and of the environment in which they live, is preceded by a short review of the sort of benefits that might be obtained. To the marine ecologist the interesting part of fisheries usually ends when the fish arrive on the deck of the fishing vessel, or perhaps his interest may go as far as the auction floor of the fishing port. To him, therefore, the benefits from a fishery involve the catch. The bigger the catch, the greater the benefits. To some extent this is, of course, true. Unless there is a good-sized catch, there cannot be many benefits—apart from exceptional cases, such as whales, where the preservation of the stock may become an end in itself. The converse, however, is not necessarily true. Catches can be high, and indeed so high that no significant sustained increases are possible, while the actual benefits to society are very much less than they might be. One possible way of classifying the variety of potential benefits that could come from better management is in fact according to the degree to which they are determined solely by the magnitude of the catch. At one end would be the contribution of fish to the world's food supply; at the other, the reduction in international and intranational conflicts. In between lie the economic and social benefits which depend as much on the reduction of costs involved in the fishery as in possible increases in the gross value of the catch.

The role of fish in feeding the world is often under-estimated by those in the richer countries. To western Europe or America fish is largely significant as supplying a little additional variety to a market which in dietary terms is already well supplied, or even over-supplied, with protein. On a world scale, fish ranks a poor second after meat as a supply of protein. Overall, fish supplies about 8% of the world total of all types of protein, and about 20% of the animal protein, compared with 40% from meat.

These global figures are based on statistics of supply, i.e. for fish, the quantities landed. To provide a proper estimate of the contribution of fish or meat to the diet they should be corrected for that proportion of fish, or carcase of the animal, that is not eaten, and for other losses that can occur between leaving the sea (or the farm) and arriving on the consumer's plate, or in his rice bowl. In the case of those fish (anchovies, capelin, etc.) which are mainly used for fish-meal the calculations then get quite complicated. In the first instance these fish do not go for human consumption, and apparently should, therefore, be omitted. However, all the fish-meal produced is used to feed farm animals (particularly broiler chickens), and the net addition to the broiler chicken production attained by including fish-meal in the diet should be credited as a contribution of fisheries to the world food supply. This is quite high. The conversion efficiency (from food to chicken) in the intensive systems in which fish-meal is mostly used is high in any case. In addition, it has been commonly stated that fish-meal contains some 'unidentified growth factor' that increases the growth rate (and the food conversion efficiency) by an amount over and above that due simply to the protein content of the meal. The fish used for fish-meal should, therefore, be added, with a certain reduction, to the total amount available for human nutrition. The reduction, representing the losses along the chain from fishing boat—fish-meal factory—poultry farm—to cooked chicken on the plate, may not be so very much more than in other types of fish use. The losses at the first stage, from fish to first consumer (chicken or other farm animal), are small. They are very much less than, for example, the simple types of fish-drying used in Africa. At best, the product from the cruder types of drying will contain much less than the protein content of the fish as it left the water. Often, attacks by beetles and other pest may mean that the final consumer gets only a third or less of the original protein.

The other comment that needs to be made about the contribution of fish to feeding the human population concerns distribution. Overall, there is no shortage in food. Current food production more than suffices to provide everyone in the world with all he needs. Unfortunately, the poorest people cannot afford to buy the food they need, while the richest eat too much. The significance of fish is that a greater proportion of the world fish catch is eaten by the poorer people than other forms of animal protein.

This is witnessed by the per capita supply of fish in different groups of countries (Table 5-30). Where imports or exports are significant, these will misrepresent the actual per capita consumption. The difference is important for the big fish-exporting countries—notably Iceland, followed by other Scandinavian countries, Canada, and the Pacific states of South America. As might be expected, the average availability of fish is much higher in the rich, developed countries than in the developing countries. However, among the latter, the countries of east and southeast Asia, from Korea to Indonesia, land more fish per capita than the United States, and not much less than the average for the developed world as a whole. In addition, there are several countries in Africa, e.g. Uganda, Tanzania, Mali, with per capita production of around 20 kg (mostly from fresh waters), which if not particularly high, represents the difference between moderate to low annual protein supply, and severe shortage.

The same is true within countries, in that fish provides food to scattered communities which have few other sources of supply. The obvious examples are small islands. How-

Table 5-30

Per capita supply of fish in different groups of countries. Reported landings divided by the total population (Based on FAO *Yearbook of Fishery Statistics*, 1979)

| Country | Developed countries: Population (millions of individuals) | Fish landings (thousand tons) | Gross supply (kg person$^{-1}$) | Country | Developing countries: Population (millions of individuals) | Fish landings (thousand tons) | Gross supply (kg person$^{-1}$) |
|---|---|---|---|---|---|---|---|
| Canada | 23·7 | 1·332 | 56·2 | N.W. Africa | 43·9 | 380 | 8·7 |
| USA | 220·3 | 3·511 | 15·9 | W. Africa | 136·7 | 1·468 | 10·7 |
| Japan | 115·9 | 9·966 | 85·9 | Central Africa | 51·7 | 458 | 8·9 |
| Oceania | 17·4 | 237 | 13·6 | East Africa | 129·6 | 896 | 6·9 |
| South Africa | 28·5 | 659 | 23·1 | Central America | 89·8 | 1·087 | 12·1 |
| EEC | 260·3 | 4·623 | 17·8 | Caribbean | 30·1 | 220 | 7·3 |
| Southern Europe | 56·3 | 1·575 | 28·0 | Pacific: South America | 36·0 | 6·959 | 193·3 |
| Iceland | 0·2 | 1·645 | 8225·0 | | | | |
| Others: Northern Europe | 17·2 | 3·249 | 188·9 | Rest of South America | 202·9 | 1·763 | 8·7 |
| Eastern Europe | 111·5 | 1·147 | 10·3 | Near East | 211·7 | 771 | 3·6 |
| USSR | 264·5 | 9·134 | 34·5 | South Asia | 874·1 | 3·481 | 4·0 |
| | | | | East and southern Asia | 341·3 | 8·586 | 25·2 |
| | | | | China | 945·0 | 4·054 | 4·3 |
| | | | | Oceania | 4·9 | 1·029 | 21·0 |
| Total* | 1·155·0 | 37·151 | 32·2 | Total* | 3·180·4 | 33·020 | 10·4 |

Grand total Population: 4 335·3; Fish landings: 71 287; Gross supply: 16·4

*Includes countries not included separately.

ever, many communities on the coasts and inland along rivers and lakes, especially those cut off from the main centres of economic activity, are in the same situation.

This brings up another important function of fishing: provision of employment. There are many communities, even in the developed world, which, without fishing and its associated shore-based trades, would have little money coming in. These include, for example, the small Scottish towns around the Moray Firth, many parts of western Norway (though in these cases the advent of offshore oil drilling is changing the situation radically) and most of Newfoundland, as well as several other parts of the Canadian maritimes.

The function of fisheries as employment provider is even more significant in many developing countries. In far too many places the lack of opportunities in the country and small towns has led to a drift to the big capital cities. This, combined with the steady rise in total population, threatens to make these big cities (Calcutta, Mexico City, Jakarta, etc.) with populations rising well into eight figures, the major disaster areas of the late twentieth century. Anything that offers counter-attractions to the big cities, and keeps or attracts people back to the other parts of the country, is extremely welcome. For example, the Indonesian government has a major programme of trans-migration, i.e. the moving of people from the over-crowded island of Java to the much less densely populated outlying islands of the vast Indonesian archipelago, for which the opportunities of fishing in the outer areas are important.

These remarks underline the fact that fishing is an economic activity. As such its success is best measured not by the gross value of the product, but by the value added, i.e. the gross value minus the outside inputs used in producing the catch. The costs of several of these, notably fuel, have increased rapidly in the last few years. This is changing the economic aspects of many fisheries, and hence the way they may need to be managed.

Fishery management and the development of national fishery policies are complicated not only because of the variety of benefits that fisheries can provide, but also by the fact that many of them are in conflict. If a group of fishermen is harvesting a given resource more or less to its biological limit with a simple type of gear, consideration of economic efficiency may point to the desirability of introducing a better type of gear. However, the use of such gear could only be accommodated within the biological constraints set by the productivity of the resource, if a proportion of the existing fishermen can be persuaded to leave the fishing. While raising the income of the individual fishermen who remain, this would conflict with the objective of maintaining employment.

An important conflict, and one that has been increasingly apparent in recent years, is that between different groups of fishermen. The conflict (real or imagined) between coastal fishermen and those coming from other countries was one of the factors that triggered the actions for a new Law of the Sea; those are now reaching their conclusion. The conflict between Iceland and England (as well as some other European countries) over cod, and between the United States and several states along the Pacific coast of central and south America over tuna, have perhaps received most attention. They are, however, far from being the only such conflicts. Where distant-water fisheries have operated there have been complaints almost everywhere from fishermen of damage to their livelihood. While these complaints were not always justified, since the two groups of fishermen may exploit different species of fish, they have nearly always gained political significance.

The change in the Law of the Sea regarding the establishment of exclusive economic zones will end many of these international conflicts by resolving them in favour of the coastal states, but many national and even international conflicts will remain. Fish will not respect the new man-made boundaries. Where the movements between adjacent EEZs is substantial there are bound to be arguments as to the shares of the catch that should be taken by each country. These arrangements get more detailed and harder to resolve when there are differences in the characteristics of the fish stocks or of the fisheries either side of the boundary. For example, one country might feel that it is entitled to a major share in the catch because the spawning grounds are contained within its EEZ. Another country might in response claim that it deserved the main share because the feeding grounds are in its EEZ, and that the bulk of the weight taken comes from the production in its waters. Similarly, arguments can be based on the relative needs of one or other country for protein, or for employment, or on the basis of the relative efficiency with which the fish can be caught.

The new Law of the Sea should resolve a few of these conflicts. In the case of salmon and other anadromous species, the state in which the fish spawn is given authority to manage the stocks wherever they might move. Bearing in mind that these states could ruin the spawning success through a variety of actions (dam construction, logging, spraying against insect pests, etc.) if these are not properly controlled, and that they would lack much of the incentive to make these controls if they did not expect to reap most of the benefits from the fish stocks, this arrangement is reasonable. It resolves conflicts such as those in the Pacific Ocean between Canada and the US coastal fishermen and Japanese high-sea fishermen, and in the Atlantic Ocean between the salmon fishermen of Canada and western Europe (among whom the anglers have been politically most powerful), and the fishermen who catch salmon on their feeding grounds, e.g. off West Greenland. The latter example is interesting in that the Greenland fishery is carried on close inshore, and is thus a case in which the coastal state does not have unlimited powers over what is caught in its waters. Apart from anadromous species, however, the new Law of the Sea is not very specific on how conflicts over shared stocks should be solved, merely calling on the states concerned to collaborate.

Conflicts within a single country can be equally difficult. The arguments in developed countries between sports and commercial fishermen are matched closely by those in developing countries between the traditional small-scale fishermen and those with newer and more powerful vessels. In both cases there is a clash between maximum economic or technical efficiency and other social objectives, recreation or maintaining the livelihood of part of the poorest section of the community.

Fishery management cannot, of course, remove all these conflicts. If more people wish to fish than the resource can support, someone is going to be disappointed. What good management can do is to ensure that the actions taken to deal with the conflict are in accordance with general national objectives, and are soundly based on a knowledge of the resource (e.g. that restrictions are not placed on industrial fishing to protect inshore fishermen when in fact the impact on the latter would be very small). This reduction of conflicts can be one of the most important functions of the fishery manager and his scientific advisers, and one that goes far beyond some of the narrow views of fishery management as being merely the conservation, i.e. preservation, of the resource.

### (g) New Approaches to Fishery Management

Several factors—the New Law of the Sea, the increasing pressure on fish stocks, and the growing public concern (at least in the richer countries) for the natural environment—are causing the question of managing fisheries to be approached in new ways.

The logical structure of fishery management has been most clearly set out in a report of a working party of FAO's Advisory Committee on Marine Resources Research (ACMRR) (FAO, 1980). In this report the similarity of the processes to those involved in managing a commercial business, or any other management, are stressed. The working party pointed out that fishery management is a co-ordinated system involving the following components: (i) setting objectives; (ii) defining boundaries; (iii) collecting data; (iv) transforming data into information; (v) formulating action; (vi) executing the chosen policies; and (vii) co-ordinating the results and the system itself.

While it cannot be claimed that the working party in itself has achieved a revolution in the ways fisheries are managed, its report does provide a milestone in the somewhat erratic progress towards better management. Looking backwards, its report, and especially the explicit recognition of management as a process of bringing together diverse and sometimes conflicting elements, represents a codification of a process that had been going for many years. Looking ahead, the report will, it is to be hoped, provide guidelines for practical management action, as well as a basis for further theoretical discussion.

#### *Objectives*

The objectives of management, especially from the ecological point of view, have recently received considerable attention (e.g. HOLT and TALBOT, 1978). More important than the specific conclusions reached in any particular study has been the realization that until the objectives are clearly stated (if only implicitly), there can be considerable confusion. Groups with different objectives will often work at cross purposes, or the management authority can come under strong criticism for failing to achieve objectives which were never aimed at. A common example is when fisheries managed for purely ecological objectives fail to achieve economically satisfactory results.

Of the ecological objectives, one that has been greatly criticized is that of maximum sustainable yield (MSY) (e.g. GULLAND, 1968b; 1978; ROEDEL, 1975; LARKIN, 1977a). Some of these criticisms and the attempts to find a single substitute for MSY have been misplaced in that the concept of MSY served a number of distinct purposes. The curve of sustainable yield, as a function of the amount of fishing or of the population abundance and its maximum (MSY) at some particular value of fishing effort or abundance, provided a useful general description for non-specialists of the ecological background for management. MSY also provided a landmark whereby the state of exploitation can be judged. If the abundance is less (or the amount of fishing more) than that producing the MSY, then the stock is clearly in some sense 'over-fished', and management is needed. Conversely, if the abundance is well above (or the amount of fishing well below) the MSY level, then probably management is not needed, at least not so far as ecological objectives are concerned. One problem is that the more the concept of sustainable yield—a simple function of abundance or amount of fishing—is modified to take account of the ecological facts of life (natural fluctuations, delays between changes in the amount of fishing, and the response of the population) the less the resulting model serves as a simple description for the non-specialist. It would seem that with all its faults MSY will remain as a useful simple description of the biological basis of management, and a landmark against which the need for ecological management can be judged.

In fact, despite the references to MSY or some equivalent words, in the text of many conventions on fishery management MSY has seldom been used in a more exact or rigid manner. One exception has been the International Whaling Commission. Especially since the adoption of its New Management Policy the quotas set for each stock of whales have been determined according to a complicated but precise formula which depends on the abundance of the stock relative to the MSY level. Only if the abundance is close to, or above, this level is any catching allowed. Though this procedure was designed to remove arguments about whale management from the political arena, and to avoid the conflict of objectives between 'conservationists' and 'exploiters', the practical effect has been to move these political conflicts into the scientific committee. The fact that a relatively minor change in the scientific conclusions can, if the stock is near the MSY level, cause big changes in the quota, has intensified the political pressures on the scientists. When such changes occur, it has caused those outside the IWC to wonder whether the Commission really knows what it is doing.

The chief contender as a replacement for MSY is MEY, the maximum economic yield. This will be taken when the difference between the gross value of the catch and the costs of harvesting is maximal. Since, as the MSY is approached, the catch (and hence value) increases much more slowly than the amount of fishing (and hence the costs), the MEY will occur at an amount of fishing that is less (often considerably less) than that giving the MSY. Similarly, stock abundance at MEY is greater than that at MSY. MEY and the state of the stock relative to the MEY does provide a broad guide as to whether or not the stock is being too heavily exploited from the economist's point of view. It is, however, certainly no better than MSY as a rigid prescription. Like MSY it takes no account of social factors. Also, the position of MEY will change as the value of fish, or the costs of fishing, alter. This is realistic, and in practical terms desirable, but is probably not a desirable feature in a standard objective.

Attempts have also been made to strengthen the biological realism of MSY. For example, HOLT and TALBOT (1978) attempted to develop new management principles, to take account of matters such as interaction between species, need to manage the whole ecosystem rather than individual species, and uncertainties involved in making assessments on which management decisions are based. This approach has been incorporated in the new Convention for the Conservation of Antarctic Marine Living Resources. Unlike most previous fishery conventions which were drawn up by those directly interested in exploitation, after it became obvious that some of the stocks were being heavily fished, the Southern Ocean Convention was established before large-scale exploitation of the species of greatest interest (krill) had begun; it was a joint effort of countries interested in exploitation (notably Japan and USSR) and those interested almost solely in conservation (notably USA). The most significant section, Article II 3(a), on the objectives of the Convention reads:

> '(a) Prevention of decrease in the size of any harvested population to levels below those which ensure its stable recruitment. For this purpose its size should not be allowed to fall below a level close to that which ensures the greatest annual increment.'

It remains to be seen how well this will work. From the conservation point of view the chief intention was to prevent exploitation of krill reaching a level at which a reduced abundance of krill could be a threat to the rebuilding of the depleted whale stocks. The extent to which the terms of the convention will achieve this objective will only become clear as and when large-scale krill fishing develops. At the moment this appears to be a somewhat distant probability (MITCHELL and SANDBROOK, 1980; see also p. 986).

One point that became clear in the background discussions concerned with this convention, and from other theoretical studies (e.g. FAO, 1978; MAY and co-authors, 1979) is that if two species interact, it is most unlikely that the MSY from both can be obtained simultaneously. In particular, a high sustainable yield from a predator (e.g. blue whales) requires a high stock (and therefore a low harvest) of prey (e.g. krill). There is some implication in the Southern Ocean Convention, and more particularly in some other documents (e.g. the US Marine Mammal Act), that—at least as far as marine mammals (especially whales) and their prey are concerned—priority should be given to maintaining marine mammals at or above the level giving the maximum net productivity (which seems to be indistinguishable from the MSY level). There seems little logical basis for this. If whales are regarded as an economic resource, maximizing their yield does not need to be given priority over increasing economic yield from their prey. In terms of conservation there is a very strong case for ensuring that their abundance does not fall to such a level that there is any risk of extinction, and a somewhat weaker case for not killing them at all, but no case for giving priority to ensuring that the greatest possible number can be killed.

*Boundaries*

These discussions over objectives, and the search for a successor to MSY, illustrates the general realization that the boundaries to fishery management have in the past been set too narrowly. The ACMRR Working Party report already referred to has listed, in tabular form, some of the activities, disciplines, and interests that are concerned in one way or another with fishery management. From the core of fishery management—the catching of fish and the control of the kinds and quantities caught—the boundaries can be explored in several directions. In the sea there is the need to consider factors, both natural and man-made, other than fishing, that can affect the stock. This means that the fishery manager must take account of changes in ocean climate, or the influence of competing interests (e.g. land reclamation which can reduce the extent of nursery areas available to several species such as shrimp or menhaden), and call upon the skills of physical oceanographers, or pollution experts. On land, what happens to the fish after it is caught—the market demand, particularly how important it is in local diet, the employment in processing and distribution, and the supply of alternatives or competing products—all can affect management decisions. These decisions (in the narrow sense of the introduction of rules and regulations such as catch quotas) are not the only factors that determine the pattern of fishing. The decisions taken by the individual fishermen, or fishing company, e.g. on buying a newer and better boat, can be affected to at least as great a degree by other aspects of government policy. Some of these, such as the degree to which profits in fishing attract less tax if reinvested in fishing, or the availability of different forms of subsidies, should be taken into account (and possibly modified) in any scheme of fishery management that takes a properly broad view of the subject.

This is not the place to discuss such aspects in detail. The main point is that the boundaries of fishery management are being considerably widened. It can no longer be considered as solely, or even primarily, the responsibility of the fishery biologist. The role of the biologist, and to some extent the nature of his work, can be expected to change accordingly.

*Data and Information*

The working group emphasized that within the broad subject described by these terms there was an important distinction between 'data'—the raw material, e.g. statistics of catch as collected from the fishery or in research work—and 'information', the result of applying suitable methods of analysis to this raw material and converting it into a form that can be understood by the ultimate user, e.g. the fishery administrator or planner. Some of the more effective, if less recognized, obstacles to fishery management have in the past been in these fields. This has mainly been a result of poor communication and understanding between the fishery administrator (the ultimate user of much of the biological information concerning fishery management, and also very often responsible for the provision of much of the data, especially that arising from the ordinary commercial fishery), and the fishery biologist (responsible for converting the raw data into usable information). One problem has in fact been that the information has not always been usable, or at least not usable without considerable difficulty. This gives rise to mutual complaints—from the biologist that the administrator (or other user) does not listen to him, and from the administrator that the biologist does not tell him what he wants to know. Similar recriminations have occurred over data—by the biologist that he is not provided with the data (e.g. detailed catch and effort statistics) which he needs to do his job, and by the administrator that the biologist usually does not make clear what he wants, and when he does, the demands are unreasonable and not accompanied by convincing reasons.

Some of the difficulties of communication were connected with the view that fishery management was solely a biological matter. This may be caricatured by the feeling of some biologists that if only they could be supplied with the perfect data, then the perfect management policy could be clearly and uniquely determined. Among the objections to this view are that complete fishery data are extremely expensive to collect (even if they could be defined). If they were collected, uncertainties in other factors (e.g. climatic variations) affecting fisheries would make the determination of the uniquely perfect management policy impossible. In any case, the difference between 'perfect' management (whatever that might be) and reasonably good management is not great, and much less than that between reasonably good management and the uncontrolled state of many present-day fisheries.

The approach to data and information is now changing with the realization that managing fisheries is a complex and difficult multi-disciplinary task, and the recognition by particular disciplines or interests that their demands for information have to be matched to the ability of other groups to provide them, and to the value of the data or information in the whole management process. This recognition is far from complete, but progress is being achieved. For example, most administrators are aware that fishery scientists must have adequate data to do their job, otherwise the information flowing back to the administrators will be poor. Equally, most biologists are aware that they can never get all the data they want.

As far as information and advice are concerned it is becoming more generally recognized that while in some ways the apparent central role of biologists (in providing an assessment of precisely what should be done) is out of date, their real importance is increasing. Only in exceptional cases in the past were the biologist's prescriptions, e.g. levels of MSY, actually followed closely; in many cases the biologist's advice was completely ignored. Now an increasing proportion of fishery administrators and other decision makers (e.g. in large fishing companies) are aware that they must have biological advice if they are not to make expensive mistakes. Biological advice is now being increasingly used as one of several, but among the more important, inputs to decisions concerning fisheries.

These changes are going to cause large changes in the way that the biologists responsible for providing management advice go about their job. Different approaches will be needed for the preparation of the standard scientific paper and advice to managers. The latter will not seek for complete proof, or unqualified conclusions. Rather, it will have to be expressed in terms of probabilities, and of the likelihood of different outcomes from various actions that might be taken.

### *Formulation and Implementation of Action*

These later stages in the management process are of less direct interest to the biologist, but are not entirely irrelevant to his work. However, there are changes in the way management measures are applied that can radically change the ability of scientists to study the stocks. The extension of jurisdiction over fisheries by most coastal states has meant that more stringent and more effective measures are being imposed. In principle the biologist should be happy that his studies are now having greater impact on the fisheries, and the well-being of the fishermen. To some extent this is so also in practice, but not wholly. The greater impact of the new measures (e.g. catch quotas) has increased the incentive to evade them, and this evasion can threaten the reliability of some of the basic data used in studying the stock. Fishery scientists have always realized that, for example, many fishermen have failed to comply with regulations on the minimum mesh size of trawls, either using smaller meshes, or adding covers or other material to reduce the effective selectivity of the legal-sized meshes. Fortunately, the effective mesh size used by the fishermen is not a vital piece of information in studying fish stocks. In contrast, knowledge of the total catch is vital. Scientists, especially in the northeast Atlantic Ocean, are therefore becoming very concerned that, because of the imposition of catch quotas or similar controls, the available statistics on the quantities of fish being caught are, for several important stocks, unreliable. This has reached the stage that some groups have had to state that they are unable to assess the state of the stocks, or to give reliable advice on further management measures. No doubt, this problem will be resolved, but in future the biologist will have to give more attention to how management measures are enforced, and to ensure that, so far as possible, enforcement does not threaten the basic supply of data.

### *Shared Stocks*

The new Law of the Sea has partitioned most of the world's fishing grounds neatly into different areas, each under the jurisdiction of a single country. This would have greatly simplified the management problems if fisheries and fish stocks could be partitioned into national boxes in this way, and if interactions across the boundaries could be ignored. This is far from being the case. Large-scale movements of many fish stocks across boundaries between national zones of jurisdiction require that most management schemes must take account of these trans-boundary movements. This was, of course, a feature of most management practices in the pre-UNCLOS era, but the details of these multinational arrangements for management will have to be greatly changed.

Under the old regime all countries had equal access to stocks on the high seas. Decisions on management action had to be taken by consensus, and this consensus could be upset by the entry of a new country into the fishery. In practice, additional entries into a fishery did not cause a major problem, since their share was almost inevitably small at first, though special provisions had often to be made for them, e.g. through earmarking a small part of the Total Allowable Catch for catches taken by non-member countries. The result of this diffuse and weak jurisdiction was that the management decisions that were reached were usually weak.

The problems under the new pattern of jurisdiction are almost the exact opposite. Rather than lack of jurisdiction, the difficulty now is too many independent jurisdictions. Each country in which a fish stock may occur during its migration has, according to the draft texts of the Law of the Sea, competence to set measures for the fisheries in its zone of jurisdiction, including determining the allowable catch that may be taken in that zone. Legally there is nothing to ensure that the sum of allowable catches set in different zones is within the capacity of the stock to sustain. Without international co-ordination it could easily happen that each country would set quotas which individually might be reasonable, but together greatly exceed the sustainable yield of the stock, leading to a collapse of the stock and loss of the fisheries.

Reaching the necessary co-ordination can be complicated (FAO, 1980). Consideration has to be given first to the choice of the general level of exploitation (e.g. should the objective be high total catches and high effort or high catch rates and lower effort); second (generally subject to more argument) to the question of how the total should be divided between the countries. In the pre-UNCLOS era recent catches could provide a reasonable first approximation to how the division should be made. ICNAF developed a complex formula for this, in which past catches were modified to take account of special interests, especially those of the coastal state. In the post-UNCLOS era past catches are only of historical interest. A country in whose waters a stock spends most of its time would not agree that it should take a small share merely because it has been slow in developing its national fisheries. The important factors are more likely to be the location (relative to national EEZs) of the spawning and feeding grounds; the relative time spent by individual fish during its life in one or other zone of national jurisdiction; and the proportion of the total catch taken in one or other national zone (which can be very different from the proportions taken by the fishermen of each country in the pre-UNCLOS era).

These changes will alter the roles of the various international bodies concerned with management. Except for the stocks which will continue to be harvested to a significant extent outside national jurisdiction (mostly whales and tunas), the direct management role of international commissions will be reduced. This has already been illustrated by the demise of ICNAF (International Commission for the Northwest Atlantic Fisheries), and its replacement by NAFO (North Atlantic Fisheries Organization), whose responsibilities for setting management measures are limited to a few minor stocks that lie outside the jurisdiction of the coastal states (on the Flemish Cap and the tip of the Grand Bank of Newfoundland). Management measures for most stocks in the northwest Atlantic Ocean are set unilaterally by a single country, if the distribution of the stock allows this, and otherwise by bilateral deals between the countries directly concerned. The trends in other parts are similar.

The scientific role of international bodies is, in contrast, at least maintained, and in many ways strengthened. While countries will wish to reserve to themselves the decisions on the development and management of the resources in the EEZs (with as much consultation as is essential with other states sharing the same stocks), they should be glad to receive scientific information and advice (which they can ignore) from any source. Thus, the activities of ICES have continued undiminished. The difference is that actions arising from deliberations of its numerous working parties concerned with different stocks, as distilled by its Advisory Commission on Fishery Management, will be taken by individual countries, or political combinations such as the European Economic Community, rather than NEAFC. At present (1981) the scientific work of NAFO suffers from the absence of the United States, but in the northern part of the region the work of NAFO's Scientific Council is very similar to that performed by ICNAF in the past. Elsewhere in the world a number of FAO regional fishery bodies—the General Fisheries Council for the Mediterranean (GFCM), the Fishery Committee for the Eastern Central Atlantic (CECAF, which has been particularly active in promoting cooperative studies of the rich resources from southern Morocco to Sierra Leone), and the Indo-Pacific Fishery Commission (IPFC)—are steadily increasing their scientific activities.

Thus mechanisms exist for countries which share stocks to work together. The mechanisms are to an increasing extent being used. However, to produce information and advice for effective joint management of these shared stocks will probably take some time. The questions to be answered include more than the single 'state-of-exploitation' questions—essentially a knowledge of the present position of the fishery on a yield curve—required for management of open-access fisheries and many fisheries restricted to a single country. It will be necessary to know such things as the position of the main spawning, nursery, and feeding areas relative to the various national EEZs. Furthermore, it is necessary for this information to be known with sufficient confidence and to have sufficient agreement between scientists of different countries, for the information to be useful in negotiations between the countries in, for example, agreeing on the share in some total allowable catch. This will take time. What will almost certainly take even more time are the later negotiations on how to make the allocations—or, more generally, to achieve acceptance. Fish move, and hence a country cannot decide alone what happens to the fisheries in its EEZ, but will have to relinquish some part of its newly acquired authority to a degree of joint decisions with adjoining states. All this means that the management of shared stocks will not be achieved easily or quickly.

### (4) Future Trends in Catches

The discussions in the immediately preceding sections on techniques and objectives likely to be pursued in planning and managing future fisheries, and the changed jurisdiction that states now have over fisheries—together with the earlier review of the resources in the regions of the world—make it possible to present some predictions of how world fisheries will develop in the future. First, however, I would like to discuss some external factors, particularly those affecting the cost of fishing and the supply of fish, that could affect future fisheries.

### (a) General Aspects

The most obvious external factors which change the cost of fishing are abrupt changes in fuel prices, particularly such as the rises in 1972/73 and 1978/80. These rises have exerted impact throughout the world. Fisheries at all levels of technology have suffered. Among the high-technology fisheries there are many, especially the trawl fisheries, in which, even before the cost increases, fuel was a high proportion of total costs. Equally hard hit have been many fishermen in developing countries who have replaced or supplemented their traditional methods with the use of small inboard or outboard motors. Here the problem is not so much the absolute magnitude of additional fuel costs, as that from the national viewpoint fuel has, for many countries, to be paid for out of a limited supply of foreign currency. From the individual fisherman's viewpoint, fuel has to be paid for in cash from a small cash income. Spiralling fuel costs therefore are threatening many programmes of upgrading traditional fisheries.

Prediction of future developments must also take account of possible technological advances that could radically reduce the costs of imposing a given level of fishing mortality. It would be bold to deny that such an advance was impossible or improbable. However, the changes in fishing techniques which have brought about the great rise in world catch between 1950 and 1970 were adaptations and improvements in techniques that already existed. In a few cases—notably the change to purse-seining—the reduction in costs compared to some of the traditional methods (e.g. drift-netting for herring) has been spectacular. In many other cases, especially in long-range fishing, the costs have remained the same. Indeed, large factory trawlers, even in the days of moderate fuel costs, were generally more costly to operate than the smaller and more traditional ways of fishing. The larger vessels have been successful because they have allowed the exploitation of hitherto unused stocks rather than because of any economic efficiency. Looking ahead, big changes in catching techniques seem unlikely. The changes that do seem probable are shifts towards simpler methods that require less energy. This could well include a return to greater use of sail power, taking advantage of the modern technology developed for yachting. Also, intermediate technology is improving some of the catching techniques, e.g. long-lining which, while demanding less fuel than, say, trawling, demands a large crew and hence high labour costs. In summary, no big changes in the costs of fishing can be expected.

On the other side of the economic equation, the demand for fish can be expected to rise. First, although there are signs that the birth-rate in many developing countries is slowing down, nearly everywhere it is still considerably higher than death-rates, and the world population, and its need for food, will continue to increase well into the next century. Increases above the global average are likely in many areas—southeast Asia, parts of Africa—where fish is the most readily available and acceptable form of animal protein. A second cause for increased demand for fish is the rise in the average income per head. The income elasticity of demand for fish does vary between countries, but overall consumption of fish per head rises with increasing income, particularly for the more valuable species (e.g. shrimp). Studies made by FAO (ROBINSON, 1979) suggest that if the real prices of fish were to remain constant (i.e. discounting the impact of inflation) the demand for fish for direct human consumption would reach some 97 million tons (including freshwater fish) by the year 2000, compared with some 67 million tons during the base years for the study (1972–1974). Demand for meal and oil is less easy to predict, but if prices of fish-meal and soya-bean meal and other competing products do not change, it is unlikely to fall below the 1972 to 1974 level of some 18 million tons of raw fish.

The actual trends in total catch, and the level of landings by the end of the century, will depend on the extent to which the supply can be increased, and the assumption of constant prices maintained. This is discussed in detail in the remainder of this section. In summary it will be shown that on the one hand the supply of the more familiar types of fish cannot be increased to the extent required and hence prices will rise; on the other hand, the supply of the less familiar types of fish could be increased, but it is less sure that they can be supplied at a price that will be both high enough to cover the cost of catching and low enough to attract potential consumers. There will probably therefore not be more than a slow growth of total world catch (unless there is a break through in the catching or marketing of unfamiliar types, such as krill or mesopelagic fish). However, there will be greater pressure on stocks already fished. This will make the rational management of these stocks more urgent and, if successful, more rewarding.

### (b) Traditional Fisheries

In attempting to predict the development of the more traditional fisheries there is a steady increase in difficulty and in the danger of the prophet being trapped. Both progress from single-species fisheries on the more stable type of species (which include most demersal species as well as tuna and crustaceans), through the single-species fisheries on the less stable species (particularly herrings and their relatives) to problems of many interacting species.

For the more stable stocks, recruitment is fairly constant. If it varies, the variation is essentially random from year to year and it is independent of any changes in the abundance of the adult stock. The response of these stocks to fishing then corresponds to the yield-per-recruit curve which can for most stocks be calculated fairly easily, especially if the ages of individual fish can be determined from scales or otoliths. Many of the stocks concerned are valuable—being either abundant, such as numerous stocks of cod and other temperate demersal species, or fetching a high price, such as tuna, lobsters, or shrimp. In either case the importance of the fishing is likely to justify enough research to provide reliable predictions of the impact of different patterns of fishing. Equally, the value of the fishery can justify substantial efforts on the administrative and legal side to apply management measures, as well as attracting enough economic resources (labour and capital) into the fishery as to make such measures highly desirable.

Thus, in these fisheries we can expect the development of sophisticated management regimes. These regimes will be concerned less with the precise level of the biological yield (which will probably be a little below the MSY) as with the attainment of a good (but again probably not the maximum possible) economic return, and with social and political ends—particularly the questions of allocation, i.e. who should get the benefits and who should be allowed to participate in the fishery.

The measures that might be introduced to increase benefits, and to achieve the desired distribution of these benefits, are the subject of lively discussion within the various professions (ecologists, economists, lawyers, etc.) concerned with the subject, and also between these professions. These discussions are largely beyond the scope of the present chapter. Convenient entries to the various non-biological aspects, and to the relevant literature, may be found in CLARK (1976); a mathematical analysis of the interaction between economics and population dynamics in CHRISTY and SCOTT (1965), one of the earliest comprehensive reviews of political and economic aspects of fishery management; and ANDERSON (1977), a paper particularly regarding current ideas in the United States in the light of the new Law of the Sea, and the US Fishery Management and Conservation Act. There is fairly a general consensus in these discussions that optimum management, while defined in different ways, will require an exploitation rate rather below that giving MSY. This will allow costs of fishing to be reduced, and also tend to reduce year-to-year fluctuations in catches. It will further almost certainly require some form of limited entry, i.e. only certain people will be allowed to fish. This favoured group might be those prepared to pay a high licence fee (in which case most benefits from management will go to the national treasury, and ultimately, to the taxpayers and population in general). It might be the fishermen (and their sons) who would then be the main beneficiaries, or it might be some other select group. Often the amateur fishermen are sufficiently numerous and politically powerful to ensure the exclusion, or severe restriction, of the professional fishermen.

For the less stable stocks (e.g. herrings and anchovies) biological consideration will carry more weight. For these stocks the ecologist is not at present in a position to say that such-and-such a level of annual catch, or of fishing mortality, will have such-and-such an effect on the stock. It is therefore not possible to determine precisely what action should be taken to attain some economic or social objective. Rather than determining some unique 'best' policy, the fishery manager will be trying to balance different risks. The more obvious risk to the outside observer is that the catches will be too great, and that the stock will collapse. However, at least as obvious to people engaged in the day-to-day business of catching, processing, and marketing fish is the risk that unduly restrictive measures will mean the loss of fish (and income) that could have been collected.

In the past the short-term views have been dominant. So long as the scientists could warn only that a collapse was possible or likely, but not that a collapse was certain, there has always been the temptation to act on the assumption that the particular stock being discussed is one of those that will not collapse, or at least not collapse this year or next. This temptation will clearly remain, but with the growing experience of stocks where the assumption has proved wrong, fishery administrations are recognizing the need to apply more conservative measures, and fishing industries are more willing to accept the necessary controls. The basis of these controls may vary. It may be to keep the fishing mortality below some determined value or to maintain at least some minimum value of spawning stock. Nevertheless, the net effect will be much the same under all systems. Catches will be kept below the theoretical MSY, and very much below the peak catches taken in many stocks before they collapsed. How much less than the potential MSY will actually be harvested will depend on how accurate is the available scientific advice, how far safety and a consistent level of catches will be preferred to a high but risky level of catches, and on the degree to which the government is able and willing to impose policies which may be unpalatable in the short run. It will be more difficult to approach full utilization when the variation is great, and in particular to take proper advantage of good year classes when they come through the fishery.

While for these stocks most attention will need to be paid to the maintenance of biological productivity, economic objectives need not be forgotten. As long ago as the 1950s, controls on the number of boats and processing plants in pelagic fisheries off South Africa (including those off Namibia, controlled by the South African authorities), combined by controls on total catch, enabled considerable economic benefits to be gained by those with licences to operate plants or boats (GERTENBACH, 1973). This situation did not last, since the current Law of the Sea enabled companies not enjoying the benefits to get around the controls by using factory ships working on the same stocks just outside the 12-mile limit.

The cautious approach is particularly needed, and likely to be used when no other species with broadly similar characteristics occur in the region. The collapse of one stock and fishery on it is then unlikely to be followed by the increase of another species (partially or wholly competing with the first) to which the fishery can turn its attention. Thus, the virtual disappearance of the Atlanto-Scandian herring from the open-ocean area between Iceland and Norway has not been followed by the increase of any other species. The Norwegian purse-seine fishery for fish-meal has only been able to maintain itself by switching both area and species—to capelin in the Barents Sea.

Elsewhere, the decline of one pelagic species has often been accompanied by the increase of another. This has been particularly well marked around Japan, with rises and declines of the individual species such as herring, sardine, squid, and mackerel. Catches of individual species have varied as much as two orders of magnitude (e.g. MURPHY, 1977), while the total catches of all species has not varied much (NAGASAKI, 1973). In this situation, a very rigid and conservative approach looking strictly at the species of original interest can give strange results. Thus, off California, management authorities are extremely concerned by the disappearance of sardine, and an increase in anchovy, while authorities in Peru are almost equally concerned by the collapse of their anchovy stock and the appearance of large quantities of sardine. (The species are not exactly the same, but their ecological relationships are probably similar.) Of course, the story is not quite as simple as suggested. The increases and decreases do not match exactly in timing and magnitude, and there are considerable arguments about what happened in each area, as well as about the degree to which the recovery of the Californian sardine would be helped by heavy fishing on anchovy (MURPHY, 1966). Nevertheless, both general ecological considerations and the history of areas where two or more similar species have occurred strongly suggest that the trends in abundance and catches of such species should be closely, and inversely, related. The practical consequences are first that the relative variation of the total catch of all species should, with proper adjustment of the effort directed to each species, be much less than that of any individual species taken alone; second, that the potential catch of all species combined that can be sustained will be less than the sum of the potential catches from each individual species, based on the periods when that species was most abundant. For example, the history of the Peruvian anchovy between 1960 and 1971 suggested that carefully managing the stock could sustain annual yields of perhaps as much as 9 million tons (ANON., 1970, 1972). By 1980 the whole system had changed, and the current stocks might suggest catches of perhaps 1 to 2 million tons of sardines and other species. Though the interactions between anchovy, sardine, and other species is not known exactly, it seems clear that it would not be possible to harvest simultaneously 9 million tons of anchovy and 2 million tons of sardines and other species. Rather, there are alternative possibilities between a purely anchoveta fishery and one predominantly on other species. With the present state of knowledge it is difficult to predict what alternatives would be available in any future year, or how these might be altered by suitable policies in preceding years. Presumably, though, these policies will tend towards encouraging catching of less preferred species or those that are relatively abundant, and restricting catches on preferred species, at least when they are relatively scarce. This will require a more flexible fishery. The vessels should be able to catch both anchovy and the more active sardines and mackerels, and the industry should be able to take advantage of the potential high market prices for sardines, etc. (canned or frozen), as well as producing fish-meal from anchovy and such quantities of other species that cannot be marketed as higher priced products.

The need to look at the interactions applies to most areas and most fisheries. In the near future it is likely to become of greatest concern in the temperate waters around the developed countries of Europe and North America. In the tropical waters, the greater number of species make the problem too complicated for easy solution. Many of the developing countries in these regions lack the sophisticated administrations to apply more than the simpler approaches to management. Even in a region like the North Sea it has so far proved difficult to apply anything except simple species-by-species management. The fact that the fish species in the North Sea do interact has long been recognized. The ICES symposium that reviewed changes in the decade 1960 to 1970 (HEMPEL, 1978) showed that there had been many changes, only some of which could be explained by the direct effect of fishing on the particular species. Others were no doubt due to natural variation, but there remained a number of changes, including the so-called 'gadoid outburst'—the series of above-average year classes of cod, haddock, and whiting—which it is tempting to ascribe at least in part to fishery-induced changes in other stocks. For example, the reduction of herring and mackerel could have reduced the predation on larval gadoids.

The practical difficulty is that there is as yet no quantitative model which will provide generally accepted estimates of, for example, how much cod year classes will increase as a result of a given decrease in mackerel. A number of models, notably that developed by Danish scientists (ANDERSEN and URSIN, 1977), attempt to produce quantitative descriptions, but suffer from the problem of fitting a large number of parameters in a set of complex equations (POPE, 1979). More general reviews of the multi-species problem have been made (e.g. FAO, 1978; MAY and co-authors, 1979). These provide good descriptions of the nature of the problem, but only qualitative estimates of the results of the interactions.

Under past patterns of management and provision of biological advice to managers, management decisions were made by consensus. The result of any uncertainties in scientific advice (e.g. the presentation of a range of possible values of the Total Allowable Catch that should be taken in the next year) was that the least restrictive action (e.g. the largest value of TAC) was taken. Hence, it was very difficult to incorporate qualitative interactions between species into scientific advice.

This situation is likely to change. On the one hand, the scientific insight in the relations between species should improve. Even if the time when accurate quantitative predictions can be made is still some way in the future, qualitative statements should become more detailed and more reliable. On the other hand, fishery administrators now have, through the extension of limits, more direct control over how fishing is carried out. They have also learnt through such painful experience as the collapse of the North Sea herring population that management of fish stocks cannot wait until all scientific arguments are resolved. The combination of these two trends means that in the reasonably near future scientific advice will be sufficiently clear, and fishery administration sufficiently willing to act on advice that still admits to some range of uncertainty, for management actions to take considerable account of likely interspecific reactions. This will presumably mean that stocks of less desirable species that compete with, or prey on, more valuable species, will be deliberately 'over-fished', while more conservative policies will be directed at the more valuable species. The result should be a significant increase in the value of the total catch, though the weight caught may not increase, but may even decrease. Thus, GULLAND (1981) has estimated that, without any increase in the weight caught, the real value of the landings from the North Sea could be increased by perhaps as much as 25%

## (c) Unconventional Resources

The likely changes outlined above, though they should lead to considerable improvements in the economic and social benefits from fishing, will not lead to any dramatic increase in the total catch. If this is going to occur it must result from harvesting of the unconventional types of fish (or other animals) which at present do not appear in large quantities on the world's fish markets.

Of these species, the one that has attracted the most attention in recent years is the krill *Euphausia superba*. Krill are distributed throughout the Southern Ocean south of the Antarctic Convergence, with particularly dense concentrations in the Atlantic sector (MARR, 1962; EVERSON, 1977). At times they occur in dense swarms which, when on the surface, can be easily recognized from the deck of a ship as large red patches.

The interest in krill as a potential resource arose in part from the obvious signs of abundance, but particularly from an awareness that the depletion of the great whales must have caused changes in the ecosystem, and there was in some sense a 'surplus' of krill corresponding to the reduction in the consumption by whales. The number of whales has dropped by some two-thirds since the beginning of the century; because the decrease has been greatest among the biggest whales, consumption has dropped even more. LAWS (1977) estimated that the reduction was nearly 150 million tons (from 190 million to a little over 40 million tons). Reference to this 150 million tons—or other similar figures that can be obtained using different estimates of the change in whale numbers, and of consumption by the average whales—as a 'surplus' has caused some confusion. It paints a picture of some separate mass of millions of tons of krill drifting in the ocean untouched. In fact, as whale predation decreases so will the mortality rate of krill and in the short run at least the abundance will increase. This can be expected to lead to various changes in the system, including increased predation by other species. Changes in several of these have been observed. The age at maturity of minke whales as estimated from the zones in the ear plugs, has decreased, starting soon after the decrease in blue whales, and well before any significant exploitation of the minke whales themselves (LOCKYER, 1979). This is presumed to have led to a big numerical increase in these whales. There is direct evidence of increases in fur seal numbers (especially around South Georgia), and of some other seals and penguins (LAWS, 1977). Hence there is probably no 'surplus' in the sense of vast quantities of unconsumed krill. What the calculations of the changes in whale consumption do show is that the ecosystem could produce this amount of krill, over and above the amount consumed by non-whale predators. If, therefore, the consumption by whales could be replaced by an equivalent harvest by man, the rest of the ecosystem remaining unchanged, then the system as a whole could remain in its original balance. Here it may be noted that whale predation is like human predation, and unlike other predation, in that the movements of whales removes material from the system. At the end of the Antarctic summer the whales migrate out of the Southern Ocean into temperate and subtropical waters, and the difference in weight between the whales moving north, with thick layers of blubber, and the survivors, with little blubber, returning south in the Antarctic spring, is very large.

Other approaches have been made to estimating the potential harvest of krill. All of these (estimations of total primary production and of the proportion going into krill; estimates of krill abundance and dynamics from catches in plankton nets; and estimates of local abundance of krill from acoustic surveys and catches in trawl nets) suggest that the biomass and reproductive rate of krill are very large. The potential harvest should be correspondingly large. At this stage, the precise value is mostly irrelevant. In any case, it can probably be estimated with reasonable precision only after large-scale harvesting has begun. What matters is that even the lower estimates are of the same order of magnitude as the present world catch of all species, and that if krill harvesting were to become a practicable economic operation, the total catch could increase dramatically.

The possibility of this has caused widespread concern among environmental interests. In the past few years few matters have aroused so much public concern as the conservation of marine mammals in general, and of the large whales in particular. Add to this the feeling that the Antarctic and its surrounding waters is the last large part of the world where the natural ecosystem has not been disrupted and polluted by man, and it is clear how the possibility that large krill catches might seriously affect the whale stocks, especially their ability to recover from their currently depleted state, has attracted considerable attention.

Whales recover because there have been favourable changes in population parameters, including pregnancy rate and age at first maturity (LAWS, 1962). The chain of cause and effect is presumably: fewer whales → less krill eaten → higher density of krill → more krill eaten by each whale (or less energy required to find and eat a given quantity of krill) → more surplus energy for growth and reproduction. Clearly, krill harvesting would disrupt this chain in the middle. In the extreme, if harvesting were so great as to reduce krill to such a low density that the whales could not get enough to eat, it would lead to starvation and ultimate extinction. In principle, any krill-catching could lead to some reduction in krill abundance and affect the whale stocks. The problem is how big the catches could be before the effect is such as to significantly change whale population parameters and reduce the net rate of increase. This raises essentially the same problems as those involved in estimating the potential yield from the krill stock. In addition it is probable that it is not so much the overall abundance of krill that affects whales, but their local density. Since krill-catching may be done at the same concentrations that whales prefer, and may involve breaking up the swarms, the impact of a given weight of harvest may be greater than if the same weight were taken randomly from the whole krill stock.

Our knowledge of the many factors involved—the feeding behaviour of whales, the swarming behaviour of krill, the type of krill concentrations most favourable for man and whales, the effect of krill abundance and density on whale growth, etc.—is mostly too poor to give precise data on how much krill can be caught before the whale stocks will be affected. However, the quantity is very large—at least some tens of millions of tons if fishing is spread evenly through the Southern Ocean, and millions of tons if it concentrated in one area. It is also clear that as harvesting increases, the first effect, well before there is any threat to the continued existence of whales, will be a change in population parameters (maturity at a higher age, lower pregnancy rates) back towards the values occurring when the whale populations were unexploited, and their abundance around the carrying capacity of the environment. This will reduce the net increase rate of the whale stocks, slowing their recovery and the level at which the population will ultimately stabilize. At this point a choice must be made between: (i) harvesting more krill, and (ii) achieving a sustained harvest of whales, a fast recovery of whale stocks and, after the period of recovery, high whale abundance. Presumably there will be some intermediate pattern of exploitation which will represent an acceptable compromise between the interests of high krill catches and high whale abundance (though it is possible that a sustained whale harvest will not be part of such a compromise).

Whether such a compromise is reached and maintained will depend on two things: a mechanism for the countries concerned to come to a suitable agreement on the patterns of krill exploitation, and sufficient scientific information to provide a sound basis to such an agreement, so that it will in fact have the effects hoped for. In the case of the krill fishery, in contrast to what happened in most other large-scale exploitations of natural resources, some of the necessary action is being taken before the resources show signs of too much exploitation.

After several years of discussion the countries with interests in the Southern Ocean, including all those party to the 1949 Antarctic Treaty and those currently carrying out research into pilot-scale harvesting of krill, met in Canberra (Australia) in May 1980 and agreed on a Convention for the Conservation of Antarctic Marine Living Resources. This convention, under which a commission will be set up in due course, manages to bring together the different interests of the conservation movements (as represented particularly from the US point of view in the negotiations), and of those interested in krill as a harvestable resource (particularly Japan and USSR). The successful signing of this convention in the face of many complicated political problems, as well as these direct differences of interest (MITCHELL and SANDBROOK, 1980), represents a big step forward, at least in intention, in the utilization of natural resources, and in taking account of the interactions between the different components of the ecosystems. How well these intentions will work out in practice depends on the quality of the scientific advice. Ultimately, the commission will have its own scientific committee, but in the immediate future it will be able to take advantage of the extensive work being carried out under the BIOMASS Programme. This Programme (EL-SAYED, 1977) is aimed at promoting and co-ordinating increased scientific research into the structure of the whole Southern Ocean. Because of the role of krill as a major consumer of phytoplankton, and the main prey of many of the larger vertebrates and invertebrates in the Southern Ocean, studies on krill play a central role in BIOMASS. A large part of FIBEX (First International BIOMASS Experiment) carried out in the Antarctic summer of 1980/81 will consist of a multi-ship survey of krill. Though the early stages of the BIOMASS work were successful mainly in showing how little was known about krill, the programme, as it progresses, will give much more knowledge of the Antarctic ecosystems. It should therefore provide a sound scientific basis for the Commission to take decisions as they become necessary.

If krill fishing does develop it may be directed to one or other of two distinct markets: a high-priced product for direct human consumption or a high-volume product for animal feeding. In either case the first requirement is a suitable catching method, and the earlier stages in the applied research into krill harvesting were concerned with finding a good catching method (EDDIE, 1977). For this, modification of the mid-water trawl used for herring has been found suitable. With a smaller mesh, and directed onto the swarms by echo sounders, catch rates of 10 tons of more in a short (half an hour or so) haul can be regularly obtained.

This is a satisfactory rate for a relatively high-priced product, but the problems start once the krill reach the deck. Compared with most fishery products, krill remain fresh only for a very short period, even in the cold of the Antarctic. If krill is to be used for human consumption it must be processed within at the most 4 h of being caught. Various forms of processing have been examined—boiled and frozen whole krill, peeled tails, and a form of mince which could be used in making surimi (a type of Japanese minced fish used in sausages, etc.) (GRANTHAM, 1977). The markets for these are not large. With good processing the tail meat has a fresh pink colour and an attractive shrimp flavour, but individual tails, after processing, are only about the size of a grain of rice. Krill tails would presumably serve much the same market as small shrimp but, because of their small size, at a lower price. The total world production of all sizes and species of shrimp is now no more than 1·5 million tons, and even if the price were low, it is difficult to see how the market for krill tails could in the near future rise to as much as that. The total market for minced fish is larger—the Japanese, the largest user, consume the equivalent of 3 million tons of fresh fish—but probably it could be harder and take longer for this market to adjust to the different characteristics of minced krill, and to absorb large quantities. Together with the small quantities (around a few tens of thousands of tons) which could be absorbed by the specialized Japanese market for whole boiled krill, it would seem unlikely that the markets for direct human consumption could take more than at most 1 or 2 million tons before the end of this century.

The big potential market is that for meal utilized for animal feeding. At present some 4 million tons of fish-meal (equivalent to 20 million tons of fresh fish) are used for feeding chickens, beef cattle, and other animals. The total amount of high-protein meal used by the animal-feedstuffs industry to supplement cereals is considerably larger than this. More fish-meal could be used, but the industry is very sensitive to price, and will only switch from the biggest current supply of protein supplement—soya meal—if the new source were cheaper than soya beans. Krill meal might find, within the general animal food market, a special role in intensive aquaculture of high-valued fish—salmon, trout, crustaceans, etc. The growth of this type of aquaculture is held back by the high costs, mainly due to the lack of good cheap food.

The key to all those developments is costs. Especially for meal, if the product can be delivered to the main markets at the right price, a substantial fishery will grow. If not, harvesting will at best continue on a pilot-scale only. The short-term indications are bad. Energy costs have been spiralling, and krill harvesting is bound to be energy intensive in the actual catching, and in transporting the catch to the markets. Except possibly in the USSR, for which the normal economic criteria do not apply, it is difficult to see large-scale krill harvesting occurring in the near future (cf. MITCHELL and SANDBROOK, 1980).

The same is true of mesopelagic fish, i.e. the large number of small-sized species, particularly myctophids and gonostomatids, which occur throughout the middle and upper layers of the open oceans (GJØSAETER and KAWAGUCHI, 1980) These fish are particularly abundant in the north Arabian Sea, in the arc from northern Somalia through Oman to Pakistan. Acoustic surveys by the Norwegian research vessel *Dr. Fridtjof Nansen* yielded high estimates of standing stock of up to 100 million tons. Since these are short-lived species, probably living for only 2 or 3 yr (though the life-span is not known accurately for any of the species), the sustainable annual yield is likely to be a significant proportion of this biomass. Including the stocks in other oceans, the potential production of these fish could easily be as high as 100 million tons. The fish are, however, very small, with only a few species reaching 10 cm, and many being much smaller. They are not attractive for direct human consumption and any fishery in the foreseeable future will have to be for fish-meal. The work of *Dr. Fridtjof Nansen* has shown that large catches can occasionally be made with mid-water trawls in the Arabian Sea, which appears to be the region where development of a fishery is most likely. With better experience of the behaviour and distribution of the fish there is little doubt that a well-equipped vessel could maintain in the favourable season a high daily catch rate.

Whether this catch rate would be economically attractive is another matter. The region lies in the monsoon zone, and there are pronounced seasonal changes in oceanographic conditions. There also seem to be exceptionally large year-to-year changes. Thus, it may only be possible to maintain good catch rates for part of the year, and then perhaps only in two years out of three. The major concentrations are often far from land, which in any case is poorly supplied with ports and shore facilities. The fishery is, therefore, likely to require considerable time steaming to and fro between port and fishing grounds, and in general to face relatively high costs. As already noted for krill, high costs are death to a potential fish-meal fishery. Though the costs are less unfavourable than for a fishery for krill-meal, a fishery for mesopelagic fish seems unlikely in the near future, unless there is some unexpected event, e.g. access to fuel supplies well below the normal world price. On the other hand, if economic conditions are favourable, a very large fishery could develop very quickly. If meal from mesopelagic fish could be produced at a price competitive with soya-bean meal, the market would be very large. The fishery could develop to 10 millions tons or more before finding sufficient markets became a problem.

*Acknowledgements.* Attempting a comprehensive review of this type would be impossible without the assistance of a large number of people with personal knowledge, published and unpublished, of individual areas. I would particularly like to express my gratitude to my colleagues in the Department of Fisheries of FAO for their assistance, in general, and in relation to particular areas.

Special thanks are due to JOHN CADDY (North Atlantic Ocean and Caribbean Sea), SHIRO CHIKUNI (North Pacific Ocean), FRANCIS CHRISTY (USA), JORGE, CSIRKE (both coasts of South America), SERGE GARCIA (West Africa), HIROSHI KASAHARA (North Pacific Ocean), ARMIN LINDQUIST (North Sea and Baltic Sea), GARY SHARP (tunas), JEAN-PAUL TROADEC (West Africa and Mediterranean Sea), and SIEBREN VENEMA (Indian Ocean). My particular gratitude is due to JUNE WERRY-LANDONE and JACKIE ELLIS for their patience and ability to decipher my writing.

## Literature Cited (Chapter 5)

ADASIAK, A. (1979). Alaska's experience with limited entry. *J. Fish. Res. Can.*, **36** (7), 770–782.

AHLSTROM, E. H. (1966). Distribution and abundance of sardine and anchovy larvae in the California Current region off California and Baja California, 1951 to 1964: a summary. *Spec. scient. Rep. U.S. Fish Wildl. Serv.*, **534**, 1–71.

AHLSTROM, E. H. and RADOVICH, J. (1970). Management of the Pacific sardine. A century of fisheries in North America. *Spec. Publs Am. Fish. Soc.*, **7**, 183–193.

ALLLEN, K. R. (1980). *Conservation and Management of Whales.* University of Washington Press, Seattle.

ALLEN, K. R. and CHAPMAN, D. G. (1977). Whales. In J. A. Gulland (Ed.), *Fish Population Dynamics.* Wiley, Chichester. pp. 335–358.

ALVERSON, D. L., PRUTER, A. T. and RONHOLT, L. L. (1964). *A Study of Demersal Fishes and Fisheries of the Northeast Pacific Ocean.* University of British Columbia, Vancouver.

ALWARD, G. L. (1911). *The Development of British Fisheries.* Grimsby News, Grimsby.

ALWARD, G. L. (1932). *The Sea Fisheries of Great Britain and Ireland*. Albert Gait, Grimsby.

ANDERSEN, K. P. and URSIN, E. (1977). A multispecies extension to the Beverton and Holt theory of fishing, with accounts of phosphorus circulation and primary production. *Meddr Danm. Fisk.-og Havunders. N.S.*, **7**, 319–435.

ANDERSON, L. G. (Ed.) (1977). *Economic Impacts of Extended Fisheries Jurisdiction.* Ann Arbor Science, Ann Arbor, Michigan, USA.

ANONYMOUS (1970). Report of the panel of experts on population dynamics of Peruvian anchoveta. *Bol. Inst. Mar Peru*, **2**, (6), 324–372.

ANONYMOUS (1972). Report of the second session of the panel of experts on population dynamics of Peruvian anchoveta. *Bol. Inst. Mar Peru*, **2**, (7), 373–458.

ANONYMOUS (1974). Informe del grupo de trabajo conjunto CAIRM/CARPAS sobre la evaluacion cientifica del estado de los stocks en el Atlantico sudoccidental. *FAO, Rome Doc.*, **CARPAS/6/74/4.**

ANONYMOUS (1977). Stock assessment of shrimp in the Indian Ocean area. Report of the ad hoc group of the IOFC special working party, Doha, Qatar. *FAO Fish. Rep.*, **193**, 1–23.

ANONYMOUS (1978a). Report of the FAO/Norway Workshop on the fishery resources of the north Arabian Sea, Rome. *Indian Ocean Programme Doc.*, **IOFC/DEV/78/43**, 1–57.

ANONYMOUS (1978b). Report of the joint meeting of the WECAF working party on assessment of fish resources, and the working party on stock assessment of shrimp and lobster resources. *FAO Fish. Rep.*, **211**, 1–132.

ANONYMOUS (1978c). Summary information on Australian resources. *Indo-Pacific Fish. Commn, Standing Committee on Resources Research and Development.* **Doc. IPFC : RRD/78/15**, 1–5.

ANONYMOUS (1978d). Proceedings of the pelagic fisheries conference, July 1977. *Occ. Publs Fish. Res. Div.*, **15**, 1–10.

ANONYMOUS (1979). Rapport du groupe de travail spécial sur l'evaluation des stocks démersal du secteur Côte d'Ivoire—Zaire. *FAO Rome Doc., COPACE/PACE/Series*, **79/14**.

ANONYMOUS (1980). State of selected stocks of tuna and billfish in the Pacific and Indian Oceans. *FAO Fish. Tech. Pap.*, **200**, 1–88.

ANONYMOUS (1981a). Review of the state of world fishery statistics. *FAO Fish. Circ.*, **710** (Rev. 2), 1–52.

ANONYMOUS (1981b). Demersal resources of the Gulf and the Gulf of Oman. *FAO*, *Rome Doc.*, **FI : DP/RAB/71/278/10**.

ANONYMOUS (1981c). Pelagic resources of the Gulf and the Gulf of Oman. *FAO, Rome Doc.*, **FI : DP/RAB/71/278/11**, 1–144.

ANONYMOUS (in press). *Report of the ICLARM/CSIRO Workshop on the Theory and Management of Multispecies Stocks, 12–23 January 1981.* Cronulla, Australia.

ANSA-EMMIN, M. (1979). Occurrence of trigger fish, *Balistes capriscus* (Gmel) on the continental shelf of Ghana. *FAO Rome Doc., COPACE/PACE/Series*, **79/4**, 20–36.

AOYAMA, T. (1973). *The Demersal Fish Stocks and Fisheries of the South China Sea*, (SCS/DEV/73/3), South China Sea Programme, Rome.

ASADA, Y. (1973). Licence limitation: the Japanese system. *J. Fish. Res. Bd Can.*, **30** (12), 2085–2095.

ATKINSON, G. T. (1908). Notes on a fishery voyage to the Barents Sea in August 1907. *J. mar. biol. Ass. U.K.*, **8**, 71–98.

BANERJI, S. K. (1973). An assessment of the exploited pelagic fisheries of the exploited fisheries of the Indian Seas. In *Proceedings of the Symposium on Living Resources of the Seas around India, 1968 (Cochin)*. Central Marine Fisheries Research Institute, Cochin (Special Publication). pp. 114–136.

BARANOV, F. I. (1918). On the question of the biological basis of fisheries. *Nauch. Issled. Ikhtiol. Inst. Izv.*, **1** (1), 81–128.

BAYAGBONA, E. O. (1965). Population dynamics: sampling the Lagos trawler croaker landings. Overfishing in Lagos: proposed cure. *Res. Rep. Fed. Fish. Serv. Nigeria*, **2**, 8–32.

BECK, V. (1974). *Bestandskundliche Untersuchungen an einigen Fischarten der Grundschleppnetz-Fischerei auf Lem Schelf rov Togo (West Africa), Dipl. Arb. Fachber. Biologie*, Universität Hamburg.

BELVEZE, H. and BRAVO DE LAGUNA, J. (1980). Les ressources halieutiques de l'Atlantique centre-est. Deuxième Partie: Les ressources de la côte ouest-africaine entre 24° N et le détroit de Gibraltar. *FAO Fish. Tech. Pap.*, (**186.2**), 1–64.

BEN-TUVIA, A. (1966). Red Sea fishes recently found in the Mediterranean. *Copeia*, (**2**), 254–275.

BEVERTON, R. J. H. (1962). Long term dynamics of certain North Sea fish populations. In L. Cren and M. W. Holdgate (Eds), *The Exploitation of Natural Animal Populations.* Blackwells, Oxford.

BEVERTON, R. J. H. and HOLT, S. J. (1957). On the dynamics of exploited fish populations. *Fishery Invest., Lond. (Series 2)*, **19**, 1–533.

BLACKBURN, M. (1960). Synopsis of biological information on the Australian and New Zealand sardine *Sardinops neopilchardus* (Steindachner). In H. Rosa and G. Murphy (Eds), *Proceedings of the World Scientific Meeting on the Biology of Sardines and Related Species.* FAO Rome. pp. 245–264.

BOELY, T. and FREON, P. (1979). Les ressources pelagique cotieres. *FAO Fish. Tech. Pap.*, (**186.1**), 13–78.

BOEREMA, L. K. (1969). The shrimp resources in the Gulf between Iran and the Arabian peninsula. *FAO Fish. Circ.*, **310**, 1–29.

BOEREMA, L. K. and OBARRIO, J. L. (1962). The case for regulations in the Panama shrimp fishery. In R. Hamlisch (Ed.), *The Economic Effects of Fishery Regulations* (FAO Fisheries Report, 5). FAO, Rome. pp. 539–548.

BOJE, R. and TOMCZAK, M. (1980). *Upwelling Ecosystems.* Springer-Verlag, Berlin.

BOONPRAKOB, V. (1976). Proposed regulatory measures for the mackerel (*Rastrelliger neglectus* (van Kampen) stocks in the Gulf of Thailand. In K. Tiews (Ed.), *Fisheries Resources and their Management in Southeast Asia.* German Foundation for International Development, Berlin. pp. 178–197.

BOONYUBOL, M. and HONGSKUL, V. (1978). Demersal fish resources and exploitation in the Gulf of Thailand 1960–1975. In *Report of the Workshop on the Demersal Resources of the Sunda Shelf* (SCS/GEN/77/13). South China Sea Programme, Manila. pp. 56–70.

BOWEN, B. K. and CHITTLEBOROUGH, R. G. (1966). Preliminary assessments of stocks of the western Australian crayfish *Panulirus cygnus* George. *Aust. J. mar. Freshwat. Res.*, **9**, 537–545.

BRANDER, K. (1981). Disappearance of the common skate (*Raja batis*) from the Irish Sea. *Nature, Lond.*, **290** (5801), 48–49.

VON BRANDT, A. (1964). *Fish Catching Methods of the World* (Revised edition 1972). Fishing News (Books) Ltd, London.

VON BRANDT, A. (1972). *Fish Catching Methods of the World* (Revised and enlarged edition). Fishing News (Books) Ltd, London.

BROWN, B. E., BRENNAN, J. A., GROSSLEIN, M. D. and HENNEMUTH, R. C. (1976). The effect of fishery on the marine finfish biomass in the northwest Atlantic from the Gulf of Maine to Cape Hatteras. *ICNAF Res. Bull.*, **12**, 49–68.

BULLIS, H. R. and FLOYD, H. (1972). Double-rig twin shrimp-trawling gear used in the Gulf of Mexico. *Mar. Fish. Rev.*, **34**, 11–12, 26–31.

BURD, A. C. (1978). Long-term changes in North Sea herring stocks. *Rapp. P.-v. Réun. Cons. int. Explor. Mer,* **172**, 137–153.

BUTTERWORTH, D. S. (1981). The value of catch-statistics-based management techniques for heavily fished pelagic stocks, with special reference to the recent decline of the southwest African pilchard stock. In K. B. Haley (Ed.), *Applied Operations Research in Fishery*. Plenum Press, New York. pp. 441–464.

CECAF (Fishery Committee for the Eastern Central Atlantic) (1979). Rapport du groupe de travail ad hoc sur les poissons pélagiques côtiers ouest-africains de la Mauritanie au Libéria (26° N à 5° N). *FAO, Rome Doc., COPACE/PACE/Series*, **78/10**, 1–165.

CECAF (Fishery Committee for the Eastern Central Atlantic) (1979b). Rapport du groupe de travail ad hoc sur les stocks côtiers démersaux vivant entre le sud de la Mauritanie et le Libéria. *FAO, Rome Doc., COPACE/PACE/Series*, **78/8**, 1–99.

CECAF (Fishery Committee for the Eastern Central Atlantic) (1979c). Rapport du groupe de travail ad hoc sur l'évaluation des stocks de céphalopodes. *FAO, Rome Doc., COPACE/PACE/Series*, **78/11**, 1–135.

CHAPMAN, D. G. (1964). Report of the Committee of Three Scientists on the Special Scientific Investigations of the Antarctic Whale Stocks. *Rep. int. Whal. Commn*, **14**, 32–106.

CHAPMAN, D. G., MYRHE, R. J. and SOUTHWARD, G. M. (1962). Utilization of Pacific halibut stocks: estimation of maximum sustainable yield. *Rep. int. Pacif. Halibut Commn*, **31**, 1–35

CHOMJURAI, W. and BUNNAG, R. (1970). Preliminary tagging studies of demersal fish in the Gulf of Thailand. In J. C. Marr (Ed.), *The Kuroshio. A Symposium on the Japanese Current.* East–West Center Press, Honolulu. pp. 517–524.

CHRISTY, F. T. (1979). Economic benefits and arrangements with foreign countries in the northern sub-region of CECAF: a preliminary assessment. *CECAF/ECAF/Series*, **79/19**.

CHRISTY, F. T., Jr and SCOTT, A. (1965). *The Common Wealth in Ocean Fisheries. Some Problems of Growth and Economic Allocation*. Johns Hopkins Press, Baltimore, Md.

CLARK, C. W. (1976). *Mathematical Bioeconomics: The Optimal Management of Renewable Resources*. John Wiley, New York.

CLARK, S. H. and BROWN, B. E. (1977). Changes in biomass of finfishes and squids from the Gulf of Maine to Cape Hatteras 1963–1974, as determined from research vessel data. *Fish. Bull. U.S.*, **75** (1), 1–21.

COPES, P. (1981). Fisheries on Canada's Pacific coast: the impact of extended jurisdiction on exploitation patterns. *Ocean Managem.*, **6** (4), 279–298.

CROSSLAND, J. and GRANDPERRIN, R. (1980). The development of deep bottom fishing in the tropical Pacific. *Occ. Pap. S. Pacif. Commn,* **17**, 1–12.

CRUTCHFIELD, J. and PONTECORVO, G. (1969). *The Pacific Salmon Fisheries: A Study of Irrational Conservation*. Johns Hopkins Press, Baltimore, Md.

CRUTCHFIELD, J. and ZELLNER, A. (1962). Economic aspects of the Pacific halibut fishery. *Fish Wildl. Serv. U.S., Fish. Ind. Res.*, **1** (1), 1–173.

CSIRKE, J. (1980). Recruitment in the Peruvian anchoveta and its dependence on the adult population. *Rapp. P.-v. Réun. Cons. int. Explor. Mer*, **177**, 307–313.

CUSHING, D. H. (1966a). *The Arctic Cod.* Pergamon Press, Oxford.

CUSHING, D. H. (1969a). Upwelling and fish production. *FAO Fish. Tech. Pap.*, **84**, 1–40.

CUSHING, D. H. (1969b). The regularity of the spawning season of some fishes. *J. Cons. int. Explor. Mer.*, **33** (1), 81–92.

CUSHING, D. H. (1977). The problems of stock and recruitment. In J. A. Gulland (Ed.), *Fish Population Dynamics.* Wiley, Chichester. pp. 116–133.

CUSHING, D. H. and DICKSON, R. R. (1976). The biological response in the sea to climatic changes. *Adv. mar. Biol.*, **14**, 1–122.

CUSHING, D. H. and HARRIS, J. G. K. (1973). Stock and recruitment and the problem of density-dependence. *Rapp. P.-v. Réun. Cons. int. Explor. Mer*, **164**, 142–155.

DAY, F. (1868). *The Fishes of India*, Vol. 2. Taylor and Francis, London.

DIETRICH, G. (1972). Upwelling in the ocean and its consequences. *Geoforum*, **11**, 3–71.

DOMAIN, F. (1979). Les ressources demersales (Poissons). *FAO Fish. Tech. Pap.*, (**186.1**), 79–122.

DRAGESUND, O. (1970). Factors influencing year class strength of Norwegian spring spawning herring. *FiskDir. Skr (Series Havunders.)*, **15**, 381–450.

EDDIE, G. C. (1977). The technology of the harvesting of Antarctic krill. *FAO/UNDP Southern Ocean Fisheries Survey Programme Report*, **3**.

EGGLESTON, D. and WAUGH, G. D. (1974). New Zealand: pelagic fisheries: their potential for development. *Proc. Indo. Pacific Fish. Counc. 15th Session*, **Sect III**, 27–37.

EL-SAYED, S. Z. (1970). On the productivity of the Southern Ocean. In M. W. Holdgate (Ed.), *Antarctic Ecology*. Academic Press, New York. pp. 119–135.

EL-SAYED, S. Z. (Ed.) (1977). *Biological Investigations of Marine Antarctic Systems and Stocks (BIOMASS)*. Scientific Council for Antarctic Research, Scientific Council for Oceanic Research, London.

ENGLISH, T. S. (1964). A theoretical model for estimating the abundance of planktonic fish eggs. *Rapp. P.-v. Réun. Cons. int. Explor. Mer*, **155**, 174–181.

EVERSON, I. (1977). *The Living Resources of the Southern Ocean* (GLO/SO/77/1). UNDP/FAO, Rome.

FAIRBRIDGE, W. S. (1952). The New South Wales tiger flathead (*Neoplatycephalus macrodon* Ogilvy). 2. The age composition of the commercial catch, over-fishing of the stocks and suggested conservation. *Aust. J. mar. Freshwat. Res.*, **3** (2), 209–219.

FOERSTER, R. E. (1968). The sockeye salmon, *Oncorhynchus nerka*. *Bull. Fish. Res. Bd Can.*, **162**, 1–422.

FONSELIUS, S. (1962). Hydrography of the Baltic deep basins. *Rep. Fishery Bd Swed. (Series Hydrogr.)*, **13**, 1–41.

FONTANA, A. (1979). Niveau d'exploitation des stock demersales du plateau continental congolais. *FAO, Rome Doc., COPACE/OPACE/Series*, **79/14**, 50–51.

FAO (Food and Agricultural Organization) (1976a). *Report of the FAO Technical Conference on Aquaculture*, 26 May–2 June 1976. Kyoto, Japan.

FAO (Food and Agriculture Organization) (1976b). Monitoring of fish stock abundance: the use of catch and effort data. A report of the ACMRR Working Party on Fishing Effort and Monitoring of Fish Stock Abundance, Rome, Italy, 16–20 December 1975. *FAO Fish. Tech. Pap.*, (**155**), 1–101.

FAO (Food and Agriculture Organization) (1978a). Some scientific problems of multispecies fisheries. Report of the Expert Consultation on Management of Multispecies Fisheries, Rome, Italy, 20–23 September 1977. *FAO Fish. Tech. Pap.*, (**181**), 1–42.

FAO(Food and Agriculture Organization) (1978b). *Mammals in the Sea*, Vol. 1. Report of the FAO Advisory Committee on Marine Resources Research Working Party on Marine Mammals, FAO, Rome.

FAO (Food and Agriculture Organization) (1980). ACMRR Working Party on the Scientific Basis of Determining Management Measures, Report of the ACMRR Working Party on the scientific basis of determining management measures, Hong Kong, 10–15 December 1979. *FAO Fish. Rep.*, (**236**), 1–149.

FAO (Food and Agriculture Organization) (1981a). Review of the state of world fishery resources. *FAO Fish. Circ.*, **710**, (Rev. 2), 1–52.

FAO (Food and Agriculture Organization) (1981b). *Atlas of the Living Resources of the Seas*. FAO, Rome.

FAO (Food and Agriculture Organization) (1981c). *Yearbook of Fishery Statistics. Catches and Landings for 1979*, Vol. 48. FAO, Rome.

FAO (Food and Agriculture Organization) (1983). *Yearbook of Fishery Statistics. Catches and Landings for 1981*, Vol. 52. FAO, Rome.

FRAZER, G. A. (1979). Limited entry: experience of the British Columbia salmon fishery. *J. Fish. Res. Bd Can.*, **36** (7), 754–763.

FUKUDA, Y. (1976). A note on yield allocation in multispecies fisheries. *ICNAF Res. Bull.*, (**12**), 83–87.

GAMBELL, R. (1976). Population biology and management of whales. In T. H. Croaker (Ed.), *Applied Biology*, Vol. 1. Academic Press, London. pp. 247–336.

GARCIA, S. (1981). Cycles vitaux, dynamique, exploitation et aménagement des stocks de crevettes Penaéides cotières. *FAO Fish. Tech. Pap.*, (**203**), 1–210.

GARCIA, S. and LHOMME, F. (1979). Les ressources de crevette rose (*Penaeus duorarum notialis*). *FAO Fish. Tech. Rep.*, (**186.1**), 123–148.

GARROD, D. J. (1961). The history of the fishing industry of Lake Victoria, East Africa, in relation to expansion of marketing facilities. *E. Afr. agric. For. J.*, **27**, 95–99.

GARROD, D. J. (1977). The North Atlantic cod. In J. A. Gulland (Ed.), *Fish Population Dynamics*. Wiley, London. pp. 216–242.

GERTENBACH, L. P. D. (1963). Regulation of the South African west coast shoal fisheries. *FAO Fish. Rep.*, **5**, 425–458.

GERTENBACH, L. P. D. (1973). Licence limitation: the South African system. *J. Fish. Res. Bd Can.*, **30** (12), 2077–2084.

GFCM (General Fishery Council for the Mediterranean) (1978). Report of the Seventh Session of the Working Party on resource appraisal and fishery statistics. *FAO Fish. Rep.*, **204**, 1–141.

GFCM (General Fishery Council for the Mediterranean) (1979). Report of the technical consultation on the assessment and management of the Black Sea turbot. *FAO Fish. Rep.*, **226**, 1–19.

GFCM (General Fishery Council for the Mediterranean) (1980). Rapport de la Consultation technique pour l'evaluation des stocks dans les divisions statistiques Baléares et Golf de Lion. *FAO Fish. Rep.*, **227**, 1–155.

GJØSAETER, J. and KAWAGUCHI, K. (1980). A review of the world resources of mesopelagic fish. *FAO Fish. Tech. Pap.*, (**193**), 1–151.

GLANTZ, M. G. and THOMPSON, J. D. (1981). Consideration of the societal value of an El Niño forecast and the 1972–1973 El Niño. In M. G. Glantz and J. D. Thompson (Eds), *Resource Management and Environmental Uncertainty*. Wiley, New York. pp. 449–476.

GORDON, H. S. (1954). Economic theory of a common-property resource: the fishery. *J. Polit. Econ.*, **62**, 124–142.

GORDON, J. D. M. and De SILVA, S. S. (1980). The fish populations of the West of Scotland shelf. Part 1. *Oceanogr. Mar. Biol. A. Rev.*, **18**, 317–366.

GRAHAM, M. (1943). *The Fish Gate*. Faber and Faber, London.

GRANTHAM, G. J. (1977). *The Utilization of Krill* (GLO/SO/77/3). UNDP/FAO, Rome.

GREEN, K. (1977). Role of krill in the Antarctic ecosystem. In *Final Draft Environment Impact Statement on the Living Resources Convention*. Department of State, Washington.

GRIFFIN, W. L., WARREN, J. P. and GRANT, W. E. (1979). A bio-economic model for fish stock management: the cephalopod fishery of northwest Africa. *CECAF Tech. Pap.*, **79/16**, 1–43.

GULLAND, J. A. (1961). Fishing and the stocks of fish at Iceland. *Fishery Invest., London.*, **23** (4), 1–52.

GULLAND, J. A. (1965). Estimation of mortality rates. Annex to Arctic Fisheries Working Group Report. *Ann. Meet. int. Coun. Explor. Sea (A.M.-I.C.E.S.)*.

GULLAND, J. A. (1968a). Recent changes in the North Sea plaice fishery. *J. Cons. int. Explor. Mer*, **31** (3), 305–322.

GULLAND, J. A. (1968b). Concept of maximum sustainable yield and fishery management. *FAO Fish. Tech. Pap.*, (**70**), 1–13.

GULLAND, J. A. (1971). Ecological aspects of fishery research. In J. B. Cragg (Ed.), *Advances in Ecological Research*. Academic Press, London. pp. 115–176.

GULLAND, J. A. (1973). Distant water fisheries and their relation to development and management. *J. Fish. Res. Bd Can.*, **30** (12), 2456–2462.

GULLAND, J. A. (1975). The harvest of the sea: potential and performance. In W. W. Murdoch (Ed.), *Environment, Resources, Pollution and Society*. Sinauer and Associates, Sunderland, Mass. pp. 167–189.

GULLAND, J. A. (1977). The analysis of data and development of models. In J. A. Gulland (Ed.), *Fish Population Dynamics*. Wiley, London. pp. 67–95.

GULLAND, J. A. (1978). Fishery management: new strategies for new conditions. *Trans. Am. Fish. Soc.*, **107** (1), 1–11.

GULLAND, J. A. (Ed.) (1979a). Report of the FAO/IOP Workshop on the Fishery Resources of the Western Indian Ocean south of the Equator, Rome. *Indian Ocean Programme*, **Doc/IOFC/DEV/79/45**, 1–99.

GULLAND, J. A. (1979b). Developing countries and the new Law of the Sea. *Oceanus*, **22** (1), 36–42.

GULLAND, J. A. (1979c). The new ocean regime: winners and losers. *Ceres, Rome*, **70**, 19–23.

GULLAND, J. A. (1980). Some problems of the management of shared stocks. *FAO Fish. Tech. Pap.*, (**206**), 1–22.

GULLAND, J. A. (1981). Long-term potential effects from management of the fish resources of the North Atlantic. *J. Cons. int. Explor. Mer*, **41**, 8–16.

GULLAND, J. A. and BOEREMA, L. K. (1973). Scientific advice on catch levels. *Fish. Bull. U.S.*, **1** (2), 325–335.

GULLAND, J. A., TROADEC, J.-P. and BAYAGBONA, E. O. (1973). Management and development of fisheries in the Eastern Central Atlantic. *J. Fish. Res. Bd Can.*, **30** (12), 2264–2275.

GULLAND, J. A. and WILLIAMSON, G. R. (1962). Transatlantic journey of a tagged cod. *Nature, Lond.*, **195** (4844), 921.

HALSTEAD, B. W. (1959). *Dangerous Marine Animals.* Cornell Maritime Press, Cambridge, Mass.

HAMRE, J. (1978). The effect of recent changes in the North Sea mackerel fishery on stock and yield. *Rapp. P.-v. Réun. Cons. int. Explor. Mer*, **172**, 197–210.

HARDIN, G. (1968). The tragedy of the commons. *Science, Wash.*, **162**, 1243–1948.

HARRIS, J. G. K. (1975). The effect of density-dependent mortality on the shape of the stock and recruitment curve. *J. Cons. int. Explor. Mer*, **36** (2), 150–157.

HEMPEL, G. (Ed.) (1978). North Sea fish stocks—recent changes and their causes. A Symposium held in Aarhus, 9–12 July 1975. *Rapp. P.-v. Cons. int. Explor. Mer, (***172**), 1–449.

HITZ, C. R. and RATHJEN, W. F. (1965). Bottom trawling surveys of the northeastern Gulf of Alaska (Summer and fall of 1961, and spring of 1962). *Comml Fish. Rev.*, **27** (9), 1–15.

HJORT, J. (1914). Fluctuations in the great fisheries of northern Europe, viewed in the light of biological research. *Rapp. P.-v. Cons. int. Explor. Mer,* **20**, 1–228.

HJORT, J., JAHN, G. and OTTESTAD, P. (1933). The optimum catch. *Hvalråd. Skr.*, **7**, 92–127.

HÖGLUND, H. (1978). Long-term variations in the Swedish herring fishery off Bohustan and their relation to North Sea herring. *Rapp. P.-v. Réun. Cons. int. Explor. Mer*, **172**, 175–186.

HOLDEN, M. J. (1978). Long-term changes in landings of fish from the North Sea. *Rapp. P.-v. Réun. Cons. int. Explor. Mer*, **172**, 11–26.

HOLT, S. J. and TALBOT, L. M. (1978). New principles for the conservation of wild living resources. *Wildl. Monogr., Chestertown*, **59**, 1–33.

HOLTHUIS, L. B. (1980). Shrimps and prawns of the world. An annotated catalogue of species of interest to fisheries, FAO, Rome. *FAO Species Catalogue,* **1**, 1–271.

HONMA, M. (1974). Estimation of overall effective fishing intensity of tuna long-line fishery—yellowfin tuna in the Atlantic Ocean as an example of seasonally fluctuating stocks. *Enyo Suisan Kenkyujo. Shimizu*, **10**, 63–85.

HOUDE, E. D. (1973). Estimating abundance of sardine-like fishes from egg and larval surveys, eastern Gulf of Mexico, preliminary report. *Proc. Gulf. Caribb. Fish. Inst.*, **25**, 68–78.

HOURSTON, A. S. (1980). The decline and recovery of Canada's Pacific herring stocks. *Rapp. P.-v. Réun. Cons. int. Explor. Mer*, **173**, 143–153.

HOUSTON, T. W. (1955). The New South Wales trawl fishery: a review of past cause and examination of present condition. *Aust. J. mar. Freshwat. Res.*, **6** (2), 165–208.

HYND, J. S. and ROBINS, J. P. (1967). Tasmanian tuna survey report of first operational period. *Tech. Pap. Div. Fish. Oceanogr. C.S.I.R.O. Aust.*, (**22**), 55.

INNIS, H. A. (1954). *The Cod Fisheries.* University of Toronto Press, Toronto.

ICNAF (International Commission for the Northwest Atlantic Fisheries) (1968). Report of the working group on joint biological and economic assessment of conservation action. *Proc. int. Commn NW. Atlant. Fish.*, (**17**), 48–84.

ICSEAF (International Communication for the Southeast Atlantic Fisheries) (1980). *Report of the Standing Committee on Stock Assessment. Collection of Scientific Papers, 7 (Part I).* Int. Commn for SE. Atlantic Fisheries, Madrid.

ICES (International Council for the Exploration of the Sea) (1980). *Report of the Advisory Committee on Fishery Management.* International Council for the Exploration of the Sea (mimeo).

IMARPE (1970a). Report of the panel of experts on population dynamics of Peruvian anchoveta. *Bol. Inst. Mar Peru*, **2** (6), 324–372.

IMARPE (Instituto Del Mar Del Peru) (1970b). Report of Expert Panel on the economic effects of alternative regulatory measures in the Peruvian anchoveta fishery. *Inst. Mar Peru, Callao*, (**34**), 1–83.

IMARPE (Instituto Del Mar Del Peru) (1972). Report of the second session of the panel of experts on the population dynamics of Peruvian anchovy, March 1971. *Bol. Inst. Mar Peru*, **2** (7), 373–458.

IMARPE (Instituto Del Mar Del Peru) (1974). Second panel of experts. Report on the economic effects of alternative regulatory measures in the Peruvian anchoveta fishery. *Bol. Inst. Mar Peru*, **3** (1), 1–40.

JENNINGS, M. G. (1980). The enforcement of fishery regulations. *CECAF Tech. Pap.*, **80/22**, 1–25.

JOHANNES, R. K. (1977). Traditional 'Law of the Sea' in Micronesia. *Micronesia*, **13** (2), 121–127.

JOHANNESSON, K. and VILCHEZ, R. (1981). Note on hydro-acoustic observations of changes in distribution and abundance of some common pelagic fish species in the coastal waters of Peru. In *Workshop on the Effects of Environmental Variation on the Survival of Larval Pelagic Fishes.* IOC Workshop Report, 28. pp. 287–323.

JOSEPH, K. M. (1974). Demersal fisheries resources off the northwest coast of India. *Bull. Explor. Fish. Proj.* Bombay 1. 45 p.

JOSEPH, K. M., RADHAKRISHNAN, N. and PHILIP, K. P. (1976). Demersal fish resources off the southwest coast of India. *Bull. Explor. Fish. Proj.* (Bombay), 1–56.

JOYNER, T. (1980). Salmon ranching in South America. In J. E. Thorpe (Ed.), *Salmon Ranching.* Academic Press, London.

JUHL, R. (1974). Economics of the Gulf of Mexico industrial and food fish trawlers. *Mar. Fish. Rev.*, **36** (11), 39–42.

JUHL, R. and DRUMMOND, S. B. (1977). Shrimp by-catch investigation in the United States of America. A status report. *FAO Fish. Rep.*, **200**, 213–226.

JUKIC, S. and PICCINETTI, C. (1979). Standing stock estimation and yield per exploitable biomass (YEB) forecast of the Adriatic edible demersal resources. *Investigación pesq.*, **43** (1), 273–282.

KASAHARA, H. (1961). *Fisheries Resources of the North Pacific Ocean*, Part I. H. R. MacMillan Lectures in Fisheries. University of British Columbia, Vancouver.

KASAHARA, H. (1964). *Fisheries Resources of the North Pacific Ocean*, Part II. H. R. MacMillan Lectures in Fisheries. University of British Columbia, Vancouver.

KASAHARA, H. (1972). Japanese distant water fisheries: a review. *Fish. Bull. U.S.*, **70** (2), 227–282.

KASAHARA, H. (1977). Management of fisheries in the North Pacific. *J. Fish. Res. Bd Can.*, **30** (12), 2348–2360.

KATONA, S. K. and WHITEHEAD, H. P. (1981). Identifying humpback whales using their natural markings. *Polar Record*, **20** (128), 439–444.

KEARNEY, R. E. (1979). Some problems of developing and managing fisheries in small island states Noumea. *Occ. Pap. Sth Pacif. Commn*, **16**, 1–18.

KEARNEY, R. E. (1981). Some economic aspects of the development and management of fisheries in central and western Pacific. South Pacific Commn. *Fish. Newsl., Noumea*, **22**, 6–15.

KERSTAN, M. and SAHRHAGE, D. (1980). *Biological Investigation on Fish Stocks in the Waters off New Zealand.* Mitteilungen aus dem Institut fur Seefischerei, Hamburg.

KESTEVEN, G. L. and STARK, A. E. (1967). Demersal fish stocks of the Great Australian Bight, as estimated from the results of operations of F.V. Southern Endeavour. *Tech. Pap. Div. Fish. Oceanogr. C.S.I.R.O. Aust.*, **24**, 4–62.

KINNE, O. (1977). Cultivation of animals: research cultivation. In O. Kinne (Ed.), *Marine Ecology*, Vol. III, Cultivation, Part 2. Wiley, Chichester. pp. 579–1293.

KLIMA, E. F. (1976). A review of the fishery resources in the western central Atlantic. *WECAF Studies*, **3**, 1–77.

KLIMA, E. F. (1977). An overview of the fishery resources of the western central Atlantic region. *FAO Fish. Rep.*, **200**, 231–252.

KONDO, K. (1980). The recovery of the Japanese sardine. The biological basis of stock size fluctuations. *Rapp. P.-v. Réun. Cons. int. Explor. Mer*, **177**, 332–354.

KRAMER, D. (1969). Synopsis of the biological data on the Pacific mackerel, *Scomber japonicus* Houttuyn. *Circ. Fish Wildl. Serv., Wash.*, **302**, (Also issued as FAO Fish. Synops. 40.)

KRISTJONSSON, H. (Ed.) (1959). *Modern Fishing Gear of the World.* Fishing News (Books) Ltd, London.

KUROGANE, K., SRIRUANGCHEEP, V., TANTISAWETRAT, C., CHULLASORN, S., SUPONGPAN, S. and BOONPRAKOB, V. (1971). On the population dynamics of the Indo-Pacific mackerel (*Rastrelliger neglectus* van Kampen) of the Gulf of Thailand. *Bull. Tokai reg. Fish. Res. Lab.*, **67**, 1–33.

LAEVASTU, T. and FAVORITE, F. (1977). *Preliminary Report on Dynamic Numerical Marine Ecosystem Model (DYNUMES II) for Eastern Bering Sea. Processed Report.* Nat. Mar. Fish Service, Northwest and Alaska Fisheries Center.

LAEVASTU, T. and FAVORITE, F. (1978a). Numerical evaluation of marine ecosystems. Part I. Deterministic bulk biomass model (BBM). *Northwest Fish. Centre Proc. Rep.*, 1–22.

LAEVASTU, T. and FAVORITE, F. (1978b). Numerical evaluation of marine ecosystem. Part II. Dynamical numerical marine ecosystem model (DYNUMES III) for evaluation of fishery resources. *Northwest Fish. Centre Proc. Rep.*, 1–29.

LARKIN, P. A. (1977a). Pacific Salmon. In J. A. Gulland (Ed.), *Fish Population Dynamics*. Wiley, Chichester. pp. 156–186.

LARKIN, P. A. (1977b). An epitaph for the concept of maximum sustainable yield. *Trans. Am. Fish. Soc.*, **106** (1), 1–11.

LARKINS, H. A. (1980). Management under FCMA—development of a fishery management plan. *Mar. Policy*, **4** (3), 170–182.

LASKER, R. (1978). The relation between oceanographic conditions and larval anchovy food in the California current: identification of factors leading to recruitment failure. *Rapp. P.-v. Réun. Cons. int. Explor. Mer*, **173**, 212–230.

LAWS, R. M. (1962). Some effects of whaling on the southern stocks of baleen whales. In E. D. Le Cren and M. W. Holdgate (Eds), *The Exploitation of Natural Animal Populations*. Blackwell Sci. Publs., Oxford. pp. 137–158.

LAWS, R. M. (1977). Seals and whales of the Southern Ocean. *Phil. Trans. R. Soc. (Series B)*, **279/963**, 81–96.

LETT, P. F. and KOHLER, A. C. (1976). Recruitment: a problem of multi-species interaction and environmental perturbations, with special reference to Gulf of St. Lawrence Atlantic herring (*Clupea harengus harengus*). *J. Fish. Res. Bd Can.*, **33**, 1353–1371.

LOCKYER, C. (1979). Changes in a growth parameter associated with exploitation of southern fin and sei whales. *Rep. int. Whal. Commn*, **29**, 191–196.

LONGHURST, A. R. (1960). Local movements of *Ethmalosa fimbriata* off Sierra Leone from tagging data. *Bull. Inst. Fond. Afr. Noire (Series A)*, **22** (4), 1337–1340.

LONGHURST, A. R. (1964). A study of the Nigerian trawl fishery. *Bull. Inst. Fond. Afr. Noire (Series A)*, **26** (2), 686–700.

LONGHURST, A. R. (1969). Species assemblages in tropical demersal fishes. In *Proceedings of the Symposium on the Oceanography and Fisheries Resources of the Tropical Atlantic*. Unesco, Paris. pp. 147–168.

MACCALL, A. D., STAUFFER, G. D. and TROADEC, J.-P. (1976). Southern California recreational and commercial marine fisheries. *Mar. Fish. Rev.*, **38** (1), 1–32.

MCGOODWIN, J. R. (1979). The decline of Mexico's Pacific inshore fisheries. *Oceanus*, **22** (2), 51–59.

MCHUGH, J. L. (1969). Comparison of Pacific sardine and Atlantic menhaden fisheries. *FiskDir. Skr. (Series Havunders.)*, **15** (3), 356–367.

MACKINTOSH, N. A. (1965). *The Stocks of Whales*. Fishing News (Books) Ltd, London.

MACPHERSON, E. (1980). Algunos comentarios sobre el canibalismo en *Merluccius capensis* de la Subarea I. *ICSEAF Collection of Scientific Papers*, **7** (Part II), 217–222.

MARR, J. C. (1960). The causes of major variations in the catch of Pacific sardine, *Sardinops caerulea*, Girard. In H. Rosa and G. Murphy (Eds), *Proceedings of the World Scientific Meeting on the Biology of Sardines and Related Species, 3.* FAO, Rome. pp. 667–791.

MARR, J. W. S. (1962). The natural history and geography of the Antarctic krill (*Euphausia superba* Dana). *'Discovery' Rep.*, (**32**), 33–464.

MATHER, F. J. (1969). Long distance migrations of tunas and marlins. *Bull. Littoral Soc.*, **6** (1), 6–14.

MAY, R. M., BEDDINGTON, J. R., CLARK, C. W., HOLT, S. J. and LAWS, R. M. (1979). Management of multispecies fisheries. *Science, N.Y.*, **205** (4403), 267–277.

MITCHELL, B. and SANDBROOK, R. (1980). *The Management of the Southern Ocean*. International Institute for Environment and Development, London.

MOISEEV, P. A. (1964). Some results of the Bering Sea expedition. *Trudÿ vses. nauchno-issled. Inst. morsk. rÿb. Khoz. Okeanogr.*, **53**, 1–21.

MONOD, T. (1967). A historical review of the marine sciences in the tropical Atlantic. *FAO Fish. Rep.*, **51**, 26–33.

MORISON, S. E. (1971). *The European Discovery of America. The Northern Voyages.* Oxford Univ. Press, New York.

MOTODA, S. and HIRANO, Y. (1963). Review of Japanese herring investigations. *Rapp. P.-v. Réun. Cons. perm. int. Explor. Mer*, **154**, 249–261.

MUNRO, J. L. (1975). The biology, ecology, exploitation and management of Caribbean reef fishes. Part 6. Assessment of the potential productivity of Jamaica fisheries. *Res. Rep. Univ. West India*, **3** (6), 1–56.

MUNRO, J. L. (1977). Actual and potential fish production from the coralline shelves of the Caribbean Sea. *FAO Fish. Rep.*, **200**, 301–321.

MURPHY, G. I. (1966). Population biology of the Pacific sardine (*Sardinops caerulea*). *Proc. Calif. Acad. Sci.*, **34** (1), 1–84.

MURPHY, G. I. (1977). Clupeoids. In J. A. Gulland (Ed.), *Fish Population Dynamics.* Wiley, London. pp. 283–308.

MURPHY, R. C. (1981). The Guano and the anchoveta fishery. In M. G. Glantz and J. D. Thompson (Eds), *Resource Management and Environmental Uncertainty.* Wiley, New York. pp. 81–106.

NAGASAKI, F. (1973). Long-term and short-term fluctuations in the catches of pelagic fisheries around Japan. *J. Fish. Res. Bd Can.*, **30** (12), 2361–2367.

NAKAMURA, H. (1969). *Tuna Distribution and Migration.* Fishing News (Books) Ltd, London.

NELSON, W. R., INGHAM, M. C. and SCHAAF, W. E. (1977). Larval transport and year-class strength of Atlantic menhaden, *Brevoortia tyrannus. Fish. Bull. U.S.*, **75** (1), 23–41.

NEMOTO, T. (1968). Feeding of baleen whales and the value of krill as a marine resource in the Antarctic. In M. W. Holdgate (Ed.), *Symposium on Antarctic Oceanography*, Santiago, Chile, 13–16 September 1966. Scott Polar Research Institute, Cambridge. pp. 240–253.

NEWMAN, G. G. (1970). Migration of the pilchard (*Sardinops ocellata*) in southern Africa. *Rep. of South Africa, Div. of Sea Fisheries, Inv. Rep.*, **86**, 1–6.

NEWMAN, G. G. (1977). The living marine resources of the southeast Atlantic. *FAO Fish. Tech. Pap.*, (**179**), 1–59.

NOSOV, E. V. (1978). *Some Biological Aspects of the New Zealand Jack Mackerel Trachurus declivis Jenyns (1842) and its Fishery Perspectives* (Translated from Russian, mimeo).

OTSU, T. and UCHIDA, R. N. (1963). Model of the migration of albacore in the North Pacific Ocean. *Fish. Bull. U.S.*, **63** (1), 33–44.

PARRISH, B. B. (Ed.) (1973). Fish stocks and recruitment: Proceedings of a symposium held in Aarhus, 7–10 July 1970. *Rapp. P.-v. Réun. Cons. int. Explor. Mer.*, **164**, 1–772.

PARRISH, B. B. and JONES, R. (1953). Haddock bionomics. 1. The state of the haddock stocks in the North Sea 1941–50, and at Faroes 1914–1950. *Mar. Res.*, **1952** (4), 1–27.

PARRISH, B. B. and SAVILLE, A. (1967). Changes in the fisheries of the North Sea and Atlanto-Scandian herring stocks and their causes. *Oceanog. mar. Biol. A. Rev.*, **5**, 409–447.

PARRISH, R. H. (1974). Exploitation and recruitment of Pacific mackerel (*Scomber japonicus*) in the north-eastern Pacific. *CalCOFI Rep.*, **17**, 136–140.

PAULIK, G. J. (1971). Anchovy, birds and fishermen in the Peru Current. In W. W. Murdoch (Ed.), *Environment, Resources, Pollution and Society.* Sinauer, Stanford, Conn. pp. 156–185.

PAULY, D. (1979). Theory and management of tropical multispecies stocks. *ICLARM Studies and Reviews*, **1**, 1–35.

PAYNE, M. R. (1977). Growth of a fur seal population. *Phil. Trans. I. Soc. (Series B)*, **279**, 67–79.

PEARSE, P. (1980). Regulation of fishing effort. *FAO Fish. Tech. Pap.*, (**197**), 1–82.

PELLA, J. J. and TOMLINSON, P. K. (1969). A generalized stock production model. *Bull. I-ATTC*, **13** (3), 421–496.

PHILLIPS, W. J. (1966). *Maori Life and Custom.* Sydney, Australia.

PICCINETTI, C., REGNER, S. and SPECCHI, M. (1979). Estimation du stock d'anchois (*Engraulis encrasicholus* L.) de la haute et moyen Adriatic. *Investigación pesq.*, **43** (1), 69–82.

POPE, J. (1979). Stock assessment in multi-species fisheries, with special reference to the trawl fishery in the Gulf of Thailand. *FAO/UNDP South China Sea Fisheries Development and Coordination Programme, Manila.* **Doc.SCS/DEV/79/19**, 1–106.

POPP MADSEN, K. (1978). The industrial fisheries in the North Sea. *Rapp. P.-v. Réun. Cons. int. Explor. Mer*, **172**, 27–30.

PRENSKI, L. B. (1980). The food and feeding behaviour of *Merluccius capensis* in Division 1.5 (with some observations on Division 1.4). *ICSEAF Collection of Scientific Papers*, **7** (Part II), 283–296.

QASIM, B. T. A. (Ed.) (1973). Proceedings of the symposium on the living resources of the seas around India. *Bull. Cent. Mar. Fish. Res. Inst., Cochin* (Special Publication), 1–748.

RAJA, A. (1973). Forecasting the oil-sardine fishery. *Indian J. Fish.*, **20** (2), 549–609.

RAJA, B. T. A. (1969). The Indian oil sardine. *Bull. Cent. Mar. Fish. Res. Inst., Cochin*, **16**, 1–128.

RANDALL, J. E. (1980). A survey of ciguatera at Enewetak and Bikini, Marshall Islands, with notes on the systematics and food habits of ciguatoxic fishes. *Fish. Bull. U.S.*, **78/2**, 201–225.

RICKER, W. E. (1954). Stock and recruitment. *J. Fish. Res. Bd Can.*, **11** (5), 559–623.

RICKER, W. E. (1958). Handbook of computations for biological statistics of fish populations. *Bull. Fish. Res. Bd Can.*, **119**, 1–300.

RITSRAGA, S. (1976). Results of the studies on the status of demersal fish resources in the Gulf of Thailand from trawling surveys 1963–1972. In K. Tiews (Ed.), *Fisheries Resources and their Management in Southeast Asia.* German Foundation for International Development, Berlin. pp. 198–223.

ROBERTSON, D. A. (1980). Hydrology and the quantitative distribution of planktonic eggs of some marine fishes of the Otago coast, southeastern New Zealand. *Fish. Res. Bull.* (N.Z.)., **21**, 1–69.

ROBINSON, M. A. (1979). Prospects for world fisheries to 2000. *FAO fish. circ.*, **722**, 1–12.

ROEDEL, P. M. (1974). The Californias and some fisheries of common concern. *CalCOFI Rep.*, **17**, 61–65.

ROEDEL, P. M. (1975). Optimum sustainable yield as a concept in fisheries management. *Spec. Publ. Am. Fish. Soc.*, **9**, 1–89.

ROTHSCHILD, B. J. and PARRACK, M. L. (1981). *The U.S. Gulf of Mexico Shrimp Fishery.* Paper prepared for the Workshop on the Scientific Basis for the Management of Penaeid Shrimp, November 1981, Key West, Florida.

ROYCE, W. F., SMITH, L. S. and HARTT, A. C. (1968). Models of oceanic migrations of Pacific salmon and comments on guidance mechanisms. U.S. Fish and Wildlife Service. *Fish. Bull. U.S.*, **66** (3), 441–462.

RYTHER, J. H. (1969). Photosynthesis and fish production in the sea. *Science, N.Y.*, **166**, 72–76.

SAETERSDAL, G., TSUKAYAMA, I. and ALEGRE, B. (1965). Fluctuaciones en la abundancia aparente del stock de anchoveta en 1959–1962. *Bol. Inst. Mar Peru*, **1** (2), 35–102.

SAVILLE, A. (1977). Survey methods of appraising fishery resources. *FAO Fish. Tech. Pap.*, (**171**), 1–76.

SAVILLE, A. (Ed.) (1980). The assessment and management of pelagic fish stocks. *Rapp. P.-v. Réun. Cons. Int. Explor. Mer*, **177**, 1–517.

SCAMMON, C. M. (1874). *The Marine Mammals of the North-Western Coast of North America.* G. P. Putnam, New York. (Reprinted by Dover Publ. Inc., 1968.)

SCAR (Scientific Committee on Antarctic Research) (1980). Report of the working group on Antarctic fish biology, Dammarie-les-Lys, France. BIOMASS *Reports*, **No. 12**, Scientific Committee on Antarctic Research, Cambridge.

SCHAAF, W. E. (1975). Status of the Gulf and Atlantic menhaden fisheries and implications for resource management. *Mar. Fish. Rev.*, **37** (9), 1–9.

SCHAAF, W. E. and HUNTSMAN, G. R. (1972). Effects of fishing on the Atlantic menhaden stock 1955–1969. *Trans. Am. Fish. Soc.*, **101** (2), 290–297.

SCHAEFER, M. B. (1954). Some aspects of the dynamics of populations important to the management of marine fisheries. *Bull. I-ATTC*, **1**, 25–56.

SCHAEFER, M. B. (1957). A study of the dynamics of the fishery for yellowfin tuna in the eastern tropical Pacific Ocean. *Bull. I-ATTC*, **2**, 245–285.

SCHAEFER, M. B. (1970). Men, birds and anchovies in the Peru Current. *Trans. Am. Fish. Soc.*, **99**, 461–467.

SHAHEEN, A. H. (1976). The sardinella fishery off the Egyptian Mediterranean coast. In The Report of the Sixth Session of the GFCM Working Party on Resources Appraisal and Fishery Statistics. *FAO Fish. Rep.*, **182**, 31–32.

SHARP, G. (Ed.) (1981). Workshop on the effects of environmental variation on the survival of larval pelagic fish. *Int. Oceanogr. Commn. Workshop Rep.*, **28**, 1–323.

SHERMAN, K., JONES, C., SULLIVAN, L., SMITH, W., BERRIEN, P. and EJSYMONT, A. (1981). Congruent shifts in sand eel abundance in western and eastern North Atlantic ecosystems. *Nature, Lond.*, **291**, 486–489.

SHOMURA, R. S. (1980). Summary report of the billfish stock assessment workshop, Pacific resources. *NOAA Tech. Memo.*, **NMFS-SWFC-5**, 1–58.

SISSENWINE, M. P., BROWN, B. E., PALMER, J. E., ESSIG, R. J. and SMITH, W. (1979). *An Empirical Examination on Population Interactions for the Fishery Resources off the Northeastern USA*. Paper presented at the workshop on multi-species fisheries management advice, November 1979, St Johns, Newfoundland.

SIVASUBRAMANIAM, K. (1981). Discussion and evaluation of the demersal resources. In Demersal Resources of the Gulf and the Gulf of Oman. *FAO, Rome Doc.*, **FI:DP/RAB/71/278/10**, 116–122.

SMITH, P. E. and RICHARDSON, S. L. (1977). Standard techniques for pelagic fish egg and larva studies. *FAO Fish. Tech. Pap.*, (**175**), 1–100.

SMITH, R. L. (1968). Upwelling. *Oceanogr. mar. Biol. A. Rev.*, **6**, 11–46.

SOUTAR, A. (1967). The accumulation of fish debris in certain Californian coastal sediments. *Rep. Calif. coop. oceanic Fish. Invest.*, **11**, 136–141.

SOUTAR, A. and ISAACS, J. D. (1974). Abundance of pelagic fish during the 19th and 20th centuries as recorded in anaerobic sediment off the Californias. *Fish. Bull. NOAA/NMFS*, **72**, 257–75.

SOUTH CHINA SEA PROGRAMME (1976a). *Report of the Workshop on the Fishery Resources of the Malacca Strait*, Part I (SCS/GEN/76/2). South China Sea Programme, Manila.

SOUTH CHINA SEA PROGRAMME (1976b). *Report of the BFAR/SCSP Workshop on the Fishery Resources of the Visayan and Sibuyan Sea Areas* (SCS/GEN/76/7). South China Sea Programme, Manila.

SOUTH CHINA SEA PROGRAMME (1977). *Report on the BFAR/SCSP Workshop on the Fishery Resources of the Sula Sea, Bohol Sea and Moro Gulf Areas* (SCS/GEN/77/11). South China Sea Programme, Manila.

SOUTH CHINA SEA PROGRAMME (1978a). *Report of the Workshop on Demersal Resources of the Sunda Shelf*, Part I (SCS/GEN/77/12). South China Sea Programme, Manila.

SOUTH CHINA SEA PROGRAMME (1978b). Pelagic resources evaluation. *Report of the Workshop on the Biology and Resources of Mackerels (Rastrelliger spp.) and Round Scads (Decapterus spp.) in the South China Sea*, Part I (SCS/GEN/78/17). South China Sea Programme, Manila. pp. 1–50.

STEELE, J. (1965). Some problems in the study of marine resources. *Spec. Publs int. Commn NW. Atlant. Fish.*, **6**, 463–476.

TEMPLEMAN, W. F. and GULLAND, J. A. (1965). Review of possible conservation actions for the ICNAF area. *A. Proc. int. Commn NW. Atlant. Fish.*, **15**, 47–56.

THOMPSON, W. F. (1950). *The Effect of Fishing on Stocks of Halibut in the Pacific.* University of Washington Press, Seattle.

THORPE, J. E. (Ed.) (1980). *Salmon Ranching.* Academic Press, London.

THUROW, F. R. (1978). The Fish Resources of the Baltic. *FAO Fish. Circ.*, **708**, 1–21.

TIEWS, K., SUCONDHAMARN, P. and ISARANKURA, A. (1967). On the change in the abundance of demersal fish stocks in the Gulf of Thailand from 1963/1964 to 1966 as a consequence of the trawl fishery development. *Contrib. Dep. Fish., Bangkok*, (**8**), 1–39.

TOWNSEND, C. H. (1935). The distribution of certain whales, as shown by the log book records of American whale ships. *Zoologica, N.Y.*, **19**, 1–50.

TROADEC, J.-P. (1979). Resource appraisal and fisheries management in the Mediterranean: the activities of GFCM in this field. *Investigación pesq.*, **43** (1), 261–271.

TROADEC, J.-P., CLARK, W. G. and GULLAND, J. A. (1980). A review of some pelagic fish stocks in other areas. *Rapp. P.-v. Réun. Cons. int. Explor. Mer*, **177**, 252–277.

TROADEC, J.-P. and GARCIA, S. (1979). Les ressources halieutiques de l'Atlantique centre-est. Première partie: Les ressources du Golfe de Guinée de l'Angola à la Mauritanie. *FAO Fish. Tech. Pap.*, (**186.1**), 1–167.

TURNBRIDGE, B. R. (1969). Pilchard survey, Nelson 1964. *Fish. tech. Rep. N.Z. mar. Dep.*, **32**, 1–23.

UCHIDA, R. (1978). The fish resources of the western central Pacific Islands. *FAO Fish. Circ.*, **712**, 1–53.

ULLTANG, Ø. (1977). Methods of measuring stock abundance other than by the use of commercial catch and effort data. *FAO Fish. Tech. Pap.*, **176**, 1–23.

VALDEZ-ZAMUDIO, F. (1973). Impacto de medidas regulatorias en la pesqueria peruana. *J. Fish. Res. Bd Can.*, **30** (12), 2242–2253.

VALDIVIA, J. E. (1978). The anchovetta and El Niño. *Rapp. P.-v. Réun. Cons. int. Explor. Mer*, **173**, 196–202.

VIRABHADRO RAO, K. (1973). Distribution pattern of the major exploited marine fishery resources of India. In *Proceedings of the Symposium on the Living Resources of the Seas around India.* Central Marine Fisheries Research Institute, Cochin. pp. 18–101.

VOSS, G. L. (1960). Potentialities for an octopus and squid fishery in the West India. *Proc. Gulf Caribb. Fish. Inst.*, **12**, 129–133.

WALSH, J. J. (1975). A spatial simulation model of the Peru ecosystem. *Deep Sea Res.*, **22**, 201–236.

WALSH, J. J. (1978). The biological consequences of interaction of the climatic El Niño and event scales of variability in the eastern tropical Pacific. *Rapp. P.-v. Réun. Cons. int. Explor. Mer*, **173**, 182–192.

WALSH, J. J. (1981). A carbon budget for overfishing off Peru. *Nature, Lond.*, **290**, 300–304.

WALSH, J. J., WHITLEDGE, T. E., ESAIAS, W. E., SMITH, R. L., HUNTSMAN, S. A., SANTANDER, H. and De MENDIOLA, B. R. (1980). The spawning habitat of the Peruvian anchovy, *Engraulis ringens*. *Deep Sea Res.*, **27** (1A), 1–28.

WANG, L. A. and PATHANSALI, D. (1977). An analysis of Penang trawl fisheries to determine the maximum sustainable yield. Ministry of Agriculture, Malaysia. *Fish. Bull. U.S.*, (**16**), 1–10.

WAUGH, G. D. (1969a). Potential of New Zealand waters for mussel culture. Report of Fisheries Committee to the National Development Conference, New Zealand. Appendix C to Annex B. Potential of New Zealand waters for rock oyster cultivation.

WAUGH, G. D. (1969b). Report of Fisheries Committee to the National Development Conference, New Zealand. Appendix D to Annex B.

WAUGH, G. D. (1979). Scientific basis for determining management measures. New Zealand experience in Interim Report of the ACMRR working party on the scientific basis of determining management measures. *FAO Fish. Circ.* **718**, 49–54.

WILLIAMS, F. (1968). Report on the Guinean trawling survey. *Publ. Organ. Afr. Unity Sci. Tech. Res. Commn.*, **99**, Vol. 1, 1–828.

WYRTKI, K. (1961). Scientific results of marine investigations of the South China Sea and Gulf of Thailand, 1959–1961. *Naga Rep.*, (**2**), 1–195.

WYRTKI, K., STROUP, E., PATZERT, W., WILLIAMS, R. and QUINN, W. (1976). Predicting and observing El Niño. *Science, N.Y.*, **191**, 343–346.

ZACHMAN, N. (1973). Fisheries development and management in Indonesia. *J. Fish. Res. Bd Can.*, **30** (12), 2335–2340.

VAN ZALINGE, N. and NAAMIN, N. (1977). *Report on the Offshore Shrimp Fishery along the South Coast of Java.* Institute of Marine Fisheries, Jakarta.

ZHU DE-SHAN (1980). A brief introduction to the fisheries of China. *FAO Fish. Circ.*, **726**, 1–31.

# AUTHOR INDEX

*Numbers in italics refer to those pages on which the author's work is stated in full.*

# TAXONOMIC INDEX

# SUBJECT INDEX